Fourth Edition

BEHAVIOURAL ECOLOGY

An Evolutionary Approach

EDITED BY

John R. Krebs
Chief Executive of the
Natural Environment Research Council
and Royal Society
Research Professor
Department of Zoology
University of Oxford

AND

Nicholas B. Davies
Professor of Behavioural Ecology
Department of Zoology
University of Cambridge

BLACKWELL PUBLISHING
350 Main Street, Malden, MA 02148-5020, USA
9600 Garsington Road, Oxford OX4 2DQ, UK
550 Swanston Street, Carlton, Victoria 3053, Australia

First published 1978
Second edition 1984
Third edition 1991
Fourth edition 1997

10 2008

Library of Congress Cataloging-in-Publication Data

Behavioural ecology : an evolutionary approach / edited by J. R. Krebs, N. B. Davies. — 4th ed.
p. cm.
Includes bibliographical references (p.) and indexes.
ISBN 978-0-86542-731-0
1. Animal behaviour. 2. Animal ecology. 3. Behaviour evolution.
I. Krebs, J. R. (John R.) II. Davies, N. B. (Nicholas B.), 1952–.
QL751. B345 1996
591.5'1 — dc20 96-30486

A catalogue record for this title is available from the British Library.

Set by Setrite Typesetters, Hong Kong

The publisher's policy is to use permanent paper from mills that operate a sustainable forestry policy, and which has been manufactured from pulp processed using acid-free and elementary chlorine-free practices. Furthermore, the publisher ensures that the text paper and cover board used have met acceptable environmental accreditation standards.

For further information on
Blackwell Publishing, visit our website:
www.blackwellpublishing.com

Contents

Contributors, v

Preface, vii

Acknowledgements, viii

Part 1: Introduction

1 The Evolution of Behavioural Ecology, 3
John R. Krebs & Nicholas B. Davies

Part 2: Mechanisms and Individual Behaviour

Introduction, 15

2 Sensory Systems and Behaviour, 19
Rüdiger Wehner

3 The Ecology of Information Use, 42
Luc-Alain Giraldeau

4 Recognition Systems, 69
Paul W. Sherman, Hudson K. Reeve & David W. Pfennig

5 Managing Time and Energy, 97
Innes C. Cuthill & Alasdair I. Houston

6 Sperm Competition and Mating Systems, 121
Timothy R. Birkhead & Geoffrey A. Parker

Part 3: From Individual Behaviour to Social Systems

Introduction, 149

7 The Evolution of Animal Signals, 155
Rufus A. Johnstone

8 Sexual Selection and Mate Choice, 179
Michael J. Ryan

9 Sociality and Kin Selection in Insects, 203
Andrew F.G. Bourke

10 Predicting Family Dynamics in Social Vertebrates, 228
Stephen T. Emlen

11 The Ecology of Relationships, 254
Anne E. Pusey & Craig Packer

12 The Social Gene, 284
David Haig

Part 4: Life Histories, Phylogenies and Populations

Introduction, 307

13 Adaptation of Life Histories, 311
Serge Daan & Joost M. Tinbergen

14 The Phylogenetic Foundations of Behavioural Ecology, 334
Paul H. Harvey & Sean Nee

15 Causes and Consequences of Population Structure, 350
Godfrey M. Hewitt & Roger K. Butlin

16 Individual Behaviour, Populations and Conservation, 373
John D. Goss-Custard & William J. Sutherland

References, 396

Index, 447

Contributors

Timothy R. Birkhead *Department of Animal and Plant Sciences, University of Sheffield, Sheffield, S10 2TN, UK*

Andrew F.G. Bourke *Institute of Zoology, Zoological Society of London, Regent's Park, London, NW1 4RY, UK*

Roger K. Butlin *Department of Biology, University of Leeds, Leeds, LS2 9JT, UK*

Innes C. Cuthill *Behavioural Biology Group, School of Biological Sciences, University of Bristol, Woodland Road, Bristol, BS8 1UG, UK*

Serge Daan *Zoological Laboratory, University of Groningen, PO Box 14, 9750 AA Haren, The Netherlands*

Nicholas B. Davies *Department of Zoology, University of Cambridge, Downing Street, Cambridge, CB2 3EJ, UK*

Stephen T. Emlen *Section of Neurobiology and Behavior, Division of Biological Sciences, Seeley G. Mudd Hall, Cornell University, Ithaca, NY 14853-2702, USA*

Luc-Alain Giraldeau *Department of Biology, Concordia University, 1455 ouest, Boulevard de Maisonneuve, Montréal, Québec, H3G 1M8, Canada*

John D. Goss-Custard *Institute of Terrestrial Ecology, Furzebrook Research Station, Wareham, Dorset, BH20 5AS, UK*

David Haig *Museum of Comparative Zoology, Harvard University, 26, Oxford Street Cambridge, MA 02138, USA*

Paul H. Harvey *Department of Zoology, University of Oxford, South Parks Road, Oxford, OX1 3PS, UK*

Godfrey M. Hewitt *School of Biological Sciences, University of East Anglia, Norwich, NR4 7TJ, UK*

Alasdair I. Houston *Behavioural Biology Group, School of Biological Sciences, University of Bristol, Woodland Road, Bristol, BS8 1UG, UK*

Rufus A. Johnstone *Department of Zoology, University of Cambridge, Downing Street, Cambridge, CB2 3EJ, UK*

John R. Krebs *Department of Zoology, University of Oxford, South Parks Road, Oxford, OX1 3PS, UK*

Sean Nee *Department of Zoology, University of Oxford, South Parks Road, Oxford, OX1 3PS, UK*

Craig Packer *Department of Ecology, Evolution and Behavior, University of Minnesota, 100, Ecology Building, 1987 Upper Buford Circle, Saint Paul, MN 55108, USA*

Geoffrey A. Parker *School of Biological Sciences, University of Liverpool, Liverpool, L69 3BX, UK*

David W. Pfennig *Department of Biology, Coker Hall, University of North Carolina, Chapel Hill, NC 27599, USA*

Anne E. Pusey *Department of Ecology, Evolution and Behavior, University of Minnesota, 100, Ecology Building, 1987 Upper Buford Circle, Saint Paul, MN 55108, USA*

Hudson K. Reeve *Section of Neurobiology and Behavior, Division of Biological Sciences, Seeley G. Mudd Hall, Cornell University, Ithaca, NY 14853-2702, USA*

Michael J. Ryan *Department of Zoology, University of Texas, Austin, TX 78712, USA*

Paul W. Sherman *Section of Neurobiology and Behavior, Division of Biological Sciences, Seeley G. Mudd Hall, Cornell University, Ithaca, NY 14853-2702, USA*

William J. Sutherland *School of Biological Sciences, University of East Anglia, Norwich, NR4 7TJ, UK*

Joost M. Tinbergen *Zoological Laboratory, University of Groningen, PO Box 14, 9750 AA, Haren, The Netherlands*

Rüdiger Wehner *Department of Zoology, University of Zurich, Winterthurerstrasse 190, CH-8057, Zurich, Switzerland*

Preface

As with each of the three previous editions of *Behavioural Ecology*, we have brought together a completely new set of chapters for this volume. This is an exciting time for the subject. Stronger links are being forged with studies of mechanisms and how they control and constrain the adaptive behaviour of individuals. New molecular phylogenies are being used to unravel the evolution of behaviour and life histories. Molecular techniques for measuring parentage have revolutionized empirical studies of breeding systems and have thrown up many surprises, especially the high levels of 'extrapair' matings in species previously supposed to be monogamous. New models are being developed to study behaviour sequences and signalling systems. There are also new ideas linking immunocompetence and mate choice. At the same time, the traditional boundaries of the subject are being extended to consider how individual decision making influences population structure and its consequences for the conservation of species and habitats.

We have tried to incorporate these developments along with the familiar topics from previous editions by organizing the book in three new sections. We have aimed for a fresh approach by choosing new authors or new subjects. We asked everyone to provide a review of the main ideas and empirical data to test them, and to focus on current controversies and unsolved questions. As with the previous editions, the book is intended for graduate and upper level undergraduate courses, where students are already familiar with the basic ideas in behavioural ecology.

We hope that these chapters will inspire the next generation of behavioural ecologists to strive for improvements in the theories and in the data to test them. Since the first edition of this book (1978), we have produced new editions at 6- to 7-year intervals (1984, 1991, 1997). The best measure of the subject's continuing success would be the need for a fifth edition in another 7 years' time.

John R. Krebs
Nicholas B. Davies

Acknowledgements

From Blackwell Science Ltd, we thank Robert Campbell and Susan Sternberg for the initial encouragement to edit a fourth edition of this book and Ian Sherman and Karen Moore for their expert advice and enthusiasm during its production. Authors of various chapters would like to thank the following for their help.

Chapter 2	Simon Laughlin.
Chapter 4	George J. Gamboa, Warren G. Holmes, Laurent Keller, Jill M. Mateo, Karin S. Pfennig.
Chapter 5	John Hutchinson, Sonia Lee, Neil Metcalfe, John McNamara, Mark Witter and the Natural Environment Research Council and Biotechnology and Biological Sciences Research Council.
Chapter 7	Rebecca Kilner and Naomi Langmore.
Chapter 9	George Chan, Tim Coulson, Jeremy Field, William Foster, Laurent Keller, Ian Owens, Yves Roisin and Roland Stark.
Chapter 12	Helena Cronin, Alan Grafen, Laurence Hurst, J. McKinnon, Naomi Pierce, S. Porter.
Chapter 16	Paul Marrow.

Once again, we thank Jan Parr for drawing the vignettes which accompany some of the figures in this book.

PART 1

INTRODUCTION

A starling Sturnus vulgaris *returns to its nest to feed its hungry brood. Why are the chicks begging so noisily? What are the proximate and ultimate factors which influence the parent's choice of prey, its provisioning rate and its clutch size?* [Photograph by Eric and David Hosking.]

Chapter 1

The Evolution of Behavioural Ecology

John R. Krebs & Nicholas B. Davies

1.1 Observations and questions

All natural history observations begin with a question. At first our curiosity may be satisfied simply by knowing the species name of the animal we are watching. Then we may want to discover what it is doing and to understand why it is behaving in a particular way. In 1978, we began the first edition of this book by asking the reader to observe a bird, such as a starling (*Sturnus vulgaris*), searching in the grass for food. The starling walks along, pausing every now and then to probe into the ground. Sometimes it succeeds in finding a prey item, such as a beetle larva, and eventually, when it has collected several prey, it flies back to the nest to feed its hungry brood.

For students of behavioural ecology, a whole host of questions come to mind as they observe this behaviour. The first set of questions concern the way the bird feeds. Why has it chosen that particular place to forage? Why is it alone rather than in a flock? What determines its choice of foraging path? Does it collect every item of food it encounters or is it selective for prey type or size? What influences its decision to stop collecting food and fly back to feed its chicks? Another set of questions emerges when we follow the starling back to the nest. Why has it chosen this site? Why this brood size? How do the two parent starlings come to an agreement over how much work each puts into offspring care? Why are the chicks begging so noisily? Are they each simply signalling their own degree of hunger or are they competing for food? Why such costly begging behaviour? If we observed the starling over a longer period of time then we may ask about what determines how much effort it puts into reproduction versus its own maintenance, about the factors influencing the timing of its seasonal activities, its choice of mate, and so on.

Behavioural ecology provides a framework for answering these kinds of questions. It combines ideas from evolution, ecology and behaviour and has emerged from five schools of thought, developed primarily in the 1960s and early 1970s. We discuss them in turn to provide a brief history of the subject and to show how the ideas have evolved in the last 20–30 years, and we point out how this book reflects recent developments.

1.2 Tinbergen's four questions

Tinbergen (1963) showed that there are four ways of answering the question 'why?' in biology. Returning to our starling, if we asked why it foraged in a particular way we could answer as follows.

1 In terms of *function*, namely how patch choice and prey choice contribute to the survival of the bird and its offspring.

2 In terms of *causation*, namely the proximate factors which caused the bird to select a foraging site or prey type. These may include cues to prey abundance, such as type of soil or vegetation, or the activities of other birds.

3 In terms of *development*. This answer would be concerned with the role of genetic predispositions and learning in an individual's decision making.

4 In terms of *evolutionary history*, namely how starling behaviour has evolved from its ancestors. This answer might include an investigation of how the starling family has radiated to fill particular ecological niches and the influence of competition from other animals on the evolution of starling behaviour and morphology (e.g. bill size, body size).

Tinbergen's studies on gulls aimed to combine the four kinds of answers and he emphasized the need to study animals in their natural surroundings, namely those where their behaviour had evolved. He championed the use of the field as a natural laboratory for observations and controlled experiments and showed how ideas can be tested by collecting quantitative data on behaviour patterns (e.g. Tinbergen, 1953, 1972). Tinbergen's legacy is evident in current field studies of behaviour, many of which use simple experiments to measure the costs and benefits of traits. A good example is the classic tail manipulation experiments by Andersson and Møller to investigate mate choice in widowbirds and swallows, discussed by Ryan in Chapter 8.

However, early studies in behavioural ecology often focused on function and tended to ignore the other three questions. A caricature of behaviour studies in the 1930s is one where researchers imagined their animals as little machines, blindly following fixed action patterns in responses to external stimuli. A caricature from the early days of behavioural ecology and sociobiology in the 1970s is the opposite extreme of regarding animals as scheming tacticians, weighing up the costs and benefits of every conceivable course of action and always choosing the best one. Current work is leading to an intermediate position. While we expect selection to favour mechanisms that maximize an individual's fitness, we must recognize that mechanisms both constrain and serve behavioural outcomes.

A good example to illustrate this point is Lotem's recent studies of how hosts come to recognize a cuckoo egg in their nest. The cuckoo, *Cuculus canorus*, is a brood parasite which exploits various small birds as hosts to raise its offspring. The female cuckoo lays just one egg per host nest. The cuckoo chick hatches first, whereupon it ejects the host's eggs over the side of the nest, so becoming the sole occupant. Given the cost of parasitism, it is not surprising that many hosts have evolved defences such as rejection of odd eggs in their nest. Nevertheless, the puzzle is that egg rejection rarely reaches 100% in the host population and, furthermore, hosts never reject the cuckoo chick. A consideration of mechanisms may help to solve both puzzles. Experiments show that the defence mechanism used by hosts involves learning the characteristics of their own eggs the first time they breed and then rejecting eggs which differ from this learned set (Lotem *et al.*, 1995). This makes it unlikely that the host population will evolve 100% rejection of cuckoo eggs because hosts which are parasitized during their first breeding attempt will learn the cuckoo egg as part of their own set. Nevertheless, the learning rule works quite well and leads hosts to reject many parasite eggs. At the chick stage, however, a learning rule does less well than a rule 'accept any chick in my nest'. This is because there is a considerable cost of misimprinting; any host parasitized in its first attempt would learn only the cuckoo chick as its own and would then subsequently reject its own young in future, unparasitized, broods (Lotem, 1993). The main message from this study is that it is not very fruitful to discuss the evolution of 'rejection' without specifying the mechanisms, because these will determine the costs and benefits involved. Studies of mechanism and function must go hand in hand.

In 1975, Wilson predicted the demise of ethology, with mechanisms becoming the domain of neurobiology, and function and evolution the domain of sociobiology. This prediction was fulfilled until recent years, when there has been a welcome renewed interest in linking mechanism and function. We have marked this change by devoting the first section of this volume to this fruitful interchange. For example, Giraldeau (see Chapter 3) shows how research on foraging behaviour has stimulated new questions about learning and memory mechanisms, and Sherman *et al.* (see Chapter 4) point out common features of recognition mechanisms of kin, mates and predators and discuss their functional significance.

1.3 Ecology and behaviour

Even before Darwin, biologists often interpreted morphological adaptations in relation to the environment in which the species lived. Darwin's achievement was to show how these could arise without a Creator. Once the early ethologists, such as Lorenz and Tinbergen, had demonstrated that behaviour patterns were often as characteristic of a species as its morphological features, attempts were made to correlate differences between species in behaviour with differences in ecological factors, such as habitat, food and predation. A pioneering study was that by Cullen (1957), who was a student of Tinbergen. She interpreted the reduced anti-predator behaviour of kittiwake gulls, *Rissa tridactyla*, compared to the ground-nesting gulls, in relation to their safer nest sites on steep cliffs. Two other early studies were those by Winn (1958), who linked the reproductive behaviour of 14 species of darter fish (Percidae) to their ecology, and by Brown and Wilson (1959), who related the colony size and structure of dacetine ants to their feeding habits.

Crook's (1964) study of weaver birds (Ploceinae) has become established as the model for this approach. Crook showed how differences between species in food and predator pressure affect a whole host of adaptations, including nesting dispersion (colonies versus territories), feeding behaviour (solitary versus flock) and mating systems (monogamy versus polygamy). This comparative method was soon extended to other groups, including primates (Crook & Gartlan, 1966), other bird species (Crook, 1965; Lack, 1968), ungulates (Jarman, 1974), carnivores (Kruuk, 1975) and coral reef fish (Fricke, 1975).

The comparative approach remains influential today, the main advances being in methodology, particularly the quantification of behavioural and ecological traits, the use of multivariate statistics to tease out confounding variables and methods for identifying independent evolutionary events (Clutton-Brock & Harvey, 1984). It is now agreed that the ideal way to carry out a comparative analysis is to reconstruct a phylogenetic tree of the group under study and to use this as the basis for independent comparisons. Phylogenies not only provide a way of identifying independent evolution but also show the sequence in which traits have evolved within a group. With molecular phylogenies there is also the potential to measure the time-scale of evolutionary change. This new approach to Tinbergen's fourth question is one of the major developments in recent years and is discussed by Harvey and Nee in Chapter 14.

1.4 Economic models of behaviour

Many early studies in ethology recognized that behaviour patterns involve costs as well as benefits. For example, Tinbergen *et al.* (1963) showed that removal of the egg-shell after hatching reduced predation of black-headed gull, *Larus ridibundus*, nests (the egg has a conspicuous white interior). But, leaving newly hatched chicks unattended is costly too, which probably explains why the parent delays egg-shell removal until the chicks have dried out and become less vulnerable to attacks from neighbouring gulls.

The pioneer in the use of mathematical models in ecology to quantify these kinds of trade-offs was Robert MacArthur, who first applied the idea of optimal choice in the context of foraging behaviour (e.g. MacArthur & Pianka, 1966; MacArthur, 1972). The argument for using optimality models in behavioural ecology is that natural selection is an optimizing agent, favouring design features of organisms which best promote an individual's propagation of copies of its genes into future generations. Behaviour patterns clearly contribute to this ultimate goal, so we expect individuals to be designed as efficient at foraging, avoiding predators, mate choice, parenting, and so on. Optimality models have three components: (i) an assumption about the choices facing the animal (e.g. prey type); (ii) an assumption about what is being maximized (e.g. rate of energy gain); and (iii) an assumption about constraints (e.g. bill size, searching speed). For example, from a knowledge of prey available and morphological constraints, we could predict how our starling should select prey so as to maximize its rate of food delivery to its brood. If the model fails to predict the observed behaviour, we can then use the discrepancies to help identify which of our assumptions was incorrect. Classic early studies include work by Schoener (1971) and Charnov (1976a,b) on prey choice and patch choice by foragers, and Parker's study of copulation time in the yellow dungfly, *Scatophaga stercoraria* (Parker, 1970a; Parker & Stuart, 1976).

The optimality approach has met with some criticism but in our view it remains the most powerful method for studying the design of behaviour (for discussion see Maynard Smith, 1978; Stephens & Krebs, 1986). Perhaps the main problem facing a behavioural ecologist is that the animals under study are clearly faced with trade-offs not just within a particular activity but also between activities. The starling has to find food, keep an eye out for predators and return to the nest to keep its brood warm, for example. In Chapter 5, Cuthill and Houston discuss techniques for considering how different activities combine to influence fitness. In particular, they show how dynamic programming can be used to model sequences of behavioural choices.

1.5 Evolutionarily stable strategies

An animal's environment does not consist solely of places to feed, nest, shelter and hide from predators. There is also a living environment of competitors. Often an individual's best choice will be influenced by what these competitors are doing. Thus, the best place for our starling to feed will depend on where the other starlings go, the best strategy to adopt in a fight will depend on what the opponent does, and the best sex ratio for an individual to produce in its offspring will depend on the population sex ratio. Early studies to recognize this important point include Fisher's (1930) explanation for why parents expend equal resources on male and female progeny, Hamilton's (1967) analysis of stable sex ratios under local mate competition, Parker's (1970c, 1974b) field study of how male dungflies distribute themselves across different mating sites and work by Fretwell and Lucas (1970) on habitat choice by birds. All these studies analysed the problem in terms of which choices would produce an equilibrium distribution in the population.

Maynard Smith's (1972, 1982) concept of the evolutionarily stable strategy (ESS) is now widely accepted as the way of analysing decision making where the payoffs are frequency dependent. A strategy is an ESS if, when adopted by most members of a population, it cannot be invaded by the spread of any rare alternative strategy. This idea has been influential in analysing many problems in behavioural ecology including fighting behaviour and communication (see Chapter 7), mating systems (see Chapter 6) and cooperation and conflict in social groups (see Chapters 9 and 11). In many cases no single strategy is an ESS, so one of the main messages for field workers has been to expect variability in behaviour. Sometimes the variability is between individuals, so there is a polymorphism in the population. Hamilton's (1979) study of dimorphic males in fig wasps provided an early classic example of stable alternative strategies within a species. For recent examples, see Shuster and Wade (1991), Lank *et al.* (1995) and Sinervo and Lively (1996). More often, individuals vary in their behaviour depending on what their competitors do. A rule 'go to the patch with the greatest number of worm casts' would be fine for a bird if it was the only forager, but in the presence of competition this may not be the best thing to do. The behavioural ecologist's task here is to consider what would be the stable decision rules (see Chapter 3 by Giraldeau for a discussion of this problem).

ESS models have been particularly useful in studies of signalling systems. In many cases animal displays seem at first sight to be unnecessarily extravagant, for example the stretching, gaping and calling of young birds as they beg for food or the energetic dancing of males on a lek as they attempt to attract a female for mating. Zahavi's (1975, 1977a) handicap principle proposed that signals are costly to prevent cheating. The key here is that there are often conflicts of interest between signallers and receivers; it pays offspring to beg for more than their fair share of food and it pays even poor-quality males to attract mates. Zahavi suggested that if signals were costly then this would enforce honesty so that, for example, only really hungry chicks would gain from begging and only the best-quality males could perform impressive displays. ESS models by Grafen (1990a,b) have confirmed that such costs produce a signalling equilibrium. This theory is now stimulating empirical work on the costs and benefits of signalling in relation to individual quality, and Johnstone (see Chapter 7) reviews these studies together with more recent theoretical developments.

1.6 Kinship, social evolution and breeding systems

Up to the mid-1960s, many interpretations of animal social behaviour were in terms of how it was of advantage to the group. Wynne-Edwards (1962) proposed that social behaviour was an adaptation for regulating animal populations and many ethologists also used group selection to explain behaviour. For example, Tinbergen (1964) interpreted the mobbing of a hawk by a group of birds as behaviour which, although of danger to the individual, was advantageous to the group. He argued that 'only groups of capable individuals survive — those composed of defective individuals do not, and hence cannot reproduce properly. In this way the result of cooperation of individuals is continually tested and checked, and thus the group determines, ultimately, through its efficiency, the properties of the individual'.

Group selection was criticized most cogently by Williams (1966) and Lack (in an Appendix to his 1966 book). They showed that clutch size and also many social interactions enhanced an individual's fitness and argued that adaptations evolve for individual benefit, not for the benefit of a group. This left the problem of altruism, behaviour which increased the fitness of others at a cost to the altruist's own personal reproduction. The key insight to understanding the evolution of altruism was provided by Hamilton (1964a,b). He argued that individuals can pass copies of their genes on to future generations not only through their own reproduction (direct fitness) but also by assisting the reproduction of close relatives (indirect fitness). Hamilton's now famous rule specifies the conditions under which reproductive altruism evolves and there is good evidence, especially from insects (see Chapter 9) and social vertebrates (see Chapters 10 and 11), that kinship provides the key to understanding altruistic behaviour. The huge interest in cooperative breeding during the last 20 years is largely inspired by Hamilton and is a good example of how empirical work is often driven by advances in theory.

Further impetus to studies of social systems came from pioneering papers by Parker and Trivers. Parker (1970a) recognized the importance of multiple mating for the evolution of reproductive behaviour and coined the term 'sperm competition' for sexual competition after mating, when sperm from different males compete for fertilization of a female's ova. Trivers laid the foundations for theories of conflicts in family groups (see Chapters 9 and 10), including male–female conflict and parent–offspring conflict (Trivers, 1972, 1974; Trivers & Hare, 1976). He emphasized the importance of the earlier conclusions of Bateman (1948) that different factors limit reproductive rates in males and females. Females tend to invest more in offspring and their reproductive rate is usually limited by resources. A male, on the other hand, has the potential to father offspring at a faster rate than a female can produce them. For males, therefore, reproductive success is limited more by access to females. Trivers argued that females should be more choosy in mating while males should practice a 'mixed reproductive strategy', both guarding their mates and also attempting to gain extrapair matings. Parker (1979, 1984b) also emphasized the conflict between the sexes, and both he and Maynard Smith (1977) used ESS models to analyse how these may be resolved.

In the last decade, detailed behavioural observations combined with molecular measures of parentage (e.g. DNA fingerprinting) have confirmed the importance of sperm competition and sexual conflict. This has revolutionized our view of mating systems. Just compare, for example, Lack's (1968) conclusion 30 years ago that monogamy predominates in birds because 'each male and each female will leave most descendants if they share in raising a brood' with the current evidence for widespread mixed paternity and sexual conflict (see Chapter 6). While it is clear that males compete for mates and females are indeed choosy, there is still vigorous debate about exactly what benefits females gain from their choice (see Chapter 8).

1.7 Critical views

Behavioural ecology has been criticized on a number of grounds during the past 20 years, the main points being as follows.

1.7.1 Determinism

Lewontin and colleagues (Lewontin et al., 1984; Lewontin, 1991) have interpreted the position of behavioural ecologists as implying genetic determinism. Statements such as '...a gene for altruism...' could be read as meaning that if an individual carries a certain gene or combination of genes it immutably and irrevocably behaves in a particular way. This would be biologically unsound, since the adult phenotype depends on complex interactions between genes and environment during development. It could also, Lewontin and colleagues argued, be open to an ideological interpretation in which social policies were built around the assumption that humans are purely products of their genetic heritage. Although it would be fair to say that behavioural ecologists have generally underplayed the complexities of behavioural development, the phrase '...a gene for...' is never used to imply genetic determinism, but rather as shorthand for '...genetic differences between individuals that are potentially or actually subject to selection'; in other words it implies gene selectionism not genetic determinism (Dawkins, 1982).

1.7.2 Panglossianism

A parody of behavioural ecology (and of neo-Darwinism in general) is that every last detail of any organism's behaviour, anatomy, physiology and so on can be explained by natural selection, for example the fact that carrots are orange and parsnips white. Gould and Lewontin (1979), in a classic article, coined the phrase 'The Panglossian paradigm', referring to Dr Pangloss in Voltaire's *Candide*, who took the view that everything was always for the best. According to its critics, the pure adaptationist approach is flawed for two reasons. First, it ignores the fact that evolution is a historical process influenced by chance, and some of the outcomes would be quite different if the video of life were played again (Gould, 1989). Differences between species or phyla may have no 'logic', they may just be the one chance outcome out of a huge range of possibilities. Second, at any one moment in time, the degree of perfection of adaptation of behaviour, physiology and so on are constrained by many factors such as developmental flexibility, historical accident and interactions between genes. While these criticisms do not undermine the value of a Darwinian framework as a powerful device for analysing and predicting behaviour, they are reflected in the fact that in this edition of the book we have included greater emphasis both on the analysis of historical events in evolution, increasingly possible because of new phylogenetic data, and on the constraints that limit adaptation here and now.

1.7.3 Anthropomorphism

Kennedy (1992), in a thoughtful and detailed critique of behavioural ecology, points to the dangers of using (often anthropomorphic) linguistic shorthand to describe functional categories of behaviour. One of his key points is that the use of functional labels such as 'foraging', 'mate searching' and 'parental allocation' tend to become substitutes for a proper analysis of what is actually going on and may even encourage anthropomorphic interpretations of behaviour. For instance, a behavioural ecologist interested in 'honest signalling' (itself a dangerously anthropomorphic term!) between nestlings and parents in a particular bird species might analyse in a model, or by experiments, whether or not '...parents allocate resources in response to the need of individual nestlings...' Kennedy would argue that, in fact, parents respond to stimuli, including those emitted by the offspring, and that this is what determines the pattern of feeding. The terms 'allocation' and 'need' are, in effect, terms related to functional considerations of optimal reproductive strategies and should not be taken to constitute causal explanations of behaviour.

Kennedy's critique recalls the distinction between two of Tinbergen's four questions, function and causation, and it also serves as a reminder that in carrying out experimental manipulations of behaviour one can normally only determine the stimuli to which animals respond, not the functional reasons for a response, which are inferred from the logic of natural selection.

The main lessons for behavioural ecologist are these.

1 That functional models should not be taken to imply particular mechanisms or decision rules (in analyses of 'tit-for-tat' as a model of cooperation, the metaphor of the model has been interpreted literally by some authors to mean that animals play the actual game originally modelled by Axelrod and Hamilton — see Chapter 11).

2 That care should be taken in experimental analyses of functional models not to conflate manipulation of the stimuli to which animals respond, for example prey size or tail length, with the Darwinian interpretation of adaptation.

3 That functional labels should not substitute for a full analysis of the causes of behaviour.

1.8 Looking ahead

Sam Goldwyn neatly summarized the dangers of predicting: 'Never prophesy, especially about the future.' However, the changes signalled by the new emphases in this edition of *Behavioural Ecology* coincide with a change in the nature of the subject. As we have mentioned earlier, Wilson's (1975) predicted fragmentation of the subject of ethology into functional and causal aspects is not happening; what is now emerging is a new form of integrated study of behaviour. In order for this to flourish, one of the keys will be to embrace the powerful armoury of techniques, from gene splicing to magnetic resonance imaging that have transformed other areas of biology.

PART 2

MECHANISMS AND INDIVIDUAL BEHAVIOUR

The desert ant Cataglyphis fortis *wanders over a circuitous path for several hundred metres, yet it can return directly to its nest. How does it achieve this amazing navigational feat?* [Photograph by Rüdiger Wehner.]

Part 2: Introduction

Visitors from another planet would find it easier to discover how an artificial object, such as a car, works if they first knew what it was for. In the same way, physiologists are better able to analyse the mechanisms underlying behaviour once they appreciate the selective pressures which have influenced its function. For example, the sensory mechanisms of an insect eye are much easier to understand once it is realized that the eye functions not only to detect food, predators and mates but also light from the sky for navigation. Just as function can help an understanding of mechanisms, so mechanism can help us appreciate adaptive design. Behavioural ecologists need to know how underlying mechanisms both serve and constrain the outcomes they measure in terms of fitness costs and benefits.

The chapters in this section are all concerned with this fruitful interchange between studies of mechanism and function. In Chapter 2, Wehner shows how behaviour is affected by constraints from both the physical environment and the animal's own body. For example, how an individual best communicates depends on whether it signals in the air or in water and certain types of behaviour are beyond an animal's capability simply because of scaling factors. A small animal is unable to generate sufficient force to kick or hit effectively whereas it can bite and crush opponents. In some cases, the sensory system used is nicely adapted to the environment; a good example discussed by Wehner is the different frequencies used by bats for echolocation in relation to foraging height and interference from vegetation. In other cases, sensory systems impose constraints. Wehner shows that the single lens eyes of spiders and vertebrates have much better resolving power than the compound eyes of insects and they are also energetically more efficient. This explains why visual communication in insects is restricted to short-range encounters and why other sensory channels (acoustic, olfactory) are used for longer distances. Have insects got stuck with an inferior design feature? Not necessarily. As Wehner argues, compound eyes may have advantages over single-lens eyes in terms of panoramic vision and navigation.

Wehner's chapter also shows how complex behaviour may be the outcome of simple subroutines. A female parasitic wasp uses a simple mechanism to measure the size of a host egg (which determines how many eggs she herself should lay); some migratory birds may use a simple sun-compass mechanism to fly the great-circle route around the globe (which minimizes the distance between summer and winter quarters); desert ants use path integration combined with landmark memory to find their way home after foraging trips; and male hoverflies use a simple movement rule to guide their flight to intercept a passing female.

Giraldeau continues the mechanistic theme in Chapter 3. To behave adaptively, individuals often need to assess the quality of their habitat or of a potential mate or competitor. How do ecological factors influence the way animals gather, store and process this information? A fundamental problem is how animals measure time intervals, for example to determine foraging rates. Giraldeau describes experiments which show that birds behave according to the predictions of scalar expectancy theory. He then goes on to discuss how individuals estimate patch quality not only on the basis of their own current gain in relation to what they expect, but also by using the 'public information' provided by the behaviour of others. Such social learning often involves acquiring information about places and objects but there is surprisingly little good evidence that individuals actually copy the behaviour patterns of others. Giraldeau provides a critical account of whether social learning can give rise to the long-term behavioural traditions which form the basis of culture, considering some classic examples such as the pecking of milk bottle tops by British titmice and the washing of potatoes by Japanese macaques.

Studies of spatial memory provide especially good evidence that ecological factors influence the design of nervous systems. Bird species which cache food and then later on recover their hoards have a larger hippocampus than non-storers. In polygynous mammals, where males patrol larger home ranges than females, males also have a relatively larger hippocampus. Giraldeau reviews experiments with titmice and corvids which show that the better spatial memory of food-storing species can also be used in non-caching contexts, so it seems to result from an improvement in general mechanisms for processing spatial information rather than a specialization for food caching per se.

In Chapter 4, Sherman, Reeve and Pfennig consider whether there may be common principles underlying recognition systems. In a wide ranging review they show, for example, that anuran tadpoles prefer to associate with siblings, that tiger salamanders devour non-relatives but avoid eating close kin, that vervet monkeys give different alarm calls for leopards, eagles and snakes, that female great tits avoid mating with males that sing like their fathers, that female ground squirrels are nicer to full-sisters than half-sisters and that female mice, and perhaps humans too, prefer mates with dissimilar major histocompatibility complex genotypes. All these examples involve recognition, whether of mates, kin, prey or predators, and Sherman *et al.* argue that all involve the same three components, namely: (i) production of a label; (ii) perception of the label; and (iii) an action (acceptance or rejection). Given errors in recognition systems, selection is expected to favour an optimal balance between acceptance errors and rejection errors. The chapter discusses whether we should expect labels to be genetic or environmental in origin and shows that perception often involves comparison of the label with a learned template. These templates can be learned by self-inspection, from nestmates or from parents, and experiments show how individuals can be tricked into forming inappropriate templates (e.g. goslings learn Konrad Lorenz as their mother or warblers treat a cuckoo chick as their own offspring). Experiments also suggest that acceptance thresholds become more permissive with increasing costs or decreasing benefits from choosiness. For example, paper wasp workers are more intolerant of unrelated kin on their nest than away from their nest, female zebra finch become choosier when they are exposed to more high-quality males and hosts vary their rejection of foreign eggs in relation to their probability of being parasitized by cuckoos.

The key to assessing costs and benefits in economic models of behaviour is the concept of trade-offs. Cuthill and Houston tackle this problem in Chapter 5, focusing on how animals can make best use of time and energy. Imagine, for example, a starling searching for food during the breeding season. How should it allocate time between feeding and other activities such as singing, mate-guarding, avoiding predators or keeping its chicks warm? How fast should it fly out to the feeding grounds? Increased flight speed reduces travel time, enabling more time for feeding, but the increased energy expended has to be recouped by increased food intake. The chapter shows that animals often cope with variations in the food available by forming fat stores as an insurance. However, these stores are costly because they reduce locomotory performance, so increasing predation risk and reducing feeding efficiency. Field observations and experiments indicate that the optimum fat reserves vary with dominance, predation pressure and patterns of food availability.

Cuthill and Houston show how dynamic programming can be used to predict optimal behaviour sequences, such as patterns of feeding and resting during the day. The technique takes account of how the best behaviour varies with the animal's state (e.g. food reserves), the payoffs from various choices and the time available (e.g. time to dusk). The chapter then considers how short-term consequences of behaviour (food gain, mate attraction) can, in principle, be related to lifetime reproductive success. These functional models should help direct future work on the physiological mechanisms which control food intake. Current textbook accounts ignore the influences of environmental stochasticity, predation and social dominance on the equilibrium level of food reserves.

Darwin coined the term 'sexual selection' for selection of traits involved in competition for mates. In a key paper, Parker (1970a) recognized that sexual competition also occurs after mating when the sperm from different males compete for fertilization of a female's ova. He called this process sperm competition. In Chapter 6, Birkhead and Parker demonstrate that sperm competition is widespread, as revealed by behavioural observations of multiple mating by females and molecular genetic analysis showing multiple paternity.

They begin the chapter by arguing that sperm competition is likely to have shaped the evolution and maintenance of the two sexes. Models show that the two sexes evolve as the stable outcome in which small gamete producers (males) parasitize the resources of large gamete producers (females). In theory, even small levels of sperm competition means that it pays a male to maximize number of gametes produced (to ensure paternity) instead of contributing resources to the zygote.

The study of sperm competition provides an excellent example of how understanding mechanisms (how sperm are stored and compete in the female tract) is vital for an understanding of the function of behaviour (the costs and benefits of multiple mating for either sex). The chapter reviews Parker's studies of the yellow dungfly and Birkhead's studies of the zebra finch, both ideal model systems for combining field observations with laboratory studies. In both species there is second male advantage in fertilizations of eggs, but the mechanisms differ. In the dungfly the second male advantage fits a model of some displacement of the first male's sperm followed by random mixing inside the female's sperm stores. In the zebra finch, there is passive sperm loss from the female's tract and the second male gains an advantage simply because fewer of his sperm are lost by the time fertilization occurs. The authors discuss the consequences for how males and females should behave to maximize their fitness. Early studies took very much a male-centred view of multiple mating, focusing on how males maximized paternity through frequent copulation, mate-guarding, the insertion of mating plugs or the production of anti-aphrodisiacs in the seminal fluid. Current work suggests that females often control sperm competition and may gain both genetic benefits and better resources by mating with more than one male. The chapter reviews the evidence for this and discusses how mating systems may arise as the outcome of sexual conflict. An unresolved issue is to what extent females can manipulate sperm inside their reproductive tracts so as to favour fertilization by particular males.

Chapter 2

Sensory Systems and Behaviour

Rüdiger Wehner

2.1 Introduction

Behavioural ecologists agree that if they metaphorically regard animals as 'decision-makers' (Krebs & Kacelnik, 1991), they do not imply that the animals decide on the basis of conscious choices or any appreciation of the computational structure underlying the problem to be solved. Instead, they assume that simple processes mediate apparently complex behavioural decisions. This assumption flies in the face of what the majority of neuroscientists have thought all along (e.g. Marr, 1982), namely that nervous systems form relatively complex internal representations of the outside world, and then use information derived from these global representations to accomplish any particular behavioural task that comes up. This conventional wisdom — the representational paradigm, supported also by cognitive scientists (Gallistel, 1990) — has been challenged recently by the notion that many behavioural tasks may not require elaborate representations of the external world. By exploiting constraints that are introduced when the animal interacts with its environment, special-purpose task-directed programmes may be able to solve a given behavioural problem more effectively (Ballard, 1991; Aloimonos, 1993; Churchland *et al.*, 1994). It is here that the ways of thinking of behavioural ecologists and physiologists converge.

This convergence, however, has not yet been put into action. Of course, the approaches of behavioural ecologists and physiologists differ in emphasis and focus. While the former — the functional or why-question approach — aims at an understanding of the fitness (and hence evolutionary) consequences of a particular mode of behaviour, the latter — the mechanistic or how-question approach — tries to understand the physiological machinery mediating that behaviour. Consider, for example, the case of a foraging honey bee, and in particular the question when the bee should stop collecting nectar and start to carry the load back to the hive. Functional analyses show that under a wide range of ecological conditions crop load can be predicted best by assuming that the bee maximizes energetic efficiency (energy gain per unit of metabolic cost) rather than net rate of gain (energy gain per unit of time), or any other more complex alternative (Schmid-Hempel *et al.*, 1985). However, economic models of this sort or another do not tell anything about how bees measure variables such as energy gain or foraging costs, and how they integrate these measures in order to compute the amount of nectar they should extract from the flowers visited during individual foraging trips. It does not even prove that the 'currency', which describes the bee's behaviour in economic terms, is actually computed by the animal in the way proposed by the model. Only a physiological analysis can tell what sensory mechanisms a bee employs and what neural computations it performs in order to arrive at what the behavioural ecologist thinks is the currency used in a particular foraging task.

Although until recently mechanistic and functional approaches have been entertained by researchers of different camps, they are in no way mutually exclusive, but complementary. In the example mentioned above, knowing physiologically that worker honey bees are constrained by a limited amount of flight performance, or flight–cost budget (Neukirch, 1982), may emphasize the economist's finding that energetic efficiency rather than intake rate is the animal's decisive currency.

It is upon constraints imposed on behaviour by various sources that this chapter concentrates. One source is the animal's physical environment (see Section 2.2). Certain habitats on the surface of our planet favour particular sensory channels, and as sensory systems differ in their potential for, say, resolving spatial detail, certain behavioural tasks can be accomplished only in one type of habitat or another. When it comes to constraints set by the organism itself, body size is an important although widely neglected factor (see Section 2.3). The kind and amount of sensory information that can be handled and used by a nervous system depends more dramatically on the size rather than the particular design of the system. The latter is responsible for what could be called the fine tuning of behavioural performances (see Section 2.4) — and it is here that behavioural ecologists with their intrinsic interest in micro-evolutionary processes become most intrigued.

2.2 Constraints imposed on behaviour by the physical environment

It goes without saying, but is not always fully appreciated, that the most fundamental functional characteristics of animal design have been shaped by very general properties of the physical world within which an animal lives, moves and behaves. For example, there are much larger fishes in water than there are birds in the air; the body of a fish is more perfectly streamlined than that of bird; for a fish it is more difficult to extract oxygen from its environment, but less costly to move through this environment than it is for a bird. A little exercise in physics and physiology will immediately show that all these functional differences between aquatic and terrestrial animals are due ultimately to the way in which water and air differ in such general properties like density, viscosity, oxygen content or gas diffusion rates.

Moreover, the fundamental physical properties of air and water are responsible not only for how animals gain and spend their energy, but also for how they gain the information to move about, to detect, localize and recognize objects of interest — in short, to explore their environment (see Dusenbery, 1992, for a thorough treatment of such topics). One simple example might help to make the point. Above the water surface one can see even the islands furthest away on the horizon, but one will not receive any sounds from there. In contrast, under water the visual scene gets blurred and obscured even a few metres ahead of the observer, but one may readily pick up the sounds produced by the engine of a ship too far away to be seen. The reasons for these striking differences are simple and straightforward.

Owing to the physical properties of light, vision is the most accurate source of spatial information that an animal can gain about the world. In both air and water, absorption and scattering of light decrease the brightness and contrast of the image, respectively, but these effects are much stronger in water than in air (Lythgoe, 1988). Marine fishes, for example, have responded to the strong selection pressures of their dim-light environment by boosting the light sensitivities of their visual systems in various ways (Locket, 1977; Munz & McFarland, 1977; Douglas & Djamgoz, 1990). As depth increases, the spherical lenses of their eyes become more powerful, the photoreceptors increase in size, are arranged in multiple layers or are combined to functional multireceptor units; screening pigments usually shielding the photoreceptors are lost, and light reflectors underlying the receptor layer are formed — until, at depths of 800–1200 m, eyes disappear altogether (and prevail only in some bioluminescent species). At this 'faunal break' quantum capture rates have become so low that any visual signal gets buried in photon and receptor noise, and finally vanishes.

While this suite of adaptations to environmental constraints tells a clear-cut story, other seemingly similar specializations are more difficult to interpret in terms of the optimization towards which any adaptation works. Take, for example, the spectral absorption characteristics of the rhodopsin photopigments built into vertebrate photoreceptors. The maximal absorption rates of the high-sensitive, dim-light receptors, the rods, are tightly clustered around 500 nm (Goldsmith, 1991). This is a reasonably good adaptation to the spectral light conditions prevailing at depths of about 100 m, but at lower depths, as well as at and above the water surface, the photon flux is greatest at much longer wavelengths (Fig. 2.1a). Why have rod photopigments, which are trimmed for high quantum capture rates, not responded to these strong selection pressures and shifted their maximal spectral sensitivities (their λ_{max}-values) to longer wavelengths? This question is all the more intriguing as the photopigments of the cone-type receptors containing the same chromophore (retinal-1) and the same opsin-type protein moiety as the rod-type receptors are usually well adapted to the colour of the water in which their owners live. This holds true even for closely related species inhabiting — like the snappers (genus *Lutjanus*) of the Australian Great Barrier Reef — different marine habitats, e.g. the clear blue water of the outer reef, the greener water of the inshore reefs or the more

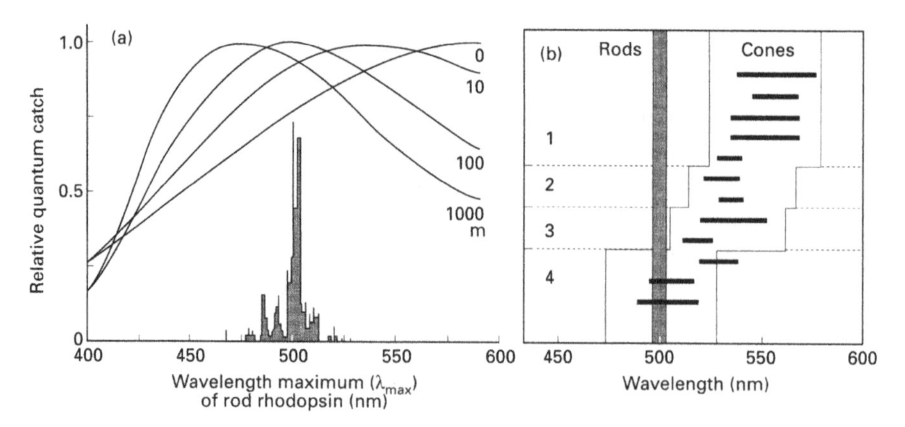

Fig. 2.1 (a) Histogram: absorption maxima (λ_{max}-values) of 274 photopigments (rhodopsins) of vertebrate rod photoreceptors. Curves: relative sensitivity (relative quantum catch) of rod rhodopsin as a function of λ_{max} (abscissa) calculated for various depths of water (0–1000 m). (Modified from Goldsmith, 1991.) (b) Absorption maxima (λ_{max}-values) of the photopigments of rods (dark grey area) and cones, i.e. double cones, (black bars) of 12 species of teleost fish belonging to the genus *Lutjanus* and inhabiting different marine habitats: 1, outer reef; 2, middle reef; 3, inner reef; 4, estuary. The left and right limitations of each bar mark the λ_{max} of the two members of the double cones present in the teleost retina. The light grey area represents the range of λ_{max}-values calculated, for each water type, to confer greater than 90% of the sensitivity of the most sensitive rhodopsin. (Modified from Lythgoe *et al.*, 1994.)

heavily stained mangrove and estuarine waters (Lythgoe *et al.*, 1994). Again, however, in all these species the absorption spectrum of the rod photopigment stays put (Fig. 2.1b).

Why does the family of rod pigments exhibit such evolutionary inertia, while that of the cone pigments does not? I ask this question, at this juncture, not in order to discuss hypotheses about the molecular biology of vision, but to caution against simplistic adaptational explanations. As this particular case shows, molecular constraints might be as significant as ecological ones. More generally, it is difficult to include in any hypothesis all the variables over which adaptation integrates. In the present context, the λ_{max}-value is certainly only one of many attributes of the rhodopsin molecule that is sensitive to natural selection. Note, for example, it might already be the high absolute sensitivity of the rods — higher by orders of magnitude than that of the cones — that limits any shift of λ_{max} to larger wavelengths. On the surface of the earth as well as in yellowish freshwater habitats, such shifts would indeed increase the number of quanta absorbed, but they would also increase the rate of dark-noise events and hence decrease the signal-to-noise ratio. In addition, rhodopsin is not only a receiver of light, but also a membrane-bound enzyme involved in the phototransduction cascade.

As mentioned before, due to the 'veil of scattered light' between the eye and the object, underwater vision is essentially a short-range affair. In contrast, underwater hearing extends into the far field. Hence, in aquatic as well as in nocturnal animals acoustic (and especially sonar) systems of orientation are much more effective than visual guidance schemes. Dolphins and bats offer prime examples.

The propagation of sound pressure waves is almost five times faster in water than in air. Furthermore, the power of sound emission depends on the product of the velocity of propagation and the density of the medium. This product — the impedance of the medium (Michelsen, 1983) — is 3500 times larger for water than for air. Consequently, low-frequency sounds as produced by baleen whales (about 20 s^{-1}) attenuate very little when travelling over large distances. Calculations of the transmission losses, which occur at various depths, especially in those layers in which sound is effectively trapped due to particular temperature conditions, show that fin whales, for instance, might hear each other over distances of several hundred kilometres (Payne & Webb, 1971). One only wonders how these whales could use an acoustic communication system in modern times, when the engines of ships produce powerful sounds exactly in the whales' frequency band.

Let us now turn to the special case of the sonar system. Whatever the medium within which such a system works, there must always be trade-off between the range and the accuracy of target detection: the higher the frequency of the emitted sound, the better the spatial resolution that can be achieved, but the stronger the attenuation of the signal as distance increases. In evolutionary terms, bats have responded to this trade-off situation by choosing their microhabitats and predatory life-styles correspondingly. For example, in comparisons across different species and genera of bats, the frequency of the echolocating sound (and, correspondingly, the best frequency of the auditory system) is inversely related to the height of the preferred foraging area (Neuweiler *et al.*, 1984). In other words, the frequency increases the closer the bats hunt to the ground or to the edges of vegetation (Fig. 2.2). Above the forest canopy, where potential prey (flying insects) are the only objects from which sounds can be reflected, a premium is paid for far-ranging (low-frequency) signals, while within cluttered environments the detection of targets against a noisy background becomes a more severe problem. In this case, high-accuracy (high-frequency) sounds are advantageous. What looks like the exception to the rule is *Megaderma lyra*, the false vampire (Fig. 2.2, M). However, this bat, which hunts close to the ground, detects its prey (beetles, birds, mice, etc.) not by using its sonar system but by listening to the sounds produced by the moving prey itself.

Due to the properties of sound transmission in air and water, some details of sonar systems should be different in aquatic and terrestrial animals. For example, sounds of constant frequency exhibit higher velocities and larger wavelengths in water than in air. Hence, the same accuracy of orientation (for data on toothed whales see Au, 1988; Würsig, 1989) requires that the echolocative sounds are of higher frequencies in aquatic than in terrestrial animals. This is indeed what occurs (Nachtigall & Moore, 1988).

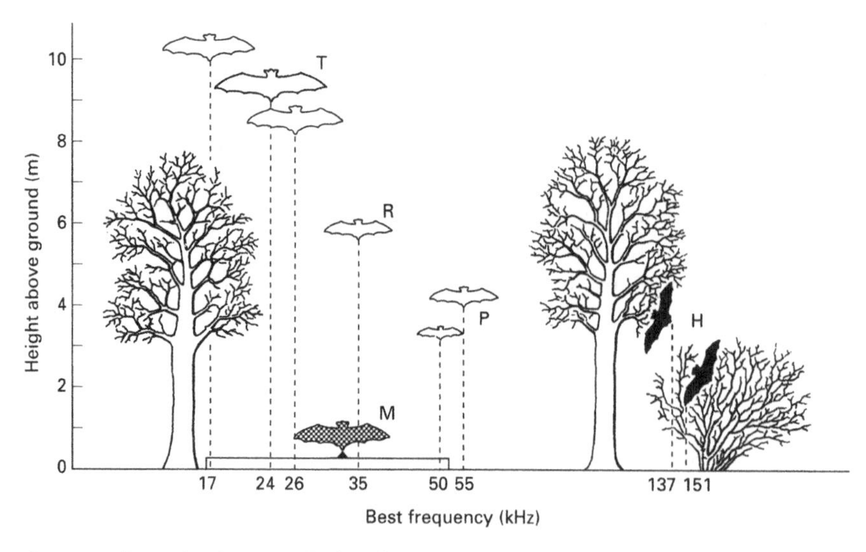

Fig. 2.2 Relationship between the best frequencies (of audiograms recorded in the dorsal midbrain) and the preferred foraging ranges of echolocating bats in southern India. The nine species of bats studied belong to the following genera: H, *Hipposideros*; M, *Megaderma*; P, *Pipistrellus*, R, *Rhinopoma*; T, *Tadarida*. (Modified from Neuweiler *et al.*, 1984.)

Similar environmental constraints apply to auditory communication, where they have been studied in both vertebrates (Wiley & Richards, 1978) and insects (Römer & Bailey, 1990). They are especially intriguing in the latter, because the frequencies of most insect songs lie in the high sonic or ultrasonic range, i.e. well above those of most vertebrates. The attenuation and degradation of these high-frequency sounds by vegetation poses intricate questions. For example, as insect habitats act as effective low-pass filters (Fig. 2.3a) and as all orthopterans studied so far have the potential for frequency analysis, the frequency-dependent attenuation of sound could by used as a means to estimate the distance between sender and receiver. However, as the amount of frequency filtering depends on the structure and density of vegetation, the frequency content of a signal does not provide an unambiguous cue. In fact, the spacing of calling bushcrickets in the field varies with the loudness of the calls (with larger animals producing more intense sounds) and the density of vegetation. As the calling males are not informed about either variable, they can maintain only acoustic rather than absolute distances to their neighbours. Males in which the sound output is experimentally reduced, or which live within denser vegetation, move closer together than the controls (Dadour & Bailey, 1990).

But, there is yet another and even more severe problem, namely to localize rather than only to detect the sound source. Due to multiple reflection and scattering of sound in vegetation, the sound field around a listening insect might become rather diffuse (Fig. 2.3b–d). Hence, nervous systems, especially

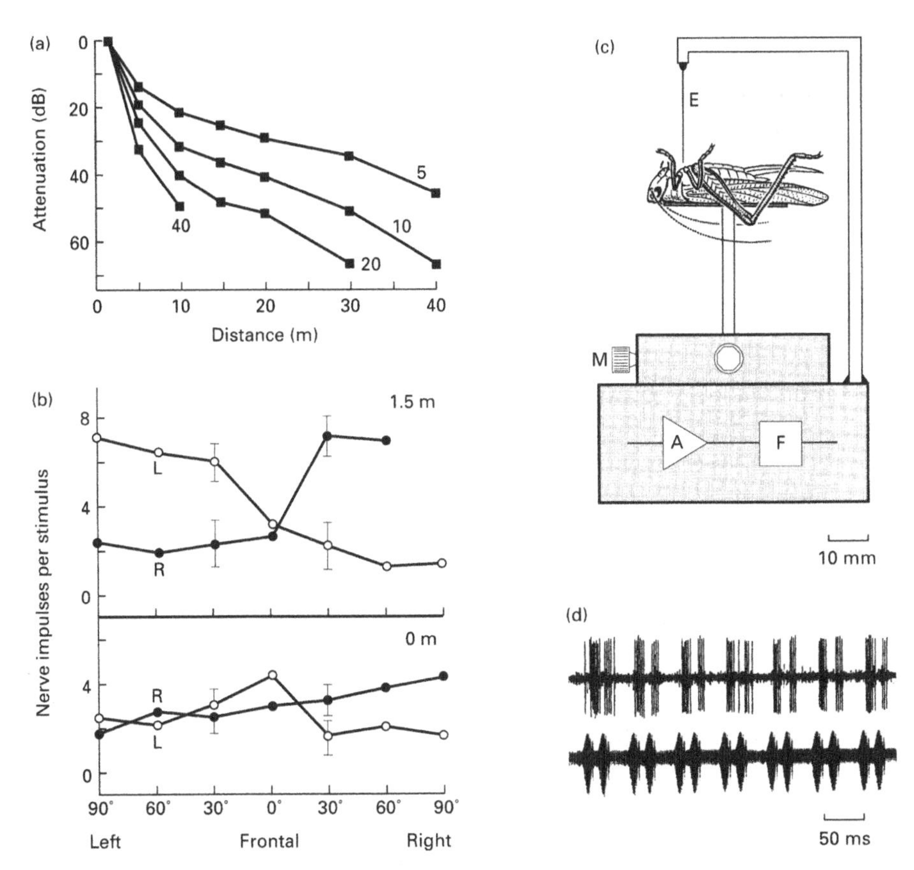

Fig. 2.3 (a) Frequency-dependent sound attenuation in a bushcricket habitat. The four curves refer to sounds of 5, 10, 20 and 40 kHz. (Modified from Römer & Bailey, 1990.) (b) Directional selectivity of a pair of auditory interneurons (T-fibres) of the bushcricket *Tettigonia viridissima*. The recordings (see c and d) were taken in dense bushland at a distance of 10 m from the sound source either 1.5 m above the ground (upper figure) or on the ground (lower figure). The solid and open circles refer to the right (R) and left (L) interneuron, respectively. (c) Recording device: portable 'biological microphone'. The device enables long-term extracellular recordings from single identified interneurons in the prothoracic ganglion. The animal is mounted with the ventral side facing upwards. A glass-insulated tungsten electrode (E) is inserted in the ganglion with a microdrive (M). The socket of the recording contains an amplifier (A), a bandpass filter (F; 0.5–5.0 kHz) and the miniature microdrive (M) by which the preparation can be moved in three dimensions relative to the fixed electrode (E). The portable recording unit is placed within the habitat at various distances, heights and directions relative to the sound source. (d) Response of an auditory interneuron (upper trace) as monitored with the portable recording unit in the field. The sound stimulus (lower trace) is the conspecific male stridulatory song. (Modified from Rheinländer & Römer, 1986.)

of small animals, face the severe problem to extract information about sound-source directions from weak directional cues. The notion 'especially in small animals' brings us to our next topic: how body size influences the way animals behave.

2.3 Constraints due to one of the most fundamental biological characteristics: body size

Animals come in various forms, but functionally even more importantly, they also come in various sizes. From the smallest to the largest, they span a range of body masses that covers more than 10 orders of magnitude (Schmidt-Nielsen, 1984). Within this range, they are not isometric, even if the organization of their bodies follows the same general pattern (or *bauplan*). Nearly all morphological and physiological variables change in proportion to each other, as body size varies: relative to body size they are scaled in non-isometric (allometric) ways. Furthermore, the constraints that pertain to body size may become so severe that they can be overcome only by a novel design. For example, unicellular organisms move by using cilia or flagella, but if animals which are only one or two orders of magnitude larger (e.g. small crustaceans) were covered with cilia, they would get nowhere. As body size increases, a new design is needed — locomotion by movable body appendages. Or, to cite another example: diffusion is an adequate mechanism for supplying oxygen to all body parts of a small organism (less than about 1 mm in diameter), but it is too slow and hence completely inadequate for oxygen supply to larger animals. A novel mechanism — transport by convection — must be added to diffusion. As can be inferred already from these two examples, size dependencies in biological phenomena are anything but trivial. In fact, the appearance of the physical environment to an organism and the organism's evolutionary response depend most strongly not on whether the organism is a bee or a bird, a worm or a whale, but on how big it is (Schmidt-Nielsen, 1984; Vogel, 1988; Pennycuick, 1992).

How does this apply to an animal's behavioural ecology? One of the most fundamental interactions between an animal and its environment is the way in which it moves about within this environment. Here, it is already as simple an aspect of locomotion as tripping that scales dramatically with body size. The bigger an animal is the harder it falls. The momentum when a (large) organism hits the ground is proportional to the fourth power of its length (note that momentum is mass times velocity and that for short falls by large creatures drag is negligible). Hence, small animals can afford to stumble, but large ones cannot (Went, 1968). If this might seem to be too trivial an example, let us turn to a more intricate mode of behaviour: the throwing of projectiles like stones or rocks, as it is practised by apes but not by smaller (although manually dexterous) animals. One might surmise that the smaller animals lack the necessary sensorimotor skills. Be this as it may, the much more fundamental reason is that they just cannot impart enough momentum to a projectile to make it an effective weapon. The momentum of a projectile thrown by an animal of proportionate mass is again proportional to the fourth power of the animal's length. By the same token, kicking and hitting can be performed only by large animals, while biting, crushing and squeezing will work for small animals as well (Vogel, 1988). In conclusion, certain types of behaviour are beyond an animal's reach for reasons not (or not only) of neural performance but simply of body size.

2.3.1 Visual acuity

When it comes to behaviour that depends on the analysis of fine spatial detail, vision provides the most accurate source of information. In accord with this potential offered by the physics of light, simple eye-spots or more advanced types of eye have evolved independently 40–60 times in almost all major groups of animals (Salvini-Plawen & Mayr, 1977), and have led to at least 10 different biological solutions to the physical problem of forming an image (Land & Fernald, 1992). Among those animals which rely most heavily on vision, two types of eye prevail: single-lens eyes and compound eyes. The former occur in a wide variety of taxonomic groups (coelenterate medusae, annelid worms, gastropod and cephalopod molluscs, insect larvae and spiders), whereas the latter are restricted, almost exclusively, to the arthropods.

It is intuitively clear — but, can be derived from optical theory as well — that visual resolution decreases with eye (and body) size, but this size dependency varies dramatically between single-lens and compound eyes. In the former the radius of the eye increases linearly with resolution, while in the latter it is proportional to the square of resolution (Land, 1981; Wehner, 1981). This prediction derived from optical analyses is confirmed by the evolutionary result (Fig. 2.4a): compound eyes are rather large and restricted to small animals. Therefore, it is costly to support them and to carry them around (Laughlin, 1995). In order to acquire a unit of visual information (a pixel), the owner of compound eyes must invest more in terms of energy expenditure than an animal equipped with single-lens eyes (Fig. 2.4b). In other words, for a given size of eye single-lens eyes offer the potential of much better resolution than compound eyes. For example, in the principal (single-lens) eyes of jumping spiders the interreceptor angle can be as small as 2.4 arc min (Williams & McIntyre, 1980) and, hence, comes close to the one found in the human fovea (0.6 arc min). As a consequence, the spider can distinguish between conspecifics and similar-sized prey at distances as great as 30 body lengths away (Jackson & Blest, 1982). An insect of about the same body size but equipped with compound eyes must get one or two orders of magnitude closer to the object to resolve the same amount of spatial detail.

Why, then, are insects and crustaceans using such an inferior optical instrument? Why have they not replaced their compound type of eye with the single-lens type? Connecting a new set of eyes to an existing neural hardware might not have been a viable option (Laughlin, in press), so that insects might have got stuck with a type of eye that worked well at low resolution, but then could not be changed. However, as many insect larvae show, insects are, in fact, able to build high-performance single-lens eyes and to use them effectively in behaviour (Wehner, 1981).

(a)

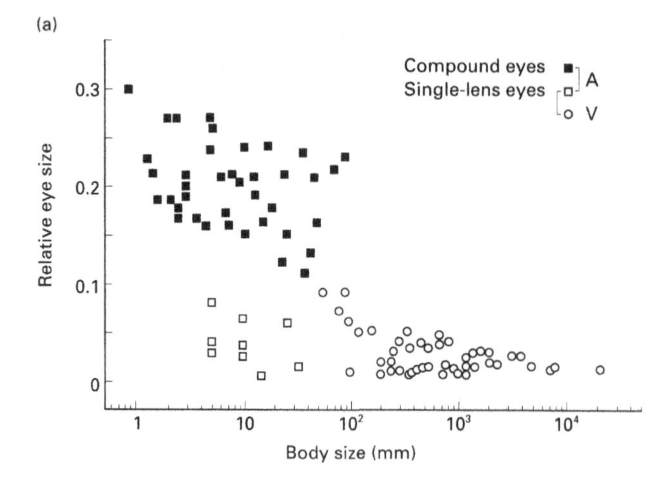

(b)

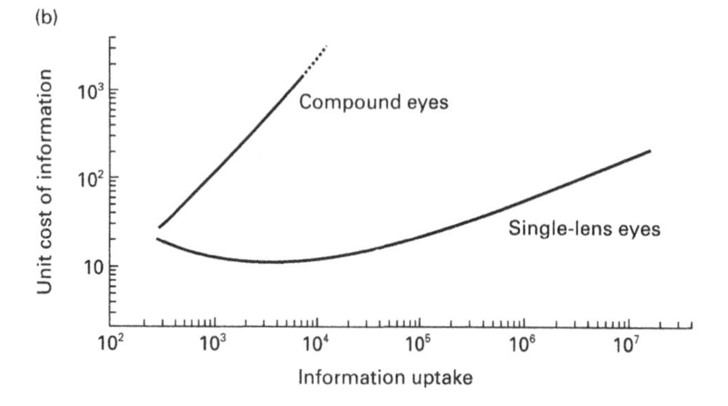

Fig. 2.4 (a) Relative size of compound eyes (filled symbols) and single-lens eyes (open symbols) in arthropods (A, squares) and vertebrates (V, circles). Body size is given in linear dimensions ('nominal length', i.e. the cube root of body mass), and so is eye size (the largest diameter of the eye). The relative size of the eye is defined as eye size divided by body size. (Modified from Wehner, 1981.) (b) The unit cost of visual information ($10^3\ \mu m^3$ transported retinal mass per pixel) plotted as a function of the total amount of information acquired (pixels per steradian of solid angle of visual space). The cost is defined in terms of the energy required to transport the volume of eye that subserves one pixel, since the metabolic cost of phototransduction is negligible if compared with the transportation costs. (From Laughlin, 1995; and personal communication.)

On the other hand, behaviour will certainly have exerted more selection pressures on eye design than merely the need for high visual acuity. For example, compound eyes may be the advantageous type of eye whenever panoramic vision becomes important. This is because compound eyes cover the surface of the head and hence create a convex — rather than concave — retina. The number of single-lens eyes a spider employs total up to eight, yet the spider cannot see all of its surroundings at once (Fig. 2.5a). On the other hand, many insects with only two compound eyes are able to view the entire visual world simultaneously, and at the same time may have at their disposal

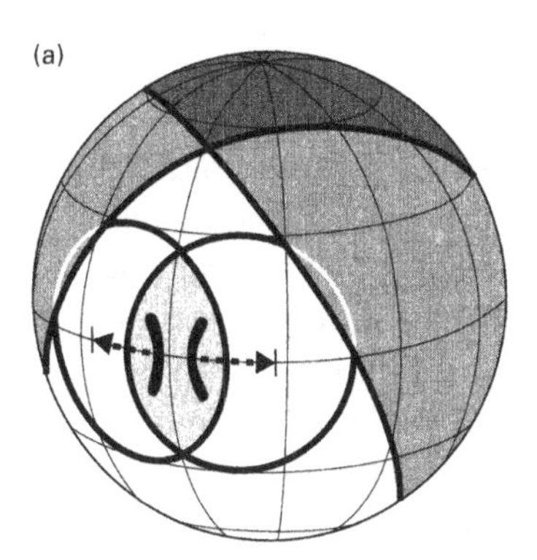

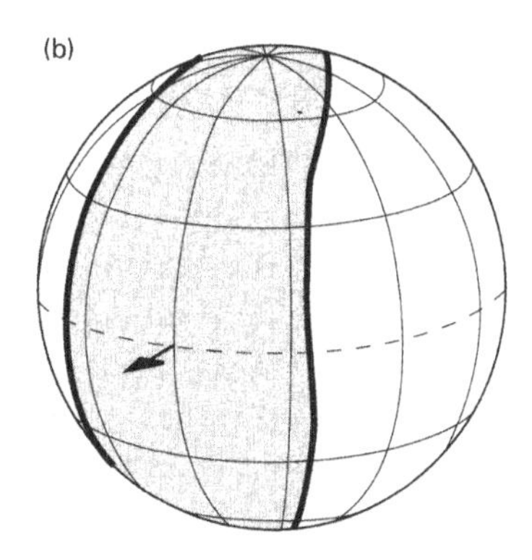

Fig. 2.5 (a) The visual field of a jumping spider, *Metaphidippus aeneolus*. The small visual fields of the anteromedian (principal) eyes shown in black can be moved over a horizontal range indicated by the heavy broken arrows. Light grey: monocular and binocular fields of view of the anterolateral eyes; dark grey: monocular and binocular fields of view of the posterolateral eyes. The visual fields of the tiny posteromedian eyes have been omitted. (From Wehner & Srinivasan, 1984; based on data from Homann, 1928; Land, 1969.)
(b) Visual field of a praying mantis, *Tenodera australasiae*. Light and dark grey indicate the monocular and binocular fields of view, respectively. The arrow points forward. The dashed line marks the horizon. (From Wehner & Srinivasan, 1984; based on data from Rossel, 1979.)

a huge binocular field to which more than 70% of all ommatidia can contribute (Rossel, 1979; Fig. 2.5b). By monitoring the apparent motion of the environment and the objects within it as the animal moves, compound eyes provide useful information as to the animal's own motion and to the landmark skyline around the moving animal. Systems of navigation and course control in which such information is used do not necessarily demand high visual acuity. One can obtain sufficiently reliable information on movement by monitoring the low spatial-frequency content of the environment. Compound eyes, viewed in this light, 'creatively' destroy unwanted information at the very first stage of vision, by using their coarse-grain optics to filter out superfluous spatial detail.

In summary, the principal design features of compound eyes — relatively poor resolution and large fields of view — enable insects to perform well in dealing with *global* aspects of their visual world. Such aspects are used in course control and navigation; these types of behaviour might have been the ones for which compound eyes have evolved primarily. For performing detailed *local* analyses, i.e. detecting and identifying objects like conspecific mates, insects must employ complicated anatomical compromises to insert acute zones in their faceted eyes (Land, 1989) and must get rather close to the object under scrutiny. This may be the reason why in insect communication visual signals are usually restricted to short-range encounters. Over larger distances, species-specific messages are conveyed through other sensory channels — olfactory, acoustic, vibrational — which thus play a more significant role in insect communication.

2.3.2 Sound-source detection

The need to localize sounds becomes most apparent in acoustic communication. In this context, only those sounds are localized and recognized that can also be produced by the animal. Here again, physical limitations abound: the smaller the animal, the higher the minimal frequency that can be generated, and the lower the distance over which sounds can be received. This is because maximum efficiency of sound emission requires that the diameter of the sound source is of the same order or magnitude as the wavelength of sound. Consequently, insects with body lengths below 1 cm are restricted generally to ultrasound, but ultrasound is a useful means of communication only in free space or at short range (Michelsen, 1983). Yet, many small insects communicate over distances of many times their body lengths, even up to metres. For doing so, they must abandon the acoustic channel and switch to substrate-borne vibrations (Markl, 1983). Such signals can travel across the surface of an insect's host plant with rather little attenuation. In addition, the efficiency of converting muscle power into vibrational power is much higher than that of the conversion into acoustic power, so that it is not only functionally more effective but also energetically less costly for a small insect to communicate through solid substrates rather than through air.

If one is small, air offers yet another 'cheap' possibility: communication by near-field air oscillations rather than sound–pressure waves. Such oscillations as caused by wing vibrations are used by *Drosophila* flies to communicate with females (Bennett-Clark, 1971) and by dancing honey bees to convey their sound message to other workers (Michelsen *et al.*, 1987). At close range these oscillations are so intense (0.5–1.0 mm peak-to-peak amplitude, 1–3 mm s^{-1}; Michelsen, 1983) that they are able to activate antennal mechanoreceptors of near-by conspecifics. As they decrease with the third power of the distance to the source, they are the signals of choice for 'private conversation'.

2.4 Constraints set by the animal's computational capabilities

In trying to understand the computational software and physiological hardware of animal behaviour, neuroscientists have often been led astray by their ideal of general, all-purpose designs. The following section of this chapter shall remind us of what we have known all along, but not always fully appreciated, namely that an animal's solution reflects a unique nervous system with adaptive limitations and particular biases. Formally similar problems may be solved by different animals in different ways depending on the animal's evolutionary history and present-day ecology. Idiosyncrasies in neural circuitry may persist as long as they do their job and as long as the animal has managed to design its way around them. Just recall the example of compound-eye vision: as the need for higher resolution increases, insects squeeze high-acuity zones into the low-acuity facet arrays of their eyes rather than exchange their type of eye for one that is intrinsically superior in terms of overall acuity.

2.4.1 Coping with spherical geometry: the egg and the globe

Ichneumonid wasps of the genus *Trichogramma* lay their eggs into the eggs of other insect species. The number of eggs which are deposited depends on the size of the host egg. In determining the volume of the spherical host, the wasp does not trace out spherical triangles and perform spherical trigonometry, but assumes a particular body posture, in which the angle between the head and the first segment of the antenna (the scapus) is related to the radius of the sphere (Fig. 2.6). This angle is probably monitored by the mechanosensory bristles located at the joint between head and scapus. Note, however, that this simple method, by which the volume of a sphere is 'computed' by relying on a simple angular measurement, works only if the wasp adjusts its body position so as to keep two other measures constant: (i) the height of the thorax above the surface; and (ii) the angle between thorax and head. Both conditions are met in the animal's behaviour.

While parasitic wasps must cope with the geometry of minute spheres, migrating birds must trace out navigational courses across the surface of the globe. On their spring and autumn migration even small birds — say, warblers and waders — can travel for several thousand kilometres non-stop over

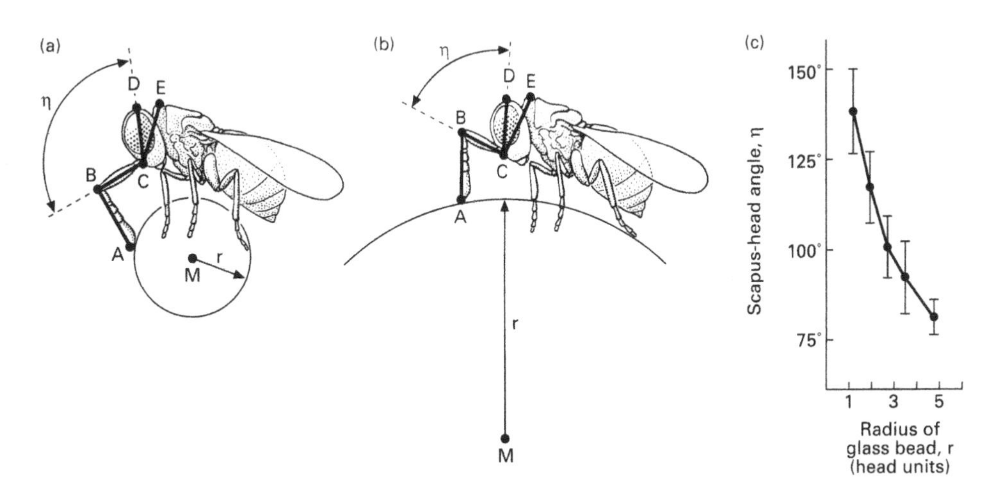

Fig. 2.6 Parasitoid wasps, *Trichogramma minutum* (Ichneumonidae), use the surface curvature of their host eggs to determine the number of progeny allocated to the host. (a, b) Female wasps examining glass-bead models of different sizes (radius r). (c) Angles between head and scapus observed for different sizes of glass-bead models. η, scapus–head angle (angle BCD); M, centre of spherical glass bead. (From Wehner, 1987; based on data from Schmidt & Smith, 1986.)

potentially hostile territory like the vast expanses of sea or desert. What are the most convenient routes these migrants should take? With all other things being equal (which they never are on the surface of our planet), the energetically least demanding way of travelling is to follow the great-circle (orthodrome) route. This route (Fig. 2.7) defines the shortest distance between two points, but is cumbersome and difficult to compute, because it intersects successive lines of longitude at different angles. However, there is a short-cut way of travelling along the great-circle route: if the bird followed a sun-compass course, but did not reset its internal clock as it moved eastward or westward, i.e. crossed different time zones, it would automatically fly along that route without having to compute it by spherical trigonometry (Alerstam & Pettersson, 1991). Radar studies suggest that this mechanism is employed by Siberian waders crossing the Arctic Ocean.

In these polar regions, where there are no topographical and ecological barriers to cross, the great-circle route is the energetically most efficient one. On the other hand, if brent geese followed the great-circle route on their way from their spring stop-over sites in Iceland to their breeding grounds in northern Canada, they would have to cross the Greenland ice cap where it is steepest and widest. Instead, they take a more circuitous route following more or less a constant-angle (loxodrome) course from Iceland to the east coast of Greenland, turn south, stay for 2–7 days within a rather delimited area, and then continue across southern Greenland on a course nearly identical to the one taken at Iceland. It is fascinating to hypothesize that the geese use their temporary halt at east Greenland to reset their internal clock from local Icelandic to local Greenlandic time, and then continue on the same sun-compass course they have followed previously (Alerstam, 1996). We know from laboratory studies in other species of birds that under exposure to a new 24-h light/dark regimen

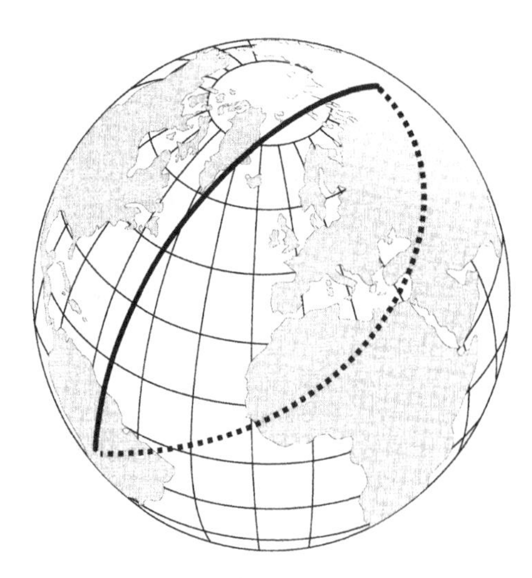

Fig. 2.7 Orthodrome (great-circle) and loxodrome (constant-angle) courses drawn on the surface of the globe. Solid line, orthodrome; dotted line, loxodrome.

it usually takes 3–6 days to recalibrate the sun compass. In conclusion, depending on the ecological needs experienced during their evolutionary history, migrating birds might take orthodrome or loxodrome courses and select and maintain either course by rather simple computational means.

Moreover, modern large-scale satellite-based radiotelemetry reveals that long-distance migrants do not travel for very long distances on either orthodrome or loxodrome courses, but seem to employ a number of navigational subroutines rather than an all-purpose system of navigation. For example, North American warblers reach their South American wintering sites by following neither orthodrome nor loxodrome courses, but by taking a wide eastward sweep across the Atlantic. Surprisingly as it might appear at first sight, this vast detour is the energetically most economic route, because it allows the birds to exploit large-scale wind and barometric pressure patterns (Williams & Williams, 1978).

In conclusion, during evolutionary time the migration routes of birds have been shaped by a number of quite different selection pressures, e.g. by synoptic weather patterns, large-scale topography, suitability of celestial or magnetic cues, etc. As Alerstam (1996) has succinctly put it, birds travel without any idea in their minds that the earth is a globe. Instead, they have responded to the selection pressures mentioned above by developing a number of sophisticated tools of migration and ways to integrate and adapt these tools in intricate ways.

2.4.2 Reading skylight patterns and landmark panoramas: the insect navigator

Insect navigation, although less impressive than bird navigation in spatial scale, is just as intriguing in terms of behavioural sophistication — all the more as in insects some of the underlying mechanisms have recently been unravelled in unprecedented detail. The best studied and, in fact, most eminent insect navigators are eusocial hymenopterans like ants and bees. These central place foragers (Stephens & Krebs, 1986) continually move to and from their central place, the site of the colony, to retrieve widely scattered food particles from the colony's environs. It is during these foraging endeavours that the spatial coherence of the superorganism — the colony — is relaxed and is re-established only by the navigational performances of the individual colony members.

An example of this performance is given in Fig. 2.8. While foraging in a circuitous way over distances of more than 200 m, *Cataglyphis* ants of the Sahara desert navigate by path integration. They continually measure all angles steered and all distances covered, and integrate these angular and linear components of movement into a continually updated vector always pointing home. This is a computationally demanding task which *Cataglyphis* must solve with its small nervous system, and — as recent research has shown (e.g. Wehner *et al.*, 1996) — it does so by relying on a number of rather simple subroutines.

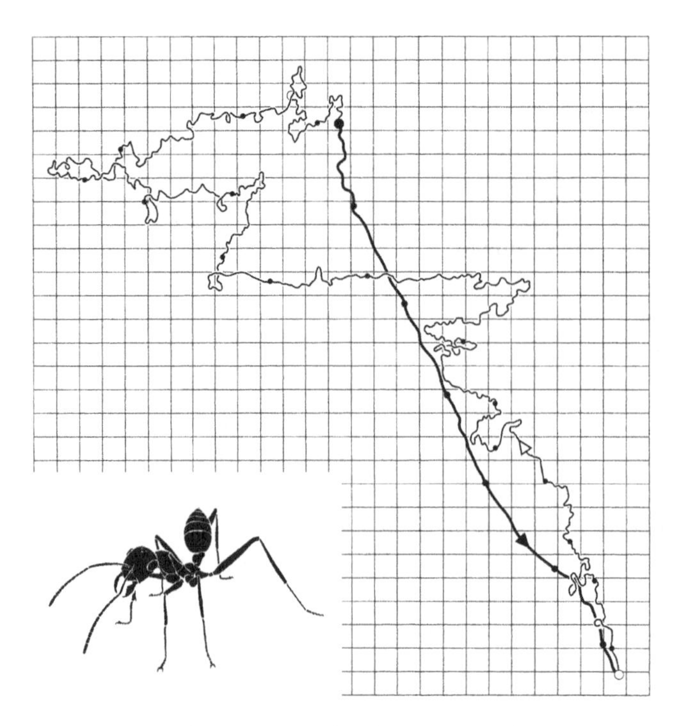

Fig. 2.8 Outward and homeward paths of an individually foraging desert ant, *Cataglyphis fortis* (see inset). The start of the foraging excursion (nesting site) and the site of prey capture are indicated by the open and the large filled circle, respectively. Time marks (small filled circles) are given every 60 s. Grid width, 5 m; length of outward path (thin line), 592.1 m; length of return path (heavy line), 140.5 m. (Modified from Wehner & Wehner, 1990.)

In the present context, let me focus on the compass used by *Cataglyphis* to monitor the angular components of its movements. This compass is a skylight compass based primarily on a peculiar straylight pattern in the sky, the pattern of polarized light (or E-vector pattern; Fig. 2.9a). At this juncture, it is not important to understand this pattern in any physical detail. Suffice it to say that in any particular pixel of sky the electric (E) vector of light oscillates in a particular direction, and that the photoreceptors in a particular region of the ant's (and bee's) eye are sensitive to these oscillations.

But, there is more to it. The skylight pattern the insect experiences is not static, but changes with the elevation of the sun above the horizon (compare left and right half of Fig. 2.9a). These dynamics notwithstanding, *Cataglyphis* can infer any particular point of the compass — say, 30° to the left of the solar meridian — from any particular point in the sky. This task must be accomplished when, for example, under cloud cover or due to experimental tricks played by the human investigator, E-vector information can be obtained only from a small gap of clear sky. If a physicist tried to solve this navigational problem from first principles, he/she would have to run a rather sophisticated series

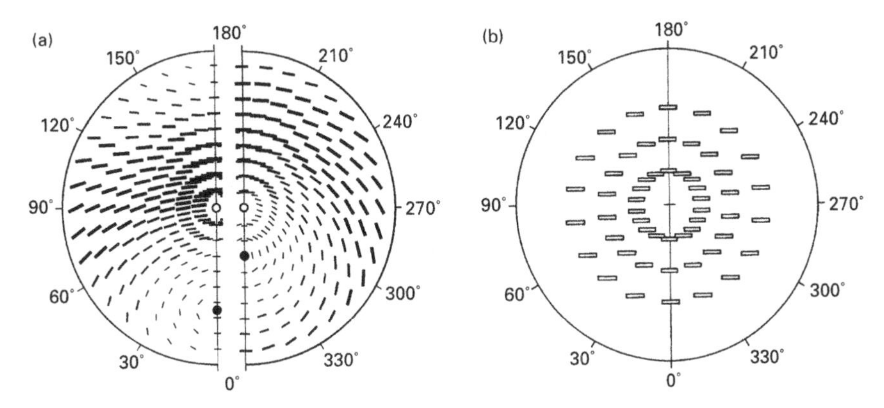

Fig. 2.9 (a) Two-dimensional representation of the E-vector pattern in the sky shown for two elevations of the sun (black disc): 25° (left) and 60° (right). The orientation of the E-vectors (the directions of polarized light) are represented by the orientation of the black bars. The sizes of the black bars mark the degree (percentage) of polarization. The zenith is depicted by an open circle. 0°, solar meridian; 180°, anti-solar meridian. (b) The ant's internal representation of the sky as derived from behavioural experiments. The open bars indicate where in the sky the insect assumes any particular E-vector to occur. This 'template' is used invariably for all elevations of the sun (for details see Wehner, 1994).

of measurements and computations, and use spherical geometry to perform elaborate three-dimensional constructions (Fig. 2.10). The insect navigator, however, comes programmed with a strikingly simple internal representation — or 'template' — of the external E-vector patterns (Fig. 2.9b). This fixed neural template resembles the skylight pattern when the Sun is at the horizon, but differs from it for all other elevations (for a review of the behavioural and neurobiological analysis of the E-vector compass see Wehner, 1994).

The tantalizing question now is this: how can *Cataglyphis* navigate correctly by using an internal representation of the sky that is not a correct copy of the external world? In the full blue sky, with the entire E-vector pattern available, the best possible match between the external pattern and the internal template is achieved when the insect is aligned with the solar — or anti-solar — meridian, the zero-point of the compass. (The distinction between these two principal meridians can be made by other means.) The match decreases systematically, as the animal rotates about its vertical body axis, i.e. selects other compass directions.

Due to the discrepancy between the internal template and the external pattern, mismatches occur whenever only parts of the skylight pattern are available. For example, an individual E-vector is matched with its corresponding detector in the template only when the animal deviates by a certain angular amount from the solar meridian, so that the zero-point of its compass gets shifted. Consequently, navigational errors arise when the foraging animal experiencing, say, the entire skylight pattern, is suddenly presented with a small patch of sky. In fact, it was from these systematic errors observed in the

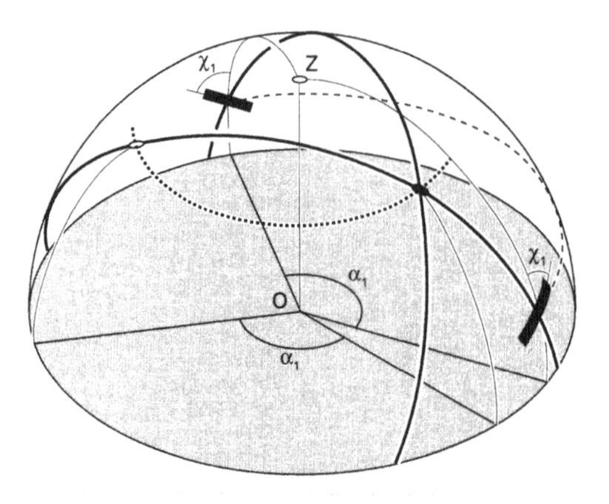

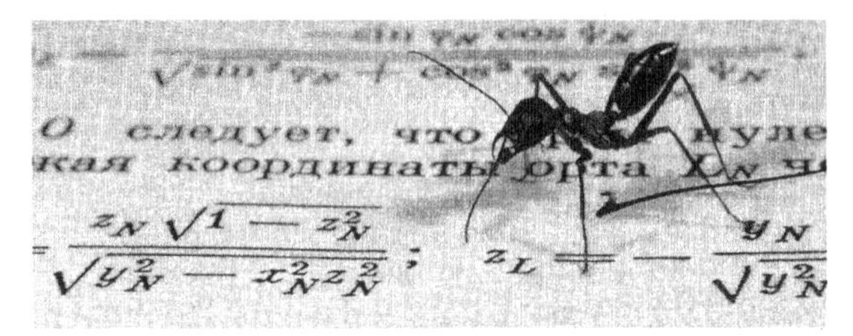

Fig. 2.10 How a physicist could infer the position of the sun from viewing at least two small patches of skylight: First, determine the E-vector direction (χ) in the two patches of sky (this is a problem in itself, which is not discussed here). Then, construct the great circles that run at right angles through the E-vectors. The position of the sun (filled circle) is defined by the intersection of the two great circles. If only one E-vector is visible, the position of the sun cannot be determined unambiguously. Provided that the elevation of the sun is known at the particular time of day the (two) intersection points of the great circle and the parallel of altitude (dotted arc) defined by the elevation of the sun yield the correct position of the sun (filled circle) and a second one (open circle) that is separated by the azimuthal distance α_1 from the correct one. *Cataglyphis* does not perform such constructions but uses a generalized template of the sky (see Fig. 2.9b). (Modified from Wehner, 1981.) In the lower part of the figure *Cataglyphis* inspects a paper of Frantsevich (1982) outlining a model of E-vector navigation.

insect's behaviour under certain experimental conditions that evidence for the internal template could be derived in the first place. Note, however, that such errors do not occur when the animal is *continuously* presented with the same patch of sky. It then always uses the same reference direction, be this the actual solar meridian or any other celestial meridian that is characterized by the currently best match between the template and the outside world. For comparison, if a human navigator used a magnetic compass in which the needle erroneously but consistently pointed towards east rather than north, this 'defective' instrument would work as a reliable compass as well.

In conclusion, evolution has managed to build into the insect navigator a nervous system that includes only some general knowledge about the geometrical characteristics of the celestial world, but this partial knowledge is sufficient if the navigator restricts its field trips to short periods of time. The insect assumes that the celestial hemisphere does not change during any of its particular foraging excursions. Given its short foraging times which lie in the range of tens of minutes rather than hours, this is generally a valid assumption.

Similarly, simplified solutions are employed by insect navigators when landmarks are used to back up the noisy path integration system. As shown in Fig. 2.8, in which an ant performed its foraging and return path within the expanses of a flat and featureless Saharan salt pan, the path-integration system worked without the aid of any landmark-based information. However, as this system is prone to cumulative errors, landmark guidance helps to reduce homing time, often substantially. The effective use bees and ants can make of landmarks as visual signposts (Wehner, 1981) has led to the assumption that insects are able to assemble map-like internal representations of the landmarks in their nest environs and then use such 'cognitive maps' to find their way to a familiar site, even from points at which they have never been before (Gould, 1986). Although this notion has generated a lot of excitement — and controversy — more recent research has shown that ants and bees are indeed able to make intensive use of landmark information in relocating nesting and feeding sites, but that they do not incorporate such sites into a map-like system of reference (Wehner & Menzel, 1990; Dyer, 1996). The strategies they employ are more straightforward, foolproof and largely sufficient for the task to be accomplished.

One task, for example, is to pinpoint the nesting site after the path integration system has led the animal into close proximity of the goal. As suggested by the experiments described in Fig. 2.11, ants seem to acquire a two-dimensional visual template — or 'snapshot' — of the three-dimensional landmark array around their nest, and later move so as to match this template as closely as possible with the current retinal image. This matching-to-memory routine can be studied best by distorting the training array of landmarks and recording the animal's responses to its altered visual world. In these experimental situations particular matching algorithms are able to indicate at which locations a better (partial) match is obtained than at any other location in their immediate neighbourhood, and it is at these locations that the local peaks in the insect's search density profile occur (Cartwright & Collett, 1983).

This snapshot-matching mechanism used in landmark guidance might have some fundamental neural traits in common with the template mechanism employed in skylight navigation. The obvious difference is that the skylight patterns are predictable, but the landmark configurations are not. Hence, the E-vector template can be hardwired, as it actually is, but the landmark snapshots must be acquired during the animal's individual foraging life.

In conclusion, the insect obtains landmark-based information not by taking a bird's eye — or a bee's eye — view of the terrain over which it travels, but gains this information successively and by egocentric perceptions during the process of path integration. This context-bound acquisition and retrieval of

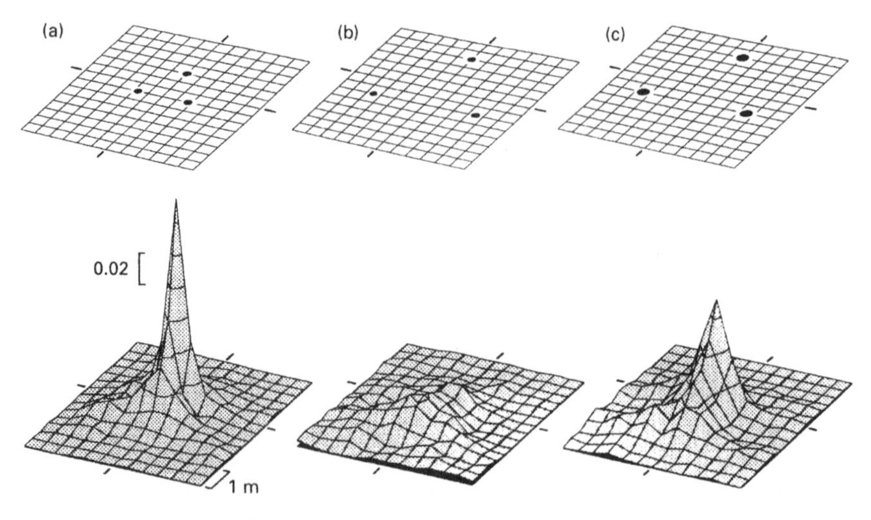

Fig. 2.11 Search density profiles of desert ants, *Cataglyphis fortis*, trained to the centre of an array of three cylindrical landmarks. The test area containing three different landmark arrays is shown in the upper figures. (a) Landmarks in the training position. In the training area (not shown) the nest is positioned in the centre of an equidistant triangle formed by the three cylinders. (b) Landmarks separated by twice the training distance. The ants behave as though lost. (c) Landmarks twice the training size and separated by twice the training distance. Again, a match between the stored image ('snapshot') and the current retinal image can be achieved when the ants are in the centre of the landmark array. However, due to the larger distance of the landmarks from the goal (as compared to the training situation shown in (a)), motion parallax cues are weaker, and hence the search density profile is broader than in (a). The results in the three experiments are in full accord with the matching-to-memory hypothesis. (Modified from Wehner *et al.*, 1996.)

landmark information reduces the danger of getting inappropriately trapped by similar landmark configurations present elsewhere in the animal's environment. For example, the snapshot-matching mechanism, by which the ant finally pinpoints its nesting site, is activated only after the path-integration system has been reset to zero (Wehner *et al.*, 1996). In addition, the insect can take snapshots at various sites and from various vantage points, and can even align them as sequences of visual images like 'beans on a string' along frequently travelled routes. As such routes can be entered — and familiar sites can be approached — from various vantage points, landmark memories are retrieved and used in quite flexible ways. This flexible use of site-recognition and route-guidance mechanisms leads to navigational performances that might give the impression of map-based behaviour.

2.4.3 Computing interception courses: male pursuits and fly-ball catching

Another — and beautifully simple — example of how a difficult computational problem is turned into a tractable one is provided by male hoverflies pursuing and finally catching passing females. As Collett and Land (1978) have shown

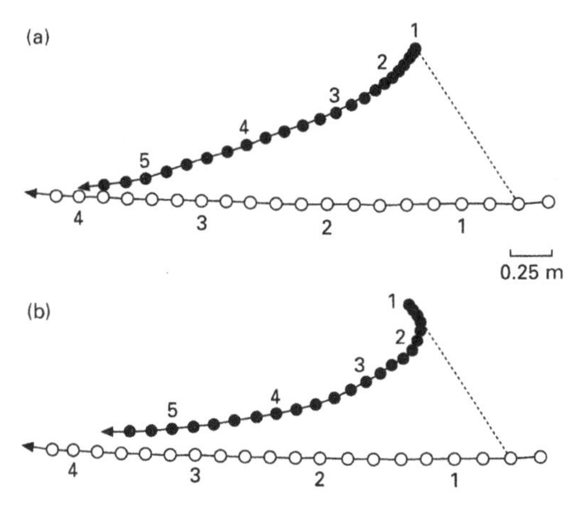

Fig. 2.12 (a) Film recording of a hoverfly male, *Volucella pellucens*, pursuing his quarry. Positions of male (filled symbols) and quarry (in this experimental case a black wooden block, 1.5 cm in diameter; open symbols) are given every 20 ms and numbered every 100 ms. The broken line indicates the line of sight between male and quarry 20 ms before the fly accelerates. As shown by the male's trajectory, the male sets out on the proper interception course. (b) Simulation of the male's behaviour on the assumption that he does not adopt an interception course but tracks his quarry, i.e. turns continuously towards it. This simulation does not describe the fly's real behaviour. (From Collett & Land, 1978.)

by filming hoverflies in the field, a male fly is able to foresee the female's flight path and to compute the proper interception course (Fig. 2.12a). The male initially does not turn towards his quarry, when the latter is first seen, but immediately sets out on an interception course. Theoretically, this task can be accomplished by a simple calculation only if the male 'knows' the absolute size and the absolute velocity of his female target (as well as his own acceleration when speeding up to catch the target) and incorporates these 'biological constants' into his neural computations. If these constants are given, the male can obey the simple rule that the size of the turn he makes ($\Delta\phi$) depends on the initial position (θ_t) and velocity ($\dot{\theta}_t$) of the target image within the male's visual field as follows: $\Delta\phi = \theta_t - 0.1\ \dot{\theta}_t \pm 180°$. The data indicate that the initial turns of the males obey this rule and lead to collisions between the male and the target, if — and only if — the target is a conspecific female. As in biological terms there is no need for a hoverfly male to chase anything other than a female, it is rather likely that natural selection has incorporated into the male's nervous system all the information about the female's flight behaviour that the chasing male needs to know.

It is not only a male hoverfly that must compute interception courses, but also a human fielder running to catch a cricket high-ball. In principle, the ball's path across the sky could be computed by a set of differential equations based on the observed curvature and acceleration. Obviously, fielders running for a high-ball do not get engaged in such intricate computations. Instead, they seem to follow one or the other simple rule. One hypothesis holds that they select a running path that maintains a linear optical trajectory for the ball relative to the wicket and the background scenery. In short, a fielder is supposed to adjust his speed and direction so that the (apparent) trajectory of the ball looks straight (linear-trajectory hypothesis: McBeath *et al.*, 1995). If the ball is hit directly at the fielder rather than at an angle to either side, another simple rule might be used, namely to select a running path that keeps the apparent speed of the ball constant (zero-acceleration hypothesis: McLeod & Dienes, 1993). Both strategies, which receive support from video recordings of running paths, do not tell the fielder where or when the ball will land, and hence he does not run to the point where the ball will fall, and then wait for it. They simply set him on a course which will ensure interception — and this is all that matters.

2.5 Outlook

Behavioural ecologists and physiologists share a mutual interest in each other's efforts. In the case of the former, this interest is obvious, because behavioural ecologists are keen to learn how neural information-processing mechanisms might have constrained the functional design of the behaviour they analyse in economic terms. It is perhaps less obvious that physiologists should be interested in knowing why it is that a particular neural subsystem mediating a particular kind of behaviour has evolved in one way rather than another.

Until recently, behavioural scientists have been preoccupied with the belief that physiological mechanisms underlying behaviour have been designed from first principles (e.g. Mittelstaedt, 1985). They have usually aimed at outlining the complete algorithmic solution to a given behavioural problem, and then asked the physiologist to discover how this solution is implemented in the hardware of the nervous system. This is the classical approach 'neuroethologists' have entertained for decades. However, neurophysiological analysis is technically demanding, and exhaustive reconstructions of entire neural subsystems are even more so. All too easily does one get lost amidst the hurly-burly of the higher nervous centres. Are such herculean efforts worth it?

In this state of affairs, physiologists have learned an important lesson: that the mechanisms they study are adaptations tailored to particular ecological needs rather than general-purpose processing devices. It is these needs that the physiologist should be concerned with, in order to be able to formulate the right questions in the first place. Let me provide an example by going back to the safe ground of my favourite organism, the desert ant *Cataglyphis*. Only after we had realized that the awe-inspiring navigational performance of a homing ant could be dissected into a number of simpler special-purpose subroutines, each responsible for a particular aspect of the task, were we able to look properly at the underlying sensory and neural mechanisms. A system must be designed to solve the problem in question — but no more; or as

Diamond (1993), while discussing design limits of physiological systems, and especially the question of how physiological capacities are matched to their expected loads, has neatly put it: 'How much is enough but not too much?'

One day, historians of science might well come to the conclusion that the recent developments in behavioural ecology have had an impact on the way physiologists started to think in evolutionary terms and have caused them to promote what could be dubbed — analogous to Huxley's (1940) connotation of 'new systematics' — 'new physiology' (*sensu* evolutionary physiology). Hence, there is hope that these recent developments in conceptual approaches will help to bridge the gap in our understanding of what is economically desirable — in terms of the functional design of a given behavioural trait — and what is, after all, physiologically feasible.

Chapter 3

The Ecology of Information Use

Luc-Alain Giraldeau

3.1 Introduction

The use of information is of central importance to a number of behavioural systems. Information, for instance, is involved in social decisions (see Chapters 4, 10 and 11), communication (see Chapter 7) and selection of mates (see Chapter 8). It can affect population structure (see Chapter 15) as well as the distribution of animals over habitats (Lima & Zollner, 1996; see Chapter 16). The goal of this chapter is to introduce behavioural ecologists to a diverse set of questions concerning information use by animals. The chapter explores how an animal's ecology affects the way it gathers, stores and processes information. It asks questions about the types of information acquired (but, see also Chapter 2) and the temporal scale of their effects on behaviour. Section 3.2 reviews information usage as it concerns resource estimation problems while Section 3.3 deals with its use in a pre-detection context. Subsequent sections deal with what some argue to be specialized forms of information use. Section 3.4 considers the extensive comparative research on spatial memory, especially in the context of avian food caching. Section 3.5 explores social learning that involves the use of information produced by other individuals (public information). Section 3.6 deals with cultural transmission of information, paying special attention to the factors that promote or retard the cultural transmission of behavioural innovations. The chapter ends with some hints of the future research directions for the emergence of a behavioural ecology of information use and cognition.

3.2 Quality estimation

This section looks at quality estimation models. The decision to settle in a habitat, forage in a food patch or escalate in combat often hinges on an individual's subjective assessment of the quality of the habitat, food patch or opponent, respectively. The models deal with how animals gain subjective assessments of the parameters required to make adaptive decisions. Most examples involve foraging, possibly because the study of learning uses food almost exclusively as a convenient reward, and because the emergence of foraging theory has underlined the importance of cognitive constraints in economic decision making. Despite an emphasis on foraging decisions, the intention remains that the topics should be of relevance to decision processes in other behavioural systems.

3.2.1 Estimating time: scalar expectancy theory (SET)

The ability to estimate short time intervals is likely to be of survival value in a number of circumstances, ranging from foraging to predator avoidance and fighting. In foraging theory, an animal's ability to measure and remember time

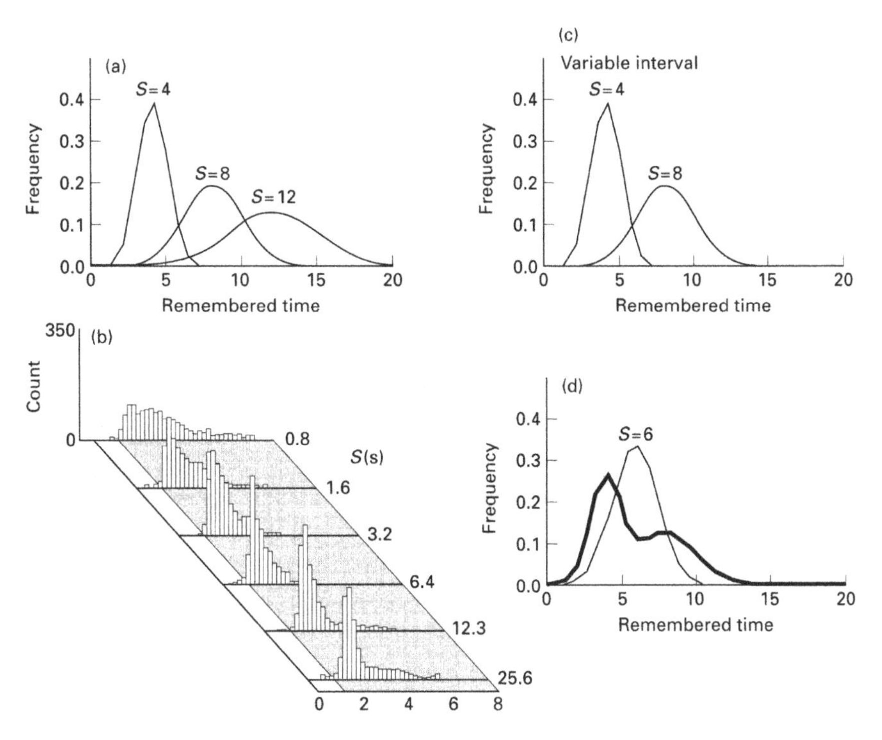

Fig. 3.1 (a) An example of three memory distributions resulting from the experience of three different time intervals (S) to reward as assumed by SET. Keeping the coefficients of variation constant as S increases forces flattening of the distributions. (b) The pooled frequency distributions of relative giving up intervals (GUI/S) for six starlings exposed to six values of S. Note that the distributions are similar in mean and variance, especially for the three largest values of S. The greater variance observed for the shortest S may indicate a lower limit to the variability of remembered distribution. (From Kacelnik *et al.*, 1990.) (c) SET memory distributions for two equiprobable intervals (S = 4 or 8 s). (d) The mixed distribution (thick line) resulting from the variable intervals in (c) is superimposed to the distribution resulting from a fixed interval of S = 6 s. Note the slight negative skew of the distribution for the variable compared to fixed interval distribution, even though both distributions have the same mean of 6 s. The negative skew makes it more likely that a shorter value of S will be remembered from the memory of variable intervals.

intervals is of particular importance to the estimation of resource quality whether it involves Bayesian (see Section 3.2.2) or linear operator (see Section 3.2.3) models. Of the many theories concerning how animals estimate time (Gibbon *et al.*, 1988; Gallistel, 1990) one, SET (Gibbon, 1977; Gibbon *et al.*, 1988; Gibbon & Church, 1990; Kacelnik *et al.*, 1990), has been used in behavioural ecology.

SET assumes that animals measure and memorize time elapsed (S) to a significant event (i.e. food reward). The memory of S, however, is imperfect and for each sample of S experienced a distribution of values centred around S is memorized. When an animal needs to remember S, it draws a value randomly from the distribution so that the distribution's variance determines the range of recall error (Fig. 3.1a). Two aspects of SET are particularly relevant to behavioural ecology:

1 the relationship between the distribution's variance and S;
2 the effect that variability in experienced S has on recall error.

Variance and interval size

SET hypothesizes that the variance of the memorized distribution increases with the magnitude of S but that the coefficients of variation of all stored distributions remain constant. Hence, the memorized distributions for different S values can all be made identical by simple scalar transformations of each other (Fig. 3.1a). For instance, starlings (*Sturnus vulgaris*) required to exploit food patches in operant devices, have a temporal memory that conforms to properties of SET (Brunner *et al.*, 1992). Starlings foraged in an experimental environment where patches provided prey at a fixed unchanging interval S, but depleted suddenly after some unpredictable number of prey. The only cue that signalled depletion was time since the last prey capture. A bird that had perfect temporal estimates of S would have left the patch once S had elapsed without a prey. Leaving sooner would have cost some prey, leaving later would have wasted time. The distribution of intervals that starlings waited before giving up a patch as empty was indeed bell-shaped as SET assumed (Fig. 3.1b). Moreover, variance in observed waiting times grew with the S experienced and most distributions were superimposable following a scalar transformation exactly as predicted by SET (Fig. 3.1b).

Variability of experienced S

Up to now we have considered cases where animals experienced constant, unchanging values of S. In most natural situations, however, intervals between significant events such as the appearance of drift prey in a brook or the courtship displays of a mate will vary. For instance, prey may be encountered at variable intervals ranging say, from 2 to 8 s. What effect will this variability have on the memory of intervals for that food patch? SET postulates that the memory distribution of time intervals is the mixture (aggregate) of the distributions corresponding to each experienced *S*. So, for instance, an option that is equally likely to provide rewards after 4 or 8 s will have a temporal memory that is the mixture of distributions around 4 and 8 s (Fig. 3.1c). An inevitable consequence of the scalar property that forces memory distributions of longer *S* to be flatter, and hence more variable than those for shorter *S*, is that the mixed distribution will be negatively skewed compared to a distribution for a fixed value of *S* with the same mean (Fig. 3.1d). It follows that when animals are confronted with a choice between a fixed and a variable option with the same mean interval, they are slightly more likely to recall a shorter interval from the aggregate than from the fixed distribution. SET, therefore, predicts preference for the variable alternative, assuming shorter delays are preferred (Fig. 3.1d).

The consequences of SET can be of broad significance to models of patch exploitation (Kacelnik *et al.*, 1990; Reboreda & Kacelnik, 1991; Kacelnik & Todd, 1992). These have just started to be considered in other foraging systems such as risk sensitivity (Reboreda & Kacelnik, 1991), diet choice (Shettleworth & Plowright, 1992) and sampling (Shettleworth *et al.*, 1988). Extending SET to estimation of reward size (Bateson & Kacelnik, 1995) as well as fighting and mate choice decisions may prove valuable.

3.2.2 Estimating quality in unchanging environments: Bayesian models

Bayesian estimation is a way of forming an estimate of an object's value on the basis of the combination of current sample information acquired from the object and a prior expectation of object value distribution in the environment. The exact way in which current and prior information are combined characterizes the Bayesian updating process (McNamara & Houston, 1980; Stephens & Krebs, 1986). Transposed to a patch foraging context, Bayesian updating involves combining the prior expectation of the distribution of prey in patches with current patch-sample information to generate an updated estimate of the number of prey currently remaining in the patch (McNamara & Houston, 1980; Green, 1980, 1984; Iwasa *et al.*, 1981). A simple graphical illustration of Bayesian updating under three different expected patch quality distributions should suffice to make the point qualitatively.

Consider three possible types of environments. In one, prey are underdispersed such that the number of prey per patch (quality) is highly variable with patches containing either many or few prey. In another, prey are overdispersed such that each patch contains relatively similar numbers of prey. In a third, prey are randomly and independently dispersed. For all three types of environment, time elapsed while in the patch and number of prey encountered are sufficient patch-sample statistics for efficient Bayesian estimation of current patch quality, and the updated estimates must decline with increasing time in the patch (Iwasa *et al.*, 1981; Fig. 3.2). Depending on the prior expected

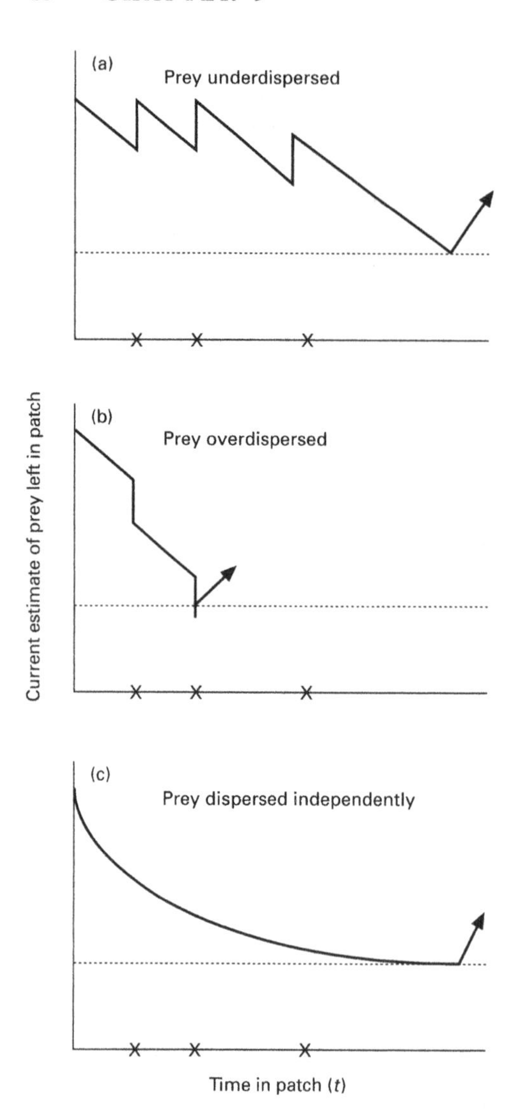

Fig. 3.2 Graphical representation of Bayesian estimation of the number of prey items left in a patch in environments that differ in the way prey are distributed. The thick lines show how the estimated number changes over time and the number of encounters (X) with prey. The dashed horizontal line gives the critical number at which an organism ought to abandon the patch (arrow). Note that in all three cases the expected number of prey remaining always declines with time spent in a patch. The difference lies in the effect of prey encounters on the estimated number remaining. When prey are underdispersed (a), each encounter with an item raises the expected number of items remaining in the patch, because patches tend to contain many or very few items. However, when prey are overdispersed (b), each encounter makes the expected number of prey remaining decline, because patches vary little in prey number. Hence, the only useful information provided here is that the patch now contains one less prey item. When prey are dispersed randomly and independently (c), prey encounters say nothing about the number of prey remaining because prey are dispersed independently. In that situation the best foragers can do is to decrease their estimate of the number of prey remaining as a function strictly of time in the patch. (Modified from Iwasa *et al.*, 1981.)

distribution of prey, however, encountering a prey has radically different consequences on the updated estimate of the number of prey remaining in the patch. For the underdispersed environment, each prey encounter increases the likelihood that the patch contains still more prey because patches contain either very many or very few prey. So, the updated estimate of the number of prey remaining in the patch increases following each encounter (Fig. 3.2a). In the overdispersed environment, where all patches are of similar quality, encountering a prey now means that the patch likely contains one less prey. So, the updated estimate of the number of prey remaining in the patch now declines with encounters (Fig. 3.2b). In the environment where prey are distributed randomly and independently, encountering prey says nothing about the number of prey remaining in the patch and so has no effect on the updated estimate. Here, the best policy is to decrease the estimation with patch time irrespective of encounters (Fig. 3.2c).

Because updated estimates increase with prey encounters in underdispersed environments, the use of Bayesian estimation in those cases predicts that the most successful individual:

1 has the highest updated estimate of the number of prey remaining in the patch;

2 exploits the patch more extensively;

3 requires longer unrewarded search before its updated estimate declines to the threshold for patch abandonment (Fig. 3.2a).

Predictions for an overdispersed environment are the exact opposite (Fig. 3.2b). The patch departure of budgerigar (*Melopsittacus undulatus*) dyads foraging in the laboratory for millet seeds hidden in fine gravel follow predictions of Bayesian updating (Valone & Giraldeau, 1993). The birds foraged in environments where seeds were either under- or overdispersed. In underdispersed environments the birds' success predicted their order of departure from the patch. A bird left first mostly when it was the least successful member of the pair and last when it was the most successful. In addition, the most successful member of the pair tolerated longer runs of unrewarded search before leaving than the less successful one. When the same birds foraged in overdispersed environments, their individual success no longer predicted the order of patch departure nor the length of unrewarded search (Valone & Giraldeau, 1993).

Behaviour consistent with Bayesian estimation has also been reported in desert granivores (both birds and mammals: Valone & Brown, 1989), inca doves (*Columbina inca*: Valone, 1991), black-chinned hummingbirds (*Archilochus alexandri*: Valone, 1992a), cranes (*Grus grus*: Alonso *et al.*, 1995) and bluegill sunfish (*Lepomis macrochirus*: Wildhaber *et al.*, 1994). One non-foraging example concerns the male amphipod (*Gammarus lawrencianus*) that is increasingly reluctant to give up a female the longer he has amplexed with her, as if upgrading her reproductive value during amplexus (Hunte *et al.*, 1985). Future work should explore how animals construct prior distributions, how and whether they update these priors, how they collect patch sample information and the frequency with which they update their estimates.

Bayesian estimation may be a powerful tool for estimating patch quality, especially when prior distributions do not change quickly over time. Changing priors makes updating patch sample information messier because individuals must alter their prior expectation as they update their current estimates. Linear operator models may offer simpler devices for tracking changing environmental conditions.

3.2.3 Tracking quality: linear operator models

Here we consider situations where resource distributions change such that no stable prior expectation can be acquired. To track change in a useful way, individuals must distinguish between local stochastic fluctuations in parameter values from more significant directional movement. They must also take into account the declining reliability of information over time. Tracking models usually reduce the influence of local irrelevant fluctuations by integrating information (i.e. averaging) over some time-span. Declining reliability is generally dealt with either by discarding or devaluating outdated information.

Discarding is used by memory window models where means of environmental parameters are estimated over a range (window) of experiences, dropping the oldest events as new ones are added to the estimate (Cowie, 1977). Despite their appeal, memory window models may be difficult to test (Cowie & Krebs, 1979; Mackeney & Hughes, 1995). The alternative approach, which involves devaluating outdated information, is more commonly invoked and forms the basis of linear operator learning rules (Kacelnik *et al.*, 1987).

Studies of linear operator models (also called exponentially moving weighted averages; Devenport & Devenport, 1993, 1994) usually involve two components: (i) an algorithm that updates information (the linear operator); and (ii) one that prescribes the behavioural decision (the decision rule) once estimates of the alternatives have been updated (Houston & Sumida, 1987). A number of different linear operator rules have been proposed as updating algorithms (reviewed by Kacelnik *et al.*, 1987). One, the relative payoff sum (RPS) (Harley, 1981), has historical importance if only because it had initially been described as a serious candidate for an evolutionarily stable learning rule, a rule that if adopted could not be outcompeted by any other updating algorithm (Harley, 1981; Maynard Smith, 1982). Whether the RPS is an evolutionary stable strategy (ESS) rule, however, remains controversial (Houston & Sumida, 1987; Tracy & Seaman, 1995). Nonetheless, the rule has been tested experimentally with some qualitative success (Milinski, 1984), so it is worth going over it in more detail, if only as an illustration of how it works.

The updating part of the RPS rule, like all linear operators, partitions the experience of each behavioural alternative into past and present. The past and the present are given a relative weight and then summed to yield an updated estimate for a given behavioural alternative. The updating process is repeated for each potential alternative. The decision component of the RPS rule uses the updated estimates to determine which behavioural alternative should be expressed in the future. For the RPS the decision rule is to use an option in proportion to its relative value, matching:

$$P_i(t+1) = \frac{V_i(t)}{\sum_{j-1}^{k} V_j(t)}$$

where $P_i(t + 1)$ is the probability of responding to option i on trial $t + 1$. The numerator gives the value of alternative i after the last trial t, while the denominator is the sum of values for all k alternative after trial t. Whether matching is an appropriate decision rule is not our concern here (Houston & Sumida, 1987). The main focus is on how RPS updates estimates of $V_i(t)$.

The RPS rule updates the value of alternative i after trial t as a linear operator of the form:

$$V_i(t) = \alpha V_i(t-1) + (1-\alpha) r_i + Q_i(t)$$

where α $(0 < \alpha < 1)$ the memory parameter, sets the weight of past (α) and present $(1 - \alpha)$ experience. $Q_i(t)$ is a measure of the rewards obtained from alternative i during trial t. r_i acts as a prior or residual value for alternative i $(V_i(0) = r_i)$. The peculiarity of RPS is that without rewards, the updated value of i converges to r_i rather than to zero as it would in all other linear operator models. The residual ensures some non-zero chance that any of the k alternatives will be used, even when one alternative is overwhelmingly more valuable, acting somewhat like an insurance against overlooking a previously depleted resource that has unexpectedly recovered. To date, empirical support for the RPS rule is weak at best (Milinski, 1984; Kacelnik & Krebs, 1985).

Some studies provide results that are partly consistent with linear operator learning rules. For instance, both tracking of travel times by foraging starlings (Cuthill *et al.*, 1990, 1994), and monitoring of interprey encounter times in pigeons (Shettleworth & Plowright, 1992) seem to follow linear operator rules. In addition, linear operator rules are also consistent with the observation that distance travelled to next flower by bumble bees (*Bombus bimaculatus*) depends on the sequence of the last three flowers visited (Dukas & Real, 1993). In all of these recent cases, foragers devalued past experience strongly, sometimes responding only to the very last event (Kacelnik & Todd, 1992; Cuthill *et al.*, 1994). Future work should explore whether the weight of past versus current experience is a fixed cognitive constraint adapted to a species' ecology (Mackeney & Hughes, 1995) or possibly a decision variable that can be adjusted to each specific tracking problem encountered by individuals (Valone, 1992a; Devenport & Devenport, 1993; 1994; Cuthill *et al.*, 1994).

3.2.4 Estimating quality in social environments: using public information

Conceivably, in social environments, the activity of other individuals can provide 'public information' (Valone, 1989) concerning food location or quality. Because the rate of Bayesian updating (see Section 3.2.2) is limited by an individual's sampling rate, a group of *G* individuals using public information can sample at up to *G* times the rate of solitary individuals, reaching reliable estimates of patch quality *G* times faster or *G* times as reliably for any given sampling period (Clark & Mangel, 1986). Public information, therefore, can provide an advantage to group foraging behaviour (Clark & Mangel, 1984).

Starlings are social foragers and sample a food patch by probing their bill into the substrate (Tinbergen, 1981). Public patch-sample information, in this context, is provided by other individuals' probes of the patch. Templeton (1993) used starlings to explore whether public information is used in a patch-sampling context. In one laboratory experiment (Templeton & Giraldeau, 1996), starlings foraged on artificial patches that offered 30 probe sites. The birds expected patches to be either partly or completely empty so they had to probe a number of empty sites before deciding that a patch was empty. To avoid effects of food depletion the birds were always tested while probing totally empty patches. The number of probes made before giving up a patch was measured for each subject as it foraged alone then in the company of a partner. Two types of partners were provided. A low public information partner had been trained to probe just a few holes while a high public information partner had been trained to probe all 30 sites. Starlings altered their probing in a way that is consistent with the use of public information. Individuals probed most when foraging alone. As expected from public information use they sampled fewer holes with the high public information than the low public information partner (Fig. 3.3). Since no food was present during testing, the difference cannot be due to faster food depletion with the high public information partner.

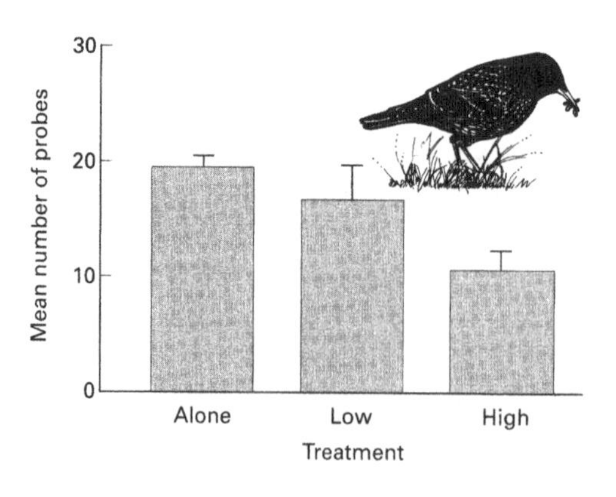

Fig. 3.3 Templeton and Giraldeau's (1996) experimental evidence of public information use by starlings faced with a patch-quality estimation problem. The mean (± SEM) number of empty sites probed by five starlings before giving up a patch is highest when tested alone, and lower when foraging with a partner. Moreover, when foraging with a partner, a subject samples more when paired with a low-information partner than with a high-information partner.

Templeton's (1993) starling work forced individuals into subject and partner roles. However, in normal circumstances sampling within groups will likely be a frequency-dependent game with some individuals exploiting the public information provided by others without contributing any in return. Indeed, when pairs of starlings are confronted with the problem of tracking the quality of a variable alternative food patch, individuals specialize into samplers and non-samplers (Krebs & Inman, 1992). Consequently, foraging groups may generate much less public information than was anticipated initially (Clark & Mangel, 1984, 1986).

3.3 Detecting cryptic prey: search image or search speed?

Prey items can pose detection problems for a number of different reasons, two of which have been the focus of considerable study: crypticity and mimetism. Research on model–mimic systems focuses mostly on information processing and decision-making aspects (Getty, 1985). Most of its development involves the use of signal detection theory, a variant of Bayesian statistical decision theory (Getty, 1985; Getty & Krebs, 1985; Getty *et al.*, 1987). Behavioural studies involving the detection of cryptic prey, on the other hand, focus mostly on the perceptual (Tinbergen, 1960; Dawkins, 1971) and attentional (Dukas & Ellner, 1993) processes as well as optimal search tactics (Gendron & Staddon, 1983; Gendron, 1986; Reid & Shettleworth, 1992; Getty, 1993; Getty & Pulliam, 1993; Dukas & Clark, 1995). This section concentrates on detection of cryptic prey.

Tinbergen (1960) suggested that foragers confronted with cryptic prey develop a 'search image' that enhances their detection. The formation of a search image, which is thought to involve only the detection process, requires successive encounters with the prey, and strongly interferes with the concurrent acquisition or use of a different search image (Pietrewicz & Kamil, 1979; Lawrence & Allen, 1983; Gendron, 1986; Guilford & Dawkins, 1987; Reid & Shettleworth, 1992).

Support for the search image hypothesis has been reported in a number of observational and experimental studies (references in Guilford & Dawkins, 1987; Endler, 1991; Reid & Shettleworth, 1992). Most evidence consists of apostatic prey selection, i.e. that a cryptic prey type is practically excluded from a forager's diet when it is rare, but is over-represented above some threshold abundance. Dawkins (1971) cautions that alternative hypotheses can account for apostatic prey selection. For instance, learning where or how to search for prey, improving capture or handling efficiency, and changing prey preference can all lead to apostatic selection without a search image. Apostatic selection can also result when cryptic prey cause foragers to reduce their search rates in order to increase the likelihood of detecting cryptic prey (Gendron & Staddon, 1983; Gendron, 1986; but see also Getty, 1993; Getty &

Pulliam, 1993). So, clear, irrefutable evidence for a search image remains elusive (Guilford & Dawkins, 1987).

The reduced search rate hypothesis has emerged as a serious challenger to the search image hypothesis (Gendron & Staddon, 1983; Guilford & Dawkins, 1987). It states that predators include more cryptic prey in their diet, not because they detect them more efficiently, but because they search more carefully. The search rate and search image hypotheses make a number of common predictions, but the crucial prediction is that reduction in search rate allows a predator to detect any other equally cryptic prey type whereas a search image supports only one prey type. So, a predator offered two equally cryptic prey types concurrently will choose only one if it forms a search image but will be equally likely to find either if it reduces its search rate.

This prediction was tested with a series of elegant experiments involving pigeons (*Columba livia*) (Reid & Shettleworth, 1992). The experiments consisted of presenting pigeons with wheat grains dyed either yellow, brown or green on a multicoloured aquarium gravel background. Preliminary experiments showed that brown and green seeds were equally likely to be selected by pigeons when presented in equal numbers on the same multicoloured background. They were, therefore, considered equally cryptic on that background. Yellow seeds, however, were conspicuous. Birds pecked in an operant device to gain access to a small piece of multicoloured substrate. First, the substrate contained only one seed. Then, after a number of single-seed presentations it contained two. If pigeons form a search image during the starting series of encounters with a prey type, then when offered a choice between two equally cryptic types they should only peck at the seed type used in the initial series. If, instead, they reduce their search rate then they should be equally likely to peck at either cryptic seed type when offered a choice. Pigeons pecked more frequently than expected by chance at the prey type used in the initial run, as predicted by the search image hypothesis. Familiarity with a prey type acquired during the initial runs cannot explain the preference because no preference was found when the same seed types were presented against a uniform beige background that made them conspicuous. As predicted by the search image hypothesis, pigeons developed a search image that was specific to a single cryptic prey type at a time and only developed such an image if the prey was cryptic.

The cognitive level at which search images are formed remains uncertain. Do search images involve perceptual or attentional mechanisms? A plausible cognitive account of search images involves predators learning to attend to a specific feature, or combination of features of a prey type that enhance detection (Reid & Shettleworth, 1992). The definition supposes that search images can allow the detection of several prey types if they happen to share the features the predator learns to attend to. The extent of interference between the concurrent operation of different search images and the factors that govern the encounter rates required for search image formation remain open questions.

3.4 Spatial memory as an adaptive cognitive specialization

Here we explore the possibility that the use of information in some cases qualifies as a specialized adaptation that involves qualitatively or quantitatively different learning and memory processes. Two distinct approaches can be used to address the issue: comparative and experimental. The comparative approach asks whether similar selection pressures gave rise, through convergent evolution, to common behavioural and neuroanatomical adaptations (Krebs, 1990; see Chapter 14). The experimental approach, on the other hand, compares the performance of different species confronted with a common laboratory problem. The objective is to establish whether different cognitive adaptations involve augmentation or reduction of common cognitive systems or, instead, are supported by separately evolved dedicated systems (Shettleworth, 1990).

3.4.1 Comparative approach

There is little doubt that an animal's ecology has placed selective pressures on its neuroanatomy, especially when sensory and motor processes are involved (Krebs, 1990). Now there is increasing evidence that selective pressures influence brain structures normally associated with higher levels of information processing. For instance, Pagel and Harvey (1989) found that insectivorous and frugivorous bats and primates tended to be more encephalized than folivorous ones. The difference is thought to be related to the larger ranges required by insectivores and frugivores coupled to the spatial and temporal unpredictability of those food types compared to leaves. Similarly, birds with larger song repertoires possess larger higher vocal centres and brain nuclei associated with song production. Crepuscular and nocturnal birds have relatively larger olfactory bulbs than diurnal birds (Healy & Guilford, 1990). Finally, birds with larger relative forebrains are more frequently reported to use novel foraging techniques (Lefebvre *et al.*, 1996a).

Of all areas comparing ecology, behaviour and neuroanatomy, the use of spatial memory for food caching represents the most extensive research programme involving the greatest diversity of taxa (e.g. Krebs *et al.*, 1989; Sherry *et al.*, 1989, 1992; Shettleworth, 1990, 1995; Clayton, 1995a; Jacobs, 1995). The following sections, therefore, focus on spatial memory in the context of retrieving cached food. Note that parallel research programmes also occur in bird song (Nordeen & Nordeen, 1990; Marler, 1991; Nottebohm, 1991; Catchpole & Slater, 1995).

Avian food caching: spatial memory and the hippocampus

Food storing has evolved in a wide range of taxonomic groups (Vander Wall, 1990), but careful comparative work in birds has focused on two groups of passerines: the Paridae (titmice and chickadees) and Corvidae (jays and nutcrackers). In parids, single pieces of food (seeds, insects) are stored for a few hours to a few weeks in hundreds of scattered cache sites that are never re-used (Krebs, 1990). Corvids store many thousands of seeds for longer times, up to 7–11 months in Clark's nutcracker (*Nucifraga columbiana*: Vander Wall, 1990).

Monocular occlusion experiments have provided elegant demonstrations of the use of spatial memory in food-storing birds. The procedure has animals storing and then recovering food while only one eye is available (Sherry *et al.*, 1981). It provides evidence for memory because, in birds, visual pathways of the two eyes cross completely at the optic chiasm such that, for the most part, each hemisphere receives input from the contralateral eye (Clayton & Krebs, 1993, 1994a,b). So, marsh tits (*Parus palustris*) allowed to store seeds while one eye is covered with a small, opaque, plastic cup fail to recover food above chance levels within the first 3 h when a different eye is available for retrieval, indicating that information stored in the other hemisphere (memory) was required to recover the seeds (Sherry *et al.*, 1981; Clayton & Krebs, 1993). Similar results were also obtained with corvids (Clayton & Krebs, 1994b).

Given that avian food caching requires remembering the spatial locations of hundreds to thousands of caches for periods ranging from days to months, the question remains whether some specialized neuroanatomical structure is required to accomplish the task. In mammals, considerable experimental evidence points to the hippocampus as a brain structure associated with higher cognitive functions such as memory and processing of spatial information (Squire, 1992). Lesion studies with black-capped chickadees (*P. atricapillus*) provide indisputable evidence that the avian hippocampus, which includes the hippocampus and area parahippocampus, are involved in spatial memory (Sherry & Vaccarino, 1989). Birds that were either unlesioned, lesioned in the hippocampal region or lesioned in the hyperstriatum ventralis, a brain area unrelated to spatial memory, stored seeds normally in aviaries containing artificial trees. However, only unlesioned birds and those with non-hippocampal lesions showed memory for cache sites. Hippocampal-lesioned birds could not recover their seed caches above chance level, although their ability to learn other tasks such as colour discriminations appeared totally unimpaired (Sherry & Vaccarino, 1989). Studies of avian hippocampus development also demonstrate involvement of the avian hippocampus in the retrieval of cached food (Clayton & Krebs, 1994c; Clayton, 1995a,b; Healy *et al.*, 1995). For instance, the volume of the hippocampal area relative to the rest of the telencephalon in the food-storing marsh tit depends upon actually retrieving stored food. The act of retrieving stimulates growth of hippocampal volumes while deprivation of opportunities to cache and retrieve induces loss of hippocampal volume (Clayton & Krebs, 1994c). There is even some suggestion that the onset of seasonal hippocampal neurogenesis described in some food-storing birds may be triggered by the seasonal need to cache and recover stored food (Krebs *et al.*, 1995).

Comparative studies involving both New and Old World avian taxa show that food caching is associated with enlarged hippocampus (Fig. 3.4). Hence, selection pressures that give rise to food caching also lead to reallocation of nervous tissue in favour of the hippocampus. Note that increased hippocampal volume does not necessarily imply an increased total brain size (Harvey & Krebs, 1990). Some unknown factor, therefore, appears to constrain a species' total brain volume. As a consequence, selection for increased hippocampus volume might have imposed some potential, yet undocumented, cost to other neural functions (Krebs, 1990). Nonetheless, these comparative avian studies show a positive relationship between relative hippocampus volume and food-storing behaviour in two phyletically different groups (Fig. 3.4).

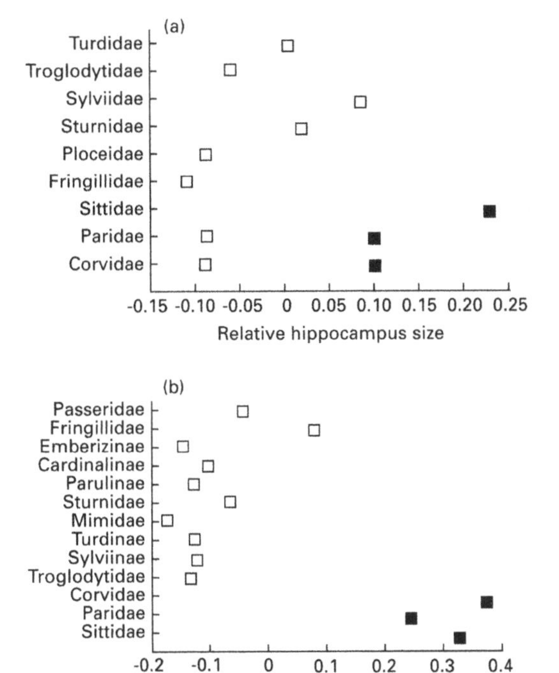

Fig. 3.4 Residual variation in hippocampal volume after body size and telencephalon volume effects have been removed through multiple regression. Dark squares are food storers, open squares non-food-storers. (Data from (a) Krebs *et al.*, 1989; (b) Sherry *et al.*, 1989. Figure taken from Krebs, 1990.)

Sex differences in spatial memory and hippocampal size

The spatial memory of voles has been studied extensively because some species show dramatic sexual dimorphism in home-range sizes depending on their mating system (Gaulin & FitzGerald, 1986, 1989). For instance, adult males of the polygynous meadow vole (*Microtus pennsylvanicus*) have home ranges in the breeding season that are four to five times larger than those of adult females or immatures (Gaulin & FitzGerald, 1989). By contrast, the monogamous prairie (*M. ochrogaster*) and pine (*M. pinetorum*) voles show no sexual dimorphism in home-range size (Gaulin & FitzGerald, 1986, 1989).

Polygynous males are sexually selected to occupy larger ranges because the greater a male's range, the more females it interacts with (Jacobs, 1995). In line with expectations, the hippocampus of males of the polygynous species are significantly larger than the female's, while no such sexual dimorphism is apparent in the monogamous species (Jacobs *et al.*, 1990). Moreover, the meadow vole's sexual dimorphism in hippocampal size disappears outside the breeding season, a time when the male's home range is of comparable size to the female's. That means that hippocampal cells are grown each year prior to the breeding season and then reabsorbed after breeding; a seasonal neurogenesis also reported for vocal centres of canaries. Increased hippocampal size, therefore, can also be a sexually selected secondary sexual character (Jacobs, 1995).

The size of the hippocampus can be an adaptation to specific ecological conditions. Size, however, is only one representation of a neuroanatomical structure's information processing and storage capabilities. Finer level cyto-architectural and neurochemical changes could also influence competence, sometimes without affecting a structure's size (Krebs, 1990). The future will no doubt reveal the extent to which ecology affects finer levels of the brain's structure.

3.4.2 Experimental approach

The spatial memory used by food-caching animals could be a cognitive specialization in two different ways. First, it can be qualitatively different from other forms of spatial learning and rely on a separate and entirely dedicated cognitive process. Alternatively, it may simply have improved the all-purpose mechanisms that are commonly used for the processing of any spatial information. Determining which type of specialization is involved requires making different kinds of comparisons. To establish that the memory is a distinct and dedicated cognitive process one must compare food-storing animals on food-storing and non-food-storing spatial tasks. Establishing whether it is a general enhanced spatial competence requires comparing storing and non-storing species on some common non-storing spatial task.

Avian food-storer spatial memory not a dedicated food-caching specialization

If spatial memory used by food-storing animals is a distinct and dedicated adaptive specialization, then food storers would only exhibit their enhanced spatial memory in the food-caching context. Any other spatial task would tap into a more mundane cognitive system probably shared with non-caching animals. Experiments designed in part to answer this question have been conducted with corvids mostly by Kamil, Balda and colleagues, while equivalent work on parids has been carried out by Krebs, Sherry, Shettleworth and colleagues (reviewed by Krebs, 1990; Shettleworth, 1990, 1993, 1995).

While several North American corvid species cache food, each relies on it to a different extent. At one extreme, Clark's nutcracker and pinyon jay (*Gymnorhinus cyanocephalus*) inhabit cold places and rely heavily on stores of pine seeds during the winter and spring. In comparison, both scrub (*Aphelocoma coerulescens*) and Mexican (*A. ultramarina*) jays inhabit warmer climes and rely much less on food stores (Balda & Kamil, 1989; Balda *et al.*, 1996). If these animals' spatial memory is specialized only for use in cache recovery, then when tested on other types of spatial tasks their ability should be uncorrelated to their rankings on a food-caching task.

In a test that involved caching food and then retrieving it within an indoor arena, nutcrackers and pinyon jays outperformed scrub jays although pinyon jays outperformed nutcrackers because they tended to hide seeds in clumps facilitating recovery (Balda & Kamil, 1989). In a non-caching task, the birds had to return to sites where they encountered food earlier within a radial-arm maze analogue (Kamil *et al.*, 1994). The task differed from caching because the birds did not store the seeds they encountered. The birds were trained to expect a buried seed in each of 12 sand-filled cups arranged in a circle on the floor of an aviary. They were then allowed into the aviary while only four randomly chosen cups were uncapped and hence available for exploitation. Following a retention interval, the birds re-entered the aviary. This time eight cups were uncapped, the same four they exploited in the initial phase plus another random set of four. The efficiency with which the subjects directed their search to the new set of cups measured their spatial memory. The ranking of spatial memory in both storing and non-storing spatial tasks were similar: nutcrackers and pinyon jays were more efficient than scrub and Mexican jays. So, the spatial memory used by food-caching corvids was unlikely to be a distinct, dedicated cognitive specialization. Instead, it was more likely supported by an enhanced general spatial cognitive system.

Parallel experimental results have been obtained for two species of food-storing parids: black-capped chickadees and coal tits (*P. ater*: Shettleworth *et al.*, 1990). Both species were as efficient at returning to locations where they encountered seeds they had not cached as they were at returning to cache locations. Hence, spatial memory used both by food-storing parids and corvids seems to be an enhanced version of a general spatial cognitive ability that can and is used in a non-caching context.

Evidence that avian food storers have superior spatial memory

Because both corvids and parids use their food-storing spatial memory for non-storing spatial tasks, it should be simple to compare storing and non-storing species on a common spatial problem. If storers have a better spatial memory than non-storers, they should exhibit any one or more of the following features: learn spatial tasks faster, memorize more spatial information and support longer retention intervals without loss of memory.

In corvids, champion storers like Clark's nutcracker and pinyon jay are compared to less assiduous storers like scrub and Mexican jays on a radial arm problem (Kamil *et al.*, 1994). As expected, nutcrackers and pinyon jays are able to remember more spatial information than Mexican and scrub jays but inexplicably they also forget it faster. After 300-min retention intervals, species no longer differ. This is especially surprising when one considers that retention intervals for Clark's nutcrackers in the wild is on the order of months.

These surprisingly small differences in spatial memory among corvid species are comparable to those found between storing and non-storing parids. The spatial task commonly used in parid comparative experiments is called 'window-shopping'. The birds are allowed into an aviary where a number of locations contain a peanut placed behind a window. Then, after a retention interval the windows are removed and the animal allowed to return. When subjected to these kinds of window-shopping tests, differences in performances between the storing coal tit and non-storing great tit (*P. major*) were small, although they consistently favoured the storing species (Krebs *et al.*, 1990).

Except perhaps for the Clark's nutcracker, the small differences observed between the spatial memory of storing and non-storing species is remarkable given that the natural history of food-storing birds suggests an ability to memorize a staggering number of distinct spatial locations (Shettleworth, 1995). A more spectacular set of quantitative differences in spatial memory may have been in order especially since food storing is associated with relatively larger hippocampus. Some factors may have prevented food-storing species from shining as well as they could when compared to non-storing species (Shettleworth, 1990, 1995). It is also possible that the major difference in the spatial cognition of storing and non-storing species is more subtle and qualitative, affecting mostly the kinds of information animals pay attention to and the way it is stored rather than the quantity (Brodbeck, 1994; Clayton & Krebs, 1994b,d; Brodbeck & Shettleworth, 1995). For instance, monocular occlusion experiments suggest that long-term spatial memory is lateralized (i.e. located in one hemisphere) in food-storing birds but not in non-storing ones (Clayton & Krebs, 1993, 1994a,b). In marsh tits, for instance, memory of food locations transfers from the left to the right hemispheres 3–7 h after storage, a time during which efficiency of cache retrieval declines. The long-term storage of spatial memory of food-storing birds, therefore, is located in the right hemisphere (Clayton & Krebs, 1994a). Moreover, the left eye system of food-storing birds appears to specialize on storing spatial information while the right eye system specializes on non-spatial cues, both in parids and corvids (Clayton & Krebs, 1994d). Research on the spatial memory of food-storing birds, therefore, suggests that selective pressures may have altered the mechanisms of memory processing and storage more than its capacity.

3.5 The ecology of social learning

The expression 'social learning' has been used to mean a wide range of behaviour (Galef, 1988, provides an exhaustive list). Here, it refers to the use of public information (see Section 3.2.4) about places, objects and behaviour: area, object and behaviour copying, respectively.

3.5.1 Area copying

Area copying, also called local enhancement (Thorpe, 1956; Galef, 1988), occurs when an individual directs its behaviour towards the place where others are currently active. In addition to birds (Krebs *et al.*, 1972; Barnard & Sibly, 1981), area copying has been reported in social spiders (Ward, 1986), fish (Pitcher *et al.*, 1982) and mammals (Galef, 1990).

Although common, area copying has generated little interest in its ecological determinants. For instance, it is conceivable that the area to which individuals are attracted depends on the size and distribution of resource clumps (Krebs, 1973; Barnard & Sibly, 1981). Moreover, how long should an individual's behaviour be influenced by the sight of another and how is this related to the animal's ecology? Krebs and colleagues' parid studies report that area copying effects are detectable for only moments, while McQuoid and Galef (1992) report that Burmese junglefowl (*Gallus gallus spadiceus*) show significant area copying effects 48 h after the initial observation of other individuals foraging in a given place. It would be useful to explore how factors such as food patch ephemerality affect the longevity of area copying.

Adaptive hypotheses

Area copying can allow animals to avoid dangerous places. For instance, in red-winged blackbirds (*Agelaius phoeniceus*), flocks that contained experienced individuals induced naive birds to forage first in places with safe food but it failed to teach them to avoid places with poisoned food (Avery, 1994). Area copying can increase foraging gains, whether in terms of foraging rate or reduced risk of starvation (Caraco, 1981; Hake & Ekman, 1988). For instance, area copying allows great tits foraging in flocks of four to find food more quickly than in flocks of two or when foraging alone (Krebs *et al.*, 1972). The same was observed for flocks of greenfinches (*Carduelis chloris*) (Hake & Ekman, 1988). Area copying is often portrayed as an information-sharing system where individuals maintain contact with their group mates as they search for their own food (Ranta *et al.*, 1993; Ruxton *et al.*, 1995). However, keeping up both activities simultaneously can be difficult under a wide range of circumstances (Vickery *et al.*, 1991). Area copying, therefore, can also be modelled as individuals alternating between searching for (producing) their own food, and exploiting (scrounging) the food discovered by others: a producer–scrounger game (Barnard & Sibly, 1981; Caraco & Giraldeau, 1991; Vickery *et al.*, 1991). Whatever the scenario, the payoff of area copying likely declines with the frequency of its use, whether benefits are expressed in terms of individual feeding rate (Giraldeau *et al.*, 1994a) or starvation risk (Koops & Giraldeau, 1996). Simulations suggest that area copying frequencies expected from producer–scrounger games lead to greater average feeding rates than those expected from information-sharing models (Beauchamp & Giraldeau, 1996). Recent experimental evidence suggests that animals can alter their frequency of area copying when local food distribution is changed (Giraldeau *et al.*, 1994a; Koops & Giraldeau, 1996; Giraldeau & Livoreil, 1997). It would be especially interesting to explore whether the updating algorithms described in Section 3.2 can usefully describe the mechanisms through which individuals adjust their area copying.

3.5.2 Object copying

Object copying, also called stimulus enhancement (Galef, 1988) or 'releaser-induced recognition' (Subowski, 1990), is similar to area copying in that it directs behaviour. However, the behaviour is directed to an object that matches the type attended by others, rather than a place. It differs from area copying in that the effect is not constrained in space but instead is generalized to all objects of that type wherever they occur. Unlike area copying, object copying has been implicated in a number of non-foraging systems. For instance, it is described in predator and enemy recognition (see Chapter 4; Curio, 1988; Mineka & Cook, 1988; Chivers & Smith, 1994) and could be implicated in mate copying (see Chapter 8; Gibson & Höglund, 1992). The focus here is on foraging examples, specifically object copying as it relates to acquisition of food preferences and the ability to forage from new types of food.

Object copying has been described in the context of food choice in many animals. For instance, red-winged blackbirds acquire food aversions and preferences socially (reviewed by Mason, 1988), pigeons avoid seed types chosen by flock mates (Inman *et al.*, 1988), while woodpigeons (*Columba palambus*) (Murton, 1971) and greenfinches (Klopfer, 1961) prefer food chosen by flock companions. The most detailed studies of object copying in the context of food choice have, no doubt, been conducted by Galef and colleagues with Norway rats (*Rattus norvegicus*) (reviewed by Galef, 1990). Rats eat copious amounts of an unfamiliar food type if they have had the opportunity of smelling that food on another colony member's breath. Object copying in rats is so strong that it can even reverse a conditioned food aversion, an effect also reported in spotted hyenas (*Crocuta crocuta*) (Yoerg, 1991).

Object copying has been implicated in a number of studies of behaviour copying (imitation) in non-human primates. Unfortunately, in those cases it is often considered a consolation prize for failure to demonstrate behaviour copying and so attracts little further study (Whiten & Ham, 1992).

Adaptive hypotheses

Group-living itself may promote social learning (Klopfer, 1961). The underlying argument implies that only social animals have the opportunity of incorporating public information, and so only social animals should have evolved the ability to use it. The hypothesis may be tenuous since, on occasion, non-social animals could also benefit from using public information. Nonetheless, it has been tested by comparing the social and non-social learning performances of taxonomically related social and non-social animals: the pigeon and the Barbados zenaida dove (*Zenaida aurita*), respectively (Lefebvre *et al.*, 1996b). Both species are dietary opportunists but pigeons are gregarious, both nesting and feeding in flocks, while Barbados zenaida doves defend year-round territories. More pigeons learned and did so more quickly in the presence of conspecific tutors than territorial zenaida doves. However, the same differences were noted for non-social tasks, so it is unclear whether sociality has selected specifically for social learning or just learning in general. Reliable conclusions will require comparisons involving more than one phyletic group.

3.5.3 Behaviour copying

Behaviour copying, also called imitation (Galef, 1988; Heyes, 1993), occurs when a topographically novel behaviour pattern is acquired by seeing another individual use it. The behaviour must be topographically novel in order to distinguish behaviour copying from situations where behaviour patterns already in an animal's repertoire are simply redirected to new objects, places or used in a different context as would result from area or object copying (Sherry & Galef, 1984). It is generally undisputed that behaviour copying occurs in the vocal learning of birds (Catchpole & Slater, 1995). In many songbirds, for example, the development of normal male song requires exposure to the song of other males and sometimes even social interaction with them. Social vocal learning, however, has almost always been regarded as a special case of behaviour copying because, among other reasons, it is dedicated to acquisition of specialized vocal motor patterns that can be acquired only within specific developmental windows (Whiten & Ham, 1992).

Visual imitation, on the other hand, is often seen as the cognitive 'Lamborghini' of the social learning world. The assumptions are that it is fast but, unlike area or object copying, requires such elaborate and costly cognitive machinery that only a few select species can afford it. Admittedly, over 100 years of research into visual behaviour copying, mostly in non-human primates, has yielded surprisingly few instances that could not be accounted for by other mechanisms such as area and object copying. If behaviour copying occurs, it is rare (Galef, 1976; Whiten & Ham, 1992; Heyes, 1993; Byrne & Tomasello, 1995). Commonly (but not universally) accepted examples of behaviour copying involve anecdotal evidence in chimpanzees (*Pan troglodytes*) (Whiten & Ham, 1992), orangutans (*Pongo pygmaeus*) (Russon & Galdikas, 1993; 1995), experimental evidence in budgerigars (Dawson & Foss, 1965; Galef *et al.*, 1986) and possibly rats (Heyes & Dawson, 1990). Outside of vocal learning in birds, therefore, it is safe to say that behaviour copying is rare.

Adaptive hypotheses

Given that accepted instances of behaviour copying are so rare, it is not surprising that adaptive hypotheses have received little formal testing. Boyd and Richerson (1985) propose that behaviour copying is a faster, more economical means of acquiring behaviour and as such is expected to occur when:

1 non-social learning is hazardous (e.g. learning to recognize poisons or predators);

2 large environmental changes are predictable;

3 social learning is more accurate than non-social learning.

An alternative view portrays behaviour copying as a specialized reward-independent learning mechanism that is likely to evolve when the rewards required for behaviour acquisition by non-social learning mechanisms are unavailable, either because other knowledgeable individuals have already depleted the reward sites, or because it is more expedient to feed from others' discoveries than to learn to discover through non-social mechanisms (Giraldeau, 1984; Beauchamp & Kacelnik, 1991; Giraldeau *et al.*, 1994b).

In conclusion, it is apparent that a number of animals are capable of area and object copying but many fewer seem capable of behaviour copying. We know some of the ecological factors that promote the use of area copying, but the ones that promote either object or behaviour copying remain uncertain. The comparative work in columbids appears to suggest that selection does not operate independently on social and non-social learning. If this is so, then social learning may not be a distinct adaptive specialization (see Section 3.4) but simply a side-effect of selection for learning in general. Naturally, many more phyletically independent comparisons are necessary to support any conclusion.

3.6 The ecology of cultural transmission

This section first looks at population studies that are meant to describe and predict how the use of public information can lead to the spread of behavioural innovations within populations. Then, it considers studies of the transmission process itself, emphasizing factors that govern the rate and efficiency of information transfer. Finally, it looks at an emerging approach that asks whether social learning mechanisms alone (see Section 3.5) can sustain long-lived behavioural traditions that form the basis of culture.

3.6.1 The population approach

Modelling the spread of behavioural phenotypes within populations can be analogous to modelling changes in gene frequencies, especially when the trait is acquired from parents (Cavalli-Sforza & Feldman, 1981). Unlike genes, however, cultural traits can be acquired horizontally within generations, or obliquely through collateral kin. In any case, they are assumed to spread within populations following a logistic progression: slowly at first when demonstrators are rare, exponentially as demonstrators become common and then slowly again as naive observers are rarer (Cavalli-Sforza & Feldman, 1981).

Re-analysis of a number of now classic studies of innovations spreading within populations provides some support for the accelerated spread (Lefebvre, 1995a,b). For instance, the number of sites where birds have been reported to open milk bottles to drink the supernatant cream spread exponentially through the UK in the first half of this century (Fisher & Hinde, 1949; Lefebvre, 1995a; Fig. 3.5a). Potato and wheat washing both spread exponentially within a provisioned population of Japanese macaques (*Macaca fuscata*) (Kawai, 1965; Lefebvre, 1995b). Potato washing involved dipping potatoes in the sea, presumably to remove sand, while wheat washing was an efficient means of separating wheat from sand by throwing a handful of mixed sand and wheat in the sea and scooping up the floating grain. Unwrapping and eating of caramels, eating beached fish and washing apples in Japanese macaques, as well as mango and lemon eating in chimpanzees, all spread exponentially (Fig. 3.5b–g). Obviously, the pattern of spread alone is only weak support for cultural diffusion. Non-social learning in a population with normally distributed learning speeds can also produce an accelerated function (Lefebvre, 1995b). Stronger support requires establishing the social learning mechanisms underlying diffusion models. Unfortunately, these mechanisms in both avian and primate examples have remained unspecified, rightfully leading some researchers to question whether social learning was involved at all (Galef, 1976; Sherry & Galef, 1984).

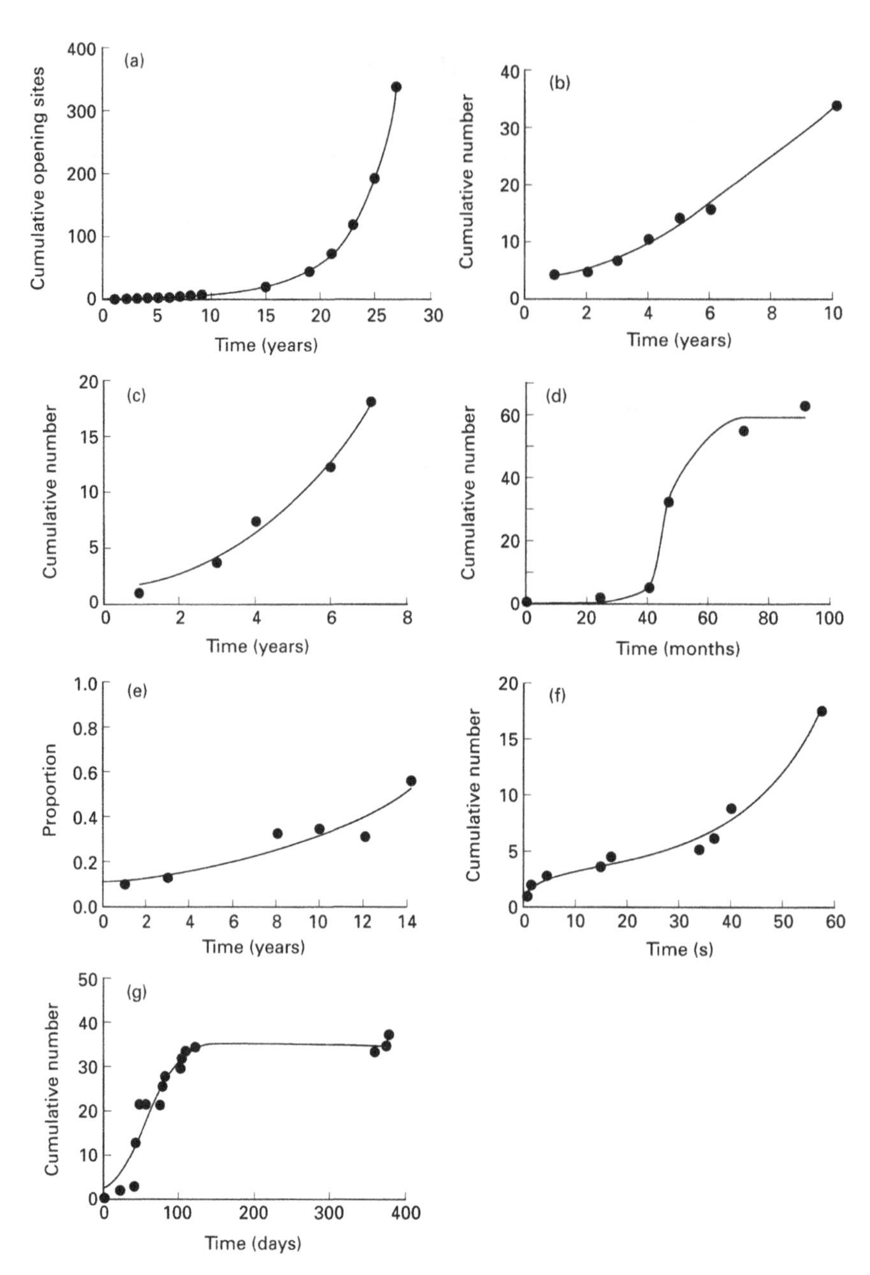

Fig. 3.5 Cumulative spread of foraging innovations over time. (a) The number of sites reporting avian bottle opening in the whole of the UK between 1921 and 1947. Cumulative number of Japanese macaques (*Macaca fuscata*) at Koshima Islet: (b) washing potatoes; (c) washing wheat; (d) eating beached fish. (e) The cumulative proportion of Japanese macaques at Takasakiyama unwrapping and eating caramels. The cumulative number of chimpanzees (*Pan troglodytes*) at Mahale (f) eating mangoes and (g) lemons. (Avian data taken from Lefebvre, 1995a, and primate data from Lefebvre, 1995b.)

3.6.2 Factors affecting the rate of spread

The population studies show that a skill may spread, sometimes slowly, other times explosively. However, the determinants of the rate remain unspecified. At least two factors have been implicated in the regulation of cultural diffusion: (i) demonstrator and observer densities; and (ii) the potential of feeding from the tutor's food discoveries (scrounging).

Demonstrator and observer densities

Logistic spread assumes that the rate of transmission of a skill is proportional to the product of the number of demonstrators and observers. Increases in either, therefore, should enhance diffusion rate. Although for pigeons the rate of acquiring a food-finding skill increases with the number of individuals simultaneously demonstrating the skill, it declines as the number of uninformative bystanders placed around a demonstrator increases (Lefebvre & Giraldeau, 1994). The observers may have difficulty identifying the individual that provides useful information so, by increasing the number of bystanders, the chance of observing the correct individual during a demonstration declines. It follows that logistic spread may not be an accurate depiction of a culturally diffused trait. Instead, compared to a logistic spread, the trait may propagate more slowly at first because of the depressing effect of bystanders while later, when demonstrators become common, the spread may be faster because the loss of bystanders is compounded to the increased number of demonstrators.

Scrounging

In a duplicate cage procedure, a caged demonstrator performs a skill while a caged subject observes. In pigeons, the procedure has consistently shown that skills are more quickly acquired when demonstrators are provided, although the exact social learning mechanism involved is rarely identified (Palameta & Lefebvre, 1985; Giraldeau & Lefebvre, 1986, 1987; Giraldeau & Templeton, 1991; Lefebvre & Giraldeau, 1994). Surprisingly, similar skills spread very little when demonstrators were available within foraging flocks (Lefebvre, 1986; Giraldeau & Lefebvre, 1987). One factor that has been implicated in the slow spread within flocks is the availability of scrounging opportunities where observers get to eat some of the food discovered by the demonstrator. To test this hypothesis Giraldeau and Lefebvre (1987) simulated the effect of scrounging in a duplicate cage procedure. They found that demonstrators failed to enhance an observer's learning when observers passively received some or most of their demonstrator's food, suggesting that scrounging may have prevented the spread of the skill within the foraging flocks (Giraldeau & Lefebvre, 1987; Giraldeau & Templeton, 1991). Whether scrounging also reduces the spread of innovations in many other species remains to be shown. For instance, it does not alter the spread of skills in domestic chickens (*G. gallus domesticus*) (Nicol & Pope, 1994). However, observations consistent with the negative effects of scrounging have been reported in capuchin monkeys (*Cebus apella*) (Fragaszy & Visalberghi, 1989, 1990), jackdaws (*Corvus monedula*) (Partridge & Green, 1987) and zebra finches (*Taeniopygia guttata*) (Beauchamp & Kacelnik, 1991).

3.6.3 Factors affecting longevity of traditions

Boyd and Richerson (1985, 1988) argue that even if social learning occurs, only behaviour copying (see Section 3.5.3) can give rise to culture. The reason is that acquisition of a trait in both other forms of social learning (area and object copying) rely strongly on an individual's rewarded performance of the behaviour, a step that allows for individual variation in its final form. Heyes (1993) goes even further, arguing that imitation itself may be insufficient to sustain cultural traditions because nothing stops individuals from subsequently modifying the socially acquired cultural phenotype through individual experience (i.e. non-social learning), hence putting an end to tradition.

Chain procedures where a previously naive observer is used as a tutor for the next naive observer have shown that in enemy recognition, cultural traditions can be maintained for several generations (Curio, 1988; Mineka & Cook, 1988; Chivers & Smith, 1994). The chain procedure may artificially enhance the fidelity of transmission because of the paucity of alternatives to copy. A more realistic means of testing fidelity was proposed by Galef and Allen (1995). They investigated experimentally whether object copying by rats in a food selection context (see Section 3.4.2) could give rise to long-lived traditions. They trained a number of individuals to avoid one of two equally unpalatable food flavours — cayenne pepper or Japanese horseradish — and used these animals to form homogeneous founder groups that all avoided one or the other flavour. The fidelity to the founders' original food preference was challenged by replacing founders, one at a time, at regular intervals by naive subjects. The founders' arbitrary preference could be maintained over generations, giving rise to long-lived traditions that survived beyond the founders' tenure within the groups, but fidelity of the cultural tradition was sensitive to the extent to which rats had access to the alternative flavour and hence the opportunity for non-social learning (Fig. 3.6).

In conclusion, therefore, research on cultural transmission remains patchy, no doubt in part because of the real problems it poses to experimentalists. Nonetheless, because cultural transmission frees behavioural evolution from slower Darwinian processes, it presents a major challenge to Darwinian-based behavioural ecology. Clearly, better documented examples of culturally transmitted traits based on solid evidence of the transmission mechanism would be very valuable. Moreover, determinants of the rate of spread and the fidelity of transmission clearly need to be explored more extensively and systematically (Galef, 1995).

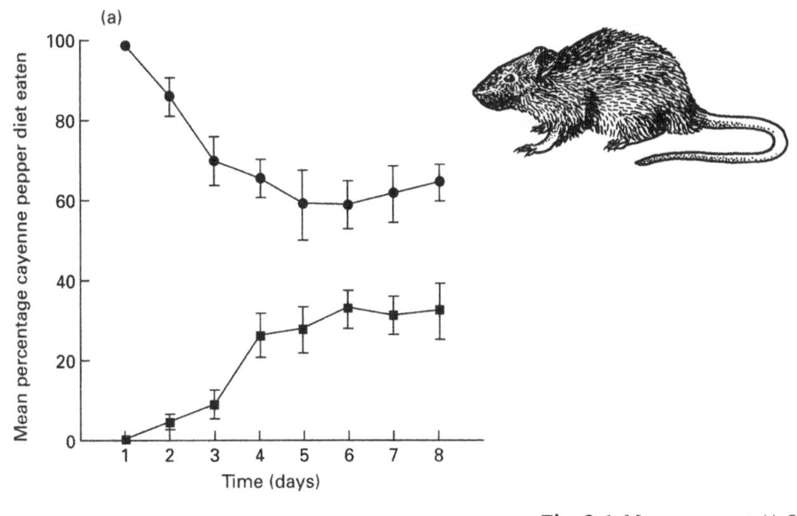

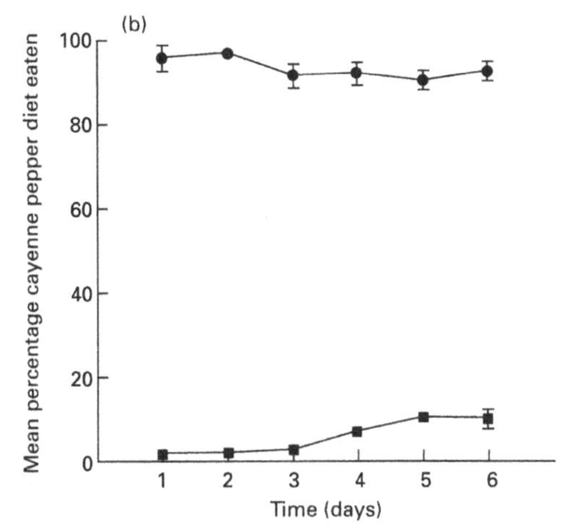

Fig. 3.6 Mean per cent (± SE) of cayenne pepper diet eaten by Norway rats (*R. norvegicus*) in groups of four individuals whose founders ate only either cayenne pepper (●) or Japanese horseradish (■) as a function of time. Each day one of the original founders is replaced by a naive individual such that from day 5 on, none of the original founders is present. (a) represents an experiment with *ad libitum* access to food while groups in (b) had access to food only 3 h a day. (From Galef & Allen, 1995.)

3.7 Future directions

The information processing ability of animals is clearly emerging as an important component of an increasing number of behavioural ecological questions. However, it will probably become an important area of behavioural ecology in its own right only once research interests switch from merely cataloguing animals as having this or that ability to investigating the ecological circumstances under which any given cognitive process is used. Only once information use is studied as a decision rather than a constraint will it be possible to trace the ecological influences on an organism's cognitive architecture (Real, 1991).

A number of studies have already engaged in this direction. For instance, instead of asking whether memory parameters should differ between species (see Section 3.2.3) one could ask whether the memory parameters of any organism should change according to the ecological problem at hand. Valone (1992b), for instance, applies a memory window model to the issue of economic patch exploitation and finds that rate maximization occurs when the size of the memory window used to estimate patch quality increases with the duration of travel between resource patches (see also Devenport & Devenport, 1994; Mackeney & Hughes, 1995). Similarly, public information should be envisaged as a process that is most profitable when individual sampling is costly and/or time available for resource exploitation is short (Clark & Mangel, 1986). Not surprisingly, in starlings it is used only when accurate personal patch-sample information is difficult to obtain and when acquiring public information does not impose any reduction in personal patch sampling (Templeton & Giraldeau, 1995, 1996).

Research has established successfully that both natural and sexual selection can promote the relative size of the hippocampus when ecological conditions call for enhanced spatial memory, but comparative experimental work has not revealed large quantitative differences in spatial memory. Differences may lie more in the quality of the information processed than its quantity. Alternatively, size of a neuroanatomical structure like the hippocampus may be a poor approximation of its processing ability. Future work, no doubt, will go both in the direction of analysing the type of information used in spatial memory (Clayton & Krebs, 1993; 1994a,b,d; Brodbeck, 1994; Brodbeck & Shettleworth, 1995) and increasing the detail with which neuroanatomical structures are studied. The picture that is emerging from the ecological study of information use opens the way to research that may reveal that animals, rather than being capable of using information or not, simply use it when it pays and in ways that are adaptive.

Chapter 4

Recognition Systems

Paul W. Sherman, Hudson K. Reeve &
David W. Pfennig

4.1 Introduction

What do these diverse situations have in common?

1 A liver-transplant patient's body rejects the new organ.

2 Before mating, male and female fireflies engage in a precisely timed flash dance.

3 A male bird disregards a singing neighbour but attacks a stranger singing from the same spot.

4 A hungry mouse and an iridescent green beetle ignore each other.

5 A carnivorous tadpole engulfs a smaller sibling, but immediately spits it out.

The answer is that each involves recognition — of self, species and mates, neighbours, prey and predators, and kin, respectively. Are there common principles governing the evolution of these varied recognition systems? This chapter explores this question and provides a unified evolutionary framework for understanding discriminative behaviour.

After briefly discussing the forms and functions of recognition systems, we show how they can be analysed in terms of three component parts. We then: (i) identify the central problem that recognition systems have been designed by natural selection to solve; (ii) discuss principles that relate the evolution of each component to the central problem; and (iii) illustrate these principles with evidence from a variety of organisms. Kin and mate recognition are highlighted, because they are major foci of mechanistic and functional research (Pfennig & Sherman, 1992, 1995). Finally, we address failures of and misunderstandings about kin recognition and suggest some profitable avenues for future research on recognition systems.

4.2 Forms and functions of recognition

We define different forms of recognition by the nature of the objects being discriminated, not by the functional significance or proximate mechanisms of recognition. Thus, *kin recognition* is differential treatment of conspecifics (including self) differing in genetic relatedness (Sherman & Holmes, 1985; Waldman *et al.*, 1988; Gamboa *et al.*, 1991c). Surprisingly, the fitness consequences of kin recognition have rarely been documented. The traditionally hypothesized benefits are dispensing nepotism (Hamilton, 1964a,b) and optimizing the balance between inbreeding and outbreeding (Bateson, 1978; Shields, 1982). However, kin discrimination also functions in other contexts. For instance, many anuran tadpoles associate preferentially with siblings (Blaustein & Waldman, 1992). In some cases this may reflect nepotism (Waldman, 1991) or learning kin phenotypes for optimal outbreeding (Waldman *et al.*, 1992), but in others it may indicate only a tendency to associate with any conspecifics, including siblings, that smell like the (safe, food-rich) natal site (Pfennig, 1990). Another function of kin recognition may be disease avoidance. Cannibalistic Arizona tiger salamander larvae (*Ambystoma tigrinum nebulosum*) feed voraciously on non-relatives but avoid eating close kin (Pfennig *et al.*, 1994). This either represents nepotism or avoidance of a deadly bacterium that is efficiently transmitted through cannibalism (Pfennig *et al.*, 1991, 1993). Infections may be especially transmissible among close relatives because they generally have similar immune systems (Pfennig *et al.*, 1994).

Mate recognition encompasses many types of recognition which differ in the nature of the potential mates being discriminated, including heterospecifics versus conspecifics (species recognition) and, among the latter, individuals differing in sex (sex recognition), relatedness (kin recognition), genetic quality or attractiveness (mate-quality recognition) or parental resources (mate–resource recognition).

There is a fundamental difference between the function(s) of mate recognition and nepotistic kin recognition. In the latter, discriminative acts increase transmission of the recognition-promoting alleles by favouring relatives likely carrying copies of those alleles. In Dawkins's (1982) terminology, kin-selected kin recognition is aimed at *detecting replicators*. In mate recognition, by contrast, discrimination increases transmission of the recognition-promoting alleles by favouring genetically compatible or superior mating partners likely to transmit beneficial alleles or resources to offspring that also receive the recognition-promoting alleles. Thus, mate recognition is aimed at *improving vehicles.*

In this chapter we use 'recognition' and 'discrimination' interchangeably, although they are not synonymous when recognition refers to an internal neural process that underlies, but can occur without, detectable behavioural discrimination (Lacy & Sherman, 1983; Byers & Bekoff, 1986). However, if discrimination never occurred, recognition (as a strictly internal process) would be an empty concept. The real point behind the distinction is that lack of discrimination in one context does not imply its absence in another.

4.3 Components of recognition systems

Any recognition process involves an *actor* and a *recipient* (Fig. 4.1); these usually are different individuals (except in self-recognition). All recognition systems can be partitioned conceptually into three component parts (Sherman & Holmes, 1985; Waldman, 1987; Reeve, 1989; Gamboa *et al.*, 1991c).

1 *Production*: the nature and development of labels (cues) in recipients that actors use to recognize them.

2 *Perception*: the sensory detection of labels by actors and subsequent phenotype matching; i.e. comparison of labels to a template (internal representation) of the phenotypic attributes of desirable (fitness-enhancing) or undesirable (fitness-decrementing) recipients; the ontogeny of templates is also part of this component.

3 *Action*: the nature and determinants of actions performed, depending on the similarity between actors' templates and recipients' labels. Discriminative actions need not be strictly behavioural, as illustrated by two examples:

(a) Arizona tiger salamander larvae transform into physically different cannibal morphs more frequently when they are reared among non-kin than among kin (Pfennig & Collins, 1993);

(b) in montane larkspurs (*Delphinium nelsonii*), genetic similarity declines with distance between plants, and pollen–pistil interactions result

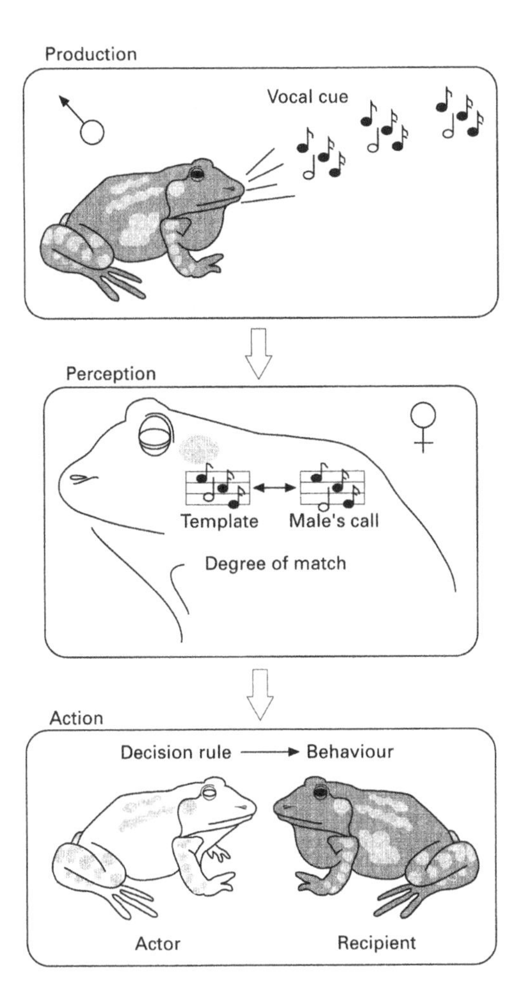

Fig. 4.1 Recognition systems comprise three components: production, perception and action (indicated by separate boxes). Any recognition process involves an actor and a recipient; here, the actor is a female frog (middle) who is attempting to recognize a potential mate (top) based on a label (cue) contained in the pattern of notes in the male's advertisement call. In this case there is a slight mismatch between the male's cue and the female's template; the degree of mismatch affects whether or not discriminative behaviour occurs.

in plants 'preferring' as mates distant relatives over close relatives or non-relatives, as indicated by the number of pollen tubes that the pistil allows to grow to its ovary (see Chapter 15; Price & Waser, 1979; Waser & Price, 1993).

Characteristics of components (1–3) should vary according to ecological and social circumstances that affect the costs and benefits of discrimination. Knowledge of the components can therefore illuminate the selective forces shaping discrimination, and vice versa. Understanding the evolution of the components requires appreciating the central problem confronting all recognition systems.

4.3.1 The central problem for recognition systems

Selection should favour individual organisms (and cells) whose recognition mechanisms classify recipients without error. In nature, however, desirable and undesirable recipients often exhibit overlapping cues (Getz, 1981, 1991; Lacy & Sherman, 1983). In addition, undesirable recipients may benefit by either mimicking the cues of desirable recipients or scrambling cues to prevent discrimination (Reeve, 1997a). Consequently, the central problem is how to optimize the balance between *acceptance errors* (accepting undesirable recipients) and *rejection errors* (rejecting desirable recipients).

Imagine graphing, for each kind of recipient, the frequency distribution of the difference between the recipients' cues and the actor's template (Fig. 4.2). The shapes and overlap between the frequency distributions for desirable and undesirable recipients depends on which cues and templates are used and how the match between them is assessed (Crozier & Dix, 1979; Lacy & Sherman, 1983; Getz & Chapman, 1987). Probabilities of each type of recognition error also depend on the position of the acceptance threshold (Reeve, 1989). Alternative designs for each recognition component affect the balance of errors differently, and the design that optimally balances these errors is predictable from knowledge of the organism's environment.

4.3.2 The production component

General principles

Actors should use recognition cues that maximize the separation of template–cue dissimilarity distributions for desirable and undesirable recipients (Fig. 4.2). Suppose the choice is between 'D-present' cues, possessed by nearly all *d*esirable recipients but also by some undesirable recipients, and 'U-absent' cues, possessed by some desirable recipients but rarely by *u*ndesirable recipients. Use of the former leads to few rejection errors but some acceptance errors, and vice versa for the latter. Accordingly, actors should use D-present cues when: (i) interactants are most often desirable recipients; (ii) fitness benefits of accepting desirable recipients are high; and (iii) costs of (mistakenly) accepting undesirable recipients are low.

For example, in the neotropical ant *Ectatomma ruidum* (Breed *et al.*, 1992) and the desert woodlouse *Hemilepistus reaumeri* (Linsenmair, 1987), nestmates share a chemical label that is readily transferred by direct contact, enabling young colony members to acquire it rapidly. Sometimes, foreigners also acquire

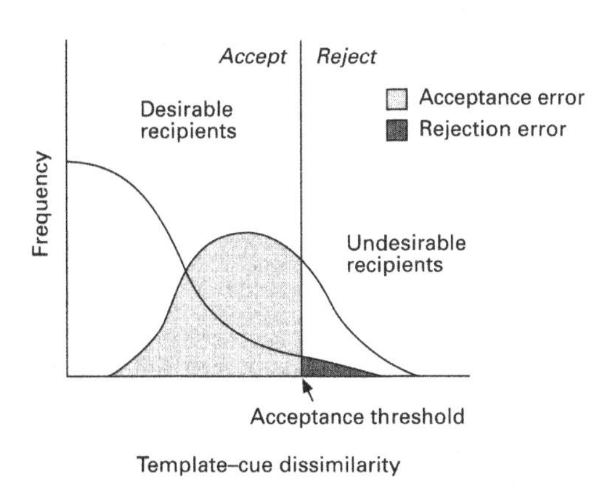

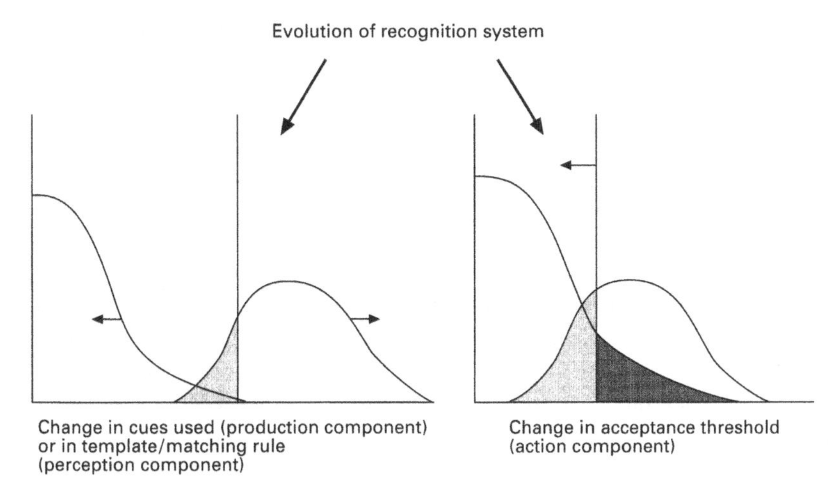

Fig. 4.2 Evolution of recognition systems. (Upper) When there is overlap in the labels of desirable and undesirable recipients, selection will favour an optimal balance between accepting undesirable recipients (acceptance errors) and rejecting desirable recipients (rejection errors). The figure illustrates the frequency distribution, for desirable and undesirable recipients, of the difference between their cues and the actor's template. For both desirable and undesirable recipients, the probabilities of each type of recognition error depend on the shapes of their frequency distributions and the position of the acceptance threshold. (Lower) Selection can alter the magnitudes of and balance between acceptance and rejection errors by changing the recognition cues used (production component), the recognition template or matching algorithm (perception component) or the decision rule — e.g. acceptance threshold, as shown here (action component); see also Box 4.1. (Modified from Reeve, 1989.)

the cue, enabling them to enter the nest to steal food. Persistent use of D-present transferable labels implies that benefits of always accepting colony mates outweigh costs of occasionally being robbed.

Actors should use U-absent cues under the converse of circumstances (i–iii). For example, to avoid accepting any pathogen-infected cells (i.e. costly acceptance errors), vertebrate immune systems use molecular markers of individuality that are possessed rarely by undesirable recipients. If antigens produced by infected (or possibly cancerous) cells are not represented among T- (immune) cell surface antigens, rejection occurs. Surface antigens are individual-specific, because loci in the major histocompatibility complex (MHC), which produces the antigens, are highly polymorphic (Brown & Eklund, 1994).

A potential cost of using U-absent cues in self-recognition is self-rejection. One clue as to why lupus and rheumatoid arthritis are so common (i.e. > 1 in 200 of the US population) is that sufferers experience reduced risk of some cancers (Gridley *et al.*, 1993; Duquesnoy & Filipo, 1994), especially those of viral origin (Kinlen, 1992). Whereas cancer represents a potentially deadly acceptance error, autoimmune diseases represent costly rejection errors. Selection should favour an optimal balance between these reciprocally related errors, but the near equivalence of slightly different balances of errors may help explain interindividual variation in susceptibilities. Elevated immune responsiveness may protect against cancers, and perhaps pathogens and environmental toxins (venoms, plant secondary compounds), but at the expense of increased likelihood of autoimmune diseases.

The production component of kin recognition

Kin-recognition cues may be any aspect of the phenotype that signifies kinship reliably. Chemical cues are widely used (Beecher, 1988; Halpin, 1991; Waldman, 1991). They are potentially information-rich (owing to their three-dimensional molecular structure), while often requiring little energy to obtain or produce (Alcock, 1993); e.g. when they are present in food or nesting materials, or generated as metabolic byproducts. Moreover, a matching system exists in most organisms to detect and decipher chemical substances (the immune system). Multiple sensory modalities may be employed in recognition, however, depending on the context. For example, as tadpoles, American toads (*Bufo americanus*) use chemical labels to school preferentially with siblings (Waldman, 1986), but as adults females may choose 'optimally related' mates based on their vocalizations (Waldman *et al.*, 1992).

Labels used in kin recognition can be of genetic or environmental origin, and may be produced endogenously by actors or acquired from their environment (Gamboa *et al.*, 1986a). As an example of genetic endogenous labels, larval *Botryllus schlosseri* tunicates settle near and sometimes fuse with individuals that carry the same allele at one histocompatibility locus (Grosberg & Quinn,

1986). Larvae settle closer to non-relatives that bear the same allele than to relatives that carry a different allele. Use of this D-present cue suggests that interactions with non-relatives are infrequent and/or costs of acceptance errors (fusing with a non-relative) are low. Because these histocompatibility loci are highly polymorphic (58–306 alleles/locus; Rinkevich *et al.*, 1995), chemically similar individuals usually are close kin.

Genetic, endogenous cues mediate self-recognition in flowering plants (Charlesworth, 1985), where proteins produced by multiallelic recognition loci in the pistil (e.g. the 'S' locus; Lewis *et al.*, 1988) are matched with pollen surface proteins. In many outcrossing species, only pollen bearing an allele that does not match the plant's own alleles will be accepted, thereby minimizing costly inbreeding (Waser, 1993). Some plants (e.g. *Collomia grandiflora*) produce both outcrossing (chasmogamous) and selfing (cleistogamous) flowers (Lord & Eckard, 1984); the former reject while the latter accept their own pollen. In outcrossing flowers, the self is the (strongly) undesirable recipient and recognition cues are U-absent, whereas in cleistogamous flowers the self is the desirable recipient and recognition cues are (presumably) D-present.

Genetic endogenous labels have been implicated in parent–offspring recognition in various birds and mammals (Beecher, 1991; Halpin, 1991), including humans (Porter, 1991; Christenfeld & Hill, 1995), and in nestmate recognition in social insects (e.g. Michener & Smith, 1987). Preferences for unfamiliar paternal half-siblings over unfamiliar non-siblings in tadpoles (*Rana cascadae*: Blaustein & O'Hara, 1982; *R. sylvatica*: Cornell *et al.*, 1989), Belding's ground squirrels (*Spermophilus beldingi*: Holmes, 1986b) and house mice (*Mus musculus*: Kareem & Barnard, 1986) also imply genetically encoded recognition labels because these kin share 50% of the male parent's genes by descent, but none of the environmental factors that full-siblings or maternal half-siblings share (e.g. in tadpoles cytoplasm, egg jelly or oviposition site cues).

Other organisms, such as paper wasps (*Polistes*: Gamboa *et al.*, 1986a; Gamboa, 1996) and honey bees (Breed *et al.*, 1995), discriminate colony mates using acquired labels that may be genetic and/or environmental in origin. Paper wasps absorb hydrocarbons from their nest at eclosion (Espelie & Hermann, 1990; Singer & Espelie, 1992), and these serve as recognition labels (Gamboa *et al.*, 1996). The relative importance of odours derived from plant fibres in the nest and genetically encoded odours applied to the paper by the wasps themselves is uncertain. However, environmental cues theoretically could rival genetic polymorphisms in diversity if different colonies use different mixtures of plants in nest construction (Gamboa *et al.*, 1986b).

Acquired, environmental cues may play a role in vertebrate kin recognition. Tadpoles of wood frogs (*R. sylvatica*) and common frogs (*R. temporaria*) prefer environmentally acquired odours to which they were experimentally exposed as embryos, sometimes even after metamorphosis (Waldman, 1991; Hepper & Waldman, 1992; Gamboa *et al.*, 1991a). These tadpoles can also use genetic endogenous labels in kin recognition (Waldman, 1981; Gamboa *et al.*, 1991a;

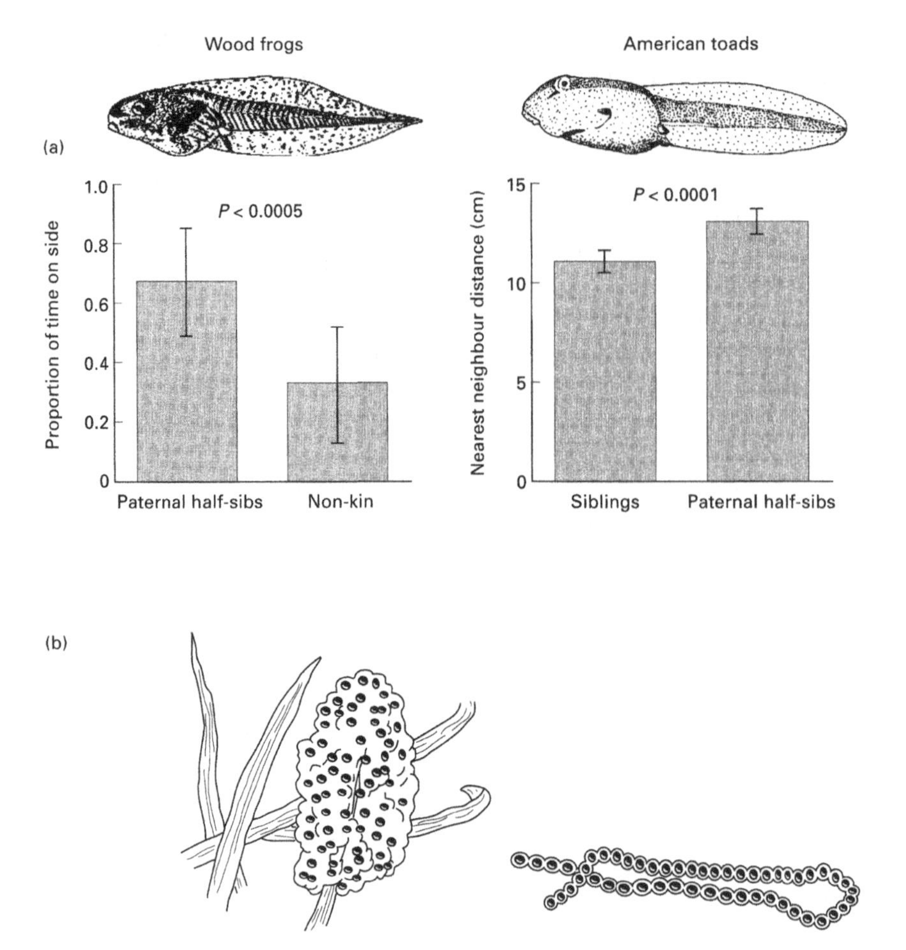

Fig. 4.3 An organism's environment can influence whether labels produced by genetic loci or environmental cues are more reliable indicators of kinship. (a) In laboratory choice tests, wood frog tadpoles (left) school preferentially with paternal half-siblings over non-kin. This implicates genetically encoded cues, because paternal half-siblings share only their father's genes by descent. (Data from Cornell *et al.*, 1989.) American toad tadpoles (right), which inhabit the same ponds, discriminate full-siblings from paternal half-siblings, which are treated like non-siblings, implying that recognition cues may come from the environment (e.g. egg jelly). (Data from Waldman, 1981.) (b) Female oviposition behaviour may explain why wood frog tadpoles, but not American toad tadpoles, rely heavily on genetically encoded cues to identify their kin. Wood frogs oviposit in communal clumps (left), presumably to insulate their eggs against the cold. Different clutches experience the same microhabitat, and only labels produced by polymorphic loci would enable different sibships to aggregate assortatively. American toads breed in warmer water and individual females often deposit strings of eggs separately (right). Each clutch experiences a slightly different chemical milieu and labels acquired from the (micro-) environment through diffusion into the egg jelly permit accurate discrimination of siblings.

Fig. 4.3). In the former experiments, environmental cues apparently overwhelmed genetic labels.

The relative importance of genetic versus environmental labels in kin recognition theoretically should optimize the balance of recognition errors (see Fig. 4.2). Relying solely on cues acquired from the environment might increase acceptance errors (mistakenly assisting non-relatives that share the same environment), whereas relying solely on genetic endogenous labels might increase rejection errors, because genetic recombination and multiple mating by females ensures that non-clonal family members will not be genetically identical for segregating alleles (Gamboa *et al.*, 1986b; Waldman, 1987; Pfennig & Sherman, 1995). The frequencies and costs of recognition errors, and thus the optimal cue system, should depend on the species' genetic system and ecology (Fig. 4.3). Endogenous genetic labels are most useful for organisms that occur in homogeneous (chemical) environments, as does a group of tunicates on a rock. For organisms that occur in more diverse environments, such as paper wasps or tadpoles, additional reliance on labels acquired from the environment may reduce rejection errors without substantially increasing acceptance errors.

Assessing the relative importance of different kinds of cues in kin recognition can be difficult because, as Breed (1983) and Carlin (1989) noted, whenever experimental subjects are reared in uniform environments, where the only detectable differences are in gene products, investigators may erroneously conclude that they use only genetic labels to recognize kin. The issue that must be addressed is whether such cues would be supplemented, or even supplanted, in nature by variable environmental cues.

The production component of mate recognition

Selection should always favour efficient recognition of mates through use of detectable, discriminable and memorable cues (Guilford & Dawkins, 1991) that reveal the properties individuals (especially females) are seeking. In sex and species recognition, females should minimize acceptance errors by using U-absent cues and minimize rejection errors by favouring males whose cues deviate least from the population mean. As an example of the former, female pied flycatchers (*Ficedula hypoleuca*) use sexually dichromatic plumage markers to recognize males (Sætre & Slagsvold, 1992); these cues are U-absent because they are restricted to one sex. Regarding the latter, species recognition in fruit flies involves female choice of males that produce stereotypic courtship sounds (Kyriacou & Hall, 1986; Hoy *et al.*, 1988), odours (Spiess, 1987) and behaviours (Hoikkala & Welbergen, 1995). These cues consist of heritable elements whose number and complexity correspond to the frequency of exposure to heterospecifics, and all of them must match the mate-recognition template for copulation to occur (i.e. they also are U-absent).

Cues used in discriminating among potential mates, by contrast, are more variable intrasexually and intraspecifically. Phonotaxis experiments on tungara frogs (*Physalaemus pustulosus*: Rand *et al.*, 1992) suggest that mate-quality recognition is based on the variable presence of one to six chucks and their pitch (females prefer the lowest frequencies, which are produced by larger males), whereas species recognition is based on a relatively invariant character, the fundamental frequency of the first 0–100 ms of the 'whine' part of the male's call.

In many mate-quality recognition systems, females prefer signals that deviate most from the population mean (Ryan & Keddy-Hector, 1992; Andersson, 1994), in contrast to sex- and species-recognition systems. Fisherian models of sexual selection (see Chapter 8) suggest that these preferences are maintained because sons of males that exhibit exaggerated traits will inherit these traits and thus be attractive to females, whereas in 'good genes' models preferences are maintained because offspring of males with exaggerated traits will be highly viable, since only males with high-viability genes can afford the costs of expressing exaggerated traits (see Chapter 8; Pomiankowski, 1988; Grafen, 1990b). In either case, females who prefer exaggerated traits minimize acceptance errors. Open-ended female preferences for exaggerated male traits should occur when the benefits of reducing acceptance errors consistently exceed the increased cumulative costs of rejection errors.

The directness of the connection between mate-recognition cues and the vehicle-enhancing qualities they signify depends on what females are attempting to obtain from males. In mate–resource recognition, stimuli from the resource itself are the cues. Thus, female blue-headed wrasses (*Thalassoma bifasciatum*) choose good spawning sites (Warner, 1987), and female hanging flies (*Hylobittacus apicalis*: Thornhill, 1981) and katydids (*Conocephalus nigropleurum*: Gwynne, 1988) choose the size and quality of males' nutritional offerings, not males' phenotypes per se. In lepidoptera, chemical mate-recognition cues often are metabolic derivatives of males' nuptial gifts (Conner *et al.*, 1990). Some female birds use cues that correlate with paternal effort, such as male body size (Petrie, 1983), courtship feeding rate (Wiggins & Morris, 1986) or colour (Hill, 1991).

When females seek high-viability genes or sexually attractive traits for offspring, they must rely on a correlation between those benefits and one or more aspects of males' phenotypes (e.g. Petrie, 1994). For example, in great-reed warblers (*Acrocephalus arundinaceus*) post-fledging chick survival is correlated with paternal song repertoire size. Females engage in extrapair copulations only with neighbouring males that have larger song repertoires than their social mate (Hasselquist *et al.*, 1996).

The connection between mate-recognition cues and benefits of discrimination is most indirect when females seek pathogen-resistant genes for their offspring, because rapid coevolution makes it unpredictable as to which genes confer resistance to prevailing pathogens. Females would frequently commit both acceptance and rejection errors by using as a recognition cue only one trait (i.e. the products of a few resistance genes). Selection has favoured two opposite solutions to this problem: (i) mating with many males, thereby increasing the likelihood that some patrilines will be resistant (e.g. social insects: Sherman *et al.*, 1988; Schmid-Hempel, 1994); or (ii) choosing a mate based on multiple condition-dependent traits that would be altered by *any* disease (i.e. cues that are connected to many potentially relevant genes). Thus, female barn swallows (*Hirundo rustica*) prefer males with long symmetrical tail feathers, cues that correlate with age and parasite resistance, which is heritable (Møller, 1994). Indeed, females in many species of birds (reviewed by Andersson, 1994, pp. 74–6) and fish (Houde & Torio, 1992; Kodric-Brown, 1993) prefer vigorously courting males with elaborate ornamentation, characteristics that are inversely related to parasitism and, presumably, genetic susceptibility to it.

4.3.3 The perception component

General principles

Selection may shape the perception component by modifying both the recognition template and the matching algorithm.

Templates

Templates are internal representations of the characteristics of desirable or undesirable recipients (see Fig. 4.1). Recognition occurs when phenotypes of recipients match these templates closely enough. Generalized templates are favoured when appropriate responses to all undesirable (or desirable) recipients are the same, despite variation in their exact cues. For example, Belding's ground squirrels usually give multiple-note trill vocalizations to terrestrial predators and single-note whistles to aerial predators (Sherman, 1977, 1985). However, the squirrels trill at *walking* hawks and whistle at *running* coyotes. Apparently the rapid approach of any large heterospecific, regardless of its exact phenotype, represents imminent danger; Evans *et al.* (1993) present similar data for chickens (*Gallus gallus*). By contrast, vervet monkeys (*Cercopithecus aethiops*) give structurally different calls to leopards, lions and hyenas, hawks, snakes, baboons and unfamiliar humans (Cheney & Seyfarth, 1990), suggesting that templates may be more specific when appropriate responses are not the same for all recipients (e.g. predators).

Template formation involves various degrees of learning. On the one hand, learning is disfavoured when recipients are not reliably present for template formation before discrimination is necessary or when template learning (particularly of undesirable recipients) might increase mortality. For example, naive (hand-reared) motmots (*Eumomota superciliosa*) are instantly repulsed by models resembling venomous coral snakes (Smith, 1975), and newborn garter snakes (*Thamnophis sirtalis*) respond strongly to odours of their primary prey; the latter preferences are heritable and differ among populations (Arnold, 1981).

Genetically encoded templates are inferred when artificial selection on male traits results in changes in female mate-selection criteria as a correlated response (reviewed by Bakker & Pomiankowski, 1995). For example, Wilkinson and Reillo (1994) selected male stalk-eyed flies (*Cyrtodiopsis dalmanni*) for large (L) or small (S) eye span relative to body size. After 13 generations, females from L-lines preferred L-males (as did unselected controls), but S-line females preferred S-males, indicating a genetic correlation between females' templates and males' traits.

On the other hand, templates must be learned when the characteristics of desirable or undesirable recipients vary over space or time. Thus, juvenile Belding's ground squirrels (Mateo, 1996), vervets (Cheney & Seyfarth, 1990) and other vertebrates (Curio, 1988; Mineka & Cook, 1988) learn the characteristics of predator contexts by observing anti-predator responses of adults (often parents).

When templates are learned, the objects or individuals that provide information about the characteristics of desirable or undesirable recipients are called *referents*. Selection favours use of referents that maximize the separation of template–cue dissimilarity distributions for desirable and undesirable recipients (see Fig. 4.2). Sometimes, actors serve as their own referents (e.g. in bat echolocation or the immune system). More often, learning occurs by: (i) associating characteristics of individuals with the consequences of interacting with them directly; (ii) observing their interactions with others; or (iii) learning the characteristics of individuals that are likely, by virtue of their location in time or space, to be desirable or undesirable recipients.

The timing of template learning depends on when the most informative referents are available and when discrimination is first adaptive. In some species, templates are learned ('imprinted') during a short sensitive period early in life, presumably because useful referents (parents) are predictably present and their characteristics reflect those of desirable recipients throughout an actor's life (Bateson, 1979). In many species, a second sensitive period occurs at reproduction, when parents imprint on their newborn offspring (Colgan, 1983).

When there is a lengthy association between individuals whose characteristics may change through time, recognition templates may be 'updated'. Template updating is what enables us to recognize our own teenage offspring regardless of how they looked as babies. In the many territorial species that discriminate neighbours from strangers (Temeles, 1994), template updating must occur whenever a new resident immigrates (Catchpole & Slater, 1995; Lambrechts & Dhondt, 1995). Sometimes, immigrants also update their song templates, because they match their new neighbour's songs (e.g. in village indigobirds,

Vidua chalybeata; Payne, 1985). Finally, females that copy the mating choices of other females (e.g. Gibson *et al.*, 1991; Dugatkin, 1992b) update their mate-quality recognition template whenever a new male is frequently chosen by conspecifics.

Matching algorithms

In general, selection should favour matching algorithms, which are internal weighting schemes, that optimally balance acceptance and rejection errors. If rejection errors are costly, cues that characterize nearly all desirable recipients (i.e. D-present cues) should be disproportionately weighted, even if some undesirable recipients also will match closely the template. Thus, myrmecophilous beetles (Vander Meer & Wojcik, 1982) and parasitic ant queens (Franks *et al.*, 1990) gain access to host ant nests because they possess a chemical cue that is heavily weighted by the host species in nestmate recognition.

Conversely, if acceptance errors are costly, actors should disproportionately weight cues that are possessed rarely by undesirable recipients (U-absent cues), even if some desirable recipients will not match closely the template. Thus, female birds avoid diseased males by scrutinizing secondary sexual characteristics associated with health, such as the colour and size of wattles, combs and snoods (Zuk *et al.*, 1992; Buchholz, 1995). Cues associated with male health are weighted so heavily that males with visible parasites (Borgia & Collis, 1989) or those that merely smell parasitized (Kavaliers & Colwell, 1995) may not match females' templates for acceptable partners.

The perception component of kin recognition

There are no clear examples of genetically encoded kin-recognition templates (Alexander, 1990; Pfennig & Sherman, 1995). This is probably because, as a result of meiotic shuffling of genetic cues and spatiotemporal variation in environmental cues, the characteristics of desirable recipients (kin) will differ for different actors, rendering genetically encoded templates unreliable. In addition, intragenomic conflict should thwart expression of selfish 'recognition alleles' (template loci linked to both recognition cue and decision rule loci: see Alexander & Borgia, 1978; Alexander, 1979; Ridley & Grafen, 1981; Chapter 12).

Organisms that live in sufficiently diverse environments (see Section 4.3.2) can learn templates partly or wholly from environmental features that reliably correlate with kinship. For example, paper wasps learn recognition odours from their nest and not directly from nestmates or themselves. Non-relatives can be fooled into accepting each other by experimentally exposing them to different fragments of the same foreign nest (Pfennig *et al.*, 1983). If wasps are isolated from their comb and nestmates when they eclose, or are exposed only to nestmates but not to nests, they later treat all conspecifics as nestmates, regardless of relatedness (Shellman & Gamboa, 1982).

Kin-recognition templates also are learned directly from parents or nestmates (Buckle & Greenberg, 1981; Blaustein *et al.*, 1987). Thus, female great tit chicks (*Parus major*) learn their father's songs and, as adults, females avoid pairing with males whose songs match this template (McGregor & Krebs, 1982). By contrast, female Belding's ground squirrels learn kin-recognition cues from nestmates. In this species, mothers and daughters and sisters behave nepotistically: they warn each other of predators and protect their own and each others' pups against infanticide by establishing and jointly defending territories (Sherman, 1977, 1981a,b). Cross-fostering studies in the field (Holmes & Sherman, 1982) and laboratory (Holmes, 1984, 1986a, 1994) indicate that nestmate females learn each other's odours just before weaning (when litters normally begin to mix), and later treat each other as siblings, regardless of their actual relatednesses. Socially learning these kin-recognition templates does not increase acceptance errors in nature because unrelated pups rarely cohabit the same nest burrow (Sherman, 1980).

Kin-recognition templates may also be learned via self-inspection. For example, the immune system learns the identity of the body's cells (von Boehmer & Kisielow, 1991). Alexander (1990, 1991) argued that genes promoting self-inspection are genetic 'outlaws' when their interests conflict with the rest of the genome. However, genes enabling formation of the molecular self-recognition template in immune recognition are not outlaws because the entire genome benefits from identifying and disabling foreign organisms (pathogens) or substances (environmental toxins). Selfish outlaw genes that caused recognition solely of products of linked genes would endanger the organism by causing serious autoimmune diseases.

Outlaw genes promoting self-inspection are more likely to spread when selfish targeting of linked production–component genes does not endanger the organism, e.g. in decisions about nepotistic allocation of resources. Even in this case, however, Alexander's argument implies only that intragenomic conflict will suppress such outlaws; it does not imply that perception–component genes causing self-inspection of the products of *unlinked* genes cannot evolve. This was Hamilton's (1987a, p. 423) point: 'so long as the genes involved in all parts of this discrimination are well spread among the chromosomes...we need not expect that other elements in the genome will evolve to suppress the effect.'

Self-inspection is favoured when template learning is desirable, but opportunities to directly assess relatives' phenotypes are limited. For example, self-inspection may occur in *Botryllus schlosseri* tunicates, the larvae of which are planktonic (Grosberg & Quinn, 1986), and in female *Gryllus bimaculatus* crickets and parasitoid wasps, which lay their eggs singly so larvae develop alone (Loher & Dambach, 1989; Ueno, 1994). In *G. bimaculatus*, females learn their own cuticular pheromones and later reject similar-smelling suitors. Unfamiliar unrelated pairs mate more quickly than do unfamiliar maternal half-siblings or full-siblings (Simmons, 1989). Since this recognition functions to improve vehicles (offspring) by reducing inbreeding, genes enabling formation of the self-recognition template are again not outlaws.

Finally, self-inspection may enable kin recognition when available relatives would yield error-prone templates as, for example, when nestmates are full- and half-siblings due to multiple mating by females (Sherman, 1991). There are two possible examples of what Dawkins (1982) dubbed the 'armpit effect'.

1 Honey bee (*Apis mellifera*) queens mate with seven to 17 drones, leading to mixed paternity (Page, 1986). Isolated workers can use their own phenotype as a kin-recognition template (Getz & Smith, 1986; Getz, 1991), and may be able to discriminate full- from maternal half-sisters (Getz & Smith, 1983). Whether nepotism occurs within the hive is controversial (see Visscher, 1986; Page *et al.*, 1989 versus Oldroyd *et al.*, 1994). A possible reason for this controversy is that discrimination may only occur when competition is severe (e.g. resources are limited) but discrimination does not jeopardize colony reproductive efficiency (Sherman, 1991; Ratnieks & Reeve, 1991).

2 Female Belding's ground squirrels mate with one to eight males and most litters are multiply sired (Hanken & Sherman, 1981). In the field, females are slightly less likely to attack full-sisters than maternal half-sisters, and more likely to share territories with full-sisters (Holmes & Sherman, 1982). Discrimination of nestmates from non-nestmates and full- from half-sisters among nestmates suggests that a female's own smell, as well as the odours of nestmates, both contribute to her kin-recognition template (Sherman & Holmes, 1985). In apparent support of self-referent phenotype matching, in laboratory studies paternal half-sisters also recognized each other (Holmes, 1986b; but see Alexander, 1991).

The perception component of mate recognition

Genetically encoded templates should be more common in species recognition than in kin recognition, because within populations there often will be little spatiotemporal variability in the characteristics of desirable recipients. Indeed, genetically encoded templates have been inferred in acoustically mediated mate recognition in crickets (Hoy *et al.*, 1977; Doherty & Hoy, 1985) and frogs (Walkowiak, 1988), based on intermediate responses of hybrids to calls of parental species. Of course, genetic encoding of species-recognition templates does not necessarily imply a lack of interpopulational variability in templates. In grey tree frogs (*Hyla chrysoscelis*) risks of species-acceptance errors have apparently influenced the evolution of female templates and matching algorithms across populations (Gerhardt, 1994). In recognizing mates, females weight the pulse repetition rate, which is unique to conspecific males, more heavily than call duration in areas of sympatry with *H. versicolor*. In areas where only *H. chrysoscelis* occurs, females weight call duration (an indication of mate quality) more heavily in mate choice. Thus, reproductive character displacement is expressed in the perception and, potentially, the action components of recognition.

By contrast, templates for mate-quality recognition should more often be learned, particularly in inbreeding avoidance, because characteristics of desirable recipients vary spatiotemporally. In house mice, the MHC genotype is expressed in urinary odours; individuals can detect the influence of even a single mutation (Brown & Eklund, 1994). Both sexes prefer to mate disassortatively (Beauchamp *et al.*, 1988; Boyce *et al.*, 1991), but preferences of males for females with MHC genotypes different from their own can be reversed by cross-fostering male pups to parents with different MHC genotypes (Fig. 4.4). This implies that males learn their template from parents.

Sex-recognition templates also are often learned from parents. Male zebra finch chicks learn their mother's plumage and bill colours and later recognize females based on those characteristics (Vos, 1994, 1995a,b). Female zebra finches, in contrast, apparently identify males by their courtship behaviour (Ten Cate, 1985; Vos, 1995a). Thus, a male's sex-recognition template directly informs him about a conspecific's sex, and indirectly informs females about sex via the males' behaviour.

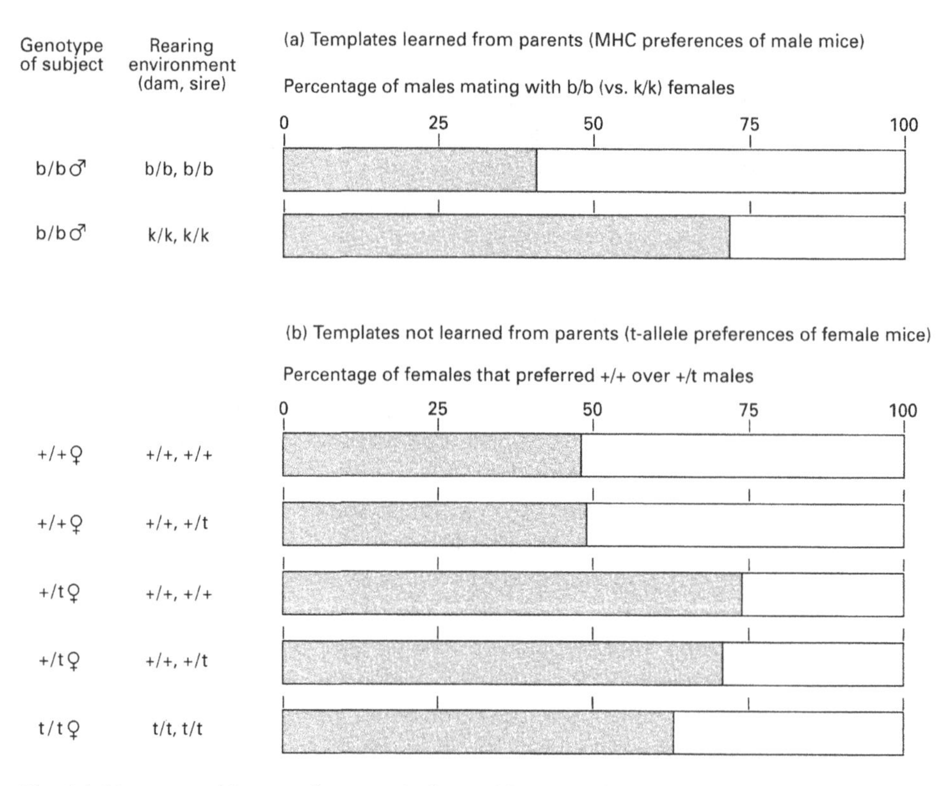

Fig. 4.4 Mate-recognition templates may be learned from social associates or by self-inspection. (a) Male house mice prefer females whose MHC genotypes differ from those of the male and female who reared each male, regardless of the male's own MHC genotype. (In this figure 50% = no preference. Data from Beauchamp *et al.*, 1988.) (b) Female house mice carrying at least one t-allele prefer males whose t-locus genotype differs from their own, regardless of the females' rearing environment. (Data from Lenington *et al.*, 1988; Coopersmith & Lenington, 1992.)

Why sex-recognition templates are learned at all is an interesting evolutionary puzzle. One possibility is that if males and females are dimorphic and typically raise young together, as in zebra finches, offspring will be favoured to use the information automatically available to them on both sex and the characteristics of successful (i.e. high-quality) mates, particularly when there is spatiotemporal variation in the latter (e.g. when polymorphisms exist within species and selection favours males pairing with females of the same morphotype as their mother). The female parent may be the better referent for sex and perhaps mate-quality recognition templates because: (i) females often have more frequent and consistent contacts with a litter or brood than do males; and (ii) due to extrapair copulations and fertilizations, the characteristics of a successful, locally adapted parent may be more accurately reflected by the mother's phenotype than that of her social mate.

When learning is favoured, but referents that are predictably available would yield inappropriate mate-recognition templates, learning via self-inspection or genetically encoded templates should occur. For example, obligate brood parasites such as brown-headed cowbirds (*Molothrus ater*) and cuckoos (*Cuculus canorus*) often interact only with their foster parents through fledging. Graham and Middleton (1989) hand-reared cowbirds and discovered that fledglings began discriminating stuffed conspecifics from heterospecifics soon after fledging. Although they concluded (p. 21) that the 'ability to recognize conspecifics is based initially upon innate responses to the appropriate visual cues', their results are also consistent with self-referent phenotype matching of species-specific plumage characteristics. Mate acquisition by male cowbirds is not based (solely) on self-learning, however. Females display ('wing stroke') in response to certain songs, and males attempt to produce songs that elicit this response (West & King, 1988). Thus, males are trained to sing songs that match the preferences of female companions, but precisely how females acquire their preferences (templates) is presently unknown.

Self-inspection may also mediate disassortative mating at the t-locus in house mice. Most t-haplotypes carry recessive lethal factors that kill homozygotes (t/t) *in utero* (Lenington, 1991). Like MHC alleles, t-alleles code for urinary chemosignals. Heterozygous (+/t) females strongly prefer wild-type (+/+) over heterozygous males, regardless of whether they were raised by two +/+ parents or one +/t and one +/+ parent, whereas homozygous (+/+) females show no preference, again regardless of rearing environment (Fig. 4.4). Lenington *et al.* (1992, p. 42) concluded that 'genes on the t-carrying chromosome modulate the cue produced by males of different genotypes as well as female preference for odours of +/+ males.' Alternatively, females may learn their own smell (t-haplotype) and then avoid males with similar t-modulated odours. Under this hypothesis, +/+ females show no preference because their own t-related odours (or lack of them) cause a permissive acceptance threshold; i.e. universal acceptance of all t-associated odours.

Do mate-quality recognition templates originate from sex- or species-recognition templates? Artificial neural networks have been used to explore the idea that accurate perception of species- or sex-recognition cues results automatically in 'hidden preferences' for the same cues in mate-quality choice contexts (e.g. Enquist & Arak, 1993, 1994; Johnstone, 1994). Based on simulation studies, these authors argued that female preferences (templates) for extravagant and symmetrical male characteristics are indeed unselected byproducts of species or sex recognition. However, as Dawkins and Guilford (1995) pointed out, these neural networks contain so few elements and connections that they bear only superficial resemblances to actual neural circuitry, and their behaviour is unlike real recognition systems in many important ways. Hidden biases may thus be artifacts of the models' simplicity, and their direction may depend on the 'training' regimen. Moreover, these models only suggest a possible *origin* for mate-quality recognition templates. Such templates will be maintained only if the resulting preferences consistently yield the predicted benefits — e.g. more attractive sons or healthier, more viable offspring (see Reeve & Sherman, 1993; Watson & Thornhill, 1994).

4.3.4 The action component

General principles

Given a particular degree of match between a recipient's cues and an actor's template, what action should the actor take? Individuals might perform a continuous range of actions (or intensities of a single action), with responses changing gradually as the cue–template match changes. Alternatively, actions may be all-or-none, as when there is a threshold above which all recipients are accepted and below which they are rejected (see Fig. 4.2). Many decision rules involve binary actions (e.g. attacking or ignoring potential prey, mating or not mating), so we restrict our attention to this conceptually simpler case.

The optimal acceptance threshold will depend on the relative rates of interaction with and the fitness consequences of accepting and rejecting desirable and undesirable recipients (Reeve, 1989). For example, the threshold should become more restrictive as the fitness costs of accepting undesirable recipients increase. However, the optimal threshold can become either more or less restrictive as the benefits of accepting desirable recipients increase, depending on whether there are limits on the number of recipients an actor can accept (Box 4.1). If acceptances are essentially unlimited (e.g. a nest-entrance guard in a social insect colony can admit all nestmate foragers), the threshold should become more permissive as the desirable recipient's value increases, because the actor should accept as many desirable recipients as possible. Conversely, if an actor can afford to accept only one or a few recipients (e.g. a female in a lek-breeding species searching for her mate for the season), the threshold should become more restrictive as the desirable recipient's value increases because a restrictive threshold enhances the probability that the

Box 4.1 Shifting acceptance thresholds in guard and search contexts

Natural selection may favour different acceptance thresholds in different contexts. Suppose $g(x)$ and $b(x)$ are the probability density functions (p.d.f.'s) describing the frequency distributions of template-recognition cue dissimilarity x for desirable (D) and undesirable (U) recipients, respectively (see Fig. 4.2). The probability of accepting a recipient given a threshold t is:

$$\int_0^t g(x)\mathrm{d}x = G(t)$$

for desirable recipients, and:

$$\int_0^t b(x)\mathrm{d}x = B(t)$$

for undesirable recipients. Let P be the fraction of Ds which each cause a fitness change F if accepted, and $1 - P$ be that for Us, which yield a fitness change $f < O$ if accepted.

The optimal acceptance threshold t^* usually cannot be explicitly obtained, because the precise shapes of the p.d.f.'s typically are unknown. However, we can predict how the threshold will shift as, say, F changes. We seek the sign of $\mathrm{d}t^*/\mathrm{d}F$, which is the same as that of $\frac{\partial w(t)}{\partial F \partial t}$ at $t = t^*$, where $w(t)$ relates the actor's fitness to t.

Case 1: guard context

Suppose an actor must decide whether to accept recipients that approach it, and multiple acceptances are possible. Assuming additive fitness changes, the actor's threshold should maximize the mean fitness increment per interaction:

$$w(t) = PFG(t) + (1 - P)fB(t)$$

Since $\frac{\partial w(t)}{\partial F \partial t} > 0$ at $t = t^*$, it must be that $\mathrm{d}t^*/\mathrm{d}F > 0$ and the threshold becomes more permissive (moves right) as F increases.

Case 2: search context

Now suppose an actor searches for a single recipient (at a cost c per search). The threshold maximizes:

$$w(t) = \frac{PFG(t) + (1 - P)fB(t) - c}{PG(t) + (1 - P)B(t)}$$

Continued on p. 88

Box 4.1 *Cont'd.*

Here, $\frac{\partial w(t)}{\partial F \partial t} < 0$ at $t = t^*$, so $dt^*/dF < 0$ and the threshold becomes more restrictive (moves left) as F increases. Thus, increasing value of Ds causes greater choosiness when only one recipient can be accepted and less choosiness when multiple recipients can be accepted (Reeve, 1989).

chosen recipient is desirable. This theoretical result supports and generalizes to all recognition systems Bateman's (1948) principle (Trivers, 1972; Hubbell & Johnson, 1987) that in mate choice, males should emphasize quantity whereas females should emphasize quality.

When the optimal acceptance threshold varies in different recognition contexts, organisms should assess environmental cues that distinguish such contexts (Hamilton, 1987b). These have been termed 'locational' (Holmes & Sherman, 1982) or 'indirect' cues in the case of kin recognition, as opposed to 'direct' or phenotypic cues (Waldman *et al.*, 1988). Non-phenotypic cues may signify *times* or *places* where desirable or undesirable recipients are likely to be found. Phenotypic and non-phenotypic cues are not mutually exclusive; spatial and temporal cues often supplement phenotypic labels.

As an example of a temporal cue, when female acorn woodpeckers (*Melanerpes formicivorus*) oviposit communally, each ejects eggs until she begins to lay (Mumme *et al.*, 1983; Koenig *et al.*, 1995); females thereby avoid destroying their own eggs. Locational cues are used by parents in some birds (e.g. bank swallows, *Riparia riparia*: Hoogland & Sherman, 1976) and mammals (Belding's ground squirrels: Sherman & Holmes, 1985) to recognize offspring. Young remain in their natal nest for several weeks after hatching (birth), so cues of the nest or burrow define a context in which the frequency of interaction with desirable recipients (offspring) is so high that universal acceptance is favoured (i.e. a maximally permissive acceptance threshold). This may seem paradoxical because parents should almost always have the opportunity to learn phenotypic cues that are probabilistically associated with their own offspring. However, selection might never favour use of these cues if they lead *even infrequently* to rejection of offspring, given the fitness cost of such rejection.

Parent swallows and mother ground squirrels stop accepting foreign young when broods begin to mix in nature. At that point, temporal cues indicate a new context in which the frequency of interaction with undesirable recipients (i.e. non-offspring) is high enough that a more restrictive acceptance threshold is optimal. Thus, parents switch to using phenotypic cues, chicks' vocalizations in the swallows (Beecher *et al.*, 1981a,b) and pup odours in the ground squirrels (Holmes, 1984). The ontogeny of sibling recognition in the latter also involves a change from universal acceptance to discrimination against non-nestmates (Holmes & Sherman, 1982; Holmes, 1986a, 1994).

The action component of kin recognition

Kin-recognition acceptance thresholds shift among contexts in the ways predicted if animals facultatively optimize the balance of acceptance and rejection errors. For example, paper wasp workers are more intolerant of unrelated non-nestmates that arrive on their nest, where brood theft and usurpation occur (Kasuya *et al.*, 1980; Klahn, 1988), than they are when they meet the same non-relatives elsewhere (Gamboa *et al.*, 1991b), probably because benefits of defensive aggression are low away from the colony (Fig. 4.5). On the nest, queens have more restrictive thresholds than workers, as evidenced by greater aggression than workers toward both nestmates and non-nestmates (Fishwild & Gamboa, 1992). Such a caste-mediated threshold shift makes adaptive sense because queens are more likely to interact with non-nestmates that intrude on the nest than workers (Fishwild & Gamboa, 1992), and queens may have more to lose reproductively than workers from accepting them.

Dramatic illustrations of context-dependent acceptance thresholds come from species that facultatively produce distinct morphs which differ in their

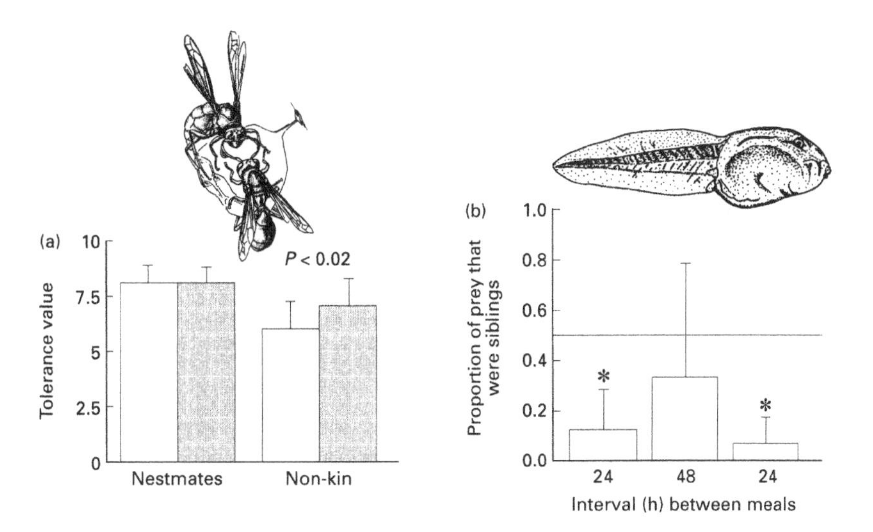

Fig. 4.5 Examples of context-dependent kin discrimination. (a) In paper wasps, resident females are equally tolerant of nestmates on (grey bars) and off (white bars) the nest comb. By contrast, resident females are significantly less tolerant of unrelated females on the comb, where brood theft and usurpation occur (Klahn, 1988), than when they meet the same non-relatives off the comb. (Data from Gamboa *et al.*, 1991b.) (b) When satiated, carnivorous Plains spadefoot toad tadpoles eat significantly fewer kin than non-kin; when they are very hungry, however, they engage in indiscriminate cannibalism. The experiment was done sequentially over 96 h, with food withheld for 24 h, 48 h, and then 24 h again. At the end of each fasting period, the carnivore's consumption of kin and non-kin was assessed. An asterisk indicates that the observed value was significantly less than random expectation (indicated by the horizontal line). (Data from Pfennig *et al.*, 1993.)

ability to harm relatives. Plains spadefoot toads (*Scaphiopus bombifrons*) breed in temporary desert pools. All tadpoles begin life as omnivores, feeding primarily on detritus. Occasionally, however, individuals eat freshwater shrimp, triggering a series of changes in morphology and dietary preference. Changed tadpoles become exclusively carnivorous, often feasting on conspecifics (Pfennig, 1992). Omnivorous and carnivorous tadpoles have different context-specific decision rules (Pfennig *et al.*, 1993). Omnivores are attracted to and school preferentially with siblings (i.e. kin are desirable recipients), while carnivores generally avoid relatives (i.e. kin are undesirable recipients of cannibalism). When carnivores are very hungry, however, they associate with and eat kin and non-kin at random — i.e. their acceptance threshold becomes more permissive because risk of starvation increases their personal cost of rejecting related prey (Fig. 4.5). Hokit *et al.* (1996) present a dramatic example of context-specific changes in aggression toward kin in larval salamanders.

The action component of mate recognition

As with kin recognition, a recipient's cues may trigger different actions by an actor depending on the mate-recognition context and its associated decision rule. There are at least three general types of mate-quality decision rules. Individuals (usually females) may choose: (i) the best mate available among those they can sample; (ii) the first mate they encounter whose characteristics exceed some threshold; and (iii) the mate chosen by other same-sex individuals (copying). Analogous decision rules apply to all types of recognition systems (Reeve, 1989).

Janetos (1980) suggested that a 'best-of-*n*' decision rule (choosing the best mate from a sample of size *n*) yields higher fitness than a 'one-step' strategy (searching terminates only when the quality of an encountered mate exceeds the expected quality of subsequent mates). Parker (1983) then showed that the optimal threshold quality for acceptance depends acutely on search costs. Real (1990) demonstrated that, when search costs are included, a sequential sampling strategy (searching ceases only when the quality of an encountered mate exceeds some minimum threshold) generally yields higher fitness than does best-of-*n*; Real's model is similar to Reeve's (1989) repeated search model, which incorporates recognition errors and makes similar predictions.

Empirical studies of mate-choice decision rules have barely begun (see Wiegmann *et al.*, 1996). Nonetheless, it is already clear that females often do sample multiple males. Thus, female great snipe (*Gallinago media*: Fiske & Kålås, 1995), cock-of-the-rock (*Rupicola rupicola*: Trail & Adams, 1989) and Australian brush turkeys (*Alectura lathami*: Birks, 1996) visit up to 12 males before returning to copulate with one. This indicates either a best-of-*n* decision rule or a sequential search in which the acceptance threshold becomes more permissive as the time spent searching increases (Fiske & Kålås, 1995). Female fruit flies require four to six bouts of courtship before becoming receptive, which enables them to sample the relative frequency of male phenotypes in the population, and to choose the rarest for mating (Spiess, 1987).

Under the models of Parker (1983), Reeve (1989) and Real (1990), optimal mate-acceptance thresholds become more permissive with increasing search costs or decreasing benefits for choosiness. In support, female guppies (*Poecilia*

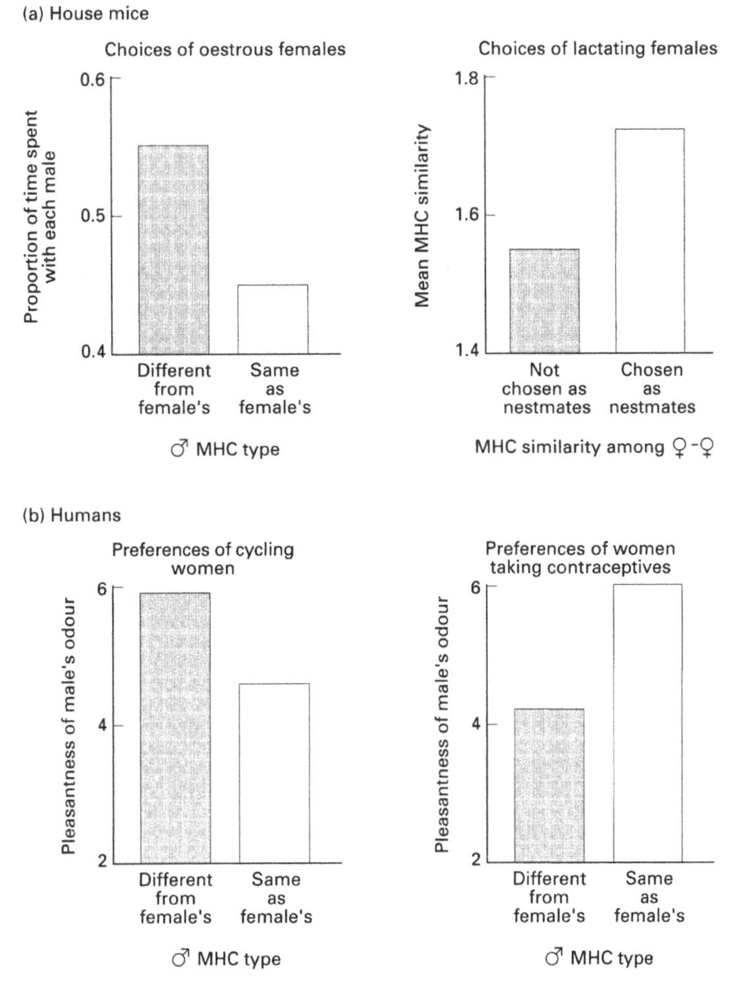

Fig. 4.6 Examples of facultative shifts between recognition systems in house mice and humans. (a) Sexually receptive female mice avoid males with similar MHC genotypes (thereby promoting outbreeding), whereas pregnant and lactating females are attracted to females with similar MHC genotypes (with whom they form nepotistic communal nesting associations). (Data from Egid & Brown, 1989; Manning *et al.*, 1992; Model 3: MHC preferences based on social learning.) (b) Women find odours of males with different MHC genotypes more pleasant than odours of males with the same genotype unless they are taking oral contraceptives; then they prefer males with similar genotypes. (Data from Wedekind *et al.*, 1995.)

reticulata) collected from a river in which predators are common reduce both their overall level of sexual activity and their preference for brightly coloured males in the presence of predators (Godin & Briggs, 1996), and female peacock wrasses (*Symphodus tinca*) are choosiest about whether they lay their eggs in nests tended by a male, and about which nests to lay in, when the fitness payoff for laying eggs in male-tended nests is highest (due to predation) and search costs are lowest (Warner *et al.*, 1995). Female zebra finches become choosier after they are exposed to males with high display rates (Collins, 1995) implying that, in accordance with theory (see Box 4.1), their mate-acceptance thresholds become more restrictive as the assessed frequency of high-quality mates increases.

The third decision rule, mate-choice copying, occurs in many vertebrates (Gibson *et al.*, 1991; Dugatkin, 1992; Höglund & Alatalo, 1995). Copiers may save time and reduce predation risks by not sampling every male; information about which males other females have accepted or rejected also enables individuals to minimize erroneous choices. Mate-choice copying also may affect the perception component of mate recognition by causing template updating in light of evidence of attractiveness to other females (see Section 4.3.3).

Clearly, mate-acceptance thresholds can be context dependent. In addition, which recognition system is in operation depends on the context. For example, female house mice prefer as mates males with dissimilar MHC genotypes (Egid & Brown, 1989; Potts *et al.*, 1991; Fig. 4.6), apparently to avoid inbreeding (Potts *et al.*, 1994). However, when females are pregnant or lactating, they form communal nesting and nursing associations with females that have similar MHC genotypes, which are usually close kin (Manning *et al.*, 1992, 1995). Interestingly, human females also prefer odours of males with dissimilar MHC genotypes, except when they are taking birth control pills; in that case, they prefer odours of males with similar genotypes (Fig. 4.6). Wedekind *et al.* (1995) suggested that the basis of the former choice is optimal outbreeding, whereas the latter indicates nepotistic kin recognition by (pseudo-) pregnant females.

4.4 Current topics in recognition research

4.4.1 Failures of kin recognition

Many organisms occasionally make acceptance or rejection errors. There are at least three evolutionary reasons why this may be so. First, circumstances favouring recognition may be rare, or may have been rare until recently. Female Belding's ground squirrels behave nepotistically only to daughters and sisters. More distant relatives (granddaughters, second cousins) are treated like non-kin (Sherman, 1980). These distant kin are infrequently alive simultaneously (Sherman, 1981b). Either selection has not favoured abilities to learn templates to recognize distant relatives (perception component), or the rate of interaction with them is so low that the optimal acceptance threshold is restrictive, excluding distant kin (action component). As another example, bird species that have long been exposed to brood-parasitic cowbirds or cuckoos consistently reject parasite eggs, whereas species whose nesting habitats have been recently invaded (e.g. due to forest fragmentation) tend to accept all eggs in their nest, including those of parasites (Mayfield, 1965; Davies & Brooke, 1989).

Second, errors may persist because the error-related costs of kin discrimination outweigh the benefits. Recall that recognition errors occur because of overlap in labels. If either rejection or acceptance errors become too costly, selection may favour universal acceptance or rejection, respectively, of kin and non-kin (Reeve, 1989). For example, male red-winged blackbirds (*Agelaius phoeniceus*: Westneat *et al.*, 1995) and house martins (*Delichon urbica*: Whittingham & Lifjeld, 1995) feed all the chicks in their nest, although 25–30% are unrelated due to extrapair copulations by their mother. Since the chicks are of the same age and species, they closely resemble each other, but they look and sound nothing like adults. Apparently it is more efficient reproductively for males to feed all the chicks than to risk rejection errors, thereby allowing their progeny to starve (Westneat & Sherman, 1993; Kempenaers & Sheldon, 1996). Interestingly, in several birds (e.g. dunnocks, *Prunella modularis*: Davies *et al.*, 1992; alpine accentors, *P. collaris*: Hartley *et al.*, 1995) males do adjust how much they feed the *entire* brood based on a non-phenotypic (time-specific) cue: the amount of exclusive sexual access they had to the female (which correlates with degree of paternity).

Brood parasitism by cuckoos also illustrates this cause of errors. Normally, first-time parents learn what their own eggs look like and reject eggs from subsequent clutches that do not match this template (Lotem *et al.*, 1995). The cost of the learning rule is that if a host is parasitized during its first breeding attempt, it will learn the parasite egg and be doomed to accept parasite eggs forever. When parasitism is sufficiently rare, the benefits of correct imprinting exceed the costs of misimprinting, so acceptance errors persist (Lotem, 1993). Hosts minimize these acceptance errors by varying their rejection rate in relation to their probability of parasitism (Davies *et al.*, 1996).

Third, when recipients benefit from the absence of discrimination they will be favoured to hide their true kinship by 'muting' or 'scrambling' recognition labels (Reeve, 1997a). For example, extrapair copulations set in motion a coevolutionary race between males attempting to discriminate their progeny and unrelated juveniles attempting to dupe the male by not revealing their genotype via their phenotype (Beecher, 1991). If young are initially uncertain about whether the resident male is in fact their father, they all should hide, change or mix their recognition cues (e.g. via rapid growth and feather development) because the fatal cost of being rejected, even if improbable, may exceed the small benefit, even if likely, of receiving extra food from a male that has identified them as his progeny. Juveniles have an upper hand in this coevolutionary struggle because, on average, they have more to lose by being recognized and rejected than males lose through misdirected nepotism.

4.4.2 Misunderstandings about kin recognition

Kin recognition, like kin selection (Dawkins, 1979), is often misunderstood. We therefore highlight and attempt to clarify several semantic and conceptual issues.

Misunderstanding 1: kin recognition favoured by kin selection must be mediated by genetic cues

The flaw in this proposition (Grafen, 1990a, 1992) becomes clear when we view kin recognition as a strategy by which recognition-promoting alleles target copies of themselves in recipients (Reeve, 1997b). The recognition-promoting allele in the actor does not directly 'see' copies of itself but nevertheless promotes its own spread through a chain of positive statistical associations linking the recognition cues in the recipient with possession of a copy of the allele in the recipient. Whether labels are produced by genetic loci or acquired from the environment, the recognition-promoting allele spreads via an *indirect* association between cues and copies of the recognition-promoting allele in the recipient. The statistical association between cues and alleles is not necessarily less (and may often be greater) for environmentally acquired than for genetically specified cues (Gamboa *et al.*, 1986b, 1991c; see also Section 4.3.2).

Misunderstanding 2: non-phenotypic recognition is not 'true' kin recognition

Halpin (1991, p. 220) suggested 'animals that rely only on "spatially-based recognition" are actually incapable of recognizing kin from non-kin, and it is precisely because of this that they are forced to rely on spatial cues to determine who will be treated *as if* they were kin.' This argument also fails to appreciate that recognition-promoting alleles spread because of indirect statistical associations between the recognition cue (i.e. the location) and presence of the recognition-promoting allele in conspecifics. When frequencies of interactions with relatives at a particular location (e.g. a nest burrow) are sufficiently high, selection may favour universal acceptance at that location (see Section 4.3.4). Rarity of acceptance errors and costs of rejection errors, not mechanistic inability to use phenotypic labels, drives the evolution of kin recognition via non-phenotypic cues. In species exhibiting location-specific behaviour, parents occasionally do rear non-kin due to mix ups (in bank swallows: Hoogland & Sherman, 1976; Belding's ground squirrels: Sherman, 1980; paper wasps: Gamboa *et al.*, 1986b), but these recognition errors are rare. Moreover, as we have seen, recognition systems based on phenotypic cues of genetic or environmental origin are not immune to errors either.

Misunderstanding 3: kin recognition is an epiphenomenon of species or group recognition

According to Grafen (1990a, 1992) most examples of kin recognition are epiphenomenal, meaning that they are not maintained by kin selection but rather by selection for species or group member recognition. There are two problems with the alternative functions Grafen identified. First, in general, kin are inappropriate templates for species recognition because if individuals learned to recognize all conspecifics based on their resemblance to kin only, they would acquire a very restrictive species-recognition template, and many rejection errors would result. Second, group member recognition is actually kin-selected kin recognition in the majority of social insects, birds and mammals because group members are close relatives (Sherman *et al.*, 1995). Grafen's general point is instructive, however, because some forms of kin discrimination function in contexts that do not involve kin selection or mate complementarity (see Section 4.2).

4.5 Conclusions

This chapter presents a synthetic framework for understanding recognition systems by conceptually partitioning them into production, perception and action components. Characteristics of these components vary predictably with the recognition context. Knowledge of each component illuminates the selective forces that shape recognition and behavioural discrimination, and vice versa, because selection favours components that enable individual cells and organisms to achieve an optimal balance between rejection errors and acceptance errors. These principles apply to all recognition systems, including recognition of mates, kin, habitats (e.g. Jaenike & Holt, 1991; Chivers & Smith, 1995), hosts (Honda, 1995), reciprocators (Reeve, 1997b), dominance (Butcher & Rohwer, 1989; Drickamer, 1992), predators and prey (Stephens & Krebs, 1986) and individuals (Caldwell, 1992; Temeles, 1994).

Understanding recognition systems is important to developing and testing hypotheses about such diverse topics in evolutionary biology as mate choice, nepotism, intragenomic conflict, immune responsiveness and learning. Exciting areas for future research include the following.

1 Quantifying the fitness consequences of recognition in nature, especially kin recognition (see Section 4.2). One promising approach is to create behavioural 'mutants' that differ in their discriminatory abilities and then to compare, within natural populations, the inclusive fitnesses of individuals that discriminate, e.g. kin (individuals that were allowed to imprint on kin) versus individuals that do not discriminate kin (because they were forced to imprint on non-kin).

Moreover, the widespread occurrence of facultative changes in the components of recognition, such as context-dependent acceptance thresholds (see Section 4.3.4, Box 4.1), has opened a new avenue of research into the functions of recognition systems. For example, the hypothesis that discrimination between desirable and undesirable recipients functions to obtain some benefit '*X*' or avoid some cost '*Y*' can be tested by assessing whether the recognition cues, template weighting scheme and/or acceptance threshold (inferred from the probability of acceptance errors) vary in predictable ways as the magnitude of *X* or *Y* changes across recognition contexts.

2 Determining the occurrence of self-referent phenotype matching in the context of mate recognition (e.g. brood parasitic birds, t-alleles in mice) and nepotism (e.g. full- versus half-sibling discrimination among nestmates in birds and mammals or within insect colonies, and paternal discrimination of offspring versus unrelated young; see Sections 4.3.3 and 4.4.1). Is the armpit effect rare, or just subtle and hard to detect? The answer will enhance our understanding of the evolution of self-inspection and intragenomic conflict.

3 Determining how mechanisms underlying perception and action components develop (see Sections 4.2, 4.3.3 and 4.3.4). Recent advances in understanding the neurophysiology of signal perception (e.g. Barlow, 1995) and processing (Schildberger *et al.*, 1989; Suga, 1995) notwithstanding, for behavioural ecologists recognition mechanisms remain 'black boxes'. Studies of the neurophysiological processes underlying the development of recognition abilities will provide deeper insights not only into the mechanisms of recognition, but also into the more fundamental processes of learning and memory.

Chapter 5

Managing Time and Energy

Innes C. Cuthill & Alasdair I. Houston

5.1 Introduction

All behaviour takes time. All behaviour consumes energy. But, the behavioural ecologist's interest in time and energy does not simply lie in the fact that behaviour utilizes these key resources. Also important is the fact that they cannot be allocated to all behaviours simultaneously. The concept of trade-offs is thus a central pillar in the evolutionary approach to behaviour. While one can see the imprint of trade-offs in observed patterns of behaviour and in comparative studies, the most direct tool for examining them is the optimality model. Economic models force one to be explicit about the nature of the trade-offs involved and to explore their putative impact on the design of the behaviour. In this chapter we concentrate on functional (ultimate) aspects of the allocation of time and energy but, as we shall see, proximate and ultimate approaches to this question can, and should, be interlinked. We start with some classic economic models of decision making that make simplifying assumptions about the value of energy, then progress to models that can accommodate changes in value as a result of current energetic state.

When should an animal feed? When energy or some other dietary component is required? When food is most abundant? Even the simplest functional hypothesis would have to take into account the fact that feeding is incompatible with many other important behaviours, all of which require energy. Even when two behaviours can be performed concurrently, there will usually be reduction in efficiency compared to when each behaviour is performed in isolation (Futuyma & Moreno, 1988; Leigh, 1990). So, an organism must acquire and store reserves not only for occasions when food is not available, but to fuel periods when it cannot feed due to the performance of other behaviours. A full analysis of daily feeding patterns is thus likely to have to take into account not only temporal variation in the costs and benefits of feeding, but temporal variation in the costs and benefits of other behaviours. But, how can we evaluate the fitness consequences of different feeding strategies?

5.2 The functional approach

The functional approach involves comparing actions in terms of their contribution to future reproductive success. In principle, this gives us a common currency for comparing actions whose consequences might differ in a variety of ways (e.g. in amount of food obtained and probability of being killed by a predator). In practice it is often difficult to assess future reproductive success, and so attention has often been focused on simple currencies which have the following properties:

1 they are easy to measure;

2 maximizing them will maximize fitness.

In the context of an animal's foraging decisions, the net rate of energetic gain γ is an obvious currency. Maximizing γ maximizes the amount of energy obtained from a given time spent foraging. Maximizing γ also minimizes the time required to obtain a given amount of energy (Schoener, 1971). We illustrate this using data on the modes of hunting of the kestrel (*Falco tinnunculus*).

Kestrels may hunt for prey by flying or by sitting on a perch. Masman *et al.* (1988) refer to these methods as flight hunting and perch hunting, respectively, and provide information on their energetic consequences. In both summer and winter, flight hunting has both a higher rate of energy intake and a higher rate of energy expenditure. In winter, the gross rate of gain b_f from flight hunting is 131.8 kJ h^{-1}, whereas the gross rate of gain b_p from perch hunting is 13.2 kJ h^{-1}. The corresponding rates of expenditure are $e_f = 52.2$ kJ h^{-1} and $e_p = 6.5$ kJ h^{-1}, respectively. Thus, flight hunting has a net rate of gain $\gamma_f = b_f - e_f = 79.6$ kJ h^{-1} and perch hunting has a net rate of gain $\gamma_p = b_p - e_p = 6.7$ kJ h^{-1}. The kestrel will maximize the energy gain in 1 day by using flight hunting for all the available time, and will minimize the time to balance its energy budget by flight hunting for a time t_f, where $t_f\gamma_f$ is the energy spent on other activities. Flight hunting is indeed the predominant foraging behaviour, but the fact that kestrels use perch hunting at all is not consistent with the simple predictions of rate maximizing or time minimizing given above. (For further discussion and parameter values for summer foraging, see Masman *et al.*, 1988). The kestrel story illustrates how a rate-maximizing model can be applied to animal decision-making. It also indicates that such models may not be able to explain all aspects of foraging behaviour.

The maximization of γ forms the basis of many optimal foraging models, including the standard models of prey choice and patch use (see Stephens & Krebs, 1986). Although net rate is an obvious and appealing currency, in some cases it has been shown that another currency gives a better description of the data. Some of these examples involve parent birds bringing food to their young. A bird's flight speed determines not only the time that the bird takes to travel a given distance, but also the rate of energy expenditure. By increasing its flight speed, a bird reduces the journey time (which increases the rate of delivery of food to the young), but will also usually increase its rate of energy expenditure. Norberg (1981) introduced a way of calculating the costs and benefits of flying at a given speed. If a bird flies a distance D at speed v, then the time to complete the journey is simply D/v. If this is the only effect that has to be considered, then the bird should fly as fast as possible (in order to reduce this time). Norberg's insight was that a bird will typically have to repay the energy that it spends on flight, and this will take time. If the bird spends energy at a rate $P(v)$ and can replace energy at a rate g, then the time to replace the energy spent is

$$\frac{P(v)D}{gv}.$$

In this model, the best speed is the one that minimizes the total time

$$\frac{D}{v}\left(1+\frac{P(v)}{g}\right)$$

of the two components. This will maximize the overall rate to the young, subject to the constraint that the parent balances its energy budget. Norberg introduced this idea in the context of a parent bird bringing food to its young (in which case D is the round-trip distance from the nest to the feeding site and back again), but the principle is general and has been used in the context of migration by Alerstam and Lindström (1990). McLaughlin and Montgomerie (1990) studying lapland longspurs (*Calcarius lapponicus*) and Welham and Ydenberg (1993) studying black terns (*Chlidonias nigra*) report that the flight speed of parent birds does not maximize overall delivery rate but is close to the speed that maximizes efficiency. Houston (1993) and Hedenström and Alerstam (1995) show that this is to be expected if the parents' energy expenditure is up against a constraint. Hedenström and Alerstam (1995) give a general review of the currencies that have been used for assessing flight speed and also establish some new general results.

Efficiency is the ratio of energy gained to energy spent and hence has no dimensions. It is obvious that if an animal has a fixed amount of energy to spend, then maximizing efficiency will maximize the amount of energy obtained. This argument has been suggested as an explanation of why honey bees (*Apis mellifera*) appear to maximize efficiency (Schmid-Hempel *et al.*, 1985), but what matters in this case is not the energy brought in by each bee over its foraging life, but the gains to the colony from energy brought in and the cost to the colony of a bee dying. Even if each bee has a fixed amount of energy that it can spend, maximizing the energy brought in by maximizing efficiency will not necessarily maximize the reproductive success of the colony (Houston *et al.*, 1988).

One reason why foragers might not maximize long-term rate depends on the idea of constraints on the amount of energy that an animal can acquire or spend in a given time. The ability of an animal to assimilate energy is likely to be limited by its physiology (Kirkwood, 1983; Weiner, 1992). In addition to this limit, it has also been argued that there is an upper limit to the rate at which an animal can spend energy (Drent & Daan, 1980). These constraints mean that an animal may not be able to forage for all the time available to it (Ydenberg *et al.*, 1994; Houston, 1995). As a result, there can sometimes be an advantage in choosing an option that does not have the highest value of γ. If there is a foraging option which also has a lower rate of energetic expenditure than the option with the highest γ, then the low-cost option can be adopted for a longer time before the constraint on energetic expenditure is reached. However, if no foraging options allow the animal to escape the constraint on energetic expenditure then it should use the option that maximizes a modified form of efficiency for as long as possible and spend the rest of its time resting (see Houston, 1995).

It is easy enough to think of biologically plausible situations when the simple currencies considered above are highly correlated with fitness. However, it is even easier to think of cases when gathering energy at the maximum possible rate, or at the lowest unit cost, are not sensible strategies. Good feeding sites may incur a high predation risk. Might not the value of energy change with time and circumstance? Is it not likely, for example, that a dominant individual, sure of displacing conspecifics from good feeding sites, might show different feeding patterns from a subordinate? And that being as fat as possible is not necessarily a good thing? How can we incorporate such biological realism into our models? To calculate the value of energy, we need to know what it is going to be used for and when. Then we need to know the costs of acquiring and maintaining energy reserves. Finally, we need a way of evaluating these benefits and costs such that we can predict the effects of changing time and circumstance on foraging patterns.

5.3 The benefits of energy reserves

Ultimately, the value of feeding is to provide the raw material for self-maintenance, growth and reproduction. In this chapter we shall confine ourselves to the use of energy, but the same principles can be applied to other dietary components. These may be required on different time scales (e.g. the season-specific requirements of small birds for calcium to produce egg-shell; Graveland & Vangijzen, 1994; Krementz & Ankney, 1995) and the optimal diet may itself involve trade-offs between different dietary requirements (Belovsky, 1978; Murphy, 1994; Raubenheimer & Simpson, 1995), but the modelling approach is the same.

No animal can fuel expenditure simply from current income. There will always be times when it is not feeding (because it cannot, or there are better things to do), so it has to store reserves for these occasions. The time-scale and magnitude of requirements can vary markedly, from sufficient fuel to survive an entire winter or to complete a migratory flight, to overnight survival. Such periods of enforced fasting are to an extent predictable, although variability in weather conditions, and changes in windspeed or direction, make the energetic requirements hard to specify exactly in advance. But, short-term changes in weather, prey behaviour and simple lack of omniscience as to prey distribution, also render moment-to-moment foraging success stochastic. Although even small animals (e.g. shrews) are unlikely to die if they fail to feed for a few minutes, it only needs several successive runs of bad luck for a disastrous cumulative effect on survival prospects. What is the evidence that animals alter foraging behaviour to anticipate either predictable shortfalls in food availability or to buffer themselves against environmental predictability?

For animals at high latitudes, winter represents a deterioration in the foraging environment for several reasons. Nights are longer with, for diurnal animals, the double penalty of reduced foraging time during the day and an increased period of enforced fasting at night. Temperatures are lower, increasing basal metabolic expenditure. Food is both lower in mean availability and, perhaps, the variance in food intake may be higher. Foraging time may be shortened and rendered more variable by periods of bad weather, such as storms or snowfall. All such effects should favour storage of greater energy reserves (Lima, 1986; McNamara & Houston, 1990; Houston & McNamara, 1993; McNamara *et al.*, 1994), but this need not imply that animals respond to these factors as proximate cues. For example, as hibernators must acquire large reserves well in advance of winter, a hyperphagic response based on decline in current food availability or temperature would be maladaptive. Thus, for example, captive alpine marmots (*Marmota marmota*) show a 100% increase in food intake even under constant feeding conditions (Kortner & Heldmaier, 1995). However, animals that remain active throughout winter could plausibly rely on direct response to local conditions, use of short-term predictive cues (such as changes in weather) or long-term predictive cues (such as photoperiod). Which strategy, or combination of strategies, is best will depend on the predictive power of the cues, the cost of tracking them, the benefit of energy storage come the change in conditions and the cost of storing energy should conditions fail to change (see also Witter & Cuthill, 1993; Rogers *et al.*, 1994).

Direct response to local deterioration in environmental conditions has been demonstrated for a variety of variables. Laboratory studies on several species of passerine bird have shown that they store more fat at low temperatures (Kontogiannis, 1967; Kendeigh *et al.*, 1969; Ekman & Hake, 1990), and Bednekoff *et al.* (1994) found higher evening masses in great tits (*Parus major*) exposed to variable overnight temperatures than to constant conditions. However, the latter result seemed to be more a response to the coldest night experienced than a response to variability per se; without a factorial design one cannot separate these effects. Bednekoff *et al.* (1994) argue that their birds use daily

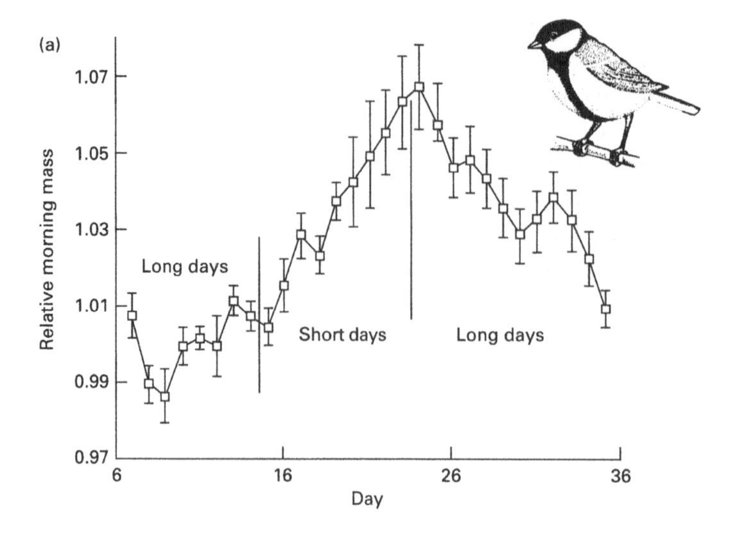

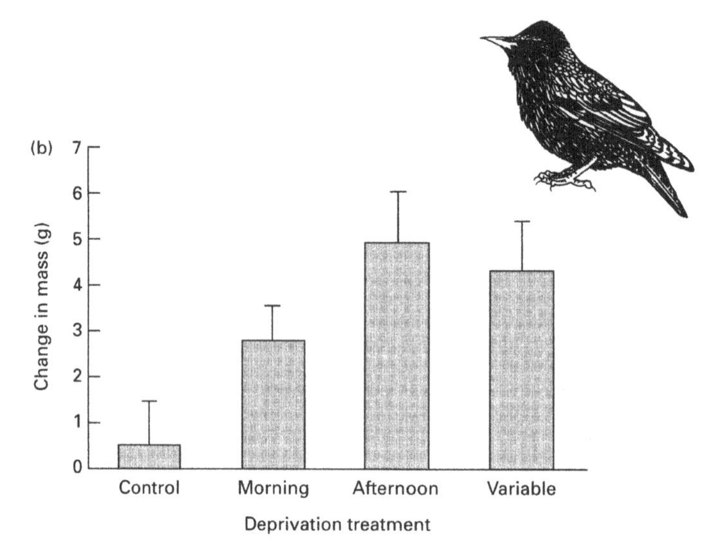

Fig. 5.1 (a) Great tits were switched from a 9-h feeding day ('long days') to a 6-h feeding day ('short days') while holding photoperiod constant. On short days, birds gained mass relative to their mass during the first phase of the experiment (scaled to a value of 1); on switching back to long days they rapidly lost mass. (From Bednekoff & Krebs, 1995.) (b) Starlings were exposed to fixed length interruptions in their feeding day. These interruptions could occur in the morning, in the afternoon, in either the morning or afternoon with equal probability ('variable') or not at all (control). Birds in all experimental groups increased mass relative to controls, with afternoon and variable interruptions producing the largest mass gains. (From Witter *et al.*, 1995.)

temperatures to predict overnight energy requirements, but that if the night differs from expectation, they regulate overnight metabolic expenditure to match their reserves.

Various forms of interruption from feeding have also been shown to produce a direct response in avian fat storage. While several field studies have indicated that photoperiod is a good predictor of body mass in birds (reviewed by Blem, 1990), Bednekoff and Krebs (1995) demonstrated a direct response to manipulation of the length of the time available to feed each day. Great tits whose feeding day was shortened by 3 h rapidly gained mass (Fig. 5.1a), probably through reducing energetic expenditure rather than eating more. In a second experiment, birds gained more mass on days of unpredictable length than days of predictable length equivalent in mean duration. The data of Bednekoff and Krebs (1995) suggest a differential response to short and long feeding days, with the former having a stronger effect; as with Bednekoff *et al.*'s (1994) temperature experiment, there is no evidence of a response to variability *per se*. Interruptions from feeding during the day have also been shown to promote increased fat storage in greenfinches (*Carduelis chloris*: Ekman & Hake, 1990) and starlings (*Sturnus vulgaris*: Witter *et al.*, 1995). Witter *et al.* (1995) sought to discriminate effects of unpredictability from the actual time of interruption: fixed length interruptions could occur in the morning, in the afternoon, either in the morning or afternoon with equal probability or not at all (control). Birds in all experimental groups increased mass relative to controls, with afternoon and variable interruptions producing the largest mass gains (Fig. 5.1b). Afternoon interruptions might be expected to have a larger effect than morning interruptions, as there is less time (until dusk) in which to replenish lost reserves (McNamara *et al.*,1994). Witter *et al.*'s (1995) results support this predicted differential response to the time at which interruptions occur but, again, no additional response to variability was found.

Species may differ in their responsiveness to manipulations of feeding conditions depending upon the predictability of the food supply to which they are adapted (Rogers & Smith, 1993; Rogers *et al.*, 1994). In meadow voles (*Microtus pennsylvanicus*) there appear to be different phenotypes with respect to response to winter conditions; some become non-reproductive and reduce body mass to save energy, others maintain body mass and breed opportunistically (Bronson & Kerbeshian, 1995). There is also evidence that the response to feeding interruptions is seasonally modulated. Witter *et al.* (1995) found that photosensitive starlings increased body mass in response to food deprivations, but that photorefractory birds maintained a constant body mass without increasing it. The difference seemed linked to photorefractoriness (lack of gonadal response to long days; Dawson *et al.*, 1985) rather than the period of moult, with which it overlaps, but the adaptive significance of such modulation is unclear (see discussion in Witter *et al.*, 1995).

Several authors have found differences in mass and fat correlated with factors expected to affect the predictability of food intake but, in light of the experimental evidence above, we should be cautious in interpreting such results as a response to variability itself. Thus, ground-feeding birds, in habitats prone to sudden snowfall, store more fat than tree-feeding species (Stuebe & Ketterson, 1982; Rogers, 1987; Rogers & Smith, 1993), but it may not be the differences in the predictability of interruption that are critical. Alternatively, birds in such conditions may have mechanisms that provide adaptive responses to feeding unpredictability without cueing on predictability itself.

More problematic is the interpretation of effects of social dominance on fat storage. Higher fat storage by subdominant than dominant great tits (Gosler, 1996) and willow tits (*Parus montanus*: Ekman & Lilliendahl, 1993) has been interpreted by these authors as a response to a less predictable food supply, due to displacement of subdominants from feeding sites. This may well be the case, and manipulations of dominance by removal of birds also produce the predicted changes in body mass (Ekman & Lilliendahl, 1993; Witter & Swaddle, 1995). However, dominance may alter not only the mean and variance in food gain, but energetic expenditure (Hogstad, 1987; Bryant & Newton, 1994) and predation risk (Ekman, 1987), all of which are predicted to affect optimal fat storage (Witter & Cuthill, 1993). The degree of competition for food has also been shown to affect the optimal fat reserves of dominants and subdominants to differing degrees (Witter & Swaddle, 1995), so in fact any relationship between social dominance and fat storage is possible, depending upon ecological circumstance (see discussion in Witter & Swaddle, 1995). Such arguments based on the dominance-dependent costs and benefits of fat storage go some way towards helping us understand the widely varying patterns seen in field data (see Witter & Cuthill, 1993; Witter & Swaddle, 1995), but we are a long way from understanding the proximate mechanisms by which such differences come about.

5.4 The costs of energy reserves

It may seem obvious to a behavioural ecologist, schooled in cost–benefit reasoning, that fat storage must have costs. But, many analyses of intra-specific variation in body fat, or mass, seem to neglect this point. Most obvious is the use of fat or mass, usually scaled to control for skeletal size variation, as a measure of condition. Animals with naturally high levels of fat, or high relative body mass, are considered to be in 'good condition' or of 'high quality'. Such data are then correlated with other fitness components, such as mate acquisition, breeding success or survival. They may also be used as a bioassay, of sorts, for habitat quality. Yet, for fat or relative mass to be positively correlated with individual or habitat quality would seem to suggest that fat storage is limited only by food availability or the ability to acquire it. Is this the case? The most extensive data sets come from avian morphometric studies and these provide substantial, if circumstantial, evidence that typical levels of energy storage are not food limited. First, temperate zone species are fatter in winter than summer, and often fattest in the coldest periods of the winter (references in Blem, 1990; Witter & Cuthill, 1993). Winter days are shorter and temperatures lower, leading to higher metabolic costs and lower food availability, so winter is precisely the time of year when we might expect birds to have lowest energy reserves. The fact that they store less fat in summer suggests that fat storage is not (usually) limited by food supply (King & Murphy, 1985). Second, body mass changes prior to migration or breeding suggest strategic regulation independent of current food supply (Mrosovsky & Sherry, 1980; Sherry *et al.*, 1980; Gwinner, 1990; Wingfield *et al.*, 1990). That energy reserves are usually maintained below the physiological maximum, or those dictated by food availability, suggests that the acquisition and storage of energy has costs.

5.4.1 Acquisition costs

There are two types of cost associated with feeding: (i) acquisition of the food; and (ii) maintenance of the energy reserve once secured. Acquisition costs intrinsic to the foraging process itself, such as search and handling time, have always been part of classical optimal foraging models (Stephens & Krebs, 1986). That different behaviours involve different metabolic costs has also been considered by simple optimality models based on maximization of net energy intake rate (see Section 5.2; also Houston, 1986). However, foraging animals are also likely to experience greater predation risk (Lima, 1986; McNamara & Houston, 1987). Through being active, a forager spends a greater amount of time exposed to predators than an inactive animal (a suggested functional explanation for the initial evolution of sleep: Meddis, 1993). Also, as foraging usually requires close attention to the task (detection of cryptic prey, stealth in approaching prey, pursuit of fleeing prey), there is liable to be a higher predation risk per unit time, due to decreased vigilance (e.g. Milinski, 1984). Although predation risk is the most widely considered cost, there may be other costs associated with heightened activity, such as risk of injury (Cuthill & Guilford, 1990) or parasitism (e.g Schmid-Hempel & Schmid-Hempel, 1993).

5.4.2 Storage costs

Food, once acquired, can either be transported to another location for subsequent processing or storage, or consumed immediately. These different immediate fates have different attendant costs, which will in turn influence the optimal acquisition strategy. Assuming that the ultimate consumer is the forager itself (foraging for dependent young or as part of a cooperative group comprises a vast literature in itself), should an animal store energy as fat or in an external larder or cache? It may be that the amount of energy required is greater than the maximum fat storage that is physiologically possible. In this case, the only options are caching or reduction of the energy requirements through torpor. Some species do not build large caches for use in subsequent months, but scatter-hoard small amounts for use on a much shorter time-

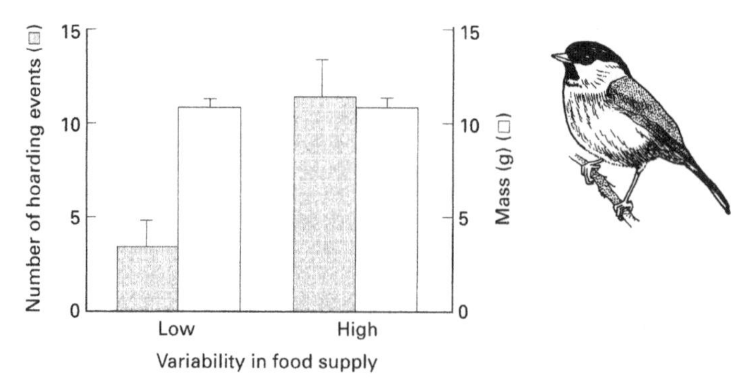

Fig. 5.2 Hurly (1992) measured seed hoarding and body mass changes in captive marsh tits by means of infrared detectors and computer-controlled balances. When an automatic feeder delivered seeds at highly unpredictable times, the birds hoarded more seeds per day, but did not increase body reserves. Greater hoarding under variable feeding conditions is predicted when birds minimize the chance of an energetic shortfall during the day or overnight. The fact that this extra energetic buffer is stored as a hoard rather than as fat, suggests that the latter is more costly than the former in this particular situation. (Replotted from data in Hurly, 1992.)

scale. Marsh tits and black-capped chickadees, for example, relocate and eat hidden food a few hours or days after storing it (Sherry, 1985; Stevens & Krebs, 1986); why not store these small amounts of energy in the body as fat? There are clearly special costs of storing energy in an external cache. These include the fact that hidden food can be pilfered, travel costs in taking the food, and subsequently returning, to the cache site, mistakes in relocating the cache (McNamara *et al.*, 1990; Lucas & Walter, 1991) and costs of memorizing the location and subsequently accessing that information (see Chapter 3). If one is considering the evolution of hoarding as a strategy, rather than the decisions of a species which already has the capacity to store food, there are the harder to quantify costs of maintaining extra neural tissue for spatial memory tasks (see Chapter 3 and Sherry *et al.*, 1992). However, storage of energy in the body, as fat for example, also has distinctive costs (see below). An examination of the relative costs of hoarding versus fat storage can shed light not only on why some species have evolved food caching as an energy storage strategy, but also the fine details of individual hoarding and retrieval decisions (McNamara *et al.*, 1990; Lucas & Walter, 1991; Hurly, 1992; Fig. 5.2). The nature of the costs of fat storage are dealt with in more detail below, as they have implications for a wider range of animals and behaviours.

5.4.3 The costs of being fat

The most obvious costs of fat storage in human life are the various pathological conditions associated with obesity: heart disease, gall and kidney stones, diabetes and various forms of cancer and arthritis (see, e.g. Bjorntörp & Brodoff, 1992). But, while human males can be considered obese if fat exceeds 20%

of their body mass, pregnant female polar bears (*Ursus maritimus*) enter their winter dens carrying 1 kg of fat for every 1 kg of lean mass (Atkinson & Ramsay, 1995) and some birds also double their mass in pre-migratory fattening (Lindström & Alerstam, 1992; Lindström & Piersma, 1993). So, a high level of fat storage is part of the life history of many animals. 'Obesity' can also arise as a byproduct of maintaining the correct nutrient balance when faced with a diet of suboptimal composition. Locusts fed on low-protein diets compensate for this deficiency by increasing intake rates, at the expense of excess carbohydrate consumption and elevated body mass (Raubenheimer & Simpson, 1995). There are almost no data on pathological effects of fat storage in wild animals (perhaps understandably: if migrating birds have heart attacks, how do we find out?), but even moderate levels of fat storage can have tangible ecological costs. These arise from the elevated body mass that fat storage entails, either directly from the adipose tissue, or indirectly through any changes in postural or locomotor muscles necessary for transporting the extra fat (e.g. Marsh, 1984; Piersma, 1990; see also Piersma, 1988).

These 'mass-dependent' costs include elevated metabolic expenditure, particularly during locomotion, and reduced locomotor performance resulting in increased predation risk and reduced foraging efficiency (reviewed in detail by Witter & Cuthill, 1993). The key issue is not whether elevated body mass has such costs, but whether they are of sufficient magnitude to be biologically interesting. For example, does the approximately 5% increase in body mass of a greenfinch in response to a temperature drop (Ekman & Hake, 1990) affect its predation risk? Conversely, would changes in predation risk be sufficient to affect patterns of fat storage? The effects of changes in body mass on energetic expenditure are derived largely from laboratory measurements of natural variation, and from theoretical calculation. That subcutaneous fat can reduce heat loss is well established for many mammals (see, e.g. Young, 1976; Pond, 1978) and various non-hibernating species may save total energy expenditure in winter by reducing body mass (Heldmaier & Steinlechner, 1981; Heldmaier *et al.*, 1989). However, adipose tissue is relatively metabolically inert, so it is active metabolism, through mass-dependent locomotor costs, that is liable to impose the most substantial costs. The metabolic costs of walking or running increase linearly with load size in mammals (Taylor *et al.*, 1982) and the cross-species relationship between body mass and energetic costs of terrestrial locomotion are similarly linear for birds and mammals (Taylor *et al.*, 1980; Peters, 1983). However, for flight, theoretical calculations suggest that the power requirements should accelerate with changes in body mass across species (to an exponent of about 1.5; Pennycuik, 1989; Rayner, 1990). But, there are few experimental data on the effects of loading on metabolic expenditure in flight, and even less on the relationship between within-individual changes in body mass and flight costs (see Bryant & Tatner, 1991; Witter & Cuthill, 1993). This is an area ripe for empirical investigation.

It seems likely that fat storage, and the increased body mass that results, may have decremental effects on locomotor performance, particularly in flying animals. Theoretical models of flight suggest reduced performance on various manoeuvres (Norberg, 1990; Alerstam & Lindström, 1990; Hedenström, 1992) and, not unreasonably, modellers of behaviour have assumed that predation risk is likely to be increased by such predicted deleterious effects on mobility (Lima, 1986; McNamara & Houston, 1990; McNamara *et al.*, 1994). But, collecting empirical evidence for mass-dependent predation risk is not straightforward. First, natural correlations between body mass (or fat) and survival probability are confounded by variation in escape ability or risk-taking behaviour (e.g. 'good' individuals store more fat, take less risks and are better at escaping predators). Second, predation rates need not be a good predictor of, or even positively correlated with, predation risk (McNamara & Houston, 1987; McNamara, 1990; Watts, 1990: Abrams, 1993a; McNamara & Houston, 1994). Animals trade off predation and starvation risk so, for example, individuals with low energy reserves may need to feed in more risky locations in order to avoid starvation, but in doing so expose themselves to higher predation risk. Fat individuals can afford to feed in safer sites, sacrificing intake rate for safety so, dependent upon the association of patch quality with predation risk, it is possible to generate a variety of relationships, positive or negative, between energy reserves and predation *rates* (McNamara, 1990; Abrams, 1993a). Thus, even field measurement of survival rates in response to experimental manipulations of body mass may give spurious estimates of predation risk. A more indirect approach is required: evidence that increased body mass alters those aspects of locomotor performance likely to affect escape ability, evidence that increased body mass is associated with increased anti-predator behaviours such as vigilance, and evidence that animals adjust body mass in response to differences in predation risk.

There are few experiments on the effects of fat storage on locomotion, but several showing that other causes of mass increase, such as carrying eggs or artificial loads, have significant effects. One must interpret these with caution, as it may not be the mass per se that is responsible for the observed changes, but such data give one confidence that fat may have qualitatively similar effects. Mass increases associated with being gravid have been shown to impair mobility and reduce escape speeds in several species of lizard and snake (Shine, 1980; Andren, 1985; Madsen, 1987; Siegel *et al.*, 1987; Cooper *et al.*, 1990). For example, gravid female *Sceloporus occidentalis* lizards show lower sprint speeds (Snell *et al.*, 1988; Sinervo *et al.*, 1991) and yolkectomies show that females with reduced broods survive better than control females (Sinervo *et al.*, 1991; Landwer, 1994). In birds, carrying eggs reduces feeding efficiency in great tits (*Parus major*; Krebs, 1970), rate of ascent in sand martins (*Riparia riparia*; Jones, 1986) and take-off angle in starlings (*Sturnus vulgaris*; Lee *et al.*, 1996). Jones (1986) demonstrated the same decreased flight performance when females were experimentally loaded by injecting an equivalent mass of water into their body cavity (the mass gain is temporary as the water is naturally expelled).

This suggests that it is the added mass of the eggs or developing follicles which is responsible for the decrease in flying ability. But, does fat storage have similar effects, given that small birds show seasonal changes in flight muscle composition during egg-laying (Jones, 1991; Houston *et al.*, 1995a,b)? Witter *et al.* (1994) manipulated both natural body mass and artificial loads to investigate the effect of fat reserves on escape performance in starlings. Ascent angle at take-off and manoeuvrability through an obstacle course were both reduced by increased body mass, whether natural or artificial (Fig. 5.3a,b). Although fat and artificial loads have different distributions around a bird's centre of gravity, the fact that the effects on flight are of similar type and magnitude suggests that there is a definite cost of increased mass per se. Strong

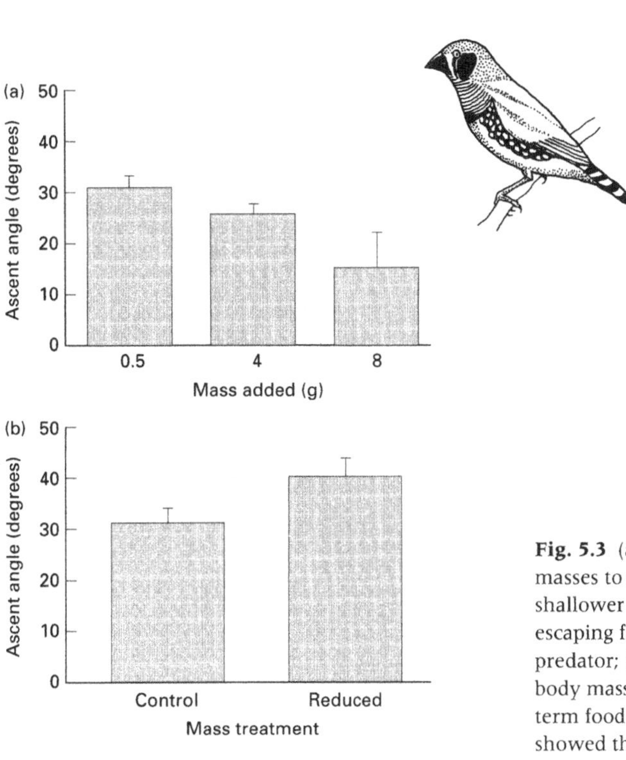

Fig. 5.3 (a) Adding artificial masses to starlings resulted in a shallower angle of ascent when escaping from a simulated predator; (b) reduction of natural body mass, by means of short-term food deprivation, also showed that heavier birds had lower take-off angles. (From Witter *et al.*, 1994.) (c) The natural daily variation in body mass of zebra finches is correlated with the time they take to ascend a given vertical height. Later in the day, birds are heavier and ascend more slowly. The data plotted are from a single individual, but the trend was similar (and highly significant) across eight replicate birds. (From Metcalfe & Ure, 1995.)

circumstantial support comes from the fact that diurnal variation in body mass is correlated with diurnal variation in take-off ability in zebra finches (*Taeniopygia guttata*: Metcalfe & Ure, 1995; Fig. 5.3c).

If escape ability is reduced by increased body mass then we might expect animals to respond by increasing compensatory anti-predator behaviours. Schwarzkopf and Shine (1992) showed that although female southern water skinks (*Eulamprus tympanum*) suffer a decrease in mobility while gravid, these females also change their anti-predatory behaviour, relying on crypsis to avoid detection rather than active escape. Relating changes in, say, vigilance to fat storage is more problematic, as manipulation of fat reserves necessarily alters not just body mass but energetic state. Thus, a fat animal may be heavier (reduced escape ability and hence higher predation risk), but it is also likely to

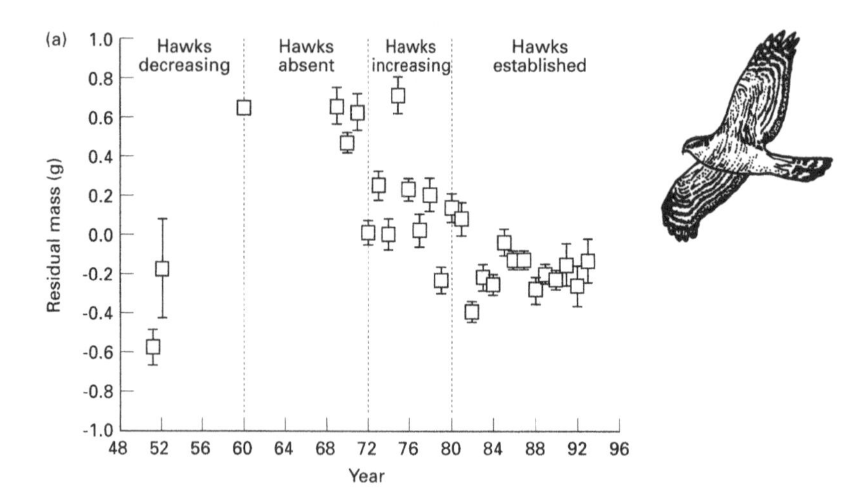

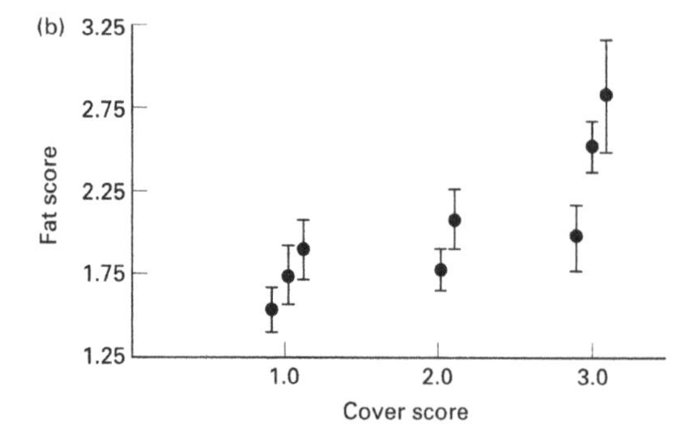

Fig. 5.4 (a) In years when sparrowhawks were absent from Wytham Woods in Oxfordshire, the resident great tits maintained higher body mass than when the predators were present. (From Gosler *et al.*, 1995.) (b) Starlings were assigned randomly to experimental aviaries differing in the amount of protective cover they contained. As would be predicted from mass-dependent predation risk, birds in aviaries with more cover, and thus a reduced predation risk, stored more fat than birds in less protected aviaries. (From Witter *et al.*, 1994.)

favour vigilance over feeding because gaining energy is relatively less important. Both the predation risk and starvation risk hypotheses predict the same change in vigilance behaviour, so the experiment in which fat levels are manipulated is worthless (but, see Witter *et al.*, 1994).

While increased anti-predator behaviours may compensate somewhat for increased vulnerability to predators (at the expense of decreased time allocated to other behaviours), animals might also regulate fat reserves in response to predation risk. Correlational data comes from changes in great tit body mass over a 40-year period in Wytham Woods, Oxfordshire (Gosler *et al.*, 1995). During years when the major predator, the sparrowhawk (*Accipiter nisus*) was absent due to organopesticide poisoning, tits maintained higher relative body masses than they do today (Fig. 5.4a). Such 'natural experiments' are open to confounding variables — population densities may have changed, and competition for food is known to affect fat reserves (Witter & Swaddle, 1995) — but, this is the best available field evidence to date. Witter *et al.* (1994) sought to manipulate predation risk in the laboratory by randomly allocating groups of starlings to aviaries with different amounts of protective cover. As predicted, birds in aviaries with more cover stored more fat (Fig. 5.4b).

5.5 Relating short-term behaviour to lifetime reproductive success

Having considered the costs and benefits of energy acquisition and storage, how do we use this information to construct more realistic models of foraging than those based on simple rate maximization? It is a central problem in behavioural ecology to evaluate the payoff from various patterns of behaviour over a short period of an animal's life. Yet, a functional explanation of behaviour is based on the contribution that behaviour makes to fitness, which can typically only be assessed by looking at an animal's lifetime reproductive success. It turns out that, at least in principle, the problem of relating short-term consequences to lifetime reproductive success is easy to solve. Each feature of an animal that is an important determinant of its reproductive success is taken to be a 'state variable'. Such variables might be internal (e.g. body temperature, parasite load) or external (e.g. territory size). An animal's behaviour will typically change its state (e.g. foraging increases energy reserves, basking increases body temperature). If we look at an animal at some time T, its expected reproductive success after this time will depend on its state at this time.

We now wish to evaluate behaviour over a period of time that ends at time T. The relationship between state and reproductive success at this final time is known as the terminal reward. This provides the link between behaviour over the short time period and lifetime reproductive success. Even if the animal is not reproducing during the period, its choice of behaviour will influence its state at T, and hence its future reproductive success. This can be illustrated with various simple examples based on foraging; having understood the principles we can then move on to more complex, and realistic, problems. We assume that the only important component of state is the animal's level of energy reserves, x. We start by assuming that foraging is deterministic and that the level of reserves has no costs in terms of metabolic rate or predation risk. We also assume that there is no predation risk while the animal forages. The animal has a range of foraging options, and under none of them will it reach the upper limit to its possible energy reserves by the final time T. For each foraging option i, there is a net rate of energetic gain γ_i. If the animal's level of reserves is x_0 at time 0, and it adopts option i throughout the time period, then its state at T will be:

$$x_0 + \gamma_i T.$$

It follows that as long as the future reproductive success increases as the level of reserves at T increases (i.e. the terminal reward is an increasing function of state) then the option with the highest net rate of gain should be chosen. We now consider various modifications.

The terminal reward is a step function and foraging is dangerous

Instead of assuming that future reproductive success always increases with reserves at T, we assume that future reproductive success is 0 if reserves are

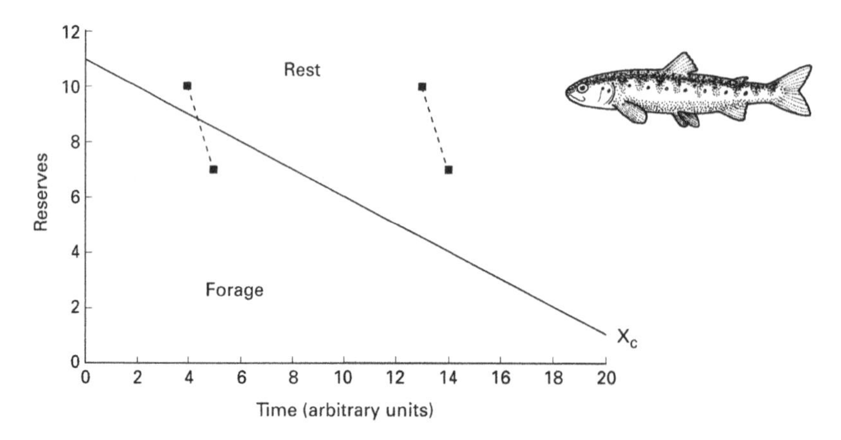

Fig. 5.5 The optimal policy for an animal over a season (arbitrarily 20 time units), where the decision whether to feed or rest depends on both energetic state and time. Here we assume that future reproductive success is 0 if reserves are less than some critical level x_c and is maximal if reserves are equal to or greater than x_c. Foraging can increase energy reserves, but exposes the animal to predation risk; resting is safe, but resting metabolism steadily burns up energy. The optimal policy for reaching the end of the season with reserves of at least x_c is to forage when reserves are below the boundary and to rest when above. Note that the shape of the optimal policy of resting and foraging is such that an animal losing energy early in the season (left-hand squares) gains from resuming foraging, while later in the season (right-hand squares) an animal with the same or lower initial reserves, losing exactly the same amount, does better by continuing to rest. Such an effect has been demonstrated experimentally in overwintering juvenile salmon (Bull *et al.*, 1996).

less than some critical level x_c and is 1 if reserves are equal to or greater than x_c (Fig. 5.5). All foraging options are dangerous, in that there is a probability that the forager will be killed by a predator. We assume that this probability per unit time spent foraging is the same for all the foraging options. If the animal decides not to forage, it can rest without incurring any predation risk. The expected future reproductive success associated with any behaviour between $t = 0$ and $t = T$ is given by the probability of surviving until T multiplied by the terminal reward. There is no advantage in having reserves greater than x_c at T, and survival probability decreases with time spent foraging. It follows that the best behaviour is to minimize the time spent foraging, subject to the condition that the level of reserves at T is x_c. This is achieved by using the foraging option with the highest possible mean net rate of gain for just long enough to ensure that reserves reach the critical level x_c at T, and spending the remainder of the interval resting. Note that this characterization does not specify the exact sequence of foraging and resting that should be observed. All sequences that result in the optimal overall allocation of time are equally good.

Animals with very high levels of reserves may be able to survive by resting throughout the foraging period. In such circumstances, reducing an animal's reserves by a given amount early in the foraging period may result in an increase in foraging compared to reducing the reserves of an animal (with the same or even lower initial level of reserves) later in the foraging period (Fig. 5.5). This effect has been found by Bull *et al.* (1996) in a study of juvenile salmon (*Salmo salar*).

In general, foraging options will differ in both their net rate of gain and their predation risk. Gilliam and Fraser (1987) analyse the optimal choice in such a case. They show that when the forager can choose between a habitat with no predation risk (a refuge) and foraging options that differ in terms of net rate of gain and their predation risk, then the optimal behaviour is to use the refuge and the habitat with the lowest ratio of predation rate to intake rate. This criterion is able to predict the behaviour of juvenile fish (*Semotilus atromaculatus*, Cyprinidae) foraging in an experimental stream.

The terminal reward is a step function, there is no predation risk but foraging is stochastic

As in the previous case, there is a critical level of reserves x_c that must be attained. Foraging is not dangerous, and there are only two foraging options. One option is deterministic with a mean net rate of gain γ. The other is stochastic, with the same mean net rate of gain.

Assuming that the animal will use one of the options for the whole time period, it is clear that it is best to use the deterministic option if the expected level of reserves at T is at least x_c i.e. if:

$$x_0 + \gamma T \geq x_c$$

and to use the stochastic option if the expected level of reserves at T is less than x_c. This is the expected energy budget rule for risk-sensitive foraging (Stephens, 1981).

When we move to considering an environment in which there is both predation risk and stochasticity (perhaps both in food intake and metabolic requirements), we rapidly move to models which are analytically intractable. Fortunately, there is a computational technique which can cope with this complexity, and capture the biological realism we require: the technique is stochastic dynamic programming.

5.5.1 Static versus dynamic models

So far we have assumed that the animal will use the same foraging option throughout the period under consideration. There are clearly cases when this sort of behaviour will be best (e.g. energy maximization), but there are other cases in which it is likely that the optimal behaviour will depend on either the animal's state, or the time remaining until final time T, or both. Such cases require an analysis that is dynamic, as opposed to the static analysis that underlies rate maximization. In finding the best behaviour, we now need to consider all possible ways in which behaviour could be specified by state or time over the time period. This is a daunting task, but there are two reasons why we need not be completely discouraged.

1 Under some circumstances, a potentially dynamic problem collapses to a static problem.

2 There is a simple and general computational technique (dynamic programming) for solving dynamic optimization problems.

The first point can be illustrated in the context of foraging. Assume that an animal can forage with intensity u. This intensity determines both the mean net rate of energetic gain and the rate of predation. Both these rates increase with u, and the predation rate is accelerating. As a result, the animal can only increase its intake rate by increasing its probability of being killed, and each increase in foraging intensity is more costly in terms of predation than the previous one. In trying to find the foraging intensity that the animal should use if it is attempting to change its state from x_0 at time $t = 0$ to x_T at time $t = T$ it turns out that under some circumstances we need only consider behaviour in which foraging intensity is constant. In particular, Houston *et al.* (1993) show that if: (i) foraging is deterministic; (ii) foraging is not subject to interruptions; and (iii) the animal's level of energy reserves does not influence its rate of gain or rate of predation, then u should be kept at the constant value that just gets reserves to x_T at T. Houston *et al.* (1993) call this the risk-spreading theorem because it states that the predation risk is spread out equally over the time available. Given a terminal reward, we can then use this result to find the best value of x at T. But, in nature foraging is likely to be stochastic, interruptions

Box 5.1 Stochastic dynamic programming (SDP)

SDP is a numerical technique for finding the optimal behavioural decision as a function of state and time. The animal is characterized by one or more state variables (e.g. energy reserves, immunocompetence, knowledge) and a set of possible behavioural options. The result of its behaviour is to change its state(s), usually in a stochastic way. For example, foraging does not always result in food. The goal is to find the optimal sequence of decisions that maximizes future reproductive success; the problem is the vast combination of possible actions as a function of state and time. SDP tackles this by working backwards from some final time (T) at which the relationship between state and future reproductive success is known. Given this relationship then, for each state at time $T-1$, the optimal choice for this final step can be found readily. In doing so, one also obtains the expected future reproductive success associated with each state at time $T-1$. One proceeds to step $T-2$ in an analogous way; one calculates the optimal choice that maximizes expected reproductive success at time $T-1$. Repeating this process for all decision times yields a matrix of decisions that constitutes the optimal policy. The optimal policy specifies the best decision for all states and times. Following this policy forward from the first time step, for given initial state(s), generates the expected behaviour. For technical details see Mangel and Clark (1988).

may be common, and predation risk is likely to be mass-dependent (see Sections 5.3 and 5.4). We tackle such a situation in the next section.

5.5.2 Dynamic programming in action

Dynamic programming is a general technique for finding a solution to a dynamic optimization problem (Box 5.1). It was first applied systematically to behavioural problems by Mangel and Clark (1986) and McNamara and Houston (1986; see also Mangel & Clark 1988; Houston *et al.*, 1988). The approach is especially useful when: (i) the relationship between behaviour and its consequences for the animal's state is stochastic rather than deterministic; and/or (ii) the animal's state influences the possible actions or their consequences. Both conditions are likely to hold for a wide variety of behaviours.

An example with both the above features is the analysis of singing versus foraging in songbirds (McNamara *et al.*, 1987). In this model, a male bird is characterized by his level of energy reserves, x. He dies of starvation if x falls to 0. During the day, the male can choose between singing and foraging; at night, he rests. Singing gives the male a chance to attract a mate, but uses up energy. By foraging the male can increase its energy reserves, but the amount of energy obtained is variable (with a specified distribution). The male's rate of energy expenditure increases with body mass and hence with energy reserves. In some runs of the model, the amount of energy needed to survive the night was fixed; in other runs it was variable. This difference turns out to be important in determining the optimal pattern of behaviour. McNamara *et al.* (1987) use dynamic programming to find the behaviour that maximizes a male's expected value of the terminal reward at the end of a period of several days. If the male is dead at this time, then the reward is zero; if he is alive but unpaired he gains a positive reward; and if he is alive and paired his reward is greater still. It turns out that the exact values given to the two cases in which the male is alive are not crucial for determining the optimal behavioural rule. The optimal rule specifies, for each level of reserves and time of day, whether the male should sing or forage. Because of the danger of starvation males tend to forage when reserves are low, but above a critical level of reserves the male sings. As the end of the day approaches, the critical level above which the male sings rises. This enables the male to build up reserves for the period of darkness when it cannot forage. When the energy used overnight is not variable males build up just enough energy to survive the night. In contrast, when the energy used overnight is variable, males need to go to roost with reserves of more than the average overnight expenditure if they are to have a reasonable chance of survival. On most nights not all of this energy will be required. As a result, the male will have energy 'left over' at the end of the night and can best utilize this energy by singing. Thus, variability in overnight energy expenditure can produce a 'dawn chorus', i.e. a burst of song at dawn. This is a striking example of how a daily routine can emerge even when the consequences of activities do not depend on time of day (for further investigations of this problem and how it relates to honest signalling, see Hutchinson *et al.*, 1993). In many circumstances, the success of a male that is calling to attract females will depend on the behaviour of other males. This requires a game-theoretic analysis (e.g. Lucas & Howard, 1995).

The importance of energy reserves for avian song has been demonstrated by Reid (1987) in the Ipswich sparrow *Passerculus sandwichensis princeps* and by Cuthill and Macdonald (1990) in the blackbird *Turdus merula*. Reid found that males sang less after cold nights, when overnight energy expenditure would have been high. Both studies found that the level of song in the morning could be increased by providing males with food. Although the model we have described is relatively simple, it illustrates some important general points. One is that the timing of an activity (song in this case) may only be understandable in the broad context of an animal's overall time budget. Another is that a selective force may be small but yet exert considerable influence. The routine of singing and foraging is driven by the threat of starvation, but under the optimal policy few birds starve. The general point is that the absolute magnitude of a cost does not tell us about the importance of the cost in determining optimal behaviour (McNamara & Houston, 1987; Abrams, 1993b; Houston &

McNamara, 1993). Thus, just because starvation is rare, it does not mean that we can understand behaviour without taking it into account.

5.5.3 Investigating behavioural routines

We have seen how dynamic programming can tackle the problem of optimal behavioural routines in the specific case of male birds singing to attract mates. The approach is, however, of general applicability and can be used to investigate routines of other behaviours and on different time-scales. For example, McNamara *et al.* (1994) consider whether a small bird in winter should forage or rest at particular times of day. The resulting optimal daily routines are driven by the conflict between keeping reserves high to avoid starvation, and keeping reserves low to avoid mass-dependent costs (see Section 5.4). If the bird is free to forage at any time of day, when food availability is low, the optimal routine is to forage fairly intensively throughout the day (Fig. 5.6). When food availability is higher, there tends to be a burst of foraging at the start of the day and also in the afternoon. Such patterns are often observed in nature (see McNamara *et al.*, 1994, for references). However, it is also realistic to consider the situation where a bird can be prevented from foraging by, for example, periods of inclement weather (see Section 5.3). In such an environment, it may be optimal to concentrate foraging at dawn and dusk, as long as an evening interruption cannot carry over to the next morning. These predictions remain untested because, as yet, we have insufficient information about environmental variation as experienced by birds in nature.

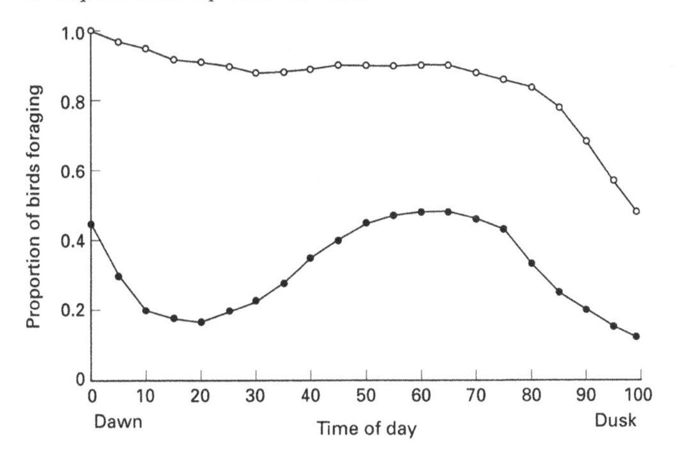

Fig. 5.6 The predicted proportion of birds foraging at different times of day, from a dynamic programming model where birds can either forage to gain energy or rest to avoid predation. In the case illustrated, foraging can occur continuously without interruption (if the bird chooses). When food availability is low, birds forage at a high level for most of the day (open circles). When food availability is higher, the overall level of foraging is lower and there is a burst of foraging at the start of the day and in the afternoon (filled circles). (Based on McNamara *et al.*, 1994.)

Other routines that have been investigated include daily patterns of food hoarding (McNamara *et al.*, 1990), daily patterns of vertical migration in fish (Clark & Levy, 1988; Burrows, 1994), seasonal body mass changes (Bednekoff & Houston, 1994) and annual patterns of moulting (Holmgren & Hedenström, 1995). The ultimate application of dynamic programming would be to consider the optimal sequence of behaviour over an entire life cycle. This is where optimality modelling meets life-history theory (see Chapter 13) and is a natural extension of the approach outlined in this chapter. Questions that are traditionally the domain of life-history theory, such as optimal clutch size, diapause and phenotypic plasticity, have all been tackled in this way (see McNamara & Houston (1992), McNamara (1994) and Houston & McNamara (1992), respectively).

The important messages from this chapter are, first, that current behaviour can have repercussions on survival prospects (or more generally fitness) at times quite removed from present actions. Second, that the best time to perform a particular behaviour can often depend on the costs and benefits of other behaviours at this and other times. Because all behaviour takes time and uses energy, an animal's energy reserve is a natural state variable to consider when modelling behavioural routines. Dynamic programming, because it allows one to model state-dependent decision-making and to link current actions to future payoffs, is a powerful tool for analysing the use of time and energy.

5.6 Final considerations

5.6.1 Is your dynamic programme really necessary?

The state-dependent approach that we have described has the advantage of producing models that can include a range of features that are biologically important. Furthermore, analysis of behaviour over a relatively short period of an animal's life is placed in the context of the whole life history (although this can be done with other methods; e.g. Abrams, 1982). The disadvantages of the approach are that quite a large amount of information may be needed in order to construct such models, and that the optimal policy will typically have to be found by a numerical technique such as dynamic programming. These advantages and disadvantages have been discussed several times (e.g. Houston & McNamara, 1988; Gladstein *et al.*, 1991; Houston *et al.*, 1992; Abrams, 1995). Instead of going over all the arguments again, we will confine ourselves to a few points which are undeniable.

1 Dynamic programming is a particular technique for finding optimal solutions. It is not always necessary to use this technique just because a simple rate-maximizing approach is inadequate. For example, the trade-off between gaining energy and avoiding predators can often be analysed with static models (e.g. Abrams, 1982).

2 Stochasticity can be analysed without using dynamic programming. When there is a single decision or a fixed sequence of decisions, it may be relatively easy to get analytic results about optimal behaviour. The daily energy budget rule (Stephens, 1981) is an obvious example. When an animal makes repeated state-dependent choices in a stochastic environment, analytic results will be difficult to obtain, even in fairly schematic models (e.g. Houston *et al.*, 1992).

5.6.2 Which state variables are important?

In principle, one of the benefits of dynamic programming is that one can explicitly link function and mechanism by incorporating realistic state variables which can be measured empirically. For an animal to make adaptive decisions based on energy and time, one needs to know how each state is represented. This is a major challenge for behavioural biologists, as experimental studies indicate that animals may respond to different types of energetic stress in different ways on different time-scales (e.g. Lucas *et al.*, 1993). However, this challenge is worthwhile as a proper understanding of behaviour must integrate functional and mechanistic accounts. We know that deterioration in the foraging environment under many, but not all, situations leads to an increased storage of energy reserves (see Section 5.3). Functional models based on dynamic programming tell us why this is adaptive, and finer details such as the predicted pattern of weight gain on time-scales ranging from the day to the entire winter (e.g Bednekoff & Houston, 1994). However, we know much less about the actual mechanisms by which 'environmental deterioration', or key factors such as predation risk, are assessed and integrated. As we have seen earlier (see Section 5.3), while theory has predicted heightened fat levels in response to unpredictability of the food supply, several experiments suggest it is the duration and timing of interruptions, as much as their unpredictability, which is important. This suggests a common mechanism. That mechanism is likely to involve the adrenal stress hormones, but very little work has been done on endocrine responses of wild animals under field conditions. Plasma corticosterone has been found to be higher in wild dark-eyed juncos (*Junco hyemalis*) after recent snowfall, as was total adrenal mass (Rogers *et al.*, 1993). As fat levels were correlated with these changes, and implanted corticosterone promotes feeding behaviour in previously food-deprived birds (Astheimer *et al.*, 1992), the implication is that corticosterone mediates the influence of nutritional stress on energy storage. Exogenous corticosterone also reduces overnight energy expenditure (Astheimer *et al.*, 1992). Elevation of plasma glucocorticoid hormones, such as cortisone and cortisol, is the main endocrine response following capture stress in vertebrates, which suggests a mechanism by which predation risk could also affect fat storage, but heightened feeding under predation risk is precisely the opposite of what functional models would predict (Lima, 1986; McNamara & Houston, 1990; McNamara *et al.*, 1994). Interestingly, the corticosteroid response to capture stress has been found to be negatively correlated with fat reserves in several species of birds (Smith *et al.*, 1994; Wingfield *et al.*, 1994a,b), but we need to know much more about the proximate factors affecting feeding behaviour and energy storage, their integration and the extent to which they are seasonally modulated (e.g. Wingfield *et al.*, 1992). Functional models, such as the dynamic ones we have discussed in this chapter, are useful in indicating the key environmental and state variables to which we should direct our attention in such a search for mechanisms. But, more than this, they help us to understand the trade-offs and constraints under which these mechanisms have evolved and are maintained.

Chapter 6

Sperm Competition and Mating Systems

Timothy R. Birkhead & Geoffrey A. Parker

6.1 Introduction

6.1.1 Definition

Sperm competition is the competition between the ejaculates of different males for the fertilization of a given set of ova (Parker, 1970a). Thus, sperm competition forms a part of sexual selection (see Chapter 8), and includes the adaptations which arises as a result of it; e.g. any behaviour, morphology or physiology associated with multiple mating by females, paternity guards and ejaculate characteristics, all viewed from both a male and female perspective (Birkhead, 1996). Over the last 25 years it has become clear that sperm competition is a remarkably powerful selective force, shaping life-history characteristics such as the body size, morphology, physiology and behaviour and even the evolution of the two sexes. Our aim in this chapter is first to consider how sperm competition may have driven the evolution of males and females (see Sections 6.1.2 and 6.1.3). Although sperm competition occurs in both external and internal fertilizers, in this chapter we will concentrate on species with internal fertilization. We discuss why females should bother to mate with more than one male, and how the resulting competition between ejaculates shapes male and female strategies (see Section 6.2 and elsewhere). In Section 6.3 we consider the taxonomic incidence of sperm competition and show that it is virtually ubiquitous. Our main focus, however, is on the underlying mechanisms that determine which copulations result in fertilization (see Section 6.4). Finally, we consider the role that sperm competition has played in sexual conflict (see Section 6.5) and the evolution of mating systems (see Section 6.6).

6.1.2 Why two sexes?

It would be hard not to notice that most organisms exist in two forms: (i) males, which produce tiny gametes (microgametes); and (ii) females, which produce much larger gametes (macrogametes). Indeed, virtually all complex multicellular organisms exist as either separate male and female individuals (dioecious species), or less frequently as male and female in the same individual (simultaneous hermaphrodites). In contrast, many unicellular forms produce just one size of gamete size (microgametes). If we define a sex in terms of the gamete size an individual produces, then there are either one or two sexes, broadly relating to the complexity of the organism. Why?

There have been various theories. The earliest was that the rate of fusion in an external medium is maximized when the fusing gametes are very different in size (Kalmus, 1932, Scudo, 1967). At least in its simplest form this requires group selection. More recently, Hurst (1990) has suggested that a strong division into two gametic sizes reduces the possibility of transfer of cellular parasites during fertilization. Here we discuss the theory of Parker, Baker and Smith (1972; referred to from now on as PBS) since it specifically deals with the role of gamete competition in the evolution of males and females. PBS showed by computer simulation that in some hypothetical externally fertilizing ancestor, coexistence of males and females would be an evolutionarily stable strategy (ESS; Maynard Smith, 1982) when there is a high advantage in provisioning the zygote, and that producing microgametes is the ESS when the advantage of provisioning is weaker. Specifically, in a population in which all individuals produce gametes of equal size (isogamy), two gamete sizes (anisogamy) are favoured if the relationship between zygote fitness and zygote size is accelerating (at least over part of its range). Large gametes result in zygotes which survive well, but can be produced only in small numbers because they are costly. Small gametes contribute little to zygote survival, but obtain vastly more fusions. Starting from an isogamous population, disruptive selection quickly produces two sexes (males and females) because small-gamete producers and large-gamete producers are simultaneously favoured; producers of intermediate-sized gametes quickly become extinct.

In a sense, males succeed by parasitizing females: sperm become smaller and smaller in order that more and more of them can be produced. In the original PBS model, fusion between gametes was random. Because of relative numbers, most fusions are between sperm, but since sperm–sperm zygotes have negligible viability, mechanisms to prevent such fusions are likely to evolve quickly, leaving a vast predominance of sperm in the gametic pool. Almost all ova are snatched immediately by the ubiquitous sperm, and sperm–ovum zygotes have good prospects. However, ova do better if they are able to fuse with other ova, since ovum–ovum fusions have much higher fitness, especially if sperm are tiny.

There is a good reason why ovum producers cannot 'retaliate'. Parker (1978b) showed that if sperm have become small (and hence numerous) enough before an ovum producer retaliates by mutating in such a way that it permits only ovum–ovum fusions, then all the wild-type, randomly fusing ova will be grabbed by sperm before they have any significant prospect of collision with mutant ova. Thus, the mutant ova remain unfertilized, or are forced eventually into self-fusions. If 'retaliation' occurs before sperm have become small, then sperm may be lost. But, renewed 'sperm drives' from within the retaliator female population would eventually proceed to a point where sperm become too small and numerous to make retaliation favourable. The ESS is stable coexistence of males and females. At this point, ova are expected to lose their motility and to channel the savings into resources for zygote survival, effectively 'settling' for sperm–ovum fusions (Parker, 1984).

Some evidence for PBS comes from algae and protozoans (Knowlton, 1974; Bell, 1978, 1982). One would expect that the importance of zygote size would increase with complexity and size; Knowlton (1974) showed that in the volvocine algae there is a general transition from isogamy in the simplest unicellular forms to anisogamy in the more complex of the multicellular forms. Bell (1978) found a similar (but less distinct) correlation in other groups of chlorophyte algae (see also Bell, 1982).

6.1.3 Sperm competition, sexual selection and sexual conflict

A primordial form of gamete competition may then account for why there are two gamete-producing morphs (males and females), not three, four or five. Furthermore, once selective fusion has evolved, PBS generates males and females in a 1 : 1 sex ratio. The reason for equal production of the two sexes is due essentially to an argument originally put forward by Fisher (1930). Suppose that all eggs are fertilized (males are never very rare). Each female gains a fixed number, n, of progeny whatever the sex ratio. But, the expected gains per male are the total progeny divided by the number of males: i.e. ($n \times$ females)/males. If males are *less* common than females, each male gains more offspring than each female by fertilizing the progeny of more than one female. If males are *more* common than females, the expected number of progeny per male is less than that per female because more than one male must share each female's progeny. The ESS — for an autosomal gene determining the sex of the progeny — is a 1 : 1 sex ratio, although different genetic mechanism and deviations from random mating can generate adaptive sex biases (Hamilton, 1967).

If a single male fertilizes a given set of ova (as may be common with internal fertilization), there is no longer any advantage in producing huge numbers of microgametes to compete for fertilizations, since there are no other microgamete-producing competitors. PBS depends upon sperm competition — it cannot generate anisogamy when gametes from just two adults fuse (although it could then arise by Kalmus' model without invoking group selection). Instead, there should be a return to isogamy with just one sex, with the two spawning individuals sharing the burden of the cytoplasmic resources in eggs, each producing gametes of 50% of the optimal size (Parker, 1982).

If this means that males cannot persist with internal fertilization, PBS is clearly wrong — because they clearly do! In fact, sperm competition is commonplace in internally fertilizing species, and the original definition of the term and development of the concept (Parker, 1970a) was actually for an internally fertilizing group — the insects. It appears that only very low levels of sperm competition are necessary to maintain males, especially if eggs and sperm are very different in size (Parker, 1982). To see why this is so, consider the cow's ovum. It is roughly 20 000 times the size of a bull's sperm. If a mutant bull were to halve its sperm number and so double sperm size, then (assuming that normal sperm contribute nothing to the zygote but deoxyribonucleic acid (DNA)) this would increase the cytoplasmic reserves of the egg by only one unit in 20 000, an entirely trivial benefit. It nevertheless generates a huge cost whenever two bulls mate with the same cow (relative fertilization probability drops from one-half to one-third). Parker (1982) has shown that it will not pay to increase cytoplasmic reserves in the sperm if the probability of sperm competition (between two bulls) is greater than (roughly) four times the ratio of minimal sperm size divided by the optimal ovum size. In most species, this ratio is a very tiny value, so that anisogamy will be highly stable (Fig. 6.1). So, sperm competition may not only have both produced the two sexes, it may also currently maintain them. This is hardly trivial — but, as we shall see, it also does much more than this.

We therefore end up with males and females in a 1 : 1 sex ratio. Darwin (1871) pointed out that this, coupled with the relative cheapness of ejaculates compared to egg batches, typically means that males compete vigorously for females, he termed adaptations arising through this process sexual selection (see also Bateman, 1948; Trivers, 1972; Chapter 8). Males may compete directly (male–male competition) or indirectly, by appealing more effectively to females (female choice). Sperm competition can be seen as a form of sexual selection occurring even after mating (Trivers, 1972). Note that the interests of the two sexes need not be coincident, and many situations involve sexual conflict, e.g. concerning whether mating should take place or how much parental care should be given by each partner (Trivers, 1972; Parker, 1979).

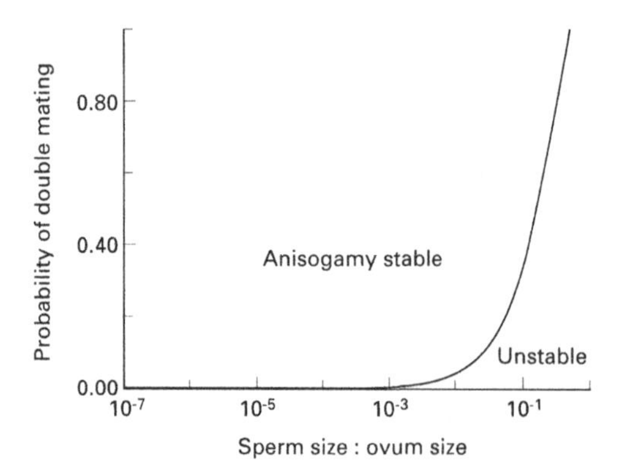

Fig. 6.1 Stability of anisogamy in relation to sperm size : ovum size ratio and the probability of double mating. Above the line, anisogamy is stable, so that in systems with a pronounced gamete dimorphism only tiny amounts of sperm competition will stabilize anisogamy. Below the line, anisogamy becomes unstable, but this requires less pronounced gamete dimorphism and higher probabilities of double mating. (Calculated from the exact equation for stability in Parker, 1982.)

6.2 Why should females copulate with more than one male?

The benefits to males of copulating with several females are obvious: they father more offspring. It is much less clear why a female should copulate with more than one male, especially since in many species a single insemination can fertilize all her ova. In some cases there may be no advantage; males may simply force copulations on females (e.g. some lizards: Olsson, 1995; ducks: McKinney *et al.*, 1983), or the costs of resistance may be greater than acquiesence (e.g. some insects). In other cases, however, females appear to actively seek multiple males, indicating that they obtain some benefit from copulating with more than one male. The benefits can be either direct or indirect (genetic) and are summarized in Table 6.1. Perhaps the most widespread explanation is that copulations with different males occur to ensure an adequate supply of sperm (Parker, 1970a), yet there is surprisingly little evidence to support this (see Woodhead, 1985). The potential indirect benefits obtained through multiple matings by females comprise: genetic diversity, attractiveness genes (Fisher's runaway sexual selection) and good genes (superior genetic quality

Table 6.1 Some of the main benefits to females of mating with multiple males.

Benefit	Examples	Reference
Direct benefits		
Fertility insurance	Insects: several	Thornhill & Alcock (1983)
	Birds: no convincing evidence	Wetton & Parkin (1991)
		Birkhead & Fletcher (1995a)
Acquisition of nutrients	Insects: many species	Thornhill & Alcock (1983)
	Birds: one example	Mills (1994)
Paternal care	Birds: cooperative polyandrous species	Davies (1992); Faaborg *et al.* (1995); Davies *et al.* (1996)
	Mammals: marmosets	Goldizen (1988)
Avoidance of harrassment	Insects: dungfly	Parker (1970a)
Change of long-term partner	Birds: oystercatcher	Ens *et al.* (1993)
Indirect benefits		
Genetic diversity	No known examples	
Genetic quality		
Attractiveness	Birds: several	Saino *et al.* (1996)
Viability	Fish: guppy	Reynolds & Gross (1992)
	Reptiles: adder	Madsen *et al.* (1992)
	sand lizard	Olsson *et al.* (1994)
	Birds: blue tit	Kempenaers *et al.* (1992)

for survivorship) (see Chapter 8). Genetic diversity is not generally regarded as a convincing explanation for multiple mating since copulation with extra males produces very little additional genetic diversity over mating with a single male (Williams, 1975).

The acquisition of attractiveness genes and good genes (see Chapter 8) is controversial since it is generally believed that traits closely associated with fitness have low heritabilities. In other words, sexual characters lack additive genetic variance. This creates the 'paradox of the lek' (see Chapter 8): why do females chose to copulate with particular males if their choice does not influence fecundity? A resolution to this paradox has been provided by Pomiankowski and Møller (1995), who showed that contrary to previous ideas, both phenotypic and additive genetic variation are actually relatively high in sexually selected traits and this is because they have been under long-term directional selection. Moreover, the results from several studies of socially monogamous birds appear to support the 'good genes' hypothesis since females of several species preferentially select males that are more attractive than their partner for extrapair copulations (e.g. Kempenaers *et al.*, 1992; Saino *et al.*, 1996). On the other hand, Sheldon (1994) has pointed out that if morphological or behavioural traits covary with ejaculate features this is also consistent with females acquiring direct fertility benefits. However, in the only test of this hypothesis to date there was no evidence that these features covary (Birkhead & Fletcher, 1995a).

In conclusion, where females obtain direct benefits, such as food, sperm or parental care as a consequence of copulating with different males, the advantage to the female are clear. However, in other cases the precise nature of the indirect benefits females obtain remain to be established, and this continues to be an important functional question.

6.3 Detection and incidence of sperm competition

Since sperm competition occurs when a female is inseminated by two or more males, sperm competition can be inferred from either the detection of mixed paternity or from direct observation of copulations. These two methods provide *estimates* of the incidence of sperm competition, because: (i) not all cases of sperm competition can be detected by paternity analysis (in some cases one male will 'win' and father all offspring); and (ii) not all copulations result in insemination. The most informative studies have been those which have used both methods in combination. This is desirable in field studies, but is often hard to achieve, whereas for laboratory studies it is much easier and such studies have been essential for revealing the mechanisms of sperm competition (see Section 6.4.2).

It is ironic that Darwin should be among the first to document sperm competition since this is a topic he assiduously avoided pursuing and hence is absent from his writings — presumably because of social mores. Darwin (1871) described how a female domestic goose *Anser anser* produced a brood comprising offspring fathered by two different males, her partner (another domestic goose) and a Chinese goose *A. cynoides*. Naturalists subsequently observed females occasionally copulating with more than one male (e.g. praying mantids: Fabre, 1897; ducks: Huxley, 1912), but such behaviour was written off as aberrant. It was not until Parker (1970a) and Trivers (1972) focused on the evolutionary implications of multiple mating that biologists started to think about the adaptive significance of sperm competition. Even before this, however, there had been experimental studies of multiple mating in insects. Many species can be reared easily in the laboratory and paternity was assigned using either genetic markers, such as a difference in colour (e.g. Hunter-Jones, 1960) or the irradiated male technique (Parker 1970a). The sperm of males subject to irradiation are capable of fertilizing ova, but have so many lethal mutations that the zygote never develops. Hence, by mating females with a non-irradiated and an irradiated male, and recording the number of developing and non-developing eggs, respectively, paternity can be assigned. Most insect studies involved reciprocal matings to quantify the effect of mating order on paternity, and most revealed a strong pattern of second (or last) male sperm precedence (see below).

Only after a substantial body of work on insects and other invertebrates was complete did biologists start to look for evidence of sperm competition in other taxa. Serious observational studies of sperm competition in birds started in the late 1970s, and somewhat later genetic evidence for multiple paternity within a brood was obtained, albeit with difficulty, using allozymes (e.g. Westneat, 1987). The discovery in 1985 of multilocus DNA fingerprinting which could assign parentage with a high degree of accuracy (Jeffreys *et al.*, 1985), was an important contributory factor in the growth of avian paternity studies (Burke & Bruford, 1987; Quinn *et al.*, 1987; Wetton *et al.*, 1987). The field of sperm competition in birds developed rapidly (Birkhead & Møller, 1992) for several reasons:

1 birds are relatively easy to observe in nature;

2 small blood samples yield DNA (from nucleated red blood cells) for paternity analyses;

3 it is relatively easy to find nests, capture the putative parents and obtain blood samples from entire families;

4 the fact that the majority of birds are socially monogamous with male parental care raises the question of the relationship between paternity and paternal care.

The extent of extrapair paternity in socially monogamous birds is remarkable, ranging from zero in a number of marine birds (e.g. Hunter *et al.*, 1992), to over 50% in some passerines (Dixon *et al.*, 1994; Mulder *et al.*, 1994). There now exists information on the level of extrapair paternity for over 100 bird species, providing a unique opportunity to explore the factors responsible for this variation (e.g. Møller & Birkhead, 1994). Evidence for sperm competition in mammals was more difficult to obtain partly because most mammals are difficult to observe and partly because molecular paternity techniques took longer to develop (Hanken & Sherman, 1981; Møller & Birkhead, 1989; Pemberton *et al.*, 1992).

Paternity assignment using increasingly refined molecular techniques (such as DNA amplification — see Westneat & Webster, 1994; Schierwater *et al.*, 1994), has revolutionized the detection of sperm competition, and techniques have been adapted for a range of taxa and there is now evidence for sperm competition in virtually every animal taxon (Table 6.2).

Although molecular techniques have revolutionized the study of sperm competition, paternity analyses tell us only about the outcome of sperm competition — the end result of a succession of behavioural and physiological processes. Behavioural observations are essential for telling us about the way multiple paternity arises. For example, not all cases of extrapair paternity in birds are the outcome of extrapair copulation: in some cases mixed paternity may arise from rapid mate switching (Pinxten *et al.*, 1993). Even in those cases where extrapair paternity results from extrapair copulations, it is important to know whether these are initiated by females, or forced on females

Table 6.2 Taxonomic occurrence of sperm competition in the animal kingdom.

Taxon	Examples	References
Cnidarians	Corals*	Levitan & Petersen (1995)
Platyhelminthes	Flatworms	Peters *et al.*(1996)
Aschelminthes	Nematodes	Barker (1994)
Molluscs	Gastropod *Arianta* spp.	Chen & Baur (1993)
Annelids	Palolo worm*	Levitan & Petersen (1995)
Arthropods		
Chelicerates	Spiders, mites	Austad (1984), Radwan & Siva-Jothy (1996)
	Horseshoe crab *Limulus**	Brockmann *et al.* (1994)
Crustaceans	Ghost crab *Inachus* spp.	Diesel (1990)
Insects	Numerous species	Thornhill & Alcock (1983)
	Dungfly *Scatophaga*	Parker 1970a (see text)
Millipedes	*Alloporus uncinatus*	Barnett *et al.* (1995)
Fish	Blue-headed wrasse *Thalossoma**	Shapiro *et al.* (1994)
Amphibians	Frog *Chiromantis**	Halliday & Verrell (1984), Jennions & Passmore (1993)
Reptiles	Adder *Vipera berus*	Madsen *et al.* (1992)
	Sand lizard *Lacerta agilis*	Olsson *et al.* (1994)
Birds	Many passerines	Birkhead & Møller (1992) Westneat & Webster (1994)
Mammals	Many species	Ginsberg & Huck (1989) Stockley & Purvis (1993)
	Humans	Baker & Bellis (1995)

* Indicates external fertilizers.

by males. A combination of behavioural observation, paternity analysis and a knowledge of mechanisms provides the best opportunity for understanding the selective forces and constraints resulting in the evolution of particular traits.

6.4 Patterns of paternity

6.4.1 General

Except for spawnings in which several males ejaculate simultaneously (e.g. in some marine invertebrates and some fish), there is a time delay between the ejaculations which compete for a given set of ova. Students of sperm competition have been much concerned with the pattern of paternity in internally fertilizing species — i.e. with the proportion of offspring fertilized by males in relation to their order of mating. When two males mate, the proportion subsequently sired by the second male is usually referred to as the P2 value (Boorman & Parker, 1976) — although P2 is often taken simply as the proportion of the progeny gained by the last male to mate.

Can we predict any general adaptive rule for P2? This may be difficult. Parker (1970a) argued that there would both be selection on males both: (i) to oust the sperm of previous males, so as to favour self's sperm; and simultaneously (ii) to prevent self's sperm being ousted by future males. These two adaptations are obviously in conflict, and female interests apart, the P2 attained can be seen as a balance between these two opposing interests.

Much will depend on the ease with which each adaptive trend can be achieved. Indeed, P2 can range from very low (i.e. first-male precedence — spiders: Austad, 1984; the Adder *Vipera berus*: Hoggren, 1995), through 'mixing', where competing males gains are roughly even (e.g. rats: Dewsbury, 1984; field crickets: Parker *et al.*, 1990), to very high (certain butterflies: Gwynne, 1984; migratory locusts: Parker & Smith, 1975; birds: Birkhead & Møller, 1992, see below). In mammals a range of precedence and mixing patterns occur because fertilization is determined by the timing of insemination relative to ovulation (Ginsberg & Huck, 1989). In insects, both sorts of adaptation ((i) and (ii) above) occur. For example, the last male to mate may displace (e.g. dungflies: Parker & Simmons, 1991) or replace (dragonflies: Waage, 1979) the previous sperm so that P2 is high. But, in some species the first male to mate leaves a plug or a part of the spermatophore within the female tract which may make sperm transfer by subsequent males difficult, resulting in lower P2 values than when no obstacle is present (see Parker & Smith, 1975). With strong constraints acting against the transfer of sperm after a previous mating, P2 may be low. But, if there is little constraint acting against displacement of previous sperm, we may expect a high P2, which is often the case, especially in insects and birds.

P2 may be high for purely mechanistic reasons. If ejaculates contain constant sperm numbers, then a high P2 can simply be the result of passive loss of the first ejaculate before the second occurs (Lessells & Birkhead, 1990). It would also occur if there is little sperm mixing, and if sperm from the last male to mate is necessarily deposited closer to the site of fertilization (Smith, 1979). There is now growing evidence that ejaculates vary adaptively depending on the risk or intensity of sperm competition (e.g. Gage, 1991): a high P2 could then be the result of an adaptive escalation in sperm numbers (Cook & Gage, 1995).

Female interests may also be important in determining paternity. If a difference in 'genetic quality' between two males can be detected, selection may favour females exerting a preference for the sperm of the best male and affect the pattern of paternity (see Chapter 5, Section 5.2). For example, if females adopt a strategy of remating only if the male is of higher quality than the one she has previously mated with, then preference for the best male would be manifest as last male sperm precedence.

6.4.2 Models and mechanisms of last male precedence

The processes underlying fertilization are typically complex and take place at a microscopic level, so it is difficult to observe what is happening directly. An alternative approach is to construct mathematical models which assume a particular mechanism of sperm competition, and predict outcomes in terms of paternity which, as outlined above, can readily be measured. We can thus attempt to deduce how ejaculates compete by comparing the observed paternity with that predicted under different mechanism assumptions. This can then form the basis of more detailed empirical investigation.

Insect models

Parker *et al.* (1990) proposed a series of linear mathematical models for analysing sperm competition data in such a way as to be able to predict the mechanism of sperm competition. If the relative number of sperm transferred by each of two males is known, then P2 can be predicted under the assumptions of:

1 the 'fair raffle', in which sperm from the two males mix randomly to generate the 'fertilization set' (the pool of sperm from which the fertilizing sperm are drawn);

2 the 'loaded raffle', in which a proportion of one or other male's sperm are discounted (see also Sakaluk & Eggert, 1996) before the fertilization set is generated, as may be the case where there is a competitive race, or passive sperm loss before the second mating;

3 sperm displacement with instant mixing (see also Parker & Simmons, 1991), in which sperm from the first male are displaced by incoming sperm with which they swiftly mix randomly to form the fertilization set;

4 sperm displacement with mixing after displacement, in which there is random mixing only after sperm from the first male are displaced by incoming sperm.

Parker *et al.* (1990) demonstrated the use of such techniques by analysing data from two insect species. In the yellow dungfly, *Scatophaga stercoraria*, the male delivers free sperm during copula. The female's sperm stores, the spermathecae, are chitinous structures of fixed volume, and the P2 data fit well with the model of sperm displacement with random instant mixing (see below). In the field cricket, *Gryllus bimaculatus*, sperm transfer does not occur during copulation itself, which serves to place an external spermatophore, from which the sperm later drain. The single spermatheca is an elastic membranous sac, and sperm from successive copulations simply add to its volume (Simmons, 1986). P2 concurs well with the fair raffle or random mixing (Parker & Simmons, 1991).

The predictions of the sperm displacement model have been tested most extensively on the yellow dungfly, a ubiquitous inhabitant of cattle pastures where male flies assemble around fresh droppings to search for and mate with the females as they arrive to lay their egg batch in the dropping. Virtually all incoming females contain enough sperm to fertilize their clutch, but nevertheless mate before ovipositing. By remating they ensure that they are guarded until the end of oviposition by the male that has just mated — a feature which greatly reduces harassment from searching males. The guarding behaviour is likely to have arisen as a paternity assurance mechanism: since the last male to mate fertilizes over 80% of the current egg batch, so the potential mating losses due to some 16-min guarding is well repaid by the protection of paternity (Parker, 1970b). However, guarding is not always effective — if an attacking male is sufficiently large a take-over and copulation may ensue (Sigurjonsdottir & Parker, 1981).

Sperm competition in dungflies seems to fit a model of constant random displacement with instant sperm mixing. To explain, imagine a tank of sperm, representing the fertilization set, which has an input pipe and an outlet pipe. During copulation, sperm flow at a constant rate into the tank through the inlet and out (by displacement) through the outlet. First imagine that the sperm entering the tank do not mix with the sperm already present, which are pushed towards and out from the outlet. The new sperm displace only the old sperm, so that the proportion of sperm from the last male (and hence P2) rises linearly at a rate equal to the input rate (volume entering per unit time) divided by the total volume of the tank. But, now suppose that there is swift random mixing of the incoming sperm with the previous sperm in the tank. At first the sperm displaced from the outlet will be only the old sperm. As the last male's sperm build up in the tank, some of the displaced sperm will be his own ('self-displacement'). By the time most of the sperm in the tank is new, most of the outflow will represent self-displacement. In fact, the proportion of sperm from the last male in the tank (and hence P2) will increase with exponentially diminishing returns (Parker *et al.*, 1990; Parker & Simmons, 1991). P2 in dungflies fits such a relationship quite closely (Parker & Stuart, 1976; Parker & Simmons, 1991, 1994); other mechanisms may generate similar results, but this is the simplest explanation of the dungfly data. What is not yet certain is how this analogue fits in with the anatomy of the female tract. It seems rather unlikely that the tank is the spermathecae, because sperm are deposited in the bursa and sperm reach the spermathecae via long thin ducts in which simultaneous inflow and outflow currents are difficult to envisage, especially for a viscous fluid.

A knowledge of the probable mechanism of sperm competition in dungflies has helped to make and test predictions about adaptation. It has long been known that the copula duration of male dungflies is close to that predicted from an optimality model. The approach used is the marginal value theorem (Charnov, 1976; Parker & Stuart, 1976). The gains to a male (P2) are plotted against the time a male spends mating and displacing rival sperm (Fig. 6.2a). The average time it takes a male to find and guard a female (the search time) is known from field work (Parker, 1970c). The optimal copula duration is given by the tangent to the P2 curve drawn from a distance on the x-axis equal to the search time; the slope of this tangent line gives the maximum rate of gain in fitness attainable from the system as a whole. The predicted copula duration is quite close to the observed average duration (Fig. 6.2a; see Parker & Stuart, 1976; Parker, 1978a), and either feeding-time costs of replenishing the sperm (Parker, 1992), or the effects of take-overs on reducing search time (Charnov & Parker, 1995) could account for the discrepancy between the observed and predicted values.

Rather than estimate the average solution, the most recent work (Parker & Simmons, 1994) has investigated how males of differing sizes should behave. Copula duration decreases with the size of the male (Fig. 6.2b). Why? First, it is not surprising to find that the bigger the male, the faster the rate of flow of sperm during copulation and the higher the displacement rate (Fig. 6.2c). Bigger males show P2 curves which rise more steeply, and we would expect that the tangent line would give shorter optimal copula durations for large males. But, there is a second reason why large males should show shorter copula durations: they experience shorter search times because they gain more take-overs of ovipositing females (Fig. 6.2d). These two effects (higher displacement rate,

Fig. 6.2 *(opposite)* Optimal copula duration in *Scatophaga*. (a) Marginal value approach. The curve is calculated from $P2 = 1 - \exp(-ct)$ where t is the copula duration and $c = 0.06$ is the best estimate of the displacement rate (Parker & Simmons, 1991). The predicted optimal copula duration (p) is approximately 42 min, while the observed duration (o) is 35.5 min (see also Parker & Stuart, 1976; Parker, 1978a). (b) Observed copula duration of males in relation to their size (measured as the cube of the hind tibia length). Dotted lines are 95% confidence intervals. The duration decreases with male size. (From Parker & Simmons, 1994.) (c) Sperm displacement rate of males in relation to size, estimated from P2 data. (From Parker & Simmons, 1994.) The displacement rate increases with male size. (d) The relative chance of gaining a female by a take-over (open circles) increases with male size and so the search and guard time (filled circles) between successive matings is reduced as males get bigger (calculated from various field data; see Parker & Simmons, 1994). (e) The effect of displacement rate (line 1) and take-over (line 2) on the predicted optimal copula duration, compared with the observed relationship (dotted line ± 95% confidence intervals).

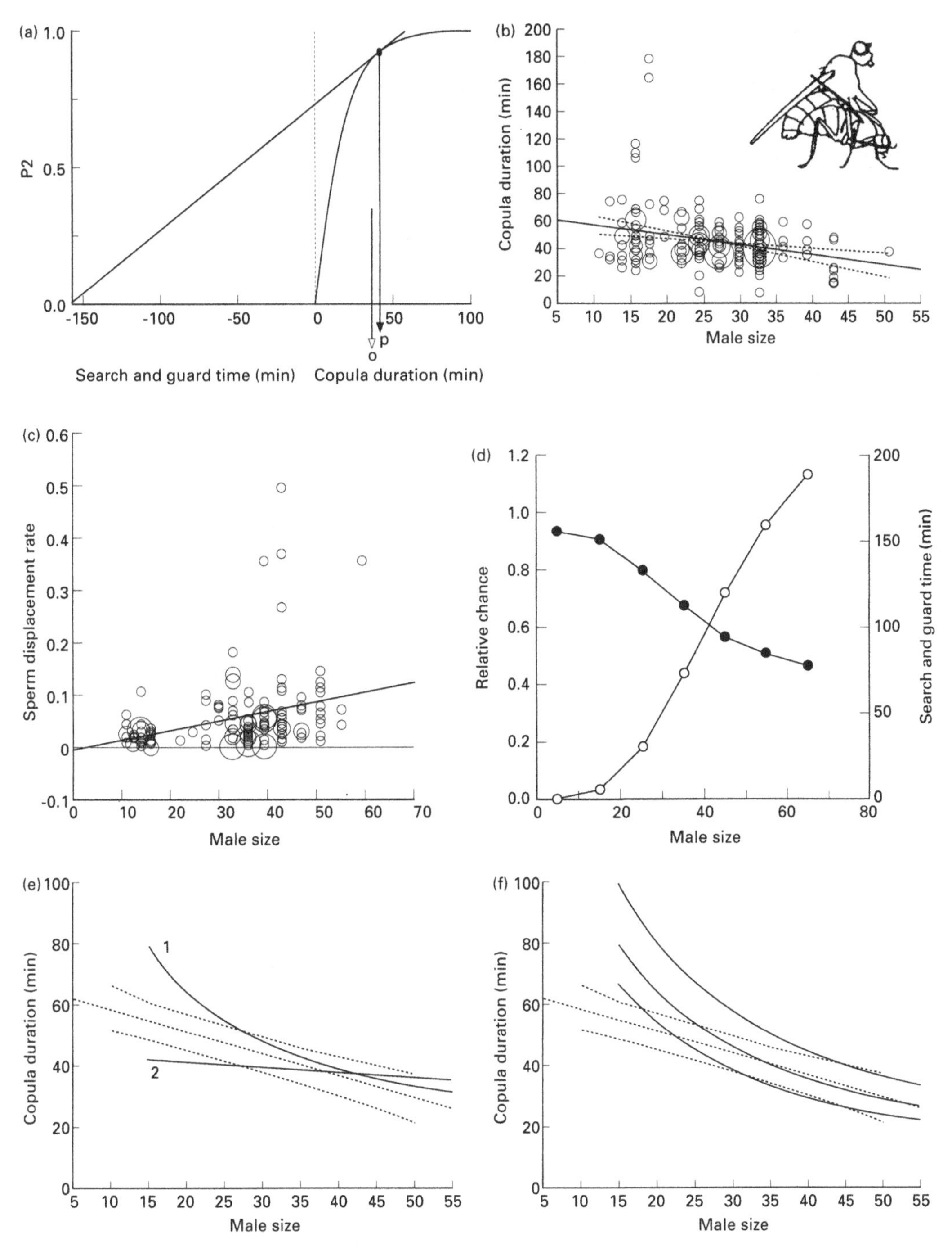

Fig. 6.2 *(continued)* (From Parker & Simmons, 1994.) In this figure the effects of displacement rate and take-over are treated separately, whereas in (f) they are combined. (f) The effect of combining the two size-dependent effects (displacement rate and take-overs) to predict the optimal copula duration (solid line ± 95% confidence intervals), compared with the observed relationship. (From Parker & Simmons, 1994.)

shorter search time) can be evaluated independently (Fig. 6.2e): optimal copula duration is much more sensitive to the variation in displacement rate. When both effects are included, the fit is a good one across most of the male size range, although the smallest 10% of males should copulate longer (Fig. 6.2f). Interestingly, it turns out that (despite larger males achieving higher gain rates in fitness) the two effects interact so that the P2 gain per mating is approximately equal for all males. Suppose there is a phenotypic characteristic that correlates with foraging efficiency (for dungflies this is male size). If uptake rate is directly proportional to the foraging efficiency characteristic, and search time is inversely proportional to it, then we get the rule that the cumulative gain from the patch should, at the optimum, be equal for all phenotypes (Charnov & Parker, 1995). Thus, in dungflies P2 is found to be constant and independent of male size as the rule predicts.

Bird models

Although there are many fewer studies of birds than insects, high P2 values appear to be the rule in birds (Birkhead & Møller, 1992). One of the most influential studies, both in terms of the magnitude of the last male effect and the mechanism by which it occurred, has been that of Compton *et al.* (1978). They used genetic markers to assign paternity in domestic fowl and artificially inseminated equal numbers of sperm from two different genotypes 4-h apart. Regardless of the order of inseminations the second male fertilized most eggs (P2 = 0.77) and they suggested that this occurred because sperm from successive inseminations remained stratified in the blind-ending sperm storage tubules (Fig. 6.3), so that a last in–first out system operated. In an earlier study Martin *et al.* (1974) found that when domestic fowl were inseminated once with a mixture of semen from two different genotypes, paternity was directly proportional to the relative number of sperm from each genotype. On the basis of these two studies the outcome of sperm competition was thought to be determined by the interval between successive inseminations: when the interval was less than 4 h, sperm mixed before going into the sperm storage tubules and paternity was proportional to the relative number of sperm from each male, but when the interval was greater than 4 h, stratification of ejaculates within the sperm storage tubules resulted in last male sperm precedence (Cheng *et al.*, 1983; McKinney *et al.*, 1984). Although this scenario seemed reasonable, a number of inconsistencies subsequently emerged, e.g. if stratification accounted for last male sperm precedence, as sperm from the second male were utilized, those of the first male should be uncovered and result in an increase his paternity. However, in Compton *et al.*'s (1978) experiment the relative success of the two inseminations remained constant over time (see Birkhead & Møller, 1992).

Lessells and Birkhead (1990) constructed a series of mathematical models to find a plausible mechanism which would account for the P2 value of 0.77

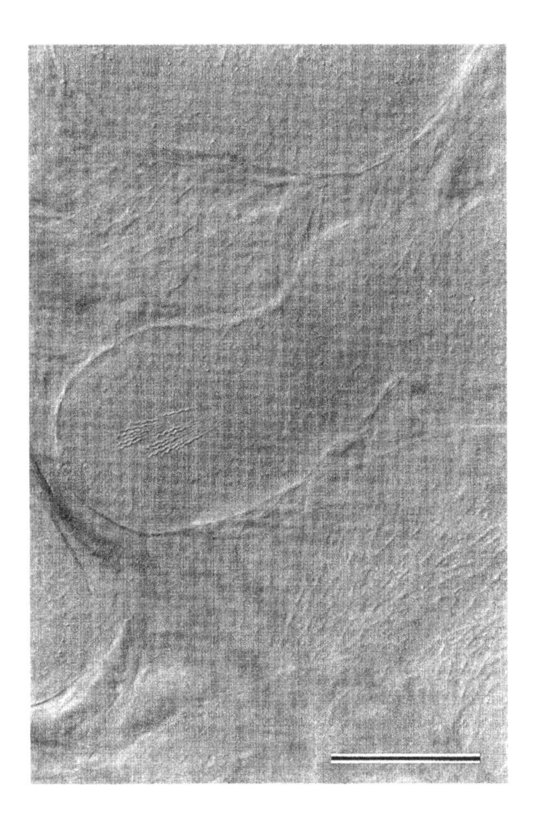

Fig. 6.3 One of 1500 sperm storage tubules from a female zebra finch. Several spermatozoa can be seen in the lumen of the tubule. The scale bar = 50 μm.

reported by Compton *et al.* (1978) in the domestic fowl. They considered three main models (Fig. 6.4).

1 Stratification in which separate ejaculates remain layered within the sperm storage tubules. This model failed because:

(a) the observed rate of sperm loss from the female tract (see Wishart, 1987) was too high to maintain the long-term levels of sperm precedence observed;

(b) the model predicts a decrease in last male precedence as sperm from the first insemination are 'uncovered', but the data showed that this did not occur and the ratio of offspring remained constant over time.

2 Passive sperm loss in which sperm are lost from the female tract at a constant rate and the second (or last) male's sperm have precedence simply because fewer of the second male's sperm are lost by the time fertilization occurs. This is similar to Parker *et al.*'s (1990) 'loaded raffle' (above). This model also failed: to achieve P2 = 0.77, the rate of sperm loss from the female tract would have to be much higher than recorded.

3 Sperm displacement in which it is assumed that space for stored sperm in the female tract is limited and that incoming sperm displace those already present. This model is similar to Parker *et al.*'s (1990) 'displacement with mixing after displacement' (above), and was able to account for Compton *et al.*'s (1978) results.

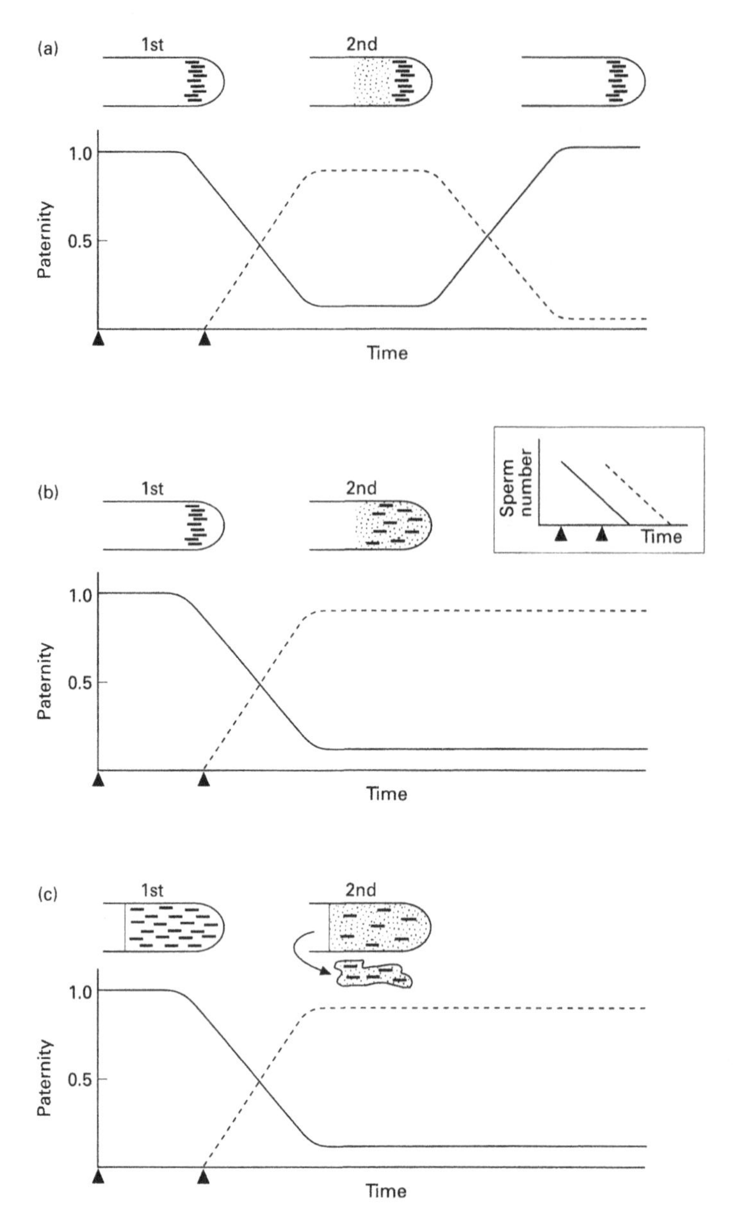

Fig. 6.4 Sperm competition in birds: three models of mechanisms that would result in last male sperm precedence. In each case there are two inseminations (black triangles) made at different times. At the top of each figure the position of sperm in a typical sperm storage tubule (SST) is shown after the first and second insemination. Lines indicate the proportion of offspring fathered by the first (solid line) and second (dashed line) inseminations. (a) Stratification. The sperm of a second insemination overlie those of the first and being closest to the SST exit the second insemination fertilizes most eggs. Once this sperm is used up, however, the first male starts to fertilize eggs again. (b) Passive sperm loss. By the time the second insemination occurs some sperm from the first have been lost. The longer the interval between inseminations the greater the last male effect. Once sperm are in the SSTs they mix, so the ratio of offspring remains constant over time. The inset shows the log number of sperm in the female tract following two inseminations (arrows); at any point in time after the second insemination there are more sperm from the second male. (c) Displacement from a limited sperm store. The second insemination displaces the sperm stored in the SSTs and the level of P2 depends on the degree of displacement. (Based on Lessells & Birkhead, 1990.)

However, when Birkhead *et al.* (1995a) repeated Compton *et al.*'s (1978) experiment, they found no marked last male precedence with a 4-h interval between inseminations. Instead, their results were more consistent with the passive sperm loss model. The discrepancy occurred because Compton *et al.*'s first inseminations took place close to the time of egg-laying when the uptake of sperm by the sperm storage tubules is greatly reduced (Brillard *et al.*, 1987), thereby giving the second insemination a greater fertilization success — hence the high P2 value (Birkhead *et al.*, 1995a). This also: (i) accounts for why Lessells and Birkhead (1990) were unable to explain Compton *et al.*'s (1978) results in terms of the passive sperm loss model; and (ii) shows that the 'success' of the displacement model was fortuitous. Indeed, empirical observations of the numbers of sperm in storage tubules now show that the displacement model's assumption that storage space is limited is biologically implausible in most situations.

The predictions of the passive sperm loss model have been tested in the zebra finch. This is a small (12–15 g), socially monogamous, colonial passerine living in the more arid parts of Australia. In the wild, extrapair courtship and copulation attempts are frequent, although extrapair paternity is relatively low: 2.4% (two out of 82 offspring, from 25 families: Birkhead *et al.*, 1988, 1990). Males attempt to protect their paternity by following their partner during the period before and during egg-laying, and to a lesser extent by copulation, which occurs about 12 times for each clutch. Using genetic plumage markers to assign paternity Birkhead *et al.* (1988) conducted two sperm competition experiments in the laboratory.

1 Mate switching experiment: the female copulated with two different males in succession and both males obtained a similar number of copulations; the second male fathered most offspring, P2 = 0.75 (95% confidence limits: 0.65–0.83).

2 Single extrapair copulation experiment: the aim was to test simultaneously the efficacy of an extrapair copulation, and the magnitude of the last male effect seen in the first experiment. Females copulated nine times on average with one male over several days and then finally once with a male of the other genotype. Overall, P2 = 0.54 (95% confidence limits: 41.6–66.1).

Thus, the extrapair copulation fertilized significantly more eggs than that expected from the 9 : 1 ratio of copulations, demonstrating both a strong last male effect and that single extrapair copulations can be disproportionately successful (Birkhead *et al.*, 1988).

To ascertain whether the passive sperm loss model (Fig. 6.4b) could account for these two sets of results required information on the relative numbers of sperm from each male in the female tract at the time of fertilization. This is determined by three factors: (i) the numbers of sperm inseminated by each male; (ii) the timing of inseminations; and (iii) the rate of loss of sperm from the female tract.

1 Ejaculate size was determined by persuading males to copulate with a freeze-dried female fitted with a false cloaca. This revealed that for males which had not copulated in the previous 7 days the initial ejaculate was relatively large (7.8×10^6 sperm), but subsequent ejaculates were much smaller (1.7×10^6 sperm) (Birkhead *et al.*, 1995b).

2 The timing of inseminations was determined directly by observation and continuous video recording.

3 The instantaneous per capita rate of loss of sperm from the female tract (0.026 sperm $h^{-1} \pm 0.007$ SE) was determined by counting the number of sperm trapped on the outer perivitelline layer of the yolk of successively laid eggs after copulations had ceased (Birkhead *et al.*, 1993).

The levels of paternity predicted by the passive sperm loss model using these variables were very similar to those observed (Fig. 6.5). The main reason the single extrapair copulation was so successful was a combination of it being last (since much of the previously inseminated sperm had already been lost from the female tract) and because it contained a relatively large number of sperm. This in turn was a consequence of the way male zebra finches produce and store sperm in their reproductive tract and the fact that the time since last ejaculations is important in determining the number of sperm ejaculated (Birkhead *et al.*, 1995b).

In the wild, female zebra finches appear to be particularly choosy about with whom they perform extrapair copulations and thus extrapair offspring are rare. Mate choice experiments in captivity have shown that female zebra finches prefer males with certain traits (e.g. high song rates: Collins *et al.*, 1994). Since they do not obtain direct fertility benefits from extrapair copulations with such males (Birkhead & Fletcher, 1995a), this indicates, by default, that they obtain indirect benefits, although there is no direct evidence for this as yet (Birkhead, 1996).

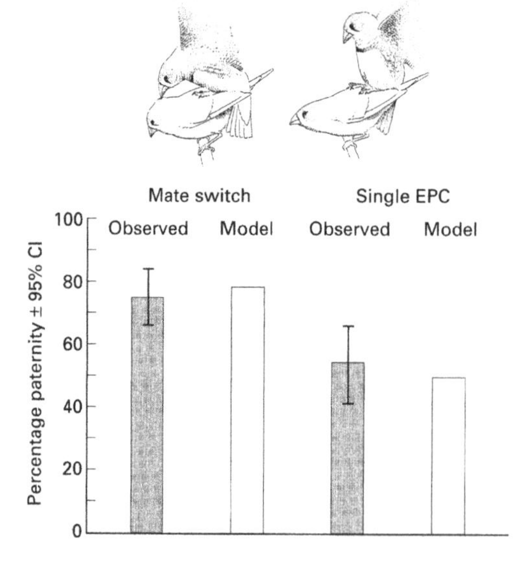

Fig. 6.5 Sperm competition in the zebra finch. Observed P2 values in two experiments and predicted P2 values derived from the passive sperm loss model. In the mate-switch experiment each male copulated several times with the female. In the single extrapair copulation experiment the first male obtained nine copulations, followed by a single copulation from an 'extrapair male'. In both experiments the observed values are similar to those predicted by the passive sperm loss model. (Redrawn from data in Colegrave *et al.*, 1995.)

6.5 Effects of mechanisms on behaviour

6.5.1 Sperm expenditure theory

Sperm expenditure is the amount of reproductive effort spent by a male on a given ejaculate. A series of 'sperm competition games' has been devised to predict patterns of sperm expenditure in relation to the intensity of sperm competition and the information available to a male at the time of mating (e.g. Parker, 1990a,b). The models assume that ejaculates are expensive (Dewsbury, 1982) in the sense that ejaculate expenditure trades off against mate-searching expenditure (Parker, 1982). Increased ejaculate effort increases the gains from a given mating if there is sperm competition, but reduces the number of matings that can be achieved. The models also typically assume that sperm competition follows some form of raffle, so that a given male's gains depend on the strategy played by his competitors.

The predictions depend on the conditions. Suppose that sperm competition is rather rare so that most females mate just once for a given set of eggs, and the rest (proportion p) mate twice. If a male has no information about p when he mates with a female, the ESS sperm expenditure (for low p), as a proportion of the total reproductive effort per mating, is roughly $p/4$ (Parker, 1982, 1990a); sperm expenditure should increase with the probability of sperm competition. The observation that, viewed across species, sperm expenditure should increase with the probability or intensity of sperm competition is a universal finding of such models. This prediction is supported by various data, although direct evidence (sperm numbers per ejaculate increase as sperm competition increases across species) is much rarer than the indirect result that across species relative testis size increases with sperm competition risk. (The first suggestion for such a correlation was made by Short (1977) for the great apes and has subsequently been shown to occur in several other taxa — see Birkhead, 1995.)

But, now suppose that a male has information that he will face sperm competition from the ejaculate of another male and, further, suppose that mating first or second carries a relative disadvantage if equal ejaculates are delivered (the 'loaded raffle'). Thus, one 'role' (first or second to mate) is favoured, and each male 'knows' which role he occupies. What is the ESS sperm expenditure pattern? If the roles are occupied randomly so that a given male is equally likely to be first or second to mate, then sperm expenditures should be equal (Parker, 1990a). But, if given males are more likely to occupy given roles, then the male in the disfavoured role should spend more on the ejaculate. Cases resembling the loaded raffle may occur in nature, and Stockley and Purvis (1993) have found support for its predictions in a comparative study across mammal species. In the 13-lined ground squirrel, *Spermophilus tridecemlineatus*, males (typically two) 'queue' for mating with an oestrous female (Schwagmeyer & Parker, 1987). There is a first-male advantage (Schwagmeyer & Parker, 1990): sperm competition may resemble a raffle loaded in favour of the first male. Although sperm numbers were not measured directly (which would have required killing the females), all behavioural measures that might correlate with sperm numbers (e.g. numbers of mounts, copulatory attempts, time of longest copulation, etc.) failed to show any differences between first and second males. The circumstances appear consistent with the loaded raffle with random roles (Schwagmeyer & Parker, 1994).

Similar sperm competition games were analysed for monogamy with extrapair copulations, and for 'sneak–guarder' situations (Parker, 1990b). In the extrapair copulations model, each male copulates with his mate, and has a low probability p of achieving an extrapair copulation with somebody else's mate. The ESS is to transfer much more sperm at an extrapair copulation (which faces certainty of sperm competition) than to one's mate (which faces sperm competition only with probability p). Indeed, bird studies show that high paternity sometimes accrues to extrapair copulation matings (Birkhead *et al.*, 1988) and that sperm numbers in extrapair copulations are likely to be higher (Birkhead *et al.*, 1995b). In the 'sneak–guarder' version, each male is either a 'sneak' (which achieves a mating with low probability p), or a 'guarder' which always mates. The same finding applies: sneaks should expend more on sperm. Evidence is found in fish, where the typically smaller sneaks produce relatively and sometimes even absolutely greater ejaculates than the guarding males (e.g. Shapiro *et al.*, 1994; Gage *et al.*, 1996). Most recently, sperm competition games have attempted to make predictions about the size of sperm under internal (Parker, 1993) or external fertilization (Ball & Parker, 1996), although why a few species should produce sperm with giant tails (such as the 58-mm longtail in the 2–3-mm male *Drosophila bifurca*) (Pitnick *et al.*, 1995) remains a mystery. Work is currently in progress to explain sperm polymorphisms such as those found in lepidopterans: the *apyrene* sperm are typically smaller and much more numerous than the *eupyrene* sperm which are transferred in the same ejaculate. Only the eupyrene sperm contains the DNA to fuse with the female pronucleus.

Most birds, many mammals and a few insects show a pattern of multiple ejaculation into a given female, rather than transferring sperm as a single ejaculate. An early sperm competition game (Parker, 1984) investigated two strategies: S (single ejaculation, in which a single large dose of sperm are transferred at the start of oestrus), and M (multiple ejaculation, in which $1/n$th the dose of sperm are transferred at n equal intervals throughout oestrus). Conception occurs randomly throughout oestrus, and sperm competition follows the passive sperm loss model: a constant proportion of each ejaculate is lost at each time unit. S always outcompetes M unless the rate of sperm loss is very high; M wins only if more than (roughly) 90% of sperm ejaculated at the start of oestrus would be lost by the end of oestrus.

6.5.2 Female behaviour

The traditional view of sperm competition has been one of sexually enthusiastic males and passive or reluctant females, mainly because selection was thought to operate more intensively on males than females (Parker, 1970a, 1984). Recent empirical observations, however, especially on birds, reveal that females sometimes actively seek additional copulation partners; this has resulted in a recent reappraisal of female roles in sperm competition (e.g. Hunter *et al.*, 1993).

In contrast to the situation where females acquire direct benefits from mating with multiple males, where indirect benefits are concerned, females seek additional partners specifically in order to have their ova fertilized by them. There are two ways that females can control the paternity of their offspring: either overtly through their behaviour, or covertly through physiological means, a process referred to as cryptic female choice (Thornhill, 1983). Intuitively, behavioural methods appear to offer the most direct way for females to control paternity. However, it may not be this simple: a female's priority must be to ensure that her eggs are fertilized. Her best strategy might therefore be to copulate with one male and then if she encounters a more preferred male, to subsequently copulate with him. In this situation females would benefit from being able to favour the sperm of the second (or last) male. In birds the last male to inseminate a female has the best chance of fertilizing her eggs and the longer the interval between the previous pair copulation and the extrapair copulation, the greater are the chances (above). This may explain why females of many bird species terminate copulations with their partner soon after egg-laying has started but before they cease to be fertile. By doing this females get the best of all possible worlds: they have sufficient sperm from their partner to fertilize the clutch, but retain the option of engaging in an extrapair copulation should they encounter an appropriate male (Birkhead & Møller, 1993).

The females of those species subject to forced copulations (above) would benefit from some physiological control over which sperm fertilize her ova. Although the idea of cryptic female choice is inherently appealing, and there are some intriguing possibilities, there is as yet very little convincing evidence for it (see Eberhard, 1996; Simmons *et al.*, 1996).

6.5.3 Males seeking extrapair copulations

Under a range of sperm competition models (above) the more sperm a male inseminates the greater are his chances of fertilizing ova. In some species males have the physiological ability to adjust the number of sperm ejaculated and ejaculate relatively large numbers of sperm in sperm competition situations (e.g. Mediterranean fruit fly *Ceratitis capitata*: Gage, 1991; and the reef fish *Thalassoma bifasciatum*: Shapiro *et al.*, 1994; Warner *et al.*, 1995). However, there is no evidence that male zebra finches can adjust the number of sperm they transfer (Birkhead & Fletcher, 1995b), instead they maximize the number of sperm they transfer through the timing of extrapair activities and seeking extrapair copulations once their own pair-copulation period is over and their sperm supplies have recovered and are maximal (Birkhead *et al.*, 1995b).

6.5.4 Paternity guards

Males exhibit a range of morphological and behavioural traits which minimize the likelihood of their female being inseminated by another male (Parker, 1970a). The most obvious of these traits is mate-guarding, in which the male remains either in physical contact or in close proximity to the female. Guarding may be pre- or post-copulatory depending on whether females store sperm and the interval between copulation and fertilization. Contact guarding occurs in the amphipod *Gammarus pulex*, in which the male carries the female for several days prior to copulation. Females do not store sperm so the first male to copulate fertilizes all eggs, hence pre-copulatory guarding (Birkhead & Pringle, 1986). Contact guarding also occurs in the dungfly (Parker, 1970b). Here, although the insemination–fertilization interval is short, males can displace previously stored sperm (see Section 6.4.2), so guarding is post-copulatory. Non-contact guarding also occurs in some insects, such as certain dragonflies (Jacobs, 1955). In birds, guarding is non-contact, and both pre- and post-copulatory: the male follows his partner from several days before she first ovulates until all her eggs are fertilized. This is necessary because females store sperm and because each egg of the clutch is fertilized separately, usually the day before it is laid (Birkhead & Møller, 1992). Male mammals also guard oestrous females, a behaviour usually referred to as consortship (Packer & Pusey, 1983; Sherman, 1989). Other paternity guards include the deposition of mating plugs (e.g. in some nematodes, arachnids, insects, snakes and mammals), anti-aphrodisiacs within the seminal fluid (Reimann *et al.*, 1967) and frequent or prolonged copulation (Birkhead & Møller, 1992).

6.6 Sexual conflict and mating systems

6.6.1 Sexual conflicts and their outcomes

Sexual conflict occurs if the evolutionary interests of males and females do not coincide. Conflicts of this sort can apply in various contexts, the most obvious being parental investment (it may pay one sex to reduce parental investment) or mating decisions. We discuss the latter here, and only briefly. If a male and female meet, it could pay both to mate, or neither to mate. There is then no sexual conflict. But, if it pays one sex to mate and the other not to mate, we can say that there will be sexual conflict over mating decision (Parker, 1979, 1983). In mating conflicts, because of the disparity in parental investment, males usually occupy the role in which mating is favourable and females the role in which it is unfavourable, although occasionally roles may be reversed. Conflict can occur over mate quality, incest, matings between ecotypes or sibling species, etc. Parker (1979) identified thresholds at which the three possible outcomes would occur — the range of conditions for conflict is likely to be greatest when the male makes no parental contribution to offspring.

What about the outcomes of mating conflicts? Clutton-Brock and Parker (1995) review game-theory models that have been applied: the general determinants of outcomes in all the games depend on the balance (see Parker, 1979) between what may be termed 'power' (which concerns the relative contest costs, the ability to inflict damage, or the relative cost of enforcing victory for the sexes) and the 'value of winning' (which concerns the relative fitness difference between mating and not mating for the sexes). Any asymmetry in the value of winning is usually in favour of the male, because the difference between mating and not mating represents n offspring, rather than just the variation in the quality of the n offspring. But, asymmetry in power may be in favour of the female — it may be easier to prevent a mating than to enforce it, unless males are much bigger. Thus, general outcomes of mating conflicts are not easy to predict. One possible way that females could avoid costs is simply to allow mating, but prevent sperm from fertilizing eggs.

6.6.2 Sexual conflict: empirical observations

The traditional view of mating systems is that 'what you see is what you get'; in other words, if a particular bird species is socially monogamous, it was also sexually monogamous and often had biparental care. But, this seems to be true in only a small proportion of cases. In most seabirds, for example, there is no conflict over copulation or parental care because the interests of the sexes are similar and partners depend upon each other to rear a single offspring. Females of these species rarely seek or engage in extrapair copulations and extrapair paternity is low (Birkhead & Møller, 1996). In other socially monogamous bird species the interests of each sex differ and paternity analyses show that both sexes routinely copulate with and are fertilized by other males (Birkhead & Møller, 1992). A conflict thus exists over parental care; being less certain of their parentage than females, a male may be less willing to invest, especially if their female has copulated with another male. In some species infidelity is costly for females because in these circumstances males reduce their care, but in others females succeed in securing their partner's full assistance in rearing offspring fathered by other males (Westneat & Sargent, 1996). All mating systems may be the outcome of intra- and intersexual conflicts (Davies, 1992; Reynolds, 1996) and the resolution of conflicts often represent a compromise, with neither of the participants achieving their preferred optimum. In socially monogamous birds, for example, this occurs because male-guarding constrains females and prevents them from timing extrapair copulations optimally and hence reducing their effectiveness (Birkhead, 1996). This in turn must also account for the considerable intraspecific variability in the outcome of extrapair copulations (e.g. Birkhead & Møller, 1992).

The conflicts and dynamic interplay between male and female decisions which result in a compromise can most clearly be seen in the polygynandrous mating system of the dunnock and alpine accentor *Prunella collaris*. In both species females obtain direct benefits in the form of male care from securing a threshold number of copulations with particular males, and to this end females solicit copulations from all males in the group. Dominant males, on the other hand, attempt to monopolize females to maximize their own reproductive success. The outcome of this conflict is reflected in the high rates of copulation solicitation by females, the refusal of many solicitations by males, but also the high rates of copulation, which for some females may number over 1000 per clutch. High copulation rates in turn are associated with relatively enormous testes and seminal glomera (sperm stores) in the males of both species (Davies *et al.*, 1996).

In some cases one sex may gain the upper hand. In the externally fertilizing blue-headed wrasse the most sexually successful males produce fewer sperm per mating, and hence fertilize a smaller proportion of each female's ova (93%) than less active males (96%). In this way males increase their reproductive output, but by choosing these particular males females suffer the cost of having fewer of their eggs fertilized (Warner *et al.*, 1995). In *Drosophila* there is last male sperm precedence ($P2 > 90\%$) and males benefit from repeated mating with different females because it maximizes the number of offspring they father (Bateman, 1948; Gromko *et al.*, 1984). For females, however, mating is costly: the more females copulate the shorter their lives. This occurs because, prior to introducing sperm during insemination, the male transfers a cocktail of other substances in their seminal fluid designed to maximize his reproductive success. These substances speed up the rate of oviposition, decrease the female's receptivity to other males and, perhaps most significantly, disable any previously stored sperm in her tract. It is thought that the proteins which disable stored sperm damage the female and significantly reduce her lifespan (Fowler & Partridge, 1989; Harshman & Prout, 1994; Chapman *et al.*, 1995; Clark *et al.*, 1995).

6.7 Conclusion

A number of fundamental questions in the field of sperm competition remain to be answered. For example, we still need to determine the function of multiple mating by females. We also need to know a great deal more about the underlying mechanisms of sperm competition, and in this respect the continued interaction between empirical studies and theoretical models will increase our understanding of sperm competition adaptations. Empirical investigations of sperm quality and the use of labels to identify the sperm of particular males promise to be particularly revealing. The continued integration of mechanism and function is important since it will allow us to make better predictions about the optimal behaviours of males and females. The field of sperm competition is a broad one spanning behaviour, physiology anatomy, molecular biology and is intimately associated with fitness. Sperm competition is such a ubiquitous and pervasive force in evolution that it will almost certainly continue to reveal exciting insights into sexual behaviour and gamete biology.

PART 3

FROM INDIVIDUAL BEHAVIOUR TO SOCIAL SYSTEMS

Male lions Panthera leo *cooperate to take over and defend prides, yet there is often a high degree of skew in their individual reproductive success. How, then, can cooperative behaviour be stable?* [Photograph by Craig Packer.]

Part 3: Introduction

Most animals spend time interacting with others of their own species. They attract mates, fight rivals, group together for safety, look after offspring, and so on. How can social groups (male–female pairs, families or colonies) be stable despite selection on individuals to maximize their own fitness, often at the expense of others in the group? The chapters in this section are all concerned with the mixture of cooperation and conflict seen in animal societies.

In Chapter 7, Johnstone explores the evolution of signals used in communication. How should we expect individuals to advertise their quality to mates or rivals? How should prey signal that they have seen a predator? How should offspring advertise their hunger to their parents? To understand signal evolution we need to consider selective pressures on both signallers and receivers and to recognize that there will often be a conflict of interest between them. There will be selection on signallers to make their signals detectable and stimulating to receivers. On the other hand, it will pay receivers to extract reliable information from the signals. The problem is that because individuals usually have conflicting genetic interests it will potentially pay signallers to misrepresent their true quality. Johnstone argues that the outcome of this conflict is the evolution of costly signals which enforce honesty, an idea originally proposed by Zahavi (1975). Honesty can be maintained by physical constraints (e.g. the carotenoid pigment used in displays by some birds and fish can be obtained only from food) or because inferior competitors find the signal more costly to produce or maintain (e.g. only good quality male swallows can bear the handicap of a long tail). The chapter reviews the theory and evidence for this conclusion and suggests two interesting problems for future work. Why do animals often use multiple signals and how does competition among signallers influence signal form and receiver response?

The most elaborate signals are concerned with mate choice, as exemplified by the tail of the peacock or the extraordinary dances of some birds of paradise. Darwin (1871) suggested these evolved through female choice, although he had little evidence for choice and did not explain why females might have such strange preferences. The last decade has seen an explosion of interest in these questions and Ryan reviews the findings in Chapter 8. Sometimes female choice is easy to understand because the female gains better resources which improve her fecundity, for example a better place to nest or a better male provider for herself and her offspring. In other cases however, the male's only contribution to the offspring is his sperm. It is these cases that have often led to the most elaborate male displays. Ryan provides a critical discussion of how female choice and male displays evolve in these circumstances. Sometimes females gain better fertilization of their eggs through choice, or they avoid males who may pass on diseases during the act of mating. In other examples, females may have a sensory bias selected for in other contexts (e.g. feeding) and males exploit this through their displays. Some phylogenetic studies show that the female preference evolved prior to the sexually selected male trait (e.g. the choice of call in the tungara frog or the female preference for swords in swordfish). In other examples, there is evidence that the female preference and the male trait have coevolved, as predicted by Fisher's 'runaway' hypothesis (coloration in sticklebacks and guppies). Finally, there is growing evidence that a female's choice for mating with particular showy males may improve the genetic quality of her offspring (Marion Petrie's study of peacocks). Perhaps the main conclusion is that there will be different benefits of choice in different cases and that several evolutionary mechanisms may interact. For example, Ryan points out that female preference evolved under sensory bias may later give rise to elaboration under a runaway process. This echoes the message from the previous chapter that we need to distinguish explanations for the origin of signals from those for their maintenance.

The next three chapters consider the evolution of social behaviour. Work in this field gained enormous impetus from Hamilton's (1964a,b) insight that individuals can pass on copies of their genes not only by producing offspring (direct fitness) but also by helping close relatives to breed (indirect fitness). This theory of 'kin selection' specifies, in Hamilton's rule, the conditions under which reproductive altruism will evolve. The social insects have provided particularly good opportunities to test this theory, especially the eusocial societies in which there is cooperative brood care with some individuals (workers) largely sacrificing their own chances of reproduction in order to help the queen produce offspring. In Chapter 9, Bourke reviews the theory and evidence that kin selection can explain reproductive altruism, focussing on both the genetic predispositions for helping behaviour (relatedness to beneficiaries) and the ecological factors which influence its costs and benefits.

Bourke discusses the famous and brilliant suggestion made by Hamilton, that haplodiploidy may explain both the prevalence of eusociality in the Hymenoptera (bees, wasps and ants) and the fact that only females in this group of insects show worker behaviour. In haplodiploidy, males develop from unfertilized eggs, and so are haploid, while females develop from fertilized eggs and are diploid. If the queen mates just once, then this genetic system creates 75% relatedness among sisters, and so seems to predispose females towards rearing reproductive sisters rather than daughters (to which they are related by 50%). The problem with this argument is that the stable sex ratio (or more strictly, investment ratio) from a worker's point of view is 3 : 1

reproductive females : males (Trivers & Hare, 1976). At this ratio, which is often achieved in nature, sib-rearing yields the same fitness returns as offspring-rearing. Therefore haplodiploidy may not, afterall, provide a special genetic predisposition for the evolution of eusociality. Bourke discusses the ecological factors which may have facilitated eusociality in the Hymenoptera, including the benefits of group-living and a large nest, which favours joining a group rather than breeding solitarily, and the possession of a sting which enables workers to contribute effectively to nest defence.

Although the cooperation in bee and ant societies is impressive, Bourke shows that conflicts of interest are also rife. Workers control sex allocation in the colony, against the queen's best interests, and recent studies show that they do this by selectively destroying male eggs or larvae. In addition, workers often lay unfertilized (male) eggs. Theory predicts when workers should police such cheating by their fellow workers, and instead favour the queen's male eggs. Recent experiments provide impressive support for the theory. The chapter also reviews 'reproductive skew' theory which attempts to explain whether dominant individuals in the colony should monopolize all the reproduction or share it with others.

In Chapter 10, Emlen discusses the social tensions in vertebrate family groups. Like the previous chapter, he uses kinship theory and reproductive skew theory to predict the conditions under which individuals should help others or attempt to breed themselves. Among the most important factors determining whether an individual should stay at home and help or disperse and breed are its relatedness to the young produced at home, its chances of successful dispersal and the quality of the territory available at home compared with that for independent breeding. Experiments with several cooperatively breeding birds (e.g. Seychelles warblers, white-fronted bee-eaters) show that young vary their decisions adaptively and are more likely to leave home when their genetic profit from breeding exceeds that from helping. Emlen discusses how the conflicts predicted by kinship theory actually occur in practice in several birds and mammals. For example: when a breeding female dies, her sons may compete with their father for matings with the new female; replacement males kill young sired by the previous male; offspring are less likely to help at home when a parent is replaced; when their father dies, his sons may evict their mother and attract a new breeding female; fathers may disrupt their son's breeding attempts to cause their son to come home to help instead. It is exciting to see the same theories applied successfully to the ants and bees of Chapter 9 and the birds and lions of Chapter 10.

Early game-theory models of contests considered how single encounters were resolved by asymmetries in fighting ability or the value of a resource. In social groups, however, individuals often meet repeatedly to contest for food, space or mates. In Chapter 11, Pusey and Packer consider how these repeated interactions may give rise to stable dominance hierarchies and cooperation. They review evidence from primates showing that alliances among both relatives and non-relatives may have various effects on the stability of social groups. Then they provide a critical discussion of the 'prisoner's dilemma' as a model for the evolution of cooperation through reciprocity. Although this model has become a favourite of theoreticians, Pusey and Packer point out that it fails to capture the key features of most real-life examples of cooperation. Real-life contests are rarely between pairs of individuals who make simultaneous decisions between just two options (cooperate versus defect), with symmetrical payoffs for each contestant. Instead, they suggest that much of the cooperation in nature reflects mutualism in which both individuals gain the highest payoff from cooperation, so there is no temptation to cheat. In other cases, some powerful individuals may be able to impose cooperation on others by the threat of punishment.

The chapter reviews three empirical studies of cooperation. In the authors' own long-term studies of African lions, individuals cooperate in hunting, care of young and territory defence but there is no hint that reciprocity is involved in this behaviour. Rather, it involves kin-selected altruism and mutualism. Predator inspection in fish has been claimed as an example of 'tit for tat' cooperation between pairs of individuals. Pusey and Packer discuss the evidence for this and the alternative hypothesis that individuals approach and retreat together simply to seek the safety of shoaling. Finally, some clever experiments with blue jays, using operant feeding devices, show that pairs readily cooperate in mutualistic games but not in a 'prisoner's dilemma'. Pusey and Packer conclude that the evidence for reciprocity in nature is not strong.

In the final chapter of this section, Haig shows that just as conflicts are rife between individuals within pairs, families and larger social groups, so they occur among the genes of an individual's genome. The common metaphor of an organism is of a well-designed machine with the genes cooperating to maximize that individual's reproductive success. Instead, Haig suggests that we might think of the genes as members of social groups with the organism's properties reflecting the same mixture of cooperation and internal conflicts which underlie the societies discussed in previous chapters. Selfish genetic elements provide a clear example of how a gene's interests may not coincide with that of the individual. A gene which submits to the meiotic lottery in reproduction will get through, as a copy, to 50% of the offspring. However, some genes (segregation distorters) kill off gametes which do not contain copies of themselves and so now get through to all the offspring. Provided they do not reduce the parent's fertility by 50%, the gene will spread. Another example of gene selfishness involves 'genomic imprinting' in which the expression of a gene varies depending on whether it is inherited from the mother or father. Paternally imprinted genes include 'growth factors' in mice which cause embryos to take more resources from the mother. This makes good sense because, given multiple paternity, paternally derived genes are less likely to be present, as copies, in siblings than maternally derived genes and so they should behave more selfishly.

Haig discusses how internal conflicts within the cell nucleus are limited by the procedures of fair segregation and recombination. Just as the rules which govern an individual's social behaviour in societies may have evolved as the evolutionarily stable ways to settle conflicts, so mechanisms which regulate the cell cycle may reflect the stable outcomes of conflicts among genes within the genome.

Chapter 7

The Evolution of Animal Signals

Rufus A. Johnstone

7.1 Introduction

This chapter is concerned with the evolution of animal signals; that is, with traits that are specialized for the purpose of communication. The diversity of signals is enormous, ranging from the bright colours typical of many birds and butterflies, to the calls of frogs and crickets, from the pheromones released by moths, ants and many other insects, to the aggressive posturing of lizards and fishes. They may serve to attract a mate or to deter a rival, to warn conspecifics of an approaching predator or alert offspring to the return of a food-bearing parent. Faced with this bewildering array of traits, so varied in their appearance and their function, any attempt to deduce general principles of signal evolution may seem doomed to failure. However, animal displays in all their variety form a coherent and distinctive class of characters, because the selective pressures that influence their design are different from those that act on other traits.

Communication occurs when the actions of (or cues given by) one animal influence the behaviour of another (Wiley, 1983; Endler, 1993). Consequently, the properties of that other individual, the receiver, exert strong selective pressures on signal design. Furthermore, signal design exerts reciprocal selective pressures on receiver behaviour. On the one hand, natural selection favours signallers who elicit favourable responses — those who are better able to intimidate opponents, for example, or to attract mating partners. On the other hand, it favours receivers who can accurately deduce the nature and intentions of signallers from their displays — those who can best determine whether an opponent's threat is real and not just a bluff, for instance, or whether a potential mate will make a good parent. To understand the process of signal evolution (commonly referred to as *ritualization*), we have to bear in mind both of these sets of selective pressures.

The structure of the chapter reflects this dual emphasis: the first section deals with the selective pressures acting on signallers; the second with the selective pressures acting on receivers. Theories of signal evolution that emphasize the former are referred to as 'efficacy based', while those that emphasize the latter are termed 'strategic'; the third and final section attempts to reconcile the apparently contrasting conclusions of the efficacy based and strategic approaches.

7.2 The signaller's perspective

From the signaller's perspective, a display is a means of manipulation. It serves to influence the receiver's behaviour in a way that benefits the signaller (whether or not the receiver also benefits is a different matter, to which I will turn in Section 7.3). Selection favours individuals whose displays are more effective at eliciting beneficial responses. At the same time, however, signalling is likely to be costly. The calls of frogs and crickets, for example, which are so effective in attracting mates, may also draw the undesired attention of predators and parasites (Cade, 1979; Tuttle & Ryan, 1981; Sakaluk & Belwood, 1984). Furthermore, they are likely to involve considerable energetic expenditure (Ryan, 1988; Prestwich, 1994). Selection thus favours individuals whose displays incur less risk and are energetically cheaper to produce. The signals we see in nature should be those that strike the optimum balance between these two conflicting pressures for greater effectiveness and lower fitness cost, i.e. those that are most efficient (Wiley, 1983, 1994; Endler, 1992, 1993).

7.2.1 Getting the message across

The first requirement of an effective signal is that it should be detectable by receivers. Communication, however, often takes place over long distances, in a noisy environment. Signals may therefore be severely attenuated and degraded by the time they reach the receiver, and may have been mixed with irrelevant stimuli. This makes it difficult for recipients to distinguish them from spurious stimulation, or from other kinds of display. A number of design features, common to many different kinds of signal, serve to increase effectiveness by making the task of detection easier. Fleishman's (1988a,b, 1992) studies of the visual 'head-bobbing' displays of anoline lizards provide a nice illustration of some of these common properties.

Territorial male anoles employ two basic types of head-bobbing display. The 'challenge' display is usually given after a direct approach toward an intruder, and is often accompanied by postural modifiers that appear to be associated with different levels of aggressiveness. The 'assertion' display, which is not accompanied by such modifications, is given spontaneously from elevated perches within a male's territory, and does not appear to be directed at specific individuals. To be noticed by viewers that are relatively far away, and unlikely to have their gaze directed at the signaller, the assertion display must be visible and attention-catching at long range. Detailed examination of this display in *Anolis auratus*, a grass anole from Panama, reveals that the signal is designed to function effectively in this way. First, as shown in Fig. 7.1, the display begins with a series of head movements that are of high acceleration, velocity and amplitude compared with the rest of the display. These movements are *conspicuous*, both because of their amplitude and speed, and because they involve frequency components distinct from those present in the common background motion

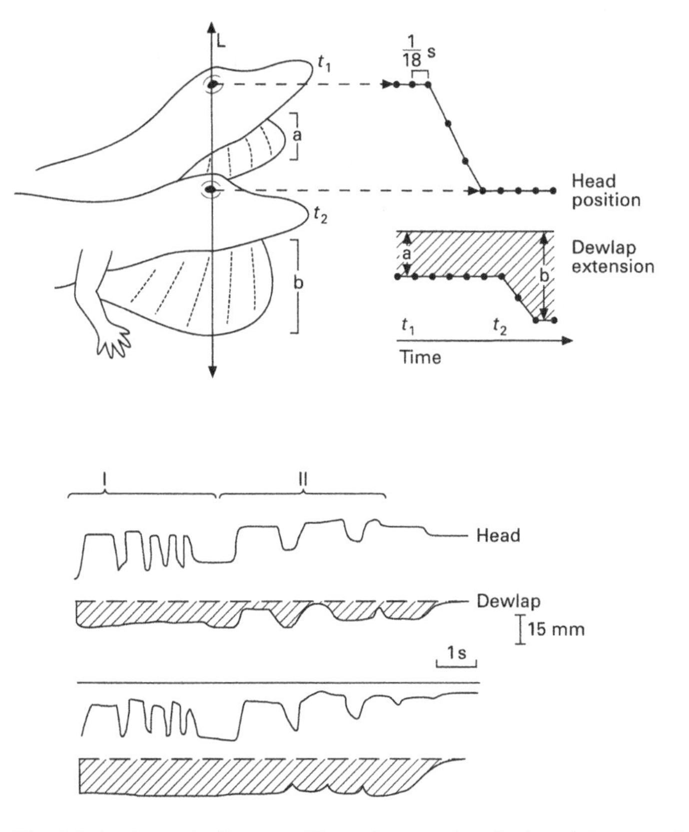

Fig. 7.1 A schematic diagram of how the assertion display of *A. auratus* is created, together with two typical examples of the display. (Modified from Fleishman, 1992.) The lizard moves its head up and down and, independently, expands and contracts its dewlap. Each display record shows the position of a fixed point on the head along the vertical line L, and the vertical distance from the bottom point of the head to the lowest portion of the dewlap, as a function of time (in the upper diagram, the lizard's head is shown at times t_1 and t_2, with the resulting record to the right of the picture). Each display is divided into a high-amplitude, introductory portion (I), and a latter portion (II).

typical of windblown vegetation. The potential for background motion otherwise to mask the display is illustrated by the cryptic movement typical of the vine snake *Oxybelis aeneus*, the anoles' major predator in Panama. While the lizards' head-bobbing contrasts with spurious background movement, the snake superimposes on its forward travel a gentle rocking motion that is similar in frequency to the swaying of a branch blowing in the wind (a type of camouflage that is rendered all the more effective because the snake moves preferentially when the surrounding vegetation has been put into motion — Fleishman, 1985).

Second, the anoles' assertion display is *repetitive*, with the head being raised and lowered several times during the initial, high-amplitude portion (Fig. 7.1). Repetition is just one form of redundancy, a more general property that can be defined as the existence of predictable relationships among different parts of a display. Redundancy helps with detection because it allows the receiver to reconstruct the correct signal from an imperfectly received one (another common instance, besides repetition, is the use of multiple signal components in parallel, a striking example of which is provided by the sexual advertisement of male Jackson's widowbirds, *Euplectes jacksoni*, which comprises an elongated tail, performance of a jump display and construction of a courtship bower — Andersson, 1989, 1991, 1993). Third, the anoles' signal is *stereotyped*, particularly in its latter part, which features the species-specific pattern of movement known as the 'signature bob'. Stereotypy reduces the difficulty of the task faced by receivers by minimizing the number of categories into which they must classify incoming signals (Cullen, 1966). Instead of producing displays that vary in complex ways, signallers increase the probability of detection by employing only one or a few standardized signals. Finally, the assertion display illustrates the use of *alerting components*. That is, it begins (as described above) with components that are relatively variable and encode little information, but which are highly detectable compared to the rest of the display. This initial section increases the chance that a recipient will notice and recognize the subsequent (information-bearing) parts of a signal, by specifying the interval of time during which it can expect to receive them.

Conspicuousness, stereotypy, redundancy and the presence of alerting components are properties common to many different kinds of signal, because they serve very generally to increase the reliability of detection (Wiley, 1983). Further consideration of the selective pressures acting on signallers can also help to account for the differences in signal design among species (or among the different displays of a single species). This is because the physical environment in which communication occurs and the audience of potential recipients differ from one signalling system to another, and these factors have an important influence on the relative efficiency of different kinds of display. What makes for an efficient signal in one context will not necessarily do so in another. Sections 7.2.2 to 7.2.4 illustrate how a consideration of the communicatory context can help to explain detailed aspects of display design, beginning with the effects of the physical environment in which signalling occurs.

7.2.2 The influence of the physical environment

The influence of the physical environment on the design of animal displays is one of the most well-studied aspects of signal evolution, particularly with regard to long-range acoustic and visual displays (for reviews of the subject, see Chapter 2 and Wiley & Richards, 1978; Gerhardt, 1983; Alberts, 1992; Römer, 1993; Forrest, 1994). Two sample studies, one of visual and one of acoustic signalling, will serve to illustrate how interspecific variation in the magnitude and form of signals can be explained by variation in environmental conditions.

Consider, first, the study by Marchetti (1993) of the plumage patterns of eight species of warblers of the genus *Phylloscopus*, which breed in the forests of Kashmir, India. All of the species are small and greenish in colour, but they possess varying numbers of pale colour patches on their wings, crown, rump and tail. These conspicuous plumage traits were found to play a role in intraspecific communication, particularly in the establishment and maintenance of territories. Experimental manipulation of male coloration in *P. inornatus* (using paint to enlarge or reduce wing-bar size or add a new colour patch) directly affected territory size, with males that were rendered more conspicuous obtaining larger territories than control individuals, who in turn obtained larger territories than males rendered less conspicuous. Comparison of all eight species then revealed that they could be unambiguously ranked from duller to brighter according to the number of colour patches they possessed (the order being: no colour patches, one wing-bar, two wing-bars, two wing-bars and a crown stripe, two wing-bars and a crown stripe and a rump patch and, finally, all of these plus white outer tail feathers). Species with more patches tended to occupy dark, dense habitats, while those with fewer patches tended to breed in open areas. In other words, there was a negative correlation between habitat brightness and species brightness (which remained significant when controlling for the effects of phylogeny). It thus appears that properties of the physical environment occupied by these species have influenced the evolution of their plumage signals. Because the ability to perceive visual displays depends on the amount of light in the environment, species that occupy darker habitats have evolved brighter coloration for intraspecific communication, while those that breed in open areas, where they are visible even without conspicuous plumage patterns, have evolved duller coloration.

The above example shows that the physical environment can influence the evolution of signal intensity (i.e. conspicuousness). A broader comparative study by Wiley (1991) of the male territorial songs of eastern North American oscine birds shows that it may also affect the particular form that a signal takes. The study considered 120 species of oscines, for each of which a number of song properties were recorded. Three of these properties concerned the temporal structure of the song, and were indicative of rapid modulation: the minimal period of repeated elements (such as syllables in a trill), the presence or absence of one or more buzzes (notes with a wide but band-limited spectrum) and the presence or absence of 'side-bands' on a song spectrogram. The habitats occupied by territorial males of the various species were then classified into six categories, three of which (coniferous forest, mixed forest and parkland) could be grouped as forested habitats, and three (shrubland, grassland and marsh) as open habitats. Statistical analysis (incorporating various measures to control for the effects of phylogeny) revealed a pronounced association between the physical environment and the temporal properties of song, with short repetition periods, buzzes and side-bands less common among birds of forested habitats. All three of these features render a song easily degradable by reverberation, which is more common in scattering environments such as forests than in open habitats. Wiley's results thus suggest that the temporal properties of the song of many oscines have evolved to reduce the effects of reverberation in forested environments.

7.2.3 The influence of the audience

While the physical environment can have strong effects on the design of animal displays, it is the influence of receiver properties on their evolution that distinguishes signals from other traits. To be effective in eliciting favourable responses, a signaller must employ displays that are well suited to detection by the sensory and neural mechanisms of intended receivers (Ryan, 1990; Guilford & Dawkins, 1991; Endler, 1992, 1993). Moreover, in order to minimize costs resulting from the attraction of predators and parasites, a signaller should employ displays that are ill-suited to detection by these unwanted eavesdroppers (ibidem). Every member of the audience of potential receivers, not just those that the signal is designed to reach, can influence the evolution of signalling behaviour (McGregor, 1993).

Studies by Endler (1978, 1980, 1991, 1992) of the coloration of male guppies (*Poecilia reticulata*) provide some of the most detailed evidence for the matching of display design to receiver sensory capacities. Wild populations of guppies have complex and polymorphic colour patterns, comprising patches of eight major types: red–orange, yellow, bronze–green, cream–white, blue, silver, brown, black and body colour. These colours, restricted to adult males, play a role in mate choice. Female preferences vary among populations, but typically favour males with more carotenoid (red, orange and yellow) patches, more structural colour (blue, green and silver) patches, and patterns that contrast in colour or patch size with the visual background (see also Kodric-Brown, 1985; Houde, 1987; Long & Houde, 1989; Houde & Endler, 1990). At the same time, conspicuous and highly colourful fish suffer greater risk of predation. Endler determined, for the various light environments typical of the species' habitat, the relative conspicuousness of a number of colour patterns, both to other guppies and to potential predators. The colour patterns considered were those typical of three different guppy populations, while the predators included the most dangerous diurnal visual predator, *Crenicichla alta* (Cichlidae), a less dangerous diurnal fish, *Rivulus hartii* (Cyprinodontidae) and the moderately dangerous diurnal freshwater prawn, *Macrobrachium crenulatum*. Endler's results suggested that selection had favoured those colour patterns that were most conspicuous to guppies while being least conspicuous to predators. For instance, in one population where prawns were the major predator, the most abundant hue was orange, a colour to which the prawns are not very sensitive. Where fish were the major predator, by contrast, all colours were more evenly distributed. Moreover, the timing of guppy courtship appeared to be adapted for efficient communication. Guppies court most frequently early and late in the day, when the colour of ambient light is different to that typical of other times. The net effect of courting under this condition was found to be better than under any midday light environment. Courting under the latter conditions would lead to decreased conspicuousness to guppies or increased conspicuousness to predators (or both).

The above studies demonstrate that the mating displays of male guppies and the sensory capacities of females are well matched, but they do not reveal whether this is the result of evolutionary change of the former to match the latter, or vice versa (or whether it is the result of coevolutionary change of both traits). Recently, however, evidence has begun to accumulate that some male mating displays have evolved to take advantage of pre-existing sensory properties of females, which evolved for reasons unrelated to communication. This possibility is referred to as 'sensory exploitation' or 'sensory trap' (Ryan, 1990; Ryan & Rand, 1990, 1993; Christy, 1995; see also Chapter 8). The courtship display of the water mite *Neumannia papillator*, studied by Proctor (1991), provides a good example. Mites of this species are ambush predators of copepods. They assume a characteristic 'net-stance' posture while hunting, in which they rest on their hind four legs on aquatic vegetation, with their first four legs held out in the water column. From this position, they orientate to and clutch at the vibrations caused by swimming prey. A male searching for mates, once he has located a female, will walk slowly around her while vibrating his legs, a display referred to as 'courtship trembling'. She, in turn, will often respond as if to prey, by orientating to the source of vibration and clutching the male in her forelegs. Male leg-trembling frequencies are well within the range of vibrations produced by copepods; moreover, experimental feeding and starvation of females revealed that hungrier individuals were more likely to orientate to and clutch at courting males. It thus appears that male mites are capitalizing on female sensory adaptations for the detection of prey. Additional support for this suggestion is provided by reconstruction of the historical pattern of preference and display evolution. Proctor (1992) has shown, by constructing a cladogram of *N. papillator* and other mites in the same family (Unionicolidae), that the female sensitivity to leg trembling (associated with the use of the net stance during hunting) may have originated prior to the evolution of the male display, in which case it cannot initially have served any role in mate choice.

Unfortunately, while studies like Proctor's show that some male displays take advantage of pre-existing female biases, there is not yet enough evidence to assess the frequency of sensory exploitation (see Johnstone, 1995a). There are good theoretical reasons to expect that sensory biases will be common. The sense organs and nervous system of an organism serve many functions besides the detection of and reaction to any particular signal, and it is unreasonable to assume that each function can always be optimized independently. Consequently, selection for sensory capacities useful in non-mating contexts (such as vibration sensitivity in water mites) is potentially a widespread source of pleiotropic mating biases. In addition, Arak and Enquist (1993) have pointed out that recognition of any given signal (such as a male sexual display) can be achieved by a large number of equally efficient mechanisms that differ only in their response to stimuli outside the normally occurring range. Because selection is blind to responses that are provoked by such stimuli, the precise mechanism used in signal recognition is thus subject to change by random drift. This may lead to a change in the display form that is optimal in stimulating a response, giving rise to hidden preferences that are open to subsequent exploitation (see also Krakauer & Johnstone, 1995). However, the fact that hidden preferences are likely to be widespread does not necessarily imply that most existing male display traits have evolved to take advantage of such preferences.

If sensory exploitation is widespread, then it may offer a new explanation for some of the common features of signal design. Both the coloration of guppies and the leg-trembling behaviour of mites are matched to unique sensory and neural properties peculiar to these species. Common display features, however, may serve to take advantage of widespread sensory properties that are shared by many different receivers. Ryan and Keddy-Hector (1992), for example, have shown (in a review of published studies) that in species where females prefer male traits that deviate from the population mean, they tend to favour traits of greater quantity, even if these lie outside the range of displays normally employed by conspecific males. In the case of visual traits, for example, females tend to prefer larger, more actively displaying, more colourful males, while in the case of acoustic display, they generally favour calls that are longer, of greater intensity and delivered at a higher rate. The exaggerated nature of many display traits might thus be attributable, in part, to a widespread bias of receivers toward greater neural stimulation. A related argument is that of Searcy (1992), who has suggested that the song repertoires of many male birds (see also Searcy & Andersson, 1986; Catchpole, 1987) may have evolved because variation in singing behaviour helps to reduce habituation on the part of the receiver, and thus to elicit a stronger response. These possibilities are discussed in more detail in Chapter 8.

7.2.4 The influence of other signallers

If a signal is to be readily detectable, it must (as previously mentioned) contrast with spurious stimulation reaching the receiver. One of the most common sources of such stimulation is the displays of other signallers attempting to communicate with other receivers in the vicinity. Species that are active at the same season and in the same habitats should employ distinctly different signals, in order to reduce the impact of this kind of interference. A recent study of reproductive character displacement (Loftus-Hills & Littlejohn, 1992) in toads provides some illustrative evidence of this possibility. The two species of narrow-mouthed toad, *Gastrophryne carolinensis* and *G. olivacea*, are widely distributed in the southern US. The former occurs in the southeast, the latter in the southwest, but there is a zone of sympatry in eastern Texas and eastern Oklahoma. Males of both species rely on an advertisement call to attract females. Recordings of calls were obtained from a number of different localities, which could be grouped into four areas: sympatric and allopatric for *G. carolinensis*, sympatric and adjacent allopatric/shallow sympatric for *G. olivacea*. Comparison of calls among these groups (taking into account the influence of body size and of water temperature on calling behaviour) revealed that while the values for the dominant frequencies of the calls did not overlap between the species, those of sympatric *G. carolinensis* were displaced away from those of both groups of *G. olivacea*, which were very similar to each other.

It is not only the displays of other species that can influence signalling behaviour. In many signalling systems, groups of conspecifics interact in their attempts to attract or deter receivers (Greenfield, 1994a,b). Examples include the interaction among males attempting to obtain mates in lekking birds and mammals or chorusing frogs and crickets, the interaction among young in a brood attempting to obtain food from their parents, or the interaction among a group of prey individuals attempting to divert the attention of predators away from themselves. Sometimes, there are clear conflicts of interest among the signallers. In other cases, the interaction appears cooperative (at least in the short term), as in the dual-male courtship of the long-tailed manakin (*Chiroxiphia linearis*), where a beta male assists with the display of an alpha male, although claiming virtually none of the resulting matings (McDonald & Potts, 1994). Whether the signallers cooperate or compete, however, the interaction can strongly influence their display behaviour.

The effects of signaller interaction (in conjunction with the properties of receivers) on display are nicely illustrated by Greenfield and Roizen's (1993) study of chorusing in the neotropical katydid, *Neoconocephalus spiza*. To attract mates, males of this species produce loud advertisement 'chirps' by forewing–forewing stridulation. An isolated male will usually maintain a regular chirp rhythm for several minutes, but those calling within 10 m of each other roughly synchronize their chirps, with the leading role in a chorusing pair usually alternating from one chirp to the next. This synchronization appears to be the result of a phenomenon known as inhibitory resetting. A male is inhibited from calling by sound initiated more than a very short time before his anticipated chirp, and this inhibition continues until the sound ends. Immediately after release from inhibition, the insect's next chirp is slightly advanced. When two males who sustain similar chirp rates call together, runs of synchrony will inevitably result (although if one male chirps faster, it will end up doing most of the calling, while the slower individual remains silent, due to repeated inhibition).

Why should males exercise inhibitory resetting and thus call in synchrony? Females of the species show a preference for the leading call in a closely synchronized sequence. This presumably imposes strong selection pressure on males to adopt a mechanism that improves the chance of calling slightly before a neighbour, and inhibitory resetting does just that. Greenfield and Roizen (1993) compared, in simulation, the performance of males employing the resetting strategy and males employing a hypothetical 'independent' strategy (in which an individual's calling behaviour is uninfluenced by his neighbours). They showed that because of the female preference for leading calls, resetting males would be favoured over independent callers, whichever strategy the two were competing with. It thus appears that synchronous chorusing in this species is the product of competitive interaction among males attempting to jam each other's signals.

7.3 The receiver's perspective

As described above, both the physical environment and the psychology of receivers exert a strong influence on the design of animal displays. While the selective pressures associated with the physical environment, however, are relatively static, those associated with receiver psychology may change in response to the evolution of signal design. This is because receivers are themselves under selection to respond appropriately to the behaviour of (and cues given by) other individuals. From the signaller's perspective, a signal may be a means of manipulation, but from the receiver's perspective, it is a potential source of information. There are many situations in which an individual stands to gain from a knowledge of the physiological or motivational state of others. When a parent bird returns to the nest with food, for example, it is faced with a brood of gaping, jostling nestlings all fighting for attention. Knowledge of which chick is hungriest would allow the parent to allocate the food where it is most needed. Similarly, when coursing predators such as wild dogs approach a group of prey, they must single out one individual to pursue. Knowledge of which prey animal is weakest would allow them to minimize pursuit costs by avoiding less vulnerable targets. Selection thus favours receivers who can better adjust their behaviour in response to the information provided by signals (which extends to ignoring displays that are uninformative). The rest of Section 7.3 deals with the implications of this selective pressure for signal evolution.

7.3.1 Cooperation and conflict in signal evolution

In a cooperative signalling system, selection acting on receivers works in concert with selection acting on signallers, in both cases favouring efficient communication. This will occur whenever the signaller benefits by eliciting a response that is also to the advantage of the receiver. One of the best-known instances of cooperative signalling is the use of a dance 'language' by honey bees to convey to fellow hive members the exact location of food. Because a worker returning from a food source to the hive benefits by directing fellow workers to the resource, just as they benefit from the information it has to provide, selection on the receivers works in concert with selection on the signaller. This is an unusually precise form of communication, in that different aspects of the dance convey information about the distance to the food source, and about the direction that must be taken to reach it from the hive (von Frisch, 1967). A simpler and more widespread example of cooperative communication is the advertisement of species identity during courtship. Both males and females share an interest in pairing with members of their own species (although, as discussed below and in Chapter 8, females may also benefit by exercising choice among conspecifics), and as a result males typically have evolved distinctive, species-specific displays, while females have evolved efficient mechanisms for their location and identification.

Although some signalling systems are cooperative, however, many kinds of communication involve a conflict of interest between signaller and receiver. Selection acting on each individual then opposes selection acting on the other, because the signaller stands to gain by provoking a response that is not to the advantage of the receiver. The exchange of threat displays during aggressive interactions provides a clear example of this kind of conflict of interest. Each participant would benefit from an accurate assessment of its opponent's fighting ability and motivational state, but each also stands to gain by misleading that opponent about its own ability, so as to more effectively deter resistance (Maynard Smith, 1974). Sexual signalling is another case in point. Although males and females share an interest in locating and pairing with conspecifics, males are typically under stronger selection to acquire many mates, while females are under stronger selection to acquire superior mates (Trivers, 1972; Clutton-Brock & Vincent, 1991; Clutton-Brock & Parker, 1992; Johnstone *et al.*, 1996). Females would thus benefit from an accurate assessment of male mate quality (allowing them to reject inferior partners), while males would benefit by misleading potential mates as to their own quality (so as to gain more matings). As a final example, although parents would benefit by allocating food in relation to the hunger or need of their young, individual offspring stand to gain by misleading the parent as to their own level of need, because they are selected to acquire more food than the parent is selected to give (Trivers, 1974; Parker & MacNair, 1978; Godfray, 1995a). In situations like these, communication is best viewed, not as an harmonious exchange of information, but as the focus of an arms race between signallers as manipulators, and receivers as 'mind-readers' (Dawkins & Krebs, 1978; Krebs & Dawkins, 1984).

Early proponents of the 'arms race' approach suggested that, on an evolutionary time-scale, informative or honest signalling was unlikely to endure for very long. Maynard Smith (1974; Maynard Smith & Price, 1973; Maynard Smith & Parker, 1976), for example, argued that threat displays conveying information about aggressiveness or level of escalation were unlikely to be stable. Suppose a population existed in which individuals did convey information about their intentions. If an individual found that its opponent was announcing a higher level of escalation than its own, it would pay to retreat at once. Consequently, a 'deceitful' mutant that invariably announced a very high level of aggressiveness, regardless of its true intentions, would be favoured by selection because its opponents would always back down. Before long, everyone would be lying, and it would then pay to ignore the signal altogether. The same argument can, with slight modification, be applied to any other situation in which there is a conflict of interest between signaller and receiver. Whenever a correlation exists between signalling behaviour and the underlying state of the signaller (i.e. whenever the signal is informative), the population appears vulnerable to invasion by a 'lying' mutant that adopts the signals typical of individuals in a different state, and thereby elicits a more favourable response.

There are, however, properties of signal design that can help to maintain honesty, even in the face of a conflict of interest, by making it impossible or unprofitable for signallers to employ a display that is not representative of their state. It is not necessary, therefore, to abandon the idea that animal signals convey information of value to receivers. Instead, one can argue that the displays we see in nature should simply be designed in such a way that they are not vulnerable to corruption by deceit. This offers a new 'strategic' perspective on signal evolution, which differs from the 'efficacy based' approach detailed in Section 7.2 (see Guilford & Dawkins, 1991). If selection favours reliability as well as efficiency, then some aspects of signal design may have evolved to ensure honesty rather than to facilitate detection. Sections 7.3.2 to 7.3.6 outline the properties that can make a display reliable, and consider the evidence that animals make use of informative signals that possess these properties.

7.3.2 Physical constraints on deceit

The simplest mechanism for the maintenance of honesty is physical constraint. If there is a direct material link between a signal and some underlying aspect of the signaller's state or condition, then lying may be physically impossible. Carotenoid-derived coloration is a good example of an 'assessment signal' of this kind. Most animals cannot synthesize carotenoids for themselves, but must ingest them. Consequently, the intensity of an animal's carotenoid colours provides information about its foraging success and nutritional status. An individual that has had poor success in finding food simply cannot produce the carotenoids necessary to maintain intense coloration. Hill and Montgomerie's (1994) study of plumage colour in the house finch (*Carpodacus mexicanus*) illustrates the effectiveness of this mechanism in maintaining honesty (see also Hill, 1992). Male house finches display ornamental carotenoid coloration that varies from pale yellow to bright red among individuals in a single population. This coloration is deposited in the feathers at the time of the annual moult. During moult (which lasts about 105 days), feathers grow in regular daily cycles, with darker material deposited at night and lighter material in the day, giving rise to 'growth bars'. The width of these bars can be used to infer the nutritional condition of an individual over the moult period, because feathers grow more slowly (yielding narrower bands) during episodes of food stress. Hill and Montgomerie examined four separate populations, and in each of them found a significant positive correlation between male plumage brightness and mean growth bar width. In other words, males that grew redder and more intensely pigmented plumage also grew feathers faster. Moreover, there was a positive correlation in all cases between the hue of growing feathers and the extent of moult of individuals examined, indicating that males that grew redder plumage tended to begin moult earlier. Both observations suggest that carotenoid-based plumage colour provides information about nutritional condition during moult. Hill (1990, 1991) has also shown that females make use of this information during mate choice, preferring to pair with redder males. In doing so they gain substantial benefits in the form of increased paternal care, because males feed their mates and offspring during incubation at a rate that was found to be positively correlated with their level of coloration. To sum up, it appears that females use a reliable assessment signal to acquire mates of superior quality.

7.3.3 Strategic constraints on deceit

While some displays appear to be honest for reasons of physical necessity, there are many signalling systems in which there is no fixed link between the signals used and any underlying aspects of the signaller's condition. An animal's choice of one threat display rather than another, for example, is not physically constrained by its fighting ability or motivational state. Equally, the length of a bird's tail is not fixed by any aspect of its condition. The maintenance of honesty under these circumstances requires a strategic justification, which can explain why 'inferior' signallers do not adopt misleadingly impressive signals even though they are physically capable of doing so. The 'handicap principle' of Zahavi (1975, 1977a,b) provides a potential explanation of this kind. Zahavi suggested that a signal could provide honest information about the 'quality' of the signaller, even in the face of a conflict of interest, provided that it was costly to produce. His argument can be illustrated by reference to sexual display, the context in which it was first discussed. Suppose that a display used by males to advertise their value to females is costly to produce, particularly for inferior males. Honesty may then be stable, because only superior individuals stand to gain a net benefit from display. A 'deceitful' mutant that opted to use the costly signal even though it was of low quality would suffer a loss of fitness, because the high cost involved would outweight the benefits to be gained from attracting more mates (see Fig. 7.2a).

The handicap principle was greeted with considerable scepticism when first proposed, but the development of formal models of honest signalling have since shown that the idea is plausible theoretically (Enquist, 1985; Grafen, 1990; Godfray, 1991; Maynard Smith, 1991; Johnstone & Grafen, 1992; Vega-

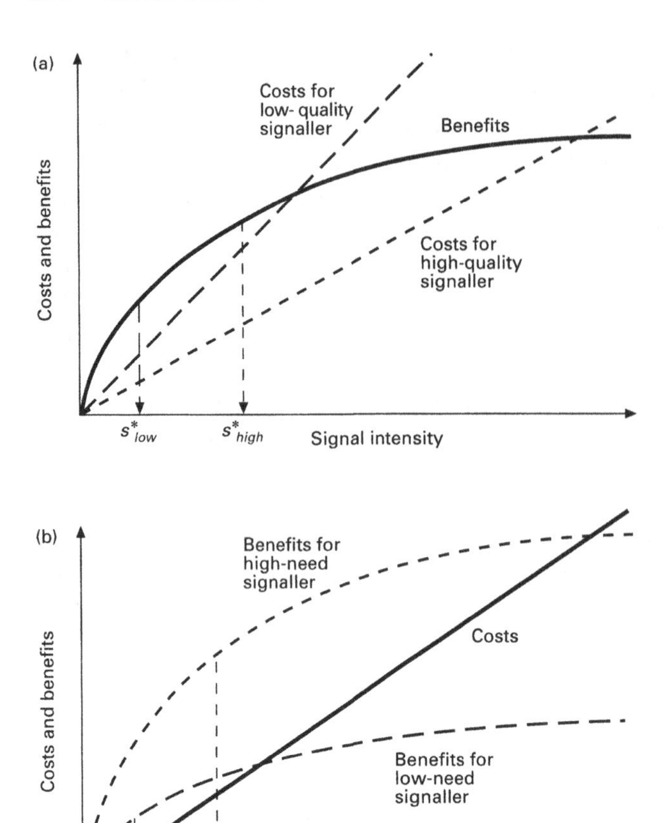

Fig. 7.2 An illustration of the way in which differential costs or benefits of display can maintain honesty. In (a) all signallers gain the same benefits, but low-quality individuals pay higher costs for a given level of signal intensity than do high-quality individuals. Under these circumstances, the optimal signal intensity (which maximizes net benefits) for the former, s^*_{low}, is less than the optimal signal intensity for the latter, s^*_{high}. In (b) all signallers pay the same costs, but low-need individuals stand to gain less benefit from a given level of signal intensity than do high-need individuals. Once again, the optimal signal intensity for the former, s^*_{low}, is less than the optimal signal intensity for the latter, s^*_{high}.

Redondo & Hasson, 1993). Moreover, the same argument can be applied to many different kinds of signal, not just to sexual display. Vulnerable prey individuals, for instance, would benefit by misrepresenting their status to predators, but a display that entails a costly increase in the risk of capture could serve reliably to advertise escape capacity, because only superior individuals could afford to employ it (Vega-Redondo & Hasson, 1993). Similarly, weaker individuals would benefit by bluffing during conflict, but a signal that increases the danger of escalated fighting could provide honest information, because only strong competitors could bear the risks involved in its use (Enquist, 1985). As a final example, offspring would benefit by simulating a dishonestly high level of hunger in order to solicit additional food from parents, but a costly begging signal could provide honest information about their need for food, because only for truly hungry individuals would the benefits of additional resources outweigh the costs involved in production of the display (Godfray, 1991). This last case differs from the previous examples in that signallers are assumed to differ, not in their ability to bear the costs of display, but in the benefits they stand to gain by eliciting a favourable response (see Fig. 7.2b). The handicap principle, in other words, allows for honest advertisement of 'need' as well as of 'quality' (see Maynard Smith, 1991; Johnstone & Grafen, 1992).

While there is now widespread agreement that the handicap principle is plausible theoretically, few empirical studies have demonstrated clearly that signal costs do play a role in the maintenance of honesty. Perhaps the best evidence of this kind is provided by Møller's (1988, 1989, 1994; Møller & de Lope, 1995) studies of the tail ornaments of male barn swallows (*Hirundo rustica*). Adult males of this monogamous species are similar to females in most respects, the exception being that their outermost tail feathers are much longer. Field experiments in which male tail length was manipulated by cutting out sections of feather and reglueing the tail together (or cutting and adding additional sections), together with simple observation, revealed that males with longer tails tend to be preferred by females during mate choice. As a result, they tend to gain more rapid access to a mate during the breeding season, and to acquire a higher quality mate, increased maternal investment and a higher frequency of extrapair copulations. Given these benefits, why do some males only grow short tail feathers? Males with experimentally elongated tails also tended to capture smaller, less profitable prey, and perhaps as a result of this exhibited lower survival rates (while males with shortened tails enjoyed improved survival prospects). Moreover, naturally short-tailed males were less able to survive with elongated tails than were naturally long-tailed males. Møller's results thus suggest that the natural tail length of male swallows reflects their ability to bear the costs of an elaborate ornament. Only superior (naturally long-tailed) males can afford to grow long tails, because the cost of doing so would be prohibitively great for inferior (naturally short-tailed) individuals. It must be admitted, however, that adding a set length to a short tail results in a proportionally greater increase in size than adding the same length to a large tail, which might lead one to expect a greater impact on flight performance in the former case even if there were no relationship between natural tail length and male 'quality'.

The above studies of swallow tail ornaments are, unfortunately, unique in their attempt to investigate whether the costs of display are greater for some signallers than others. There is considerable evidence, from a wide range of different signalling systems, that animal displays are expressed in a condition-dependent manner even when they are not physically constrained to be so (see Johnstone, 1995a, 1996a for reviews). However, while many of these displays appear to be costly, experimental evidence that this is the case is less common, and evidence for condition-dependent cost almost completely lacking. A study by Cresswell (1994) of singing by skylarks (*Alauda arvensis*) in response to attack by merlins (*Falco columbarius*) serves to illustrate the kind of evidence that is available generally. Observation of natural attacks revealed that merlins were probably not using the occurrence of song to choose which skylarks to attack, but that they did chase non- or poorly singing skylarks for longer periods compared to skylarks that sang well (this finding being based on a comparison of attacks that were not terminated by capture). In addition, song appeared to provide honest information about the birds' escape ability, since a merlin was more likely to catch a non-singing than a poorly singing than a full-singing skylark. The predator's decision as to whether or not it should continue pursuit thus seems to be based on a reliable signal of prey vulnerability. Reliability may be ensured by the energetic cost of singing while being chased, which is likely to be prohibitively great for birds that are in poor condition, and hence more vulnerable. However, there is no definite evidence that the display does involve high levels of energetic expenditure, or that it would have a more detrimental effect on the performance of non-singing than of fully singing birds. One cannot firmly conclude, therefore, that honesty is maintained by signal cost in this case.

7.3.4 The possibility of rare deception

While most studies inspired by the handicap principle have focussed on reliable signalling, and the costs that may be necessary for its maintenance, it is worth noting that models of communication based on Zahavi's theory do not always predict universal honesty (Dawkins & Guilford, 1991; Johnstone & Grafen, 1993; Adams & Mesterton-Gibbons, 1995). For a signal to be evolutionarily stable it is necessary, as discussed above, that it should provide sufficient information for selection to favour response by receivers. If it were always misleading, the argument runs, then it would soon fall into disuse, as receivers would evolve to ignore it. Provided, however, that the signal is honest 'on average', this does not rule out the possibility of occasional deceit. For example, if certain signallers find display cheaper (for a given level of quality or need) than do other individuals, then they will be able to afford to produce the display typical, in most cases, of a higher level of quality or need than their own. In doing so, they will elicit more favourable responses from receivers than would otherwise be the case. At an evolutionary equilibrium, receivers will devalue their responses to take into account what fraction of individuals signalling at any particular level belong to this class. However, as long as such individuals are rare enough, and the disadvantage (from the receivers' point of view) of responding to them inappropriately is slight enough, they need not disrupt the signalling system altogether (Johnstone & Grafen, 1993). Evidence for occasional deception can thus provide support for the handicap principle, if it can be shown that deceitful signallers do differ from others in the net benefit they stand to gain from display. A good example is provided by Redondo's studies (Redondo & Castro, 1992; Redondo, 1993) of begging by magpie (*Pica pica*) and great spotted cuckoo (*Clamator glandarius*) chicks.

The intensity with which magpie nestlings beg can be estimated by combining measures of the duration, latency and posture of the begging display, and the emission and duration of begging calls (all of which are strongly correlated with each other). Redondo and Castro first used artificial feeding to equalize the begging intensity of chicks in natural broods of four or five individuals, and then recorded both their mass gain due to parental feeding over the next hour, and their begging intensity at the end of this period. The chicks that begged most intensely were found to be those that had received smaller quantities of food from parents (relative to their body mass), indicating that the begging display of nestling pagpies provides reliable information about hunger. Moreover, a parallel experiment suggested that parents make use of this information when distributing food. After artificial feeding had been used to enhance within-brood differences in begging, those chicks that displayed most intensely were found to receive more food (relative to their body mass) during the following hour. Honesty may be maintained in this case by the energetic costs of begging, as the signals employed by hungrier nestlings involved a higher degree of muscular activity.

The above reliable system of parent–offspring communication, however, is exploited by the chicks of great spotted cuckoos, obligate brood parasites that are raised alongside host chicks, and severely depress the breeding success of magpies. Formal models of signalling suggest that individual young who are less closely related to others in the brood, or to parents, should be selected to use more costly begging behaviour, because they have less to lose by depriving nestmates of additional food (Godfray, 1995b). In accordance with this prediction, laboratory experiments conducted with magpie and great spotted cuckoo chicks of a similar developmental stage, kept in isolation under controlled conditions of food supply, revealed that cuckoo chicks exhibit exaggerated begging behaviour. For a similar degree of need, the parasite nestlings begged for longer, and gave more calls per unit time than host chicks. Furthermore, field experiments revealed that magpie parents, when given a choice between a cuckoo and a magpie chick, tended to preferentially feed the cuckoo. It thus appears that cuckoo chicks 'cheat' magpie parents by means of dishonest begging signals. This kind of deceit is, however, prevented from completely disrupting the signalling system by constraints on the frequency of brood parasitism (e.g. high rates of cuckoo egg mortality, in part due to nest defence and rejection of foreign eggs by hosts).

7.3.5 Strategic explanations for signal diversity

Studies of the kind described in Sections 7.3.2 to 7.3.4 have shown that animal signals often do convey reliable information about the state or condition of the signaller (at least 'on average'), and that in some cases they are designed in such a way as to prevent this reliability being compromised by deceit. There have been few attempts, however, to account for differences in signal design among species (or among the different displays of a single species) in terms of selection for reliability. Zahavi (1977b, 1987) has argued that costs of signal production can only help to maintain honesty if the ability to bear those costs depends on the aspects of signaller condition that are being advertised. Consequently, he suggests there is a necessary link between the form of a display and the information that it conveys. Honest advertisement of nutritional condition and energy reserves, for example, should require a signal that is energetically costly, while advertisement of escape ability should require a signal that entails a costly increase in the risk of capture by predators. However, this approach has not been used yet to generate testable predictions, for two main reasons.

1 The requirement that a signal should entail a certain type of cost does not impose much of a restriction on its design, and so does not allow one to draw detailed conclusions about its likely form. Advertisement of nutritional condition, for example, may require an energetically costly signal, but this could take the form of vocal display, strenuous posturing, or many other types of behaviour.

2 The costs involved in the production of many signals may be imposed by the behaviour of receivers, rather than being physically concomitant upon display. In these cases, the design of the signal is 'conventional', because the appropriate receiver response (imposing whatever kinds of cost are necessary for the maintenance of honesty) can be elicited by a display of any kind (Guilford & Dawkins, 1995).

A more successful attempt to explain signal diversity in strategic terms is that of Briskie *et al.* (1994), who investigated the begging behaviour of 11 species of passerine birds, which differed in their frequency of extrapair paternity. A higher level of mixed parentage means that young in a brood are, on average, less closely related to each other. Consequently, greater costs should be required to maintain honest signalling of offspring need in these species (see Godfray, 1991, 1995b; Johnstone & Grafen, 1992). Just as a cuckoo chick begs more loudly than a host chick because it has nothing to lose by depriving its nestmates of food, so chicks in a species with low relatedness among brood members should beg more loudly than chicks in a species with high relatedness. In accordance with this prediction, comparative analysis (using various methods to control for the effects of phylogeny, brood size and body mass) revealed that the loudness of nestling begging calls among the sample species was positively correlated with the percentage of extrapair young, as illustrated in Fig. 7.3. It thus appears that interspecific differences in signal intensity can, in this case, be explained in strategic terms, as a consequence of differences in the degree of relatedness among competing signallers.

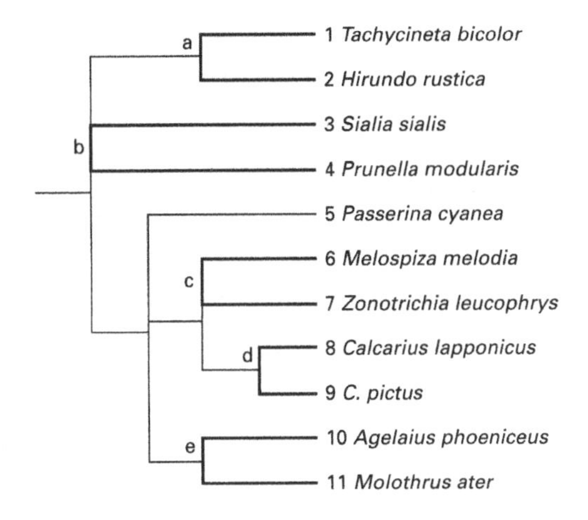

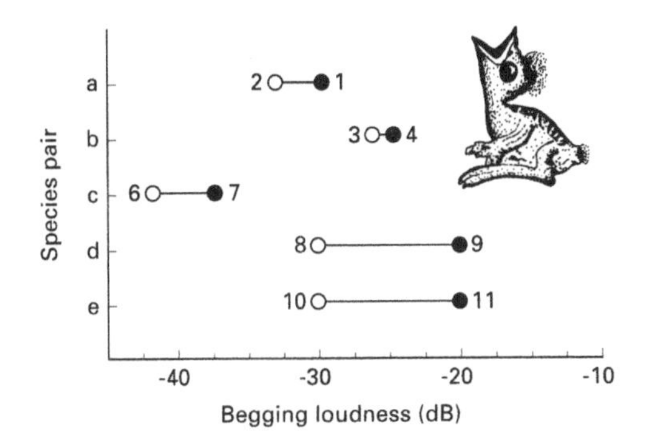

Fig. 7.3 A phylogeny of the bird species considered by Briskie *et al.* (1994), together with a paired comparison of begging loudness between species with high (open circles) and low (filled circles) levels of relatedness within a brood. (From Briskie *et al.*, 1994.)

7.4 Conclusions and prospects

Sections 7.2 and 7.3 outlined two different perspectives on signal evolution. The efficacy based theories described in Section 7.2 suggest that as a result of the selection pressures acting on signallers, animal displays will be designed so as to efficiently elicit a response. The strategic theories described in Section 7.3, by contrast, suggest that as a result of selection pressures acting on receivers, displays will be designed so as to prevent their corruption by deceit. Many common display properties can thus be explained in two different ways. The striking and costly nature of many animal signals is a case in point. On the one hand, extravagance may serve to make a display more easily detectable or stimulating to receivers; on the other hand, it may make a display more reliable by ensuring that only superior (or more needy) signallers can afford the cost of its production. How can these two different views of signal evolution be reconciled?

From the evidence presented, it is clear that neither approach can be entirely abandoned in favour of the other. As Marchetti's (1993) study of plumage brightness in warblers (described in Section 7.2.2) showed, some of the variation in signal form seen in the natural world can be explained only in terms of efficacy. *Phylloscopus* species living in darker habitats have not evolved brighter coloration as a result of selection for reliability. Equally, the study by Briskie *et al.* (1994), described above, showed that some differences in signal form can be explained only in strategic terms. Chicks of species with higher levels of extrapair paternity do not beg more loudly because of selection for detectability. To date, however, the efficacy based perspective has proved far more successful in accounting for the diversity of natural signal form. This suggests that although selection for reliability may favour the evolution of more costly, elaborate displays, it has little influence on the detailed design of a signal. Rather, it is the environment in which communication takes place, and the audience to which a display is directed, that exert the strongest influence on its appearance (Guilford & Dawkins, 1991). Animal signals, in other words, must be both efficient and reliable, but it is the former condition that places the greatest constraint on their design.

The issue is, however, complicated by the fact that different selective pressures may come into play at different times during the evolution of a trait. Studies of sensory exploitation, for example, have revealed that signals can arise as a means to exploit a pre-existing bias in receivers (see Section 7.2.3). In such cases, strategic considerations are irrelevant to the origin of the display. Once signallers have begun to take advantage of the bias, however, receiver selection will start to play a role. Strategic considerations may thus be of great importance with regard to the maintenance of the display, as a preference that leads to inappropriate responses could prove short-lived (see Arak & Enquist, 1995; Krakauer & Johnstone, 1995). Consider, for example, Ryan's (1983, 1985, 1990; Ryan & Rand, 1990, 1993) studies of sexual communication in the túngara frog, *Physalaemus pustulosus* (discussed in more detail in Chapter 8). Females of this species prefer male calls containing low-frequency chucks, a bias that appears to have evolved prior to the chuck itself. Although the preference cannot originally have served an adaptive role in mate choice, Ryan has shown that in current populations it results in females mating with larger males that fertilize more of their eggs. The maintenance of the bias may thus depend on the fact that the preferred trait provides reliable information about size.

7.4.1 Unanswered questions

As the above discussion indicates, there are many aspects of signal evolution that are still poorly understood. It seems appropriate, therefore, to end this chapter by drawing attention to some of the questions that remain unanswered. Two issues that clearly illustrate the wealth of possibilities for future research, both empirical and theoretical, are the use and evolution of multiple signals, and the consequences of competition between signallers for information transfer during communication.

Multiple signals

Why are animal displays so complex? Many displays, in contexts ranging from courtship and mating (Møller & Pomiankowski, 1993; Johnstone, 1995a) to parent–offspring communication (Kilner, 1995), involve a number of distinct signal components, and may often combine different sensory modalities. One striking example was mentioned earlier: male Jackson's widowbirds employ a suite of sexual signals comprising an elongated tail, performance of a jump display and construction of a courtship bower (Andersson, 1989, 1991, 1993). To date, the evolution of multiple signals like these has received little attention. Models of signalling, for example, have typically assumed that only a single form of display is available to signallers; it is only recently that they have begun to address the possibility of multicomponent displays (Schluter & Price, 1993; Johnstone, 1995b, 1996b; Iwasa & Pomiankowski, 1995). Equally, empirical data regarding the way in which receivers assess multiple signals are scarce (although see Zuk *et al.*, 1990, 1992; Andersson, 1991; Kilner, 1995; Scheffer *et al.*, 1996 for examples).

Explanations for the evolution of complex displays fall into two main categories: (i) those which suggest that multiple display components serve to provide the receiver with additional information about the signaller; and (ii) those which suggest that they serve to facilitate detection, or to take advantage of arbitrary preferences. Theories of the former kind include the 'back-up signal' hypothesis, which proposes that multiple signals allow more accurate assessment of one aspect of the signaller's condition (Schluter & Price, 1993; Iwasa & Pomiankowski, 1995; Johnstone, 1996b), and the 'multiple message' hypothesis, which proposes that different signals convey information about different aspects of condition (Johnstone, 1995b, 1996b). Theories of the latter kind include the suggestion that multiple displays have evolved to exploit pre-existing sensory biases (Ryan & Rand, 1993), or (in the context of sexual signalling) that they are the product of Fisherian runaway evolution (Heisler, 1985; Tomlinson & O'Donald, 1989; Pomiankowski & Iwasa, 1993; see also Chapter 8). Unfortunately, while theoretical analyses suggest that all these explanations are plausible, there is little empirical evidence against which to test them.

Møller and Pomiankowski (1993), in a comparative study (based on measurement of museum specimens) of avian taxa with and without apparent multiple male feather ornaments, found that ornament size was negatively correlated with asymmetry only in those species with a single display trait. They interpreted this as evidence that single ornaments are expressed in a condition-dependent manner, whereas multiple ornaments are not, and suggested that the latter were most likely to be the product of Fisherian runaway evolution. However, although their results provide some support for 'uninformative' theories of multiple signal evolution, evidence based on correlations between measures of quality (such as ornament symmetry) and display can be problematic (see

Johnstone, 1995a). Furthermore, experimental studies of individual species seem to paint a more complex picture.

In some cases, experimental evidence suggests that multiple signal components provide little information, but serve instead to render a display more detectable or stimulating to receivers. Consider, for instance, a recent study by Scheffer *et al.* (1996) of sexual signalling in the brush-legged wolf spider *Schizocosa ocreata* (Araneae: Lycosidae). The male courtship display of this species involves both visual and vibratory elements: synchronized tapping, waving and arching of the forelegs, which are decorated with a conspicuous tuft of bristles, and production of substratum-coupled vibrations by stridulatory organs in the pedipalps. Mating experiments in which a female was offered a choice between two males, one whose foreleg tufts had been shaved off and one who was intact, suggested that the visual display component does not play a role in mate assessment. Females were most likely to mate with the individual that first captured their attention, regardless of his condition. However, the tufts may serve to render the courtship display more detectable in situations where vibratory communication is constrained. When females were paired with single males in an artificial arena that eliminated vibratory communication, they responded less often to individuals whose tufts had been removed. Since the complex leaf-litter habitat of *S. ocreata* is ill-suited to the transmission of vibratory signals, the visual display elements may often be necessary for effective communication at a distance. In this regard, it is interesting that males of the congeneric species *S. rovneri*, which occupies compressed leaf-litter habitats better suited to vibratory communication, lack foreleg tufts and have a much simpler visual display than *S. ocreata*.

In other cases, however, multiple display components do seem to provide information about the condition of the signaller. A good example is provided by the courtship display of male guppies (*Poecilia reticulata*). As discussed in Section 7.2.3, males of this species exhibit complex polymorphic colour patterns that play an important role in mate choice, with females typically favouring males that have more carotenoid and structural colour patches, and those that contrast more strongly with the visual background (Kodric-Brown, 1985; Houde, 1987; Long & Houde, 1989; Houde & Endler, 1990). Colour is not, however, the sole determinant of male mating success. Several studies have found that male display rate also influences mate choice (Farr, 1980; Kodric-Brown, 1993; Nicoletto, 1993), with females favouring males that exhibit higher rates of sigmoid display (a rapid, highly energetic 'S-shaped' movement of the male's body). Moreover, both ornamentation and display activity are related to male condition, as measured by sustained swimming performance, by the relationship of body mass to length and by parasite load (Kennedy *et al.*, 1987; Kodric-Brown, 1989; McMinn, 1990; Houde & Torio, 1992; Frischknecht, 1993; Nicoletto, 1993). Experimental infection of males with the (naturally occurring) gut nematode *Camallanus cotti*, for instance, resulted in lowered display rates and reduced attractiveness to females in choice tests (Kennedy *et al.*, 1987;

McMinn, 1990), while temporary infection with the (naturally occurring) monogenean ectoparasite *Gyrodactylus turnbulli* led to the orange carotenoid spots of male hosts becoming paler and less saturated, with similar consequences for their attractiveness (Houde & Torio, 1992). In guppies, therefore, multiple sexual display components do appear to provide females with information about the displaying male (see Johnstone, 1995a, for additional examples of this kind).

Given the constrasting results of different studies, more data is needed regarding the occurrence of complex displays, and the way in which receivers assess their various components. Only with additional information of this kind will it prove possible to test the predictions of the various models and hypotheses discussed above.

Multiple signallers

A second topic that is poorly understood at present is the influence of competition among signallers on information transfer during communication (see Godfray, 1995a,b). The strategic view discussed in Section 7.3 suggests that receivers can acquire reliable information about signallers by focusing on costly displays that only superior or more needy individuals can afford to use. However, where several signallers vie for attention, competition has the potential to interfere with receiver assessment of signaller condition. To date, this issue has received most attention in the context of chick begging (see, e.g. Godfray, 1995b; Kilner, 1995; Kacelnik *et al.*, 1995; Price *et al.*, 1996). As discussed in Sections 7.3.3 and 7.3.4 the solicitation behaviour of nestling birds appears (at least in some cases) to vary in relation to their level of hunger, and hence to provide parents with information that allows them to distribute food adaptively among the brood (see, e.g. Redondo & Castro, 1992). Competition between brood members may, however, confound the relationship between signal and state, a possibility that is nicely illustrated by a recent study of begging in canary (*Serinus canarius*) chicks (Kilner, 1995).

Observation of visits by parent canaries to artificial nests revealed that nestling position relative to the parent strongly influenced food distribution, with the chick whose mouth was closest to the parent tending to be fed the most. Proximity, in turn, was influenced by size and by hunger. Because of a hatching spread of approximately 36 h, chicks in the experimental broods (comprising three individuals) could be ranked in a clear size hierarchy, and larger nestlings tended both to maintain a position closer to the parent and to obtain more food. Experimental manipulation of the level of food deprivation, however, could temporarily over-ride the effects of the size hierarchy, and led to the most deprived chick being fed most often. Chick position thus appears to provide parent canaries with some information about the condition of their young, although the effects of hunger are partially masked by the competitive advantage that larger chicks enjoy in jostling for position. Possibly as a result of this masking, it appears that parents also respond to other cues in addition to chick position when allocating food. When the positions of the canary chicks were controlled experimentally (by use of Plexiglas partitions within the nest), food-deprived nestlings were still favoured, suggesting that parents responded to additional factors such as chick posturing and possibly calling.

Given the potential impact of sibling competition on parental allocation of resources, one might expect that individual nestlings should adjust their behaviour in response to the level of competition with which they are faced. A recent analysis of competitive solicitation by Godfray (1995b), for instance, which considers the division of a fixed quantity of resources between two chicks, suggests that each should adjust its level of solicitation in relation to both its own condition and that of its nestmate. The model predicts, for example, that an individual should beg more intensely when paired with a hungrier competitor (who should tend to beg more than a well-fed chick). A study by Price *et al.* (1996) of the begging tactics of yellow-headed blackbirds (*Xanthocephalus xanthocephalus*) provides empirical support for this idea. Individual chicks (of similar size) were paired, in artificial nests, with nestmates of four types: (i) a big satiated nestmate; (ii) a big hungry nestmate; (iii) a small satiated nestmate; and (iv) a small hungry nestmate (hunger levels having been experimentally manipulated). Both the size and the hunger of the nestmate were found to influence various aspects of the vocal begging behaviour of the experimental subjects. Chicks begged significantly more loudly, more vigorously and for longer when their nestmate was big rather than small, and for longer when their nestmate was hungry rather than satiated (and the nestmates themselves begged more vigorously when hungry).

Given the evidence that competition can affect both the begging behaviour of chicks and their success in obtaining food from parents, models of parent–offspring communication (and of other forms of signalling) need to follow Godfray's (1995b) lead in allowing for interaction between signallers who may differ in competitive ability, as well as in their need for additional resources. Only by taking into account the complexities revealed by empirical studies of communication will it prove possible to obtain testable predictions that apply to real biological signalling systems.

Chapter 8

Sexual Selection and Mate Choice

Michael J. Ryan

8.1 Introduction

Sexual selection poses a simple but central question that Darwin addressed initially in *On the Origin of Species by Means of Natural Selection* (1859) and more extensively in *The Descent of Man and Selection in Relation to Sex* (1871): why do the sexes differ and why is the male usually the more elaborate sex? Darwin was concerned about the evolution of sexual differences other than those of reproductive organs, which he and others referred to as secondary sexual characters (e.g. see Fig. 8.1). Darwin was convinced that natural selection alone could not bring about such differences, but instead posed an alternative selection force, sexual selection, which '... depends on the advantage which certain individuals have over others of the same sex and species solely in respect of reproduction' (1883, p. 209, 3rd edn). Sexual selection, Darwin suggested, can favour the evolution of traits useful in combat between members of the same sex, or of traits that increase the attractiveness of individuals to members of the opposite sex. It is the latter subject I treat here — sexual selection and mate choice.

Darwin's simple enquiry has given rise to one of the most active disciplines in behavioural ecology and evolutionary biology (Gross, 1994). This field has attracted the attention of behavioural ecologists as well as those working in population genetics theory, parasitology, neurobiology, molecular genetics, phylogenetics and artificial intelligence. Indeed, one of the more recent developments in this field is its increasingly integrative texture (see Chapter 1).

A central concern in sexual selection is understanding how female mating preferences evolve. In many instances it is clear, as a female's mate choice influences her immediate reproductive success, i.e. her fecundity. Other times, however, the logic of female mate choice is more opaque. This is especially true in lek or lek-like mating systems. In these systems males aggregate when advertising for females, females are unimpeded in their choice of mates, and males are thought to make no contribution to the females' effort besides their sperm. Since males make no contribution to parenting, it had appeared that the choice of a mate would not influence a female's fecundity. Thus, the paradox of the lek — how can female mating preferences evolve if there is no effect of preference on female fitness (Kirkpatrick & Ryan, 1991; Ritchie, 1996)?

Until recently, much of the debate in sexual selection has centred on two hypotheses, runaway sexual selection and good genes, as the only alternatives for the evolution of female mating preferences in lek-like mating systems. Closer attention to the natural history of sexual selection, and a better understanding of the neural mechanisms underlying preferences, however, have both shown that selection can have direct, although previously unappreciated effects on preference evolution. Furthermore, phylogenetic reconstruction, laboratory experiments that probe preferences with novel stimuli and the use of neural networks to model preferences suggest that pre-existing or hidden preferences might play an important role in generating selection on male display traits. Finally, rigorous field studies, selection experiments and investigation of molecular recognition systems have all combined to provide strong empirical support for the role of good genes in the evolution of female preferences.

The field of sexual selection has grown so large we can review neither its breadth nor depth. Fortunately, Andersson (1994) has recently published an excellent, detailed review of most aspects of the field in his *Sexual Selection*. Any serious student of sexual selection must read this book. Also, the historical development of sexual selection and the debate surrounding female choice are recently chronicled in *The Ant and the Peacock* (Cronin, 1991).

8.2 Sexual selection by female mate choice

Although the role of female mate choice in the evolution of male traits was controversial (Wallace, 1889; Noble & Bradley, 1933; Huxley, 1938; Trivers, 1972; Cronin, 1991), a large number of studies in the past two decades have demonstrated that female mate choice is influenced by variation in male traits (Ryan & Keddy-Hector, 1992; Andersson, 1994). Traits as diverse as the elaborate trains of peacocks (Petrie, 1994), the hyperdeveloped tail fins forming the sword of swordtails (Basolo, 1990a), the intricately constructed bowers of bower birds (Borgia, 1985), the huge eye spans in stalk-eyed flies (Wilkinson & Reillo, 1994; Fig. 8.1), the complex acoustic repertoires of songbirds (Searcy, 1992) and túngara frogs (Ryan, 1985), and pheromones of moths (Eisner & Meinwald, 1995) are all traits that females attend to in deciding their future mates. Once the phenomenon of female mate choice became firmly established, however, there immediately raged debates as to the forces that have led to the origin and maintenance of such preferences, especially when these preferences are for traits that increase male mortality. For many years, much of the debate contrasted Fisher's (1930) theory of runaway sexual selection with various manifestations of 'good genes' hypotheses. There has been no final resolution, and additional hypotheses have further complicated matters.

As a preview to a more detailed discussion of sexual selection and mate choice, I will summarize some of the major components of the phenomenon in the context of the different approaches that have been used to study it (Fig. 8.2).

Fig. 8.1 A lek of stalk-eyed flies. A male (top) and a group of three female stalk-eyed flies roosting on a root hair. There is clear sexual dimorphism in eye span. (From Wilkinson & Reillo, 1994.)

Functional approaches are concerned with the behavioural interactions that bring about mate choice and how this generates selection on male traits and female preferences. Female preferences can guide mate choice towards males with certain traits and thus generate direct selection on these traits (Fig. 8.2(i)). If there is a genetically heritable component to trait variation, there will be an evolutionary increase in the preferred trait (Fig. 8.2(ii)). Female preferences can also be under direct selection if the preference increases the female's reproductive success either by increasing her survivorship or fecundity (Fig. 8.2(iii)). This could result directly from effects of mate choice, or from pleiotropic effects, such as sensory adaptations for predator avoidance or prey detection, which can then affect mating preferences. Alternatively, preferences can evolve by indirect selection because they are genetically correlated with the male trait under direct selection (Fig. 8.2(iv)).

Mechanistic studies attempt to identify the physiological processes underlying mating preferences (Fig. 8.2(v)). Many of these studies have concentrated on the neural mechanisms, including peripheral, central and cognitive processes, that guide female mating preferences. Although physiological processes might regulate preferences they probably serve other functions as well. Thus, certain biases in sensory systems can be favoured by selection outside of the context

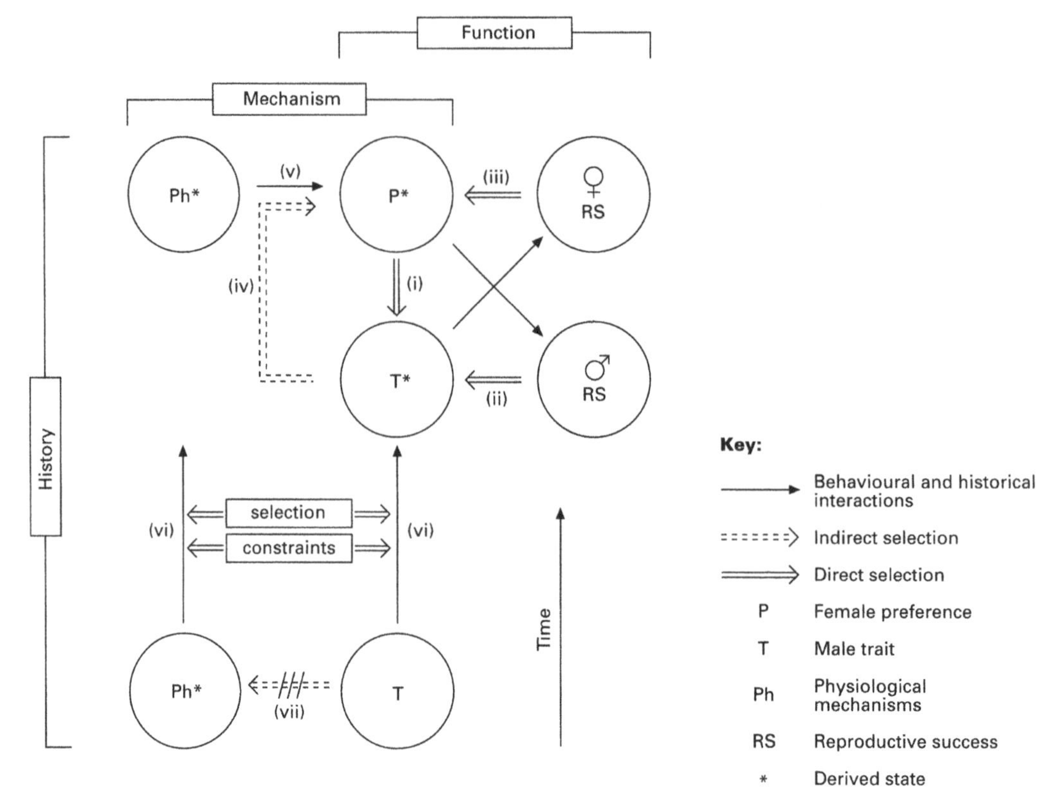

Fig. 8.2 Some components of sexual selection and approaches to their study. Roman numerals refer to explanations in text.

of mate choice because they increase female survivorship and thus reproductive success (e.g. Fig. 8.2(iii)), but these biases will also have incidental influences on mating preferences (Fig. 8.2(i)).

Historical approaches acknowledge that extant traits and preferences have experienced a long history of selection and constraints that influence their current expression (Fig. 8.2(vi)). By attempting to reconstruct the historical pattern by which traits and preferences have evolved, these studies strive to determine if preferences and traits coevolve due to genetic correlations or if traits evolve to exploit pre-existing sensory biases. The latter case is illustrated in Fig. 8.2(vii), in which the physiological mechanisms responsible for generating the preference existed prior to the evolution of the preferred trait. In such a case the sensory biases did not coevolve with the trait.

8.3 The null model for the evolution of female mating preferences

Kirkpatrick and Ryan (1991) reviewed the theoretical models for the evolution

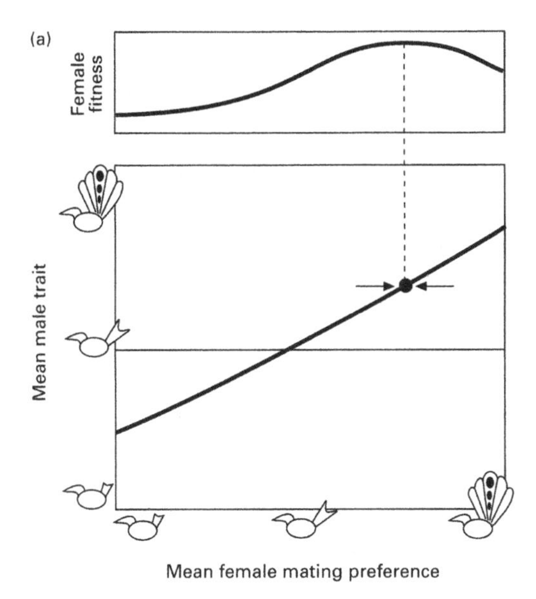

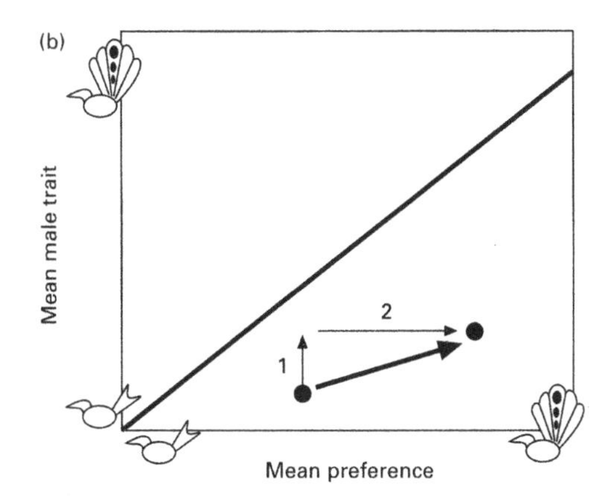

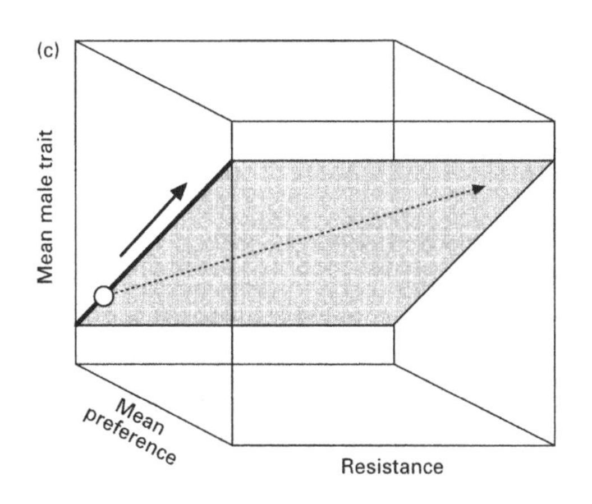

Fig. 8.3 Three hypotheses for the evolution of female mating preferences. Under direct selection (a), mating preference genes also affect female survival or fecundity. This determines the equilibrium for the preference and, through it, the male trait. In a runaway process (b), the equilibrium curve becomes unstable. Selection on the male trait (1) also causes the evolution of the female preference through a genetic correlation (2) with the result that the trait and the preference are exaggerated indefinitely as they evolve away from the equilibrium curve. In the parasite hypothesis (c) a genetic correlation is established between resistance to parasites and female preferences for a male display trait. Directional evolution of resistance (dashed arrow) results in evolution of the preference, and this causes the preference and trait to coevolve towards greater exaggeration along the equilibrium curve (solid arrow).

of female mating preferences, and that general outline is followed here (Fig. 8.3). The assumptions of these models include that fitness is measured as the number of offspring, and traits and preferences are heritable and thus evolve in response to selection. In the simplest or null model, female preferences bias mate choice towards males with certain traits. Preferences are often for traits that reduce male survivorship, such as the complex calls of túngara frogs which increase predation by the frog-eating bat (Ryan, 1985). But, preferences can counter this natural selection cost by increasing the mating success of males bearing these traits. In this null model the male trait evolves to some compromise between the different optima favoured by natural and sexual selection; that is, a compromise between maximizing a male's own survivorship and his attractiveness to females. Assuming a constant natural selection cost to the male trait, there will be a line of equilibrium that predicts the male trait which evolves when a given preference is exhibited in a population. This line represents the balance between natural and sexual selection and is a crucial component of all the models of preference evolution (Fig. 8.3). The null model assumes, as did Darwin (1871), that the female preference is a constant which predicts trait evolution; neither the null model nor Darwin addresses the evolution of these preferences. A large number of sexual selection studies now address that question.

A female's preference might influence her fecundity, and thus her fitness (Fig. 8.3). If so, one would expect the preference to evolve under direct selection. Alternatively, there could be no influence of mate choice on fecundity, but genetic variation in the male trait and the female preference might be correlated statistically, i.e. in linkage disequilibrium. Such a situation would allow both characters to evolve as a correlated response to the evolution of the character with which it is correlated. Since the trait is under direct selection by the preference, genetic correlations between trait and preference would cause preferences to evolve under indirect selection. Considering how female preferences can evolve under direct and indirect selection is a useful way to proceed.

8.4 Direct selection on female mating preferences

8.4.1 Fecundity effects as a direct consequence of mate choice

Given the debate surrounding the evolution of female choice, it is easy to miss the consensus for the controversy. In many cases a female's mate choice has an immediate effect on her reproductive success; when males offer parental care, defend young or feed their mates, prudent mate choice can be rewarded by increased fecundity. In some insects, for example, females feed on the male's spermatophore, and female choice of males with superior spermatophore quantity and quality have increased fecundity (Thornhill, 1976, 1980; Gwynne, 1984; Wedell, 1994). In resource-based mating systems, direct selection on preferences is expected to be the rule rather than the exception. When under direct selection, we expect the female preference to evolve to an optimum that maximizes her fecundity (Fig. 8.3a).

The paradox of the lek is that a female's choice of mates does not influence her fecundity, but female mating preferences have evolved under these circumstances. Recent studies have shown, however, that this crucial assumption about leks might not be true. Mate choice on the lek can have an immediate effect on female fecundity, thus preferences might be under direct selection, although in subtle ways. These cases are especially interesting because they suggest that often the lek might not be a paradox, only that researchers have not examined it closely enough.

One set of examples suggesting direct selection on mating preferences in leks involves variation in the male's ability to fertilize the female's complete compliment of eggs. Robertson (1990) and Bourne (1993) both offer compelling evidence for two frog species (*Uperolia laevigata*, and *Ololygon rubra*, respectively) that females select mates of a size that maximizes fertility rates. This is probably due to the size difference between the male and female affecting the mechanical efficiency of external fertilization.

Transmission of parasites or venereal diseases could also generate direct selection on mating preferences if male phenotype reveals parasite load or disease condition. A number of studies addressing the good genes–parasite hypothesis (see Section 8.4) have shown that females can sometimes assess male parasite load (e.g. Møller, 1990a; Milinski & Bakker, 1990), but fewer studies have shown that females actually increase their reproductive success by avoiding parasitized males. One such example is the bush cricket (*Requena verticalis*). Simmons *et al.* (1994) showed that the male's ability to donate nutrients through the spermatophore was influenced by the presence of gut parasites and, in turn, females mating with more parasitized males had reduced fecundity.

Even resource-based mating systems can generate selection on preferences in subtle ways. Eisner and colleagues (reviewed in Eisner & Meinwald, 1995) have shown that in a moth, *Utethasia*, both adults and eggs are protected against predation by pyrrolizidine alkaloids (PAs). These chemicals are not synthesized by the moth but instead are sequestered from the plants it feeds upon as a larva. When males mate with females they transfer some of their PAs to the female via the spermatophore, and these alkaloids are eventually transmitted to the eggs. Males use pheromones to attract females, the pheromone is derived from these alkaloids, and the presence of this pheromone is necessary for mate attraction. Thus, males advertise their alkaloid load to females and, it is speculated, female choice might be influenced by alkaloid levels. Although there is yet no evidence that females actively choose mates, there is some evidence that multiply inseminated females preferentially utilize the sperm of large males.

The cost of searching for a mate on the lek is an additional opportunity for direct selection on mating preferences (Parker, 1983). In this case, female survivorship rather than fecundity is affected (see Fig. 8.2(iii)). Some studies have shown that females assess mates in a manner consistent with attention to search costs. Female pied flycatchers are more likely to mate with preferred (as shown in other studies) males when males are more clumped in space (Alatalo *et al.*, 1988), and female crickets show increased choosiness in an arena when there are more refugia from predators (Hedrick & Dill, 1993).

8.4.2 Pleiotropic effects on mating preferences

A mate preference can evolve under direct selection, but selection might not be involved in mate choice. Pleiotropy can have important effects on female mate preferences, and sensory systems used in mate choice seem to manifest such effects sometimes. Sensory systems have undoubtedly evolved under both selection and constraints that can be independent of mate choice (see Chapter 2). In some cases, sensory systems might be used to locate food, assess habitat potential, detect predators and migrate. Selection to enhance performance of any of these functions could have effects on female preference for stimuli associated with mate choice that are best viewed as incidental consequences rather than evolved functions (Williams, 1966). Although such concepts have predecessors in ethology (Ryan, 1990, 1994), the interaction of selection forces and constraints that influence the performance of the sensory system in different situations has recently been termed sensory traps (West-Eberhard, 1979), sensory bias (Endler, 1992), and also has been discussed in the context of 'receiver psychology' (Guilford & Dawkins, 1993). All of these hypotheses recognize the variety of factors that influence sensory systems and emphasize that sensory perceptions implicated in mate choice, or in other functions, might be influenced by sensory biases that evolved in other contexts (see also Kirkpatrick, 1982). The specific hypothesis that males evolve traits to exploit such pre-existing biases or preferences is known as sensory exploitation (Ryan, 1990).

An example of mating preferences being influenced by selection in contexts besides mate choice is perhaps best illustrated by water mites. They locate small prey items by detecting water–surface vibrations, but this results in the females being mated by males that mimic these vibrations (Proctor, 1991). This mating preference for males producing vibrations is an incidental consequence of selection for prey recognition (see also Chapter 7). It has been suggested that selection on anolis lizards (Fleishman, 1992) to be sensitive to patterns of prey movement, and selection on fiddler crabs to detect predators (Christy & Salmon, 1991; Christy, 1995) biases the traits of males that successfully attract females. General properties of nervous systems might further affect mating preferences. For example, sensory systems can habituate to repeated stimulation, and some have suggested that songbirds have evolved complex song repertoires to release both male (Hartshorne, 1956) and female (Searcy, 1992) receivers from such habituation.

Mate choice for the correct species can also have incidental effects on mate choice within the species. Reproductive character displacement has been proposed to explain the evolution of enhanced specific differences in male courtship in areas of sympatry relative to areas of allopatry. Preferences as well as traits, however, could exhibit character displacement. Recently, Gerhardt (1994) argued that female mating preferences have been displaced in sympatry in grey treefrogs. Such an effect should reduce the error rate of females mating with heterospecifics but, as Gerhardt showed, this can have an incidental effect of generating sexual selection on conspecific calls.

8.4.3 Historical studies of sensory biases and sensory exploitation

The sensory exploitation hypothesis predicts that the evolution of sexually selected traits is influenced by these pre-existing sensory biases. This generates a prediction about the historical pattern of trait–preference evolution — preferences evolved prior to the sexually selected traits — that is distinct from the prediction of good genes and runaway sexual selection in which the preference and trait evolve in concert (see Figs 8.2(vi, vii) & 8.3c). To the extent that this historical pattern can be reconstructed, one can discriminate between these two sets of hypotheses: sensory exploitation versus indirect selection (good genes and runaway; Ryan, 1990, 1996; Shaw, 1995), although these two processes could interact (see Section 8.7).

Several studies offer strong support for the sensory exploitation hypothesis. The túngara frog, *Physalaemus pustulosus*, produces a call consisting of a whine and a chuck. The whine is always present and is necessary and sufficient to elicit mate recognition from females. The chuck is not always produced, but when added to the whine it increases the attractiveness of the call to females as well as to frog-eating bats (Ryan, 1985). This species is a member of the *P. pustulosus* species group which contains six species, three constitute one monophyletic group on the western side of the Andes and three another monophyletic group in Amazonia and Middle America (Fig. 8.4). In the Amazonian–Middle American clade, *P. pustulosus* and an undescribed species that is sister to *P. petersi* add suffixes, while *P. petersi* does not. It is not known if these suffixes are homologous; it is equally parsimonious to assume either that these suffixes were derived independently in the two species, or that the suffix was derived once at the base of this clade and subsequently lost in *P. petersi*. But, any reasonable interpretation suggests the ability to add the suffix was derived after the divergence of the two clades within the species group.

Male *P. coloradorum* do not produce chucks, but when chucks are digitally added to their calls females prefer the conspecific call with chucks to the normal conspecific whine. The most parsimonious explanation for the occurrence of a chuck preference in both clades within the species group is that the preference is shared from a common ancestor, which would have had to exist prior to the

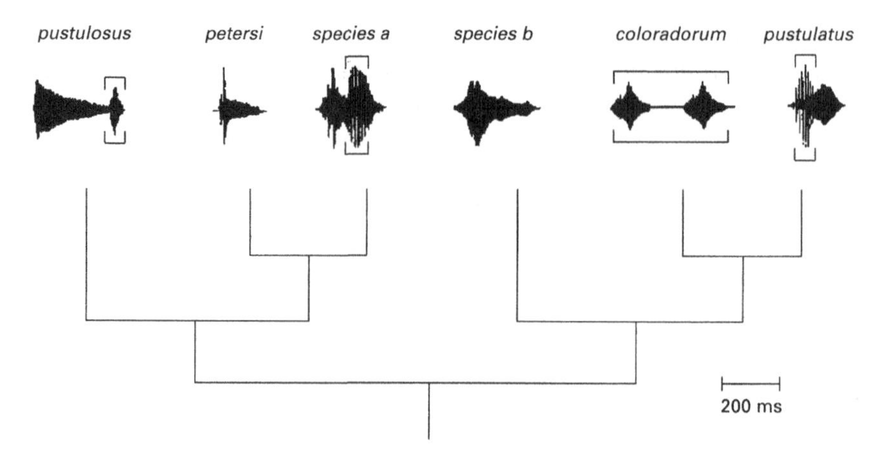

Fig. 8.4 Phylogenetic relationships of frogs of the *P. pustulosus* species complex showing the advertisement calls of each species. Call attributes unique to each taxa are indicated in brackets. (From Ryan & Rand, 1993, 1995.) This phylogeny, based on morphology, allozymes and deoxyribonucleic acid (DNA) sequence of the 12S mitochondrial gene (Ryan & Rand, 1995), differs in some details from a previous preliminary phylogeny (Ryan & Rand, 1993), although the interpretations are the same.

divergence of the two clades (Ryan & Rand, 1993). If so, the preference for chucks existed prior to the chucks, and thus males that evolved chucks were favoured by a pre-existing preference. There also are call components that occur in the western Andes clade that are absent in the other clade: a strongly amplitude-modulated prefix in *P. pustulatus* and the ability to rapidly produce calls in groups of two or three in *P. coloradorum* (Fig. 8.4). Although male *P. pustulosus* lack these call components, females prefer conspecific calls to which these components are added, offering additional evidence for the role of sensory exploitation. Furthermore, the sensory tuning that results in female túngara frogs preferring lower frequency calls of larger males did not evolve to guide female túngara frogs to larger males but, instead, appears to be an ancestral trait (Ryan *et al.*, 1990a), and even 'species-recognition' preferences are strongly influenced by what are estimated to be calls of ancestral species (Ryan & Rand, 1995).

The series of studies by Basolo on pre-existing preferences for swords in fishes also offers strong support for the sensory exploitation hypothesis. Swordtails (genus *Xiphophorus*) are characterized by an extension of several lower rays of the caudal fin which gives the appearance of a sword. Female *X. helleri* prefer males with longer tails (Basolo, 1990a). Platyfish are members of the same genus but tend to lack elaboration of the caudal rays. When artificial swords are appended to *X. maculatus* and to *X. variatus* (Basolo, 1990b, 1995a) females preferred the artificially adorned conspecific males. Since swordtails were thought to be monophyletic, Basolo reasoned that the shared preference for swords between both swordtails and platys resulted from a pre-existing bias for swords. Meyer *et al.* (1994) challenged this interpretation when their molecular phylogeny of the genus suggested that swordtails were not monophyletic and thus the sword evolved more than once. But, this criticism now seems irrelevant for two reasons. Basolo (1995b) showed preference for swords in the closely related but swordless genus *Priapella*, and a recent molecular phylogeny by Borowsky *et al.* (1995) supports the classical interpretation of the monophyly of swordtails and rejects the hypothesis offered by Meyer *et al.* (1994).

Although there are other cases in which a phylogenetic interpretation suggests pre-existing preferences for sexually selected traits (Ryan & Wagner, 1987; Proctor, 1992; Christy, 1995) the phenomenon is certainly not universal. For example, Hill (1991) had shown that female preferences for plumage coloration in house finches is under direct selection as brighter males are better fathers. (Hill also speculated that females might gain a genetic benefit for their offspring but this was not documented.) There is population variation in the size of a patch of ventral coloration; a population phylogeny suggests that the larger patch size is ancestral, but this reduction in patch size is independent of any change in preference for larger patch size and brighter males (Hill, 1994). This is similar to a situation in swordtails (*Xiphophorus*). Female *X. nigrensis* and *X. multilineatus* prefer the large-size genotype of males (Zimmerer & Kallman, 1989; Ryan *et al.*, 1990b); large size has been lost in *X. pygmaeus* but females retain the preference for large size (Ryan & Wagner, 1987). In both the finches and swordtails the loss or reduction of a sexually selected trait probably results from increased natural selection against the trait or through genetic drift. In both of these cases the stage would be set for sensory exploitation; if males were to evolve the lost trait again it would be favoured now by the pre-existing preference; of course, natural selection against the trait could still be sufficiently strong to keep it from evolving.

Enquist and Arak (1993) argued that general properties of learning and natural selection could result in 'hidden preferences' which could give rise to sensory exploitation. They used neural net models in which the network was trained to recognize a 'species-specific' shape. After the model was trained, subsequent simulations revealed a variety of shapes with which the model had no previous experience that were equally or more attractive than the initial shape (but, see Dawkins & Guilford, 1995). Staddon (1975) and Weary *et al.* (1993) also argued that a general property of some kinds of learning might result in greater preferences for unknown stimuli due to a phenomenon called 'peak shift learning'.

Direct selection on female mating preferences in resource-based mating systems has never been controversial and had tended to be overshadowed by the paradox of the lek. A growing number of studies, however, now illustrate that direct selection can be more subtle than previously thought, involving fertility effects, avoidance of sexually transmitted parasites and diseases, cross-generational transmission of chemical resources, search costs and pleiotropic effects of sensory biases. Unlike good genes and runaway sexual selection, there has never been any debate as to whether direct selection could operate. There is a consensus among theoreticians that it can, and there now appears to be a growing body of empirical evidence that it probably does in many situations not previously appreciated.

8.5 Indirect selection on mating preferences: Fisher's runaway sexual selection

A lack of direct selection on a character does not exclude that character from evolving. Besides stochastic processes such as drift, characters can also evolve under indirect selection if they are correlated with traits under direct selection.

Fisher's (1930) brief discussion of runaway sexual selection is notorious for both its insight and impenetrability. O'Donald (1962), Lande (1981) and Kirkpatrick (1982) presented formal models of Fisher's theory that clearly elucidated some of its basic principles. Again, the question is how do female preferences evolve if the preferences do not result in variation in female fecundity? All of the models of runaway assume that there is heritable variation in both traits and preferences, and assortative mating generates a statistical association between trait and preference. Initially, the assortative mating could come about by chance or, as Fisher suggested, by a fecundity advantage to females exercising the preference. In a simple example (Arnold, 1983), assume that males have short or long tails and females either prefer long tails or ignore tail length when choosing their mates (see Fig. 8.3b). Individuals have genes for both the trait and the preference but only express the character appropriate for their sex. After an episode of mating, the relative mating success of long-tailed males will be greater than that of short-tailed males due to the female preference for long tails. Furthermore, alleles that determine tail length and preference for long tails will be more likely to be found together in the same offspring than other allelic combinations of the male trait and female preference; assortative mating results in a genetic correlation of trait and preference. Thus, as the frequency of tail length evolves due to female preference for longer tails, so will the preference itself evolve as a correlated response to evolution of the male trait. The stronger the preference, the greater the evolution of the trait and, through linkage disequilibrium, the greater the evolution of the preference (see Fig. 8.3). This is why the term 'self-reinforcing choice' is often used to describe the runaway process (Maynard Smith, 1978). The term runaway should be reserved for this specific model of trait-preference evolution, and should not be used to merely define the elaboration of male traits.

Fisher's theory of runaway sexual selection has been a difficult one to evaluate empirically. Ideally, empirical support would come from data showing that male traits evolve under the opposing forces of natural and sexual selection and this results in the correlated evolution of the female preference (see Figs 8.2(iv) & 8.3b).

There is no empirical support for Fisher's hypothesis of runaway sexual selection. This might be because the runaway process is very rapid and thus unlikely to be observed. Also, in many of the systems in which sexual selection by female choice has been well documented, it is not feasible to conduct the experiments necessary to validate runaway. In many cases, runaway sexual selection is merely considered the null hypothesis and is accepted if a good genes effect can not be demonstrated (see above; Møller & Pomiankowski, 1993b); this should not be considered adequate proof of runaway because runaway and good genes are not the only forces that can influence female preferences.

8.5.1 Genetic correlation of trait and preference

Most empirical studies that have attempted to assess Fisher's theory have tested the assumption of a genetic correlation between trait and preference, either by examining patterns of geographical variation in trait and preference or selecting on the trait and measuring a correlated evolutionary response in the preference. Guppies are one of the most polymorphic vertebrates and much of this variation can be partitioned among high- and low-predator populations (Endler, 1983). Stoner and Breden (1988) (see also Breden & Stoner, 1987), Long and Houde (1989), Houde and Endler (1990) and Endler and Houde (1995) all showed that there is a correlation between the average amount of orange in males in the population and the strength of female preference for orange. Houde (1993) has pointed out that other processes besides runaway sexual selection can result in spatial covariation of trait and preference; for example, habitat differences that favour transmission of different light spectra. Nevertheless, these studies of geographical variation are at least consistent with a prediction of runaway.

Selection experiments have also been used to demonstrate a genetic correlation between trait and preference, and some have shown that evolution of a trait can result in an evolutionary response in the preference (see Fig. 8.2(iv)). The bright red nuptial coloration of male sticklebacks has been part of ethological lore since Tinbergen's (1953) account of red postal vans acting as sign stimuli releasing male aggression. Milinski and Bakker (1990) documented female preference for redder males in sticklebacks, employing some clever experiments in which the male's apparent nuptial coloration was manipulated by changing ambient light conditions. There is considerable geographical variation in the intensity of the male's nuptial coloration (McLennan & McPhail, 1990; Milinski & Bakker, 1990), and this variation in male ornament might be correlated to ornament preference. Bakker (1993) crossed wild-caught male sticklebacks whose nuptial coloration was either intense red or dull. The sons' intensity of red was correlated with that of their fathers'. Daughters of red males preferred red over dull males, while daughters of dull males tended to show no preference for nuptial coloration. Thus, there was a positive genetic correlation between trait and preference.

A genetic correlation between trait and preference was also shown in studies of stalk-eyed flies (see Fig. 8.1). These small flies have huge eye spans that can exceed their body length. They form nocturnal roosts which consist of a single male and several females perched on root hairs. Males with relatively larger eye spans are accompanied by more females, suggesting the action of female choice on eye span. Wilkinson and Reillo (1994) conducted bidirectional selection experiments on eye span. After 13 generations, females from long eye-span lines and from unselected lines both preferred long eye-span males in mating experiments, but females from the short eye-span lines preferred males with short eye spans.

8.6 Indirect selection on female mating preferences: good genes hypotheses

An alternative to the arbitrary female preferences postulated by Fisher is a more utilitarian function of female choice. Zahavi's (1975) handicap principle popularized the notion that females prefer to mate with males who have demonstrated their superior genetic quality for survivorship. Females can evaluate a male's survival ability, Zahavi suggested, by assessing the magnitude of the handicap with which he is able to survive. There is no disagreement that many sexually dimorphic traits can increase male mortality, and Zahavi suggested that these handicaps to survival evolve as honest signals, allowing females to assess male genetic quality; thus, the term female choice for 'good genes'.

There are several variations of Zahavi's original handicap proposal (Maynard Smith, 1991; Harvey & Bradbury, 1991). In the original model, the handicap is fixed and it acts as a filter to survival. If a male has a large handicap and has survived, he is likely to have high intrinsic survival ability (as opposed to merely good luck). Males without handicaps cannot be judged, and thus are assumed to be ignored by females. A more plausible variant of the original handicap model is the condition-dependent handicap. Here, the investment in the handicap varies with the male's condition such that he is optimizing the trade-off between mate attraction and survivorship to maximize his reproductive success.

When Zahavi first suggested the handicap model it was enthusiastically embraced by many field biologists but harshly criticized by most population geneticists. Early population genetics models were unable to demonstrate the spread of alleles for preferences under the original formulation of Zahavi's model. At the risk of oversimplifying a quite substantial, technically complicated and emotionally charged area of research, it appears that good genes hypotheses for the evolution of female preferences gained wide support when it was shown that preferences could evolve if they were genetically correlated to the good gene (Pomiankowski, 1988; Grafen, 1990).

The good genes hypothesis, like Fisher's theory of runaway sexual selection, also relies on a genetic correlation, but in this case between the preference and the male's 'good genes', which is taken to mean some heritable component for viability (see Figs 8.2(iv) & 8.3c). Many studies that have alleged to demonstrate female choice for good genes do so based merely on the logic that if females prefer traits that are costly to male survival, these traits are both handicaps and reliable indicators to females of a male's genetic constitution for survival. Costly displays by themselves should not be taken as evidence for female choice for good genes. Most male ornaments are used in communication. A purpose of any signal is to increase the conspicuousness of the signaller against background noise (see Chapter 7). Conspicuousness often incurs an immediate viability cost resulting, for example, from unintended advertisement to predators and parasites. Furthermore, all signals involve physiological costs, either in the growth of a structure used as a signal or in displaying the signal — it costs to grow a long tail and to make a loud call. Since mate choice involves communication it involves costs. The existence of a costly ornament is hardly serious evidence that female preferences have evolved to assess a male's genes for survival.

Testing the hypothesis of good genes requires documenting the effect of a female's mating preference on offspring viability. Ideally, such a study would demonstrate a genetic correlation between a female preference and male genes that influence viability, evolution of the viability genes and correlated evolution in the female preference. This is a tall order that is clearly not possible in most systems. There have been a number of studies recently, however, that have documented a relationship between female mating preferences and offspring viability that appears to result from a paternal genetic effect.

The focus of most good genes studies can be classified into four categories: (i) the general relationship between ornament elaboration and offspring viability; (ii) the hypothesis of female choice for parasite resistance genes; (iii) female choice of trait asymmetry as an indicator of genetic quality; and (iv) choice for genetic complementarity. The first and fourth categories offer some of the most compelling evidence for good genes.

The great tits of Wytham Woods near Oxford have been cooperative subjects of a large number of studies in behavioural ecology. Norris (1993) took advantage of this well-known system to gather some strong empirical evidence for the operation of good genes. Female tits prefer to mate with males having larger black breast stripes. Cross-fostering experiments revealed that stripe size was heritable and there was a significant relationship between stripe size of the putative father and the proportion of male offspring surviving from a brood; no such relationship existed between stripe size of the foster father and survivorship. Surprisingly, the survivorship benefits were not shared by female offspring. Møller (1994) also showed that offspring viability was correlated to the degree of male ornamentation in barn swallows. His study was conducted with a free-ranging population and the lack of any influence of parental care on offspring survival could only be argued on statistical grounds and thus is somewhat less compelling than if this variable were experimentally controlled.

A potential problem with the studies of tits and barn swallows is that high-quality females might have had preferential access to the more attractive males, and thus viability differences among offspring might have been due to maternal rather than paternal effects, although Norris and Møller were unable to detect assortative mating. Petrie (1994) controlled for maternal effects in her study of mate choice in peacocks. Previous studies had shown that the number of eye-spots in the train predicted a male's mating success, and experimental reduction of spot number was shown to influence male mating success between years (Petrie *et al.*, 1991). In her study, Petrie randomly assigned females their mates, offspring were raised under common conditions and then were released in a park, alleged to reflect almost natural conditions. Male attraction was assessed as the mean area of the father's eye-spot (studies have not shown that this character per se was under selection). There was a significant association between this measure of male attractiveness and size of offspring at 84 days, and also with the survivorship of a sample of these offspring after 24 months. These studies of tits and peacocks offer some of the strongest evidence, to date, that females obtain heritable viability benefits for their offspring through mate choice.

Other studies of good genes have not fared so well. For example, many species exhibit a large-male mating advantage (Ryan & Keddy-Hector, 1992; Andersson, 1994). In ectothermic vertebrates, growth in males is often indeterminate, thus larger males might be older and better able to accrue resources. Howard *et al.* (1994) tested the hypothesis that larger male toads (*Bufo americanus*), which are preferred by females, sire offspring with superior larval characteristics — age and size at metamorphosis. Although both traits showed significant heritability among sires, there was no effect of sire size on either age or size of their offspring at metamorphosis. Toads might have good genes but females are not selecting them.

Other tests of the good genes hypothesis have concerned more specific effects. Hamilton and Zuk (1982) suggested that bright plumage in birds indicates genetic resistance to parasites and that by preferring brighter males females increase offspring viability due to this inherited resistance. Furthermore, coevolutionary cycles of hosts and parasites should maintain genetic variation for fitness. Initial tests of this hypothesis used interspecific comparisons, testing the prediction that plumage dimorphism (and song complexity) should be greater in species with higher parasite loads. The results were equivocal; some data sets initially supporting the hypothesis did not properly control for phylogenetic effects and failed to support the hypothesis once proper phylogenetic controls were used (e.g. Read & Harvey, 1989; reviewed by Møller, 1990b; Kirkpatrick & Ryan, 1991).

Many studies of single species have demonstrated or suggested that male parasite loads can influence female mating preferences (e.g. Clayton, 1990;

Houde & Torio, 1992; Zuk *et al.*, 1993). Since parasite load can have a variety of effects on male morphology, physiology and behaviour, it is hardly surprising that parasites influence a male's ability to attract mates; in the most extreme, females do not mate with dead males. Furthermore, female preference for less parasitized males could be under direct selection, as discussed above. Studies have not shown that female preference for less parasitized males increases offspring survivorship, but Møller (1990a) has conducted a series of elegant studies with barn swallows strongly suggesting such an effect. Previous studies had shown that females prefer males with longer tails (Møller, 1988). In nature, males with longer tails have fewer parasites. In cross-fostering experiments the number of mites of a putative male parent is correlated with the number of mites on his offspring, whether the offspring are raised by him or by a foster father. There was no correlation between mites on the male parent and other 'unrelated' offspring in his nest. Furthermore, by artificially manipulating mite loads, Møller (1990a) showed a detrimental effect of mites on growth rate. This study supports some of the major assumptions of the Hamilton–Zuk hypothesis: parasites are detrimental, parasite resistance is heritable and parasite load is correlated to traits preferred by females.

Plumage brightness might indicate the presence of one type of good gene, parasite resistance, but some researchers have suggested that a single phenotypic measure might indicate overall genetic quality. Fluctuating asymmetry (FA) is the deviation from symmetry in otherwise bilaterally symmetrical traits. Soulé (1982; Soulé & Cuzin-Roudy, 1982) theorized that FA is a measure of an individual's ability to develop in the face of environmental and genetic stresses. Thus, it has been suggested that FAs might be reliable indicators used by females to assess a male's overall genetic quality (Møller, 1993; Møller & Pomiankowski, 1993a; Watson & Thornhill, 1994). Female mating preferences are sometimes correlated with male FA. In barn swallows, tail symmetry and tail length interact in influencing male attractiveness, and parasites can increase tail FA (Møller, 1992). Scorpion flies are attracted to a pheromone, some quality or quantity of which is correlated with FAs of various morphological traits (Thornhill, 1992). Such a correlation between attractive traits and morphological FAs did not hold in a study of mating calls and FAs in cricket frogs (Ryan *et al.*, 1995).

If FAs are indicative of male quality then one would expect this information to be especially reliable in sexually selected traits. Møller and Höglund (1991) compared patterns of sexually selected characters with other morphological characters. Sexually selected characters tended to have higher levels of FA. Also, sexually selected characters showed a negative relationship between character size and FA, e.g. longer tails were more symmetrical. This differed from the relationship in other characters in which average size was more symmetrical and asymmetry increased with both positive and negative departures from average. The differences between sexually selected traits and other traits is taken to suggest that FAs are reliable indicators of genetic quality.

The studies cited above show that there can be preference for symmetry, but that result per se might not be unexpected. It is possible that there is a sensory bias to prefer more symmetrical structures (Jennions & Oakes, 1994; Ryan *et al.*, 1995). Symmetry is one of the 'laws' of Gestalt perception in humans, and symmetrical patterns are more likely to be identified accurately than asymmetrical ones (Pomerantz & Kubovy, 1986). The bias toward symmetry might be even more general. In a pair of studies in which neural networks were trained for arbitrary pattern recognition, there emerged an overall preference for symmetry (Johnstone, 1994; Enquist & Arak, 1994; but, see Dawkins & Guilford, 1995, for a cautionary note). Furthermore, no studies showed either that the preference for more symmetrical mates is under direct selection due to a positive influence on female fecundity, or that it is under indirect selection via good genes due to more symmetrical males fathering more viable offspring. Møller *et al.* (1995) have shown that fluctuating asymmetry in human breasts is correlated to fecundity — women with more symmetrical breasts are more likely to have children. But, there is only anecdotal evidence that breast symmetry influences male choice of mates.

Most studies of good genes are based on females identifying absolutely superior genotypes in males. But, a series of recent studies has examined mate preference based on genetic complementarity. The major histocompatibility complex (MHC) exhibits high levels of heterozygosity — up to 60 alleles at some loci — and this heterozygosity appears to be maintained by selection (Brown & Eklund, 1994). These genes influence immune recognition by coding for cell-surface glycoproteins. The molecules are characterized by a small cleft or basket that contains a peptide or antigen of about 10 amino acids in length. If the cell presents a 'self-peptide' in this basket it is recognized as self by T cells, but if the cell is infected by a virus a small peptide from this virus is placed in the basket and the cell is then attacked by T cells. Brown (1983) speculated that this cell recognition system might also function at the behavioural level in influencing kin-directed behaviours, including inbreeding avoidance in mating (see Chapter 4).

In 1976 Yamakazi *et al.* showed that mate choice in mice was mediated by variation in MHC; there was disassortative mating by haplotype. A series of studies has further confirmed that mice mate disassortatively by MHC haplotype and this preference is mediated by olfactory cues in the urine (reviewed in Brown & Eklund, 1994). Nor are these results restricted to the congenic laboratory strains used in the initial studies. For example, Potts *et al.* (1991) showed a 27% deficiency of homozygous MHC offspring in mice under seminatural conditions; at least some of this deficiency could be explained by females settling on territories of males with different MHC haplotypes, but not by selective fertilization or abortion. In a recent study, Potts *et al.* (1994) suggest the major advantage of disassortative mating by MHC genes was due to the fitness decline associated with inbreeding rather than resistance to infectious diseases.

The *t*-complex is another well-known, highly polymorphic genetic system in mice that has been implicated in mate choice. The allelic variants are referred to as either wild type (+) or '*t*-' haplotype (*t*); there are at least 15 *t*-haplotypes. Animals that are homozygous for the same *t*-haplotype die before birth. Animals that have two different *t*-haplotypes have more or less decreased survival and males are sterile. Lenington *et al.* (1994) showed that females mate disassortatively by haplotype and their degree of discrimination is related to the fitness advantage of disassortative mating (see Chapter 4).

These detailed studies of mate choice for genetic complementarity are possible in rodents because of the wealth of information on the molecular genetics of the MHC and *t*-complexes. Other systems, however, as diverse as cnidarians (Grossberg, 1988) and toads (Waldman *et al.*, 1992) suggest similar patterns.

8.7 Interacting forces

Direct selection, runaway selection and selection for good genes are often portrayed as mutually exclusive hypotheses for the evolution of females preferences. This need not be the case. In fact, Fisher (1930) suggested that the runaway process could be initiated by females preferring males with higher survival abilities.

There are numerous scenarios by which selection forces could interact to influence the evolution of female preferences. In most species, females have strong preferences for conspecifics over heterospecifics. This is because direct selection will favour conspecific preferences given the reduced fecundity typically resulting from hybrid matings. These conspecific mating preferences could also result in sexual selection on males. Females might prefer those males that least resemble heterospecifics, despite any genetically based differences in the viability of conspecific males that would be correlated with their resemblance to heterospecifics. This could lead to assortative mating, linkage between preference and trait genes and, ultimately, runaway sexual selection. Thus, direct selection for species recognition could generate a bout of runaway (Ryan & Rand, 1993).

Searcy (1992) provides another, more specific example of how different forces can interact to influence preference evolution. Because of inherent processes of the auditory system, such as habituation, Searcy posited that male songbirds have exploited a pre-existing preference for complicated acoustic stimuli. Thus, preference for song repertoires did not coevolve with repertoires, as would need to be the case if these preferences evolved under indirect selection, as by runaway or good genes processes. Among songbirds studied, however, there is a positive correlation between strength of preference and repertoire size. Searcy's conclusion is that song repertoires were initially favoured by a pre-existing preference for complex auditory stimulation, but larger repertoires and the preference for them coevolved through a runaway or a good genes processes.

In the above examples, different forces act sequentially on preference evolution. Some population geneticists have recently attempted to ascertain the relative strengths of these different forces when they act simultaneously. Kirkpatrick and Barton (in press) provide an equation that allows an estimate of the impact of good genes preferences on the evolution of the overall mating preferences. Evolution is quantified as the change in mean preference as measured in phenotypic standard deviations (ΔI). The parameters of interest are the square root of the heritability of the trait (h_T), the heritability of the preference ($h^2{}_P$), the phenotypic correlation between the female's preference and the trait of the male she actually mates in nature (ρ_{PT}), the correlation between the male's trait and his genetic quality for survival (r_{TW}) and a measure of the genetic variance for total fitness (G_w):

$$\Delta I = {}^1/_2 \rho_{PT} r_{TW} h_T h^2{}_P \sqrt{G_w}$$

There are no single mating systems in which there are sufficient data to solve this equation. But, some of these parameters have been measured, and these can be used to estimate the upper limits of indirect effects (e.g. Bakker, 1990; Bakker & Pomiankowski, 1995; Pomiankowski & Møller, 1995). Kirkpatrick and Barton (1996) used estimates of $\sqrt{G_w}$ (0.25), h_T (0.7) and $h^2{}_P$ (0.4) from the literature. Furthermore, they assumed that the trait was a perfect indicator of male fitness ($r_{TW} = 1.0$) and that females always mated with males that perfectly matched their preference ($\rho_{PT} = 1.0$). Given these estimates, the change in preference is only 3.5% of the preference's standard deviation per generation. This effect is small relative to the effect that other selection forces are known to have on phenotypic change (Burt, 1995). Realistically, it will be even smaller in nature since it is unlikely that the male trait will indicate perfectly his fitness and that female preferences and phenotypes of mated males are not compromised by less than perfect female discriminability or alternative male mating strategies. This preliminary analysis suggests that preferences for good genes work, but when competing with direct selection they do not work very well. This analysis also provides a tractable method for measuring the strength of indirect and direct selection on preferences in real mating systems.

Although debates about preference evolution have previously contrasted mutually exclusive hypotheses, there is little doubt that multiple forces interact in their influence on the evolution of preferences. Two challenges are to ascertain how these forces might act sequentially through a species' history and the relative importance of each force when they act simultaneously.

8.8 Multiple traits and their preferences

A limitation of most treatments of sexually selected traits is the focus on single traits. In many cases, however, males have suites of sexually dimorphic traits that appear to be under sexual selection, such as the myriad of different colours and shapes of spots than adorn male guppies (Endler & Houde, 1995). Some studies have demonstrated empirically that females do attend to more than one trait (Zuk *et al.*, 1990; Kodric-Brown, 1993; Collins *et al.*, 1994; Brooks & Caithness, 1995; Endler & Houde, 1995).

Møller and Pomiankowski (1993b) indicated that multiple sexual ornaments are more likely in lek breeding and other polygynous species of birds while single sexual ornaments seem to be the rule in monogamous species. To what extent might runaway sexual selection and good genes preferences account for these differences? Møller and Pomiankowski measured the length of sexually dimorphic feathers in both lekking and monogamous birds to estimate the degree of sexual ornamentation. They assumed that traits with larger asymmetries were honest signals of good genes, and preferences for such traits evolved for choosing good genes. Furthermore, they assumed that if a trait had a low level of FA then both the trait and the preference for the trait evolved as a result of runaway sexual selection; in that sense, runaway is the null hypothesis. Their results showed that single sexual ornaments had higher levels of FAs while multiple sexual ornaments had lower levels, suggesting that in monogamous species preferences evolved for indicators of good genes while in lekking and other polygynous species traits and preferences evolved by runaway sexual selection.

One potential problem with this interpretation of multiple trait evolution is the assumption that preferences can only evolve for good genes or by runaway sexual selection. It is known, however, that a variety of other forces such as direct selection on preferences and pleiotropic effects can influence preference evolution. Another problem is that FAs are greater in monogamous species, in which males contribute substantially to offspring survival. Thus, FAs might be more indicative of male paternal quality rather than genetic quality. If so, the preference is under direct selection.

The patterns of feather FAs and numbers of sexual ornaments found by Møller and Pomiankowski is in accord with recent theoretical predictions by Pomiankowski and Iwasa (1993; Iwasa & Pomiankowski, 1994). They modelled the evolution of preferences for multiple sexual ornaments under a good genes and a Fisherian process. Under good genes processes, multiple preferences can evolve only if the joint cost of choice based on multiple traits is small. If there is an increased cost to assessing multiple traits then only a single preference is stable. Interestingly, the establishment of a single preference can preclude the later evolution of preferences for traits that are even better indicators of male genetic quality — preference for good genes precludes preferences for better genes. The cost of choice does not have such an effect on preferences that evolve by runaway sexual selection. These theoretical results thus suggest that preferences for good genes is unlikely to explain the evolution of multiple sexual ornaments, and thus do not explain preferences in lek-like mating systems.

Both the empirical and theoretical treatments suggest a dire need for more data. Although the existence of multiple sexual ornaments suggests multiple preferences, this crucial assumption needs to be determined empirically. Multivariate statistical analysis is one approach, but a more satisfactory one is experimental manipulation of independent traits. This has always been tractable in studies of acoustic signals (e.g. Gerhardt, 1992), but recent advances in video animations show this is also feasible in visual communication systems as well (Clark & Uetz, 1993). Such studies would not only allow precise demonstration of multiple preferences, but would also show how strengths of preferences might vary as a function of traits that have been hypothesized to evolve under different types of preferences. The assumption that trait asymmetry reveals why the preference for such traits evolved is also an assumption that needs to be evaluated critically. Understanding the evolution of preferences requires study of not only traits but traits and preferences (see Chapter 7 for further discussion of the evolution of multiple signals).

8.9 Mate copying: a new horizon?

Of the newest developments in sexual selection studies, many of which could not be covered in this chapter, mate copying might be one of the more exciting. Most considerations of mate choice assume that a female's mating preference is intrinsic, either because it has a strong genetic component (Bakker & Pomiankowski, 1995) or because it is learned in some early, sensitive period of development (Marler, 1991), and that its expression is independent of the choice of others. Recent studies have now shown that mating preferences can be influenced by social cues; specifically, it has been suggested that females copy the mate choice of others (Gibson & Höglund, 1992).

Mate copying has been supported by a variety of observations of birds (Höglund *et al.*, 1990; Pruett-Jones, 1992), but it has been perhaps most clearly demonstrated in guppies. In a series of laboratory studies Dugatkin (Dugatkin & Godin, 1992, 1993; Dugatkin, 1996) has allowed a female guppy to choose between a pair of males. This focal female is then constrained in a situation in which she observes the previously unpreferred male consorting with a model female and the male she preferred without a female. When the choice test is repeated, females show a significant change in their preference, increasing the time spent with the previously unpreferred male. Dugatkin showed that younger females are more likely to copy the choice of older females, and that to some extent female copying can reverse genetically determined preferences for the amount of orange coloration.

Mate copying also can occur between species. Schlupp *et al.* (1994) studied mating preferences in a sexual–asexual species complex of mollies, the sailfin molly, *Poecilia latipinna* and the Amazon molly, *P. formosa*. The Amazon molly is an all-female gynogenetic species; it reproduces clonally but must rely on sperm from heterospecific males, in this case sailfin mollies, to initiate embryogenesis. Why should male sailfin mollies mate with Amazon mollies? The assumption is that such a mating incurs only costs and no benefits for the male. Schlupp *et al.* showed, however, that female sailfin mollies exhibit mate copying even if the female is an Amazon molly. So, by mating with Amazon females, sailfin mollies increase their attractiveness to their own females.

Studies of mate copying are in their infancy and a number of questions demand attention. Although mate coping occurs in the laboratory how common is it in nature? Why do females copy; is there an advantage to copying per se (e.g. reduced search costs) or is this a specific example of a more general social facilitation of behaviour (e.g. conspecific cueing; Stamps, 1991)? To what extent can copying reverse strong intrinsic preferences among conspecifics and between conspecifics and heterospecifics? Is there a copying 'preference function' that can evolve in response to selection? How does copying influence variance in male mating success and correlated evolution of trait and preference (e.g. see Pruett-Jones & Wade, 1990; Kirkpatrick & Dugatkin, 1994)?

8.10 Summary

Sexual selection studies have moved from the single debate over Fisher's runaway sexual selection versus selection for good genes. Direct selection had always been considered an important influence on preferences in resource-based mating systems, but this fact was not relevant to attempts to understand the extreme dimorphism found in lek-like species. Now it is clear that direct selection acts in lekking species as well, either by subtle but important influences of males on female fecundity or by direct selection on sensory systems that have pleiotropic effects on mate choice. The former effect has been revealed by more detailed appreciation of the natural history of mating systems and the latter by increasing our understanding of mechanisms guiding mate choice. Both of these avenues of research offer promise for future studies. Recent studies have probably only scratched the surface in revealing the subtle influences that mate choice has on female fecundity. Future studies should continue to investigate in great detail the variety of means by which female fecundity can be subtly affected by mate choice; parasite and disease transmission are likely to have important consequences. Also, as future studies utilize more sophisticated analyses of the sensory and cognitive processes underlying mate choice, we will have a better notion as to the degree to which selection in other contexts influences mating preferences, or to which mating preferences and other functions can be compartmentalized and thus each possibly optimized.

There is now a consensus that in theory both runaway sexual selection and good genes can work, with both relying on genetic correlations between male characters and females' preferences. There is some strong support for the genetic correlations between trait and preference, and a major advance is the demonstration that male traits can influence heritable variation for viability in offspring, either because of the male's superior or complementary genotype. An interesting new avenue of research is quantifying the relative consequences of direct and indirect selection when the two interact in preference evolution. For much of the history of sexual selection, theory and empirical studies have been conducted with little interaction. There would be great benefits derived from an empirical system that would allow, for example, the precise measurements required to estimate the interaction of direct and indirect selection on the evolution of traits and preferences. Although there have been a few empirical studies demonstrating selection for good genes, such analyses would also be facilitated by more tractable model systems. Development of such empirical systems should be a high priority.

Finally, historical approaches have introduced a new dimension in attempting to reconstruct the history of traits and preferences. These studies suggest that in some cases female preferences are broader than the phenotypes exhibited by their males, and thus represent a selection force that will act on novel male traits, either positively or negatively, as they arise. The modern phylogenetic approach (see Chapter 14) has had an important influence on much of biology and this will continue to prove true in sexual selection as well. Future studies in sexual selection, and in behavioural ecology in general, should continue to add a historical analysis using the most sophisticated analyses available. This will undoubtedly require extensive collaboration with phylogeneticists.

The major advance in sexual selection is one of approach. There is a growing recognition that mate choice is a complex phenomenon that not only can influence male and female fitness, but also is guided by sensory mechanisms whose history has been characterized by a variety of selection forces and constraints. This integration is leading to a more general appreciation of the biology of sexual selection.

Chapter 9

Sociality and Kin Selection in Insects

Andrew F.G. Bourke

9.1 Introduction

This chapter deals with what the social insects can teach us about social evolution and kin selection, since it is in these areas that social insect research has made some of its most fundamental and exciting contributions to behavioural ecology as a whole. To begin with, it examines the classical problem of the origin of *eusocial* societies. In the traditional definition, such societies are those which, alongside cooperative care of the brood and an overlap of adult generations, exhibit a reproductive division of labour (Wilson, 1971). This means that some members are specialized for reproduction (queens or kings), whereas others devote themselves to foraging, nest construction, defence and brood-rearing (workers). Since workers sacrifice their own offspring production to help boost that of others, they are said to exhibit *reproductive altruism* (e.g. Trivers, 1985).

The second topic covered by this chapter is the evolution of a stable *reproductive skew* (sharing of reproduction in multiple-breeder groups). The final part discusses sex ratio evolution and conflicts of interest within eusocial colonies. Further coverage of the perennially fascinating subject of insect sociality may be found in the reviews by Wilson (1971), Hamilton (1972), Lin and Michener (1972), Alexander (1974), West-Eberhard (1975), Crozier (1979), Starr (1979), Andersson (1984), Brockmann (1984), Alexander *et al.* (1991), Seger (1991), Krebs and Davies (1993) and Crozier and Pamilo (1996).

A specific reason for concentrating on these three topics is that kin-selection theory may be applied and tested in all of them. In addition, each is relevant to the general study of behavioural ecology. For example, the evolution of social behaviour and the sex ratio represent active areas of study in all sorts of organisms (e.g. see Chapters 10 and 13). Moreover, theories of the evolution of a stable reproductive skew have been applied to birds and mammals with cooperative breeding (see Chapters 10 and 11; Keller & Reeve, 1994). Kin conflict is likewise a universal concept, applying across all scales of organization from the intragenomic level upwards (e.g. see Chapter 12; Ratnieks & Reeve, 1992; Godfray, 1995a; Maynard Smith & Szathmáry, 1995). Another reason for focusing on kin conflict is to stress that kin selection underpins competition between relatives as well as cooperation among them (Trivers, 1974; Trivers &

Hare, 1976; Seger, 1991). According to the theory, kin groups, like a family gathered for Christmas, are forever uneasily poised between twin impulses of cooperation and conflict.

9.1.1 Which are the eusocial insects?

Most eusocial insects belong to two orders, the Hymenoptera (ants, bees and wasps) and the Isoptera (termites). All ants and termites are eusocial, but many bees and wasps are solitary or exhibit other grades of social organization (Wilson, 1971). A number of aphids (Hemiptera) are arguably also eusocial (Aoki, 1977; Benton & Foster, 1992), as are some beetles (Kent & Simpson, 1992) and thrips (Crespi, 1992). How to define eusociality has become a source of controversy (Gadagkar, 1994; Crespi & Yanega, 1995; Sherman *et al.*, 1995). However, the essential point from the viewpoint of kin-selection theory is that all eusocial societies, whether broadly or narrowly defined, exhibit some degree of reproductive altruism. Some of the other key features of the eusocial insects are outlined in Table 9.1.

9.1.2 Kin selection and Hamilton's rule

The idea at the heart of the modern understanding of the evolution of altruism and sociality is Hamilton's (1964a,b) theory of *kin selection* (West-Eberhard, 1975; Michod, 1982; Trivers, 1985). The theory (also known as *inclusive fitness theory*) concerns the conditions under which genes for social actions spread through populations. Social actions are ones that increase or decrease the offspring production of conspecifics. Hamilton realized that their evolution would be affected by *relatedness*. Imagine an altruist helping to rear b (for *benefit*) extra offspring of another individual (the *beneficiary*), while incurring a loss of c (for *cost*) of its own offspring. According to Hamilton's rule, the gene for altruism undergoes selection if the condition $r_1 b - r_2 c > 0$ is satisfied, where r_1 and r_2 are the altruist's relatedness to, respectively, the beneficiary's offspring and its own offspring. West-Eberhard (1975) shows how this form of Hamilton's rule easily converts to another common version, 'help if $rb - c > 0$', where r is the relatedness of the altruist to the beneficiary.

Relatedness, under its formal definition, is a regression coefficient (Grafen, 1985; Table 9.2). Informally, it measures the probability, over and above the average probability (which is set by the gene's average population frequency), that a gene in one individual is shared by another. This allows us to see intuitively why Hamilton's rule works. Altruism can evolve because an altruistic individual can more than make up for losing c offspring with a chance r_2 of bearing the gene for altruism, if it adds to the population b individuals with a chance r_1 of sharing the gene, where $r_1 b$ exceeds $r_2 c$ (or $r_1 b - r_2 c > 0$). In short, the gene for altruism spreads because it promotes aid to copies of itself (Dawkins, 1979; Grafen, 1991).

Table 9.1 The major groups of eusocial insects and their traits.

Group	Number of eusocial species	Range	Typical genera	Nest type	Mode of colony founding	Degree of queen-worker caste dimorphism	Number of queens in mature colony	Colony size to nearest order of magnitude	Life cycle	Totally sterile workers in some species?
Hymenoptera										
Sphecid wasps (Sphecidae)	1	New World tropics	*Microstigmus*	Sac of silk and plant fibre	Haplometrosis, pleometrosis	Very low	One–few	10	Perennial	No
Hover wasps (Stenogastrinae)	50	Tropical Asia and New Guinea	*Parischnogaster, Stenogaster*	Paper and mud cells	Haplometrosis, pleometrosis	None–low	One–few	10	Perennial	No
Independent-founding paper wasps (Polistinae)	632	Worldwide	*Mischocyttarus, Polistes, Ropalidia*	Paper comb	Haplometrosis, pleometrosis	None–low	One–few	10–100	Annual, perennial	No
Swarm-founding paper wasps (Polybiini)	400–500	Tropics	*Parachartergus, Polybia*	Paper comb	Budding	None–low	Several–many	10–10 000	Perennial	No
Yellowjackets and hornets (Vespinae)	78	Asia, Europe, North America	*Vespa, Vespula*	Paper comb	Haplometrosis	High	One	10–1000	Annual	No
Sweat bees (Halictinae)	> 400	Worldwide	*Halictus, Lasioglossum*	Tunnels in soil	Haplometrosis, pleometrosis	None–medium	One–several	10–100	Annual, perennial	No
Allodapine bees plus relatives (Xylocopinae)	?	Worldwide, especially tropics	*Allodape, Exoneura, Xylocopa*	Plant stems	Haplometrosis, pleometrosis	None–low	One–few	10	Annual	No
Bumble bees (Bombinae)	200	Americas, Asia, Europe	*Bombus*	Wax cells in soil or grass	Haplometrosis	Medium	One	10–100	Annual	No
Honey bees (Apini)	5	Native to Africa, Asia, Europe	*Apis*	Wax comb in or on tree	Fission	High	One	10 000–100 000	Perennial	No

continued on p. 206

Table 9.1 *(Continued)*

Group	Number of eusocial species	Range	Typical genera	Nest type	Mode of colony founding	Degree of queen-worker caste dimorphism	Number of queens in mature colony	Colony size to nearest order of magnitude	Life cycle	Totally sterile workers in some species?
Stingless bees (Meliponinae)	277	Tropics	*Melipona, Trigona*	Wax and resin comb in tree	Fission	High	One	100–100 000	Perennial	No
Ants (Formicidae)	8804–20 000	Worldwide	*Atta, Camponotus, Eciton, Formica, Lasius, Leptothorax, Linepithema, Myrmica, Oecophylla, Pheidole, Solenopsis*	Soil mounds; tunnels in soil and wood; plant stems; carton	Haplometrosis, pleometrosis, budding, fission	High	One–many	10–10 000 000	Perennial	Yes
Isoptera										
Termites	2200	Tropics	*Amitermes, Kalotermes, Macrotermes, Nasutitermes*	Soil mounds; tunnels in soil and wood; carton	Haplometrosis, pleometrosis	High	One–many	100–1 000 000	Perennial	Yes
Hemiptera										
Social aphids (Hormaphididae, Pemphigidae)	43	Southeast Asia, Northern hemisphere	*Colophina, Pemphigus, Pseudoregma*	Plant galls	Haplometrosis	High	One	10–100 000	Annual, perennial?	Yes

Colonies may be founded by a single queen (*haplometrosis*), by multiple queens (*pleometrosis*), by a single queen plus workers (*fission*), or by multiple queens plus workers (*budding*). Paper and carton are nest materials manufactured from chewed plant matter; a comb is an array of cells. (From: W.A. Foster, personal communication (social aphids); Wilson, 1971; Brockmann, 1984; Winston, 1987; Bourke, 1988; Myles & Nutting, 1988; Engels, 1990; Hölldobler & Wilson, 1990; Seger, 1991; Ross & Matthews, 1991; Keller & Vargo, 1993; Stern & Foster, 1996. One volume references on the basic biology of some major social insect groups include Hölldobler and Wilson (1990) on ants, Ross and Matthews (1991) on wasps, Roubik (1989) on tropical bees and Winston (1987) on honey bees.)

Table 9.2 Relatedness levels in a social insect colony.

Actor, recipient	Haplodiploids: Regression relatedness	Haplodiploids: Life-for-life relatedness	Diploids: Regression and life-for-life relatedness
Queen, daughter	0.5	0.5	0.5
Female, son	1.0	0.5	0.5
Father, daughter	0.5	1.0	0.5
Queen's mate, queen's son	0.0	0.0	0.5
Sister, sister	0.75	0.75	0.5
Sister, brother	0.5	0.25	0.5
Brother, sister	0.25	0.5	0.5
Brother, brother	0.5	0.5	0.5
Queen, grandson	0.5	0.25	0.25
Female, nephew	0.75	0.375	0.25

These relatedness values are for colonies with single, once-mated, outbred queens (Fig. 9.1). Imagine a population of pairs or groups of social interactants (potential donors of a social action, or actors, and potential recipients of it). *Regression relatedness* at a gene locus is formally obtained when the gene frequency in the potential recipients is regressed on the gene frequency in the potential actors across all the pairs or groups. The slope of the resulting regression line (the regression coefficient) equals regression relatedness. By definition, it gives the probability that a recipient and actor share the focal gene independently of their average probability of doing so (which is set by the gene's average frequency in the population). *Life-for-life relatedness* equals regression relatedness multiplied by *relative sex-specific reproductive value*, which is the sex-specific reproductive value of the recipient divided by the sex-specific reproductive value of the actor. Reproductive value measures the contribution of a class of individuals to the future gene pool in the absence of selection, mutation and drift (Grafen, 1986). In haplodiploids, females have twice the sex-specific reproductive value of males because females are diploid and males are haploid. Put simply, females are intrinsically twice as reproductively valuable as males because they carry twice the number of genes. So, for example, in haplodiploids sister–brother life-for–life relatedness (0.25) equals sister–brother regression relatedness (0.5) × relative sex-specific reproductive value (0.5). In diploids, the female : male ratio of sex-specific reproductive values is 1, because both sexes have identical ploidies. Therefore, in diploids, life-for-life and regression relatednesses are the same. Several classic studies of kin selection in social insects used life-for-life relatedness (e.g. Hamilton, 1972; Trivers & Hare, 1976). However, the regression definition of relatedness needs considering for two reasons. The first is that it is the relatedness deduced from formal population genetic proofs of Hamilton's rule (Grafen, 1985). The second is that some social traits (for example, extensive worker male production) alter relative sex-specific reproductive value but not relatedness (e.g. Boomsma & Grafen, 1991), meaning that these two quantities are best kept conceptually distinct. (After Grafen, 1986, 1991; Bourke & Franks, 1995.)

As well as being the guiding principle for the evolutionary study of social behaviour, Hamilton's insight also solved the historical 'problem of altruism' (Cronin, 1991; Alexander *et al.*, 1991). Under an individual-based view of natural selection, the evolution of a sterile, altruistic caste is a paradox. How can reproductive restraint ever evolve if individuals are selected to maximize their offspring output? This difficulty, which puzzled Darwin, disappears under the gene selectionist view embodied in kin-selection theory. For, according to the theory, altruism appears at the individual level because of self-interest at the gene level. This is why kin selection still represents the flagship of the 'selfish gene' view of adaptive evolution (Dawkins, 1976).

9.2 The origin and evolution of eusociality

9.2.1 The necessity of kin selection

Kin-selection theory states that the altruism of helpers in all eusocial animals must have evolved under the conditions set out in Hamilton's rule. A form of altruism may, in some instances, evolve by reciprocity (an individual performs costly aid today in the expectation of receiving such aid tomorrow) (Trivers, 1971; see Chapter 11). But, reciprocal altruism is not 'true' altruism. For it to evolve the net benefit in terms of personal offspring number must eventually fall on the 'altruist' itself, which is why aid is delivered to begin with. True altruism, defined as occurring when the altruist suffers a net loss of offspring over its lifetime, requires kin selection. As far as can be judged, eusociality usually entails true altruism (and must do so if workers are totally non-reproductive). Hamilton's theory therefore specifies that relatedness (between the altruist and the beneficiary) must have been greater than zero in groups evolving eusociality. However, relatedness could have been quite low, provided the benefit of helping (b) was high and the cost (c) was low. In addition, as Hamilton (1972, p. 198) put it, 'insisting on the necessity of relatedness in no way precludes other factors as necessary or contributory.' In fact, Hamilton's rule shows that both genetic factors (affecting relatedness) and ecological ones (affecting benefit and cost) must be important in promoting eusocial evolution (e.g. West-Eberhard, 1975).

The kin-selectionist interpretation of eusocial evolution has traditionally been contrasted with two additional ideas. The *mutualism hypothesis* emphasized that social groups may have originated as mutualistic aggregations (Lin & Michener, 1972; Itô, 1993; see Chapter 11). The *parental manipulation hypothesis* proposed that workers were initially forced to act altruistically by a dominant parent (Alexander, 1974). However, neither of these ideas represents a genuine alternative to kin selection. The mutualism hypothesis helps explain the early stages of group-living. But, since both parties in a mutualistic interaction must, by definition, eventually gain in average offspring number, it cannot explain altruistic behaviour (Crozier, 1979). Parental manipulation, on the other hand, involves interactions among kin and so falls within kin-selection theory. In addition, there is no universal reason why workers cannot resist manipulation if it pays them to do so (e.g. Craig, 1979; Michod, 1982; Keller & Nonacs, 1993). Parental manipulation is therefore just one type of kin conflict.

9.2.2 Hymenopteran eusociality and the haplodiploidy hypothesis

The frequency of eusocial evolution in the Hymenoptera

Eusociality has evolved independently many times among the insects. For example, Wilson (1971) estimated that there have been 11 origins of eusociality in the Hymenoptera, and one in the termites. However, subsequent phylogenetic analysis has indicated that in termites reproductive altruism evolved at least twice, once in the forerunners of the worker caste and once in the defensive caste of soldiers (which, unlike ant soldiers, are not simply modified workers) (Noirot & Pasteels, 1987, 1988). Similarly, a phylogeny based on mitochondrial DNA sequences suggests that the aphid family Hormaphididae evolved a soldier caste at least five times (Stern, 1994; Stern & Foster, 1996). Therefore, eusociality has had multiple origins in groups outside as well as inside the Hymenoptera.

This finding means that unusual relatedness levels in the Hymenoptera cannot be essential for the evolution of eusociality, as Hamilton himself cautioned (Hamilton, 1972). However, high relatedness levels should certainly facilitate eusocial evolution. For example, social aphids probably live in clones (or mixtures of clones) of genetically identical individuals, because foundress females reproduce asexually (Stern & Foster, 1996). So, high relatedness presumably underpins the ease with which non-reproductive soldiers have evolved in this group (e.g. Andersson, 1984; Hamilton, 1987b). In fact, sterile soldier aphids in a clone would be almost exactly analogous to the defensive polyps of colony-living, clonal marine invertebrates (Stern & Foster, 1996).

Relatedness levels in the Hymenoptera

In the Hymenoptera, males are haploid and develop from unfertilized eggs, but females are diploid and develop from fertilized eggs. As is well known, this *haplodiploidy* leads to unusual relatedness levels among members of a family (Table 9.2). Specifically, sisters share any gene borne by their father, but have a chance of 0.5 (as in diploids) of sharing a maternal gene (Fig. 9.1). So, relatedness among sisters (calculated by averaging across the paternal and maternal halves of the genome) is $(1 + 0.5)/2$, or 0.75. By contrast, sibling–sibling relatedness in diploids is 0.5 (see Table 9.2). As well as devising kin-selection theory, Hamilton (1964a,b) suggested that the three-quarters relatedness among sisters in the Hymenoptera could promote eusocial evolution in the group. Other things being equal, a female would be more strongly selected to raise a sister (of relatedness 0.75) than a daughter (of relatedness 0.5). This would explain both the relatively high prevalence of eusocial origins in the Hymenoptera, and the restriction of worker behaviour to females (in the entirely diploid termites, workers are of both sexes).

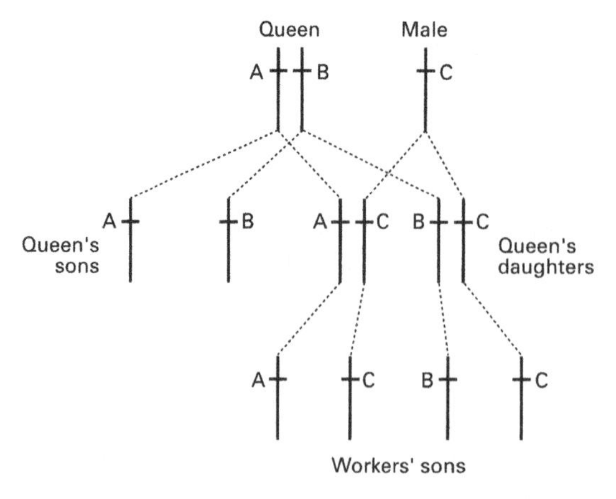

Fig. 9.1 Sisters are highly related in haplodiploids because they share all their father's genes. The figure shows the pedigree of a Hymenopteran colony, in which a single queen mates with a single male and produces sons and daughters (workers and new queens), with the workers in turn producing sons. A, B and C are alleles on representative chromosomes (vertical bars). (From Bourke & Franks, 1995.)

The *haplodiploidy hypothesis* generated a massive amount of interest because it seemed a triumphant testament to the explanatory power of kin selection. To some, however, the idea seemed too mathematical and simplistic (and too clever by three-quarters, perhaps). The whole issue has been reviewed by many authors, including Wilson (1971), Hamilton (1972), Alexander (1974), West-Eberhard (1975), Andersson (1984), Seger (1991) and Bourke and Franks (1995). Here, just a few crucial points need making. The first is that the haplodiploidy hypothesis is a subset of kin-selection theory (e.g. West-Eberhard, 1975). Hamilton's rule can easily hold without a potential helper being more closely related to siblings than offspring. Furthermore, the association of haplodiploidy with eusociality could be coincidental. In short, the validity of kin selection does not rely on the haplodiploidy hypothesis being correct.

Split sex ratios and the origin of eusociality

Another important point about the haplodiploidy hypothesis, which for a long time eluded recognition, is that, according to genetic models, it is not correct to say without qualification that haplodiploid relatedness levels promote the evolution of female workers in the Hymenoptera. Trivers and Hare (1976) showed that the evolutionarily stable population sex ratio for workers (assuming a simple colony kin structure) is 3 : 1 females : males. At this sex ratio, the *mating success* of females (their per capita number of mates) equals one-third that of males (if females have one mating on average, the average male must have three matings). This exactly cancels out the benefit that workers receive from rearing greater numbers of their highly related sisters.

In fact, assuming a simple colony kin structure, equal efficiency at raising brood and a uniform population sex ratio, sibling-rearing yields the same fitness payoff as offspring-rearing in the Hymenoptera (Craig, 1979; Grafen, 1986; Bourke & Franks, 1995).

In response to this finding, theoreticians sought conditions under which a high relatedness among sisters might promote worker behaviour. Such conditions arise if there are *split sex ratios*. These occur when separate classes within a population contribute broods with systematically different sex ratios to the same generation (Grafen, 1986). Specifically, sibling-rearing females have higher fitness than offspring-rearing ones if, at the outset of eusocial evolution, they raise broods whose sex ratio is more female-biased than the population sex ratio (Seger, 1983; Grafen, 1986; Pamilo, 1991a; Krebs & Davies, 1993). To see why, consider a population of solitary bees in which mothers rear daughters that mate then disperse. The population sex ratio is assumed to be the mothers' stable value of 1 : 1. Imagine that a few mutant colonies arise in which daughters stay at home and help rear a female-biased brood of siblings. The mating success of females and males will be unchanged because the mutant colonies are too rare to affect the population sex ratio. Consequently, female mating success no longer offsets the fitness gain to workers of rearing a bias of their highly related sisters, so worker behaviour will spread.

Other factors in the origin of Hymenopteran eusociality

Although it is theoretically sound, whether the haplodiploidy hypothesis correctly accounts for the nature and frequency of Hymenopteran eusociality is uncertain. Confirming that *facultatively social* Hymenoptera (those whose populations contain both solitarily nesting individuals and colonies with workers) exhibit split sex ratios would strengthen the view that females can exploit their high relatedness with sisters to evolve worker behaviour. The problem is to disentangle such effects from others that could be involved. For one thing, the Hymenoptera share features aside from haplodiploidy that may have facilitated eusocial evolution. These include the habit of solitary wasps and bees of building nests (providing a resource worth inheriting), and the possession by females of a sting (a useful weapon for a defensive caste) (Alexander, 1974; Andersson, 1984). Several authors have laid great stress on such non-genetic predisposing factors (e.g. Evans, 1977; Alexander *et al.*, 1991; Crespi, 1994). In addition, haplodiploidy may have genetic consequences helpful to social evolution independently of its effects on relatedness. For example, a model by Reeve (1993) showed that dominant genes for sibling care are less likely to be lost from small populations by genetic drift (random loss of alleles) under haplodiploidy than under diploidy (the *protected invasion hypothesis*).

A number of authors have proposed that life-history effects also favour worker behaviour in both diploids and haplodiploids (Queller, 1989, 1994; Gadagkar, 1990, 1991; Nonacs, 1991). The essential idea is that early mortality damages the fitness gain of a solitary breeder more than that of a worker. If a worker dies early in its active life, it may still have accrued some fitness. One reason is that a worker can begin rearing brood the instant it becomes active, because larvae requiring aid (produced by the queen) are likely to be present on the nest already. This is Queller's (1989) 'reproductive head start' advantage. Another is that, even after the focal worker's death, other workers can raise partly-reared brood to adulthood. In Gadagkar's (1990) phrase, a worker therefore has 'assured fitness returns'. By contrast, a solitary breeder that dies early may not yet have started rearing brood (for example, if it dies while nest-building); or, if it has started but not finished, will leave no companions to complete the job.

This wealth of possible influences on eusocial evolution in the Hymenoptera makes testing the haplodiploidy hypothesis rather intractable. So, perhaps the time has come to abandon the attempt. Instead, it seems more fruitful to refocus on the deeper issue — the application of Hamilton's rule, and its more sophisticated variants incorporating sex ratio and life-history effects, to natural populations.

An example of the application of Hamilton's rule in a facultatively social insect

Hamilton's rule has previously been applied to the evolutionary decisions of *Polistes* wasp queens over whether to join other queens in nest-founding (e.g. Metcalf & Whitt, 1977a; Noonan, 1981; Grafen, 1984). But, polistine wasps are eusocial; after the founding stage, queens produce worker broods. To apply Hamilton's rule in the context of the *origin* of eusociality requires studies of the decisions of females in populations of facultatively social bees and wasps. Surprisingly, there have been few studies of this type. However, a beautifully revealing example comes from the work of Roland Stark on the large carpenter bee *Xylocopa sulcatipes* (Xylocopinae) in Israel. Females nest singly or in pairs in hollow plant stems. From long-term observations of marked individuals, Stark (1992a) deduced that pairs consist of mother and daughter, two sisters or two individuals from different broods. Behaviour inside the nest was observed by the ingenious method of fixing small lead shapes to the bees and watching them through the nest wall with an X-ray machine. This way, Stark *et al.* (1990) and Stark (1992a) found that one bee of a pair is reproductively dominant and eats the eggs of the other bee. The dominant bee is also the chief forager, whereas the subordinate acts as a nest guard and so is effectively a non-reproductive helper. Observations also revealed that *X. sulcatipes* nests risk usurpation, or takeover by a foreign conspecific bee, who if successful destroys all existing brood (Stark, 1990).

Stark's (1992a) data allow Hamilton's rule to be applied to a bee's decision over whether to nest alone or be a guard (Table 9.3). The analysis followed here differs slightly from Stark's own. Only mother and daughter pairs are

Table 9.3 Nesting success and relatedness in the carpenter bee *X. sulcatipes*. (After Stark, 1992a.)

	1986		1987	
	Single female	Mother plus daughter	Single female	Mother plus daughter
Number of nests observed (N_1)	37	18	33	14
Number not usurped (N_2)	18	17	26	12
Number of offspring per surviving nest (N_3)	3.0	5.8	5.8	6.5
Average number of offspring per foundress ($N_4 = [N_2 \times N_3]/N_1$)	1.5	5.5	4.6	5.6
Relatedness to offspring reared (r_1)	—	0.5	—	0.5
Benefit (b) (= pair's N_4 − singleton's N_4)	—	4.0	—	1.0
Relatedness to offspring lost (r_2)	—	0.5	—	0.5
Cost (c) (= singleton's N_4)	—	1.5	—	4.6
Helper's net payoff ($= r_1b - r_2c$)	—	1.25	—	−1.80

In two-female nests, number of offspring per foundress refers to the offspring of the mother, since she founds the nest and daughters produce no adult offspring. The r_1 term is calculated as the average of a female's relatedness with full-sisters (0.75) and brothers (0.25), which is 0.5 (mothers mate only once). Stark (1992a) weighted the guard–sibling relatedness by the population sex ratio, which was female-biased (and equal to the colony sex ratio in all types of nest) (Stark, 1992b). However, a female-biased population sex ratio means that the males' higher mating success cancels out the advantage of rearing more sisters (see Section 9.2.2, 'Split sex ratios and the origin of eusociality'). Since the net effect on the fitness payoff is as if the sex ratio were unbiased, unweighted relatedness values are used here. This does not affect Stark's conclusions, since the female bias was slight.

considered in detail, because they are the most common type (in such pairs, the guard is always the daughter). A guard's benefit term is calculated easily as the number of extra offspring her presence confers on a pair. Since guards produce no adult offspring themselves, her cost term equals the number of offspring she could have had as a single nester. The benefit and cost terms are then weighted by the guard's relatedness to, respectively, siblings and potential offspring (Table 9.3).

The analysis shows that selection favoured helping (guarding) by daughters in the first year of the study but not in the second (Table 9.3). The same was also true for guarding by sisters (Stark, 1992a). It was unclear if joining a bee from a different nest was ever favoured, because the relatedness of these pairs was unknown. However, even helping an unrelated bee could sometimes have been profitable, if the guard stood a chance of eventually inheriting the nest (Stark, 1992a; Hogendoorn & Velthuis, 1995).

These results help explain why both helpers and single nesters coexisted in the population (Stark, 1992a). They also show exactly why helping paid off. Clearly, the guard's presence meant that pairs of bees suffered far fewer usurpations than singletons (Table 9.3). In addition, having a guard boosted the brood size of non-usurped nests (Table 9.3), almost certainly because the dominant bee could spend longer foraging (Stark, 1989; Hogendoorn & Velthuis, 1993, 1995). The conclusion is that sociality can both protect the nest against intraspecific attack and allow an efficient division of labour. These represent two very general promoters of social evolution (Lin & Michener, 1972; Oster & Wilson, 1978). Moreover, helping was more profitable in the first year because single nesters were usurped more often that year and, if they survived, were relatively unproductive (Table 9.3). Therefore, the selective pressures maintaining sociality could be those that make solitary nesting costly as much as those affecting groups directly (e.g. Emlen, 1991; see Chapter 10).

This exemplary study could have been extended in several ways. One would involve confirming the inferred levels of relatedness among pairs of bees using molecular techniques, for example allozyme analysis (e.g. Metcalf & Whitt, 1977b), DNA fingerprinting (e.g. Mueller *et al.*, 1994) or the analysis of microsatellite DNA variation (Queller *et al.*, 1993). Another line of investigation would be to test the assumption that guard bees would have as many offspring as single nesters if they nested alone. Perhaps guards are intrinsically less fertile, making helping more likely to evolve (West-Eberhard, 1975). The issue could be settled by measuring costs and benefits experimentally, for example by forcing helpers to nest alone (Queller & Strassmann, 1989). Finally, Stark's (1992a) findings pose the question of precisely why single nesters were so unproductive in the first year, suggesting a need for the measurement of ecological variables.

9.2.3 The origin of eusociality in the termites

The termites have long been recognized as providing a testing ground for theories of social evolution wholly independent from that represented by the Hymenoptera. Entomologists used to believe that termites evolved from cockroaches, and pointed to the woodroach *Cryptocercus*, a group-living cockroach found in logs, as a likely approximation to the termites' ancestor (Seelinger & Seelinger, 1983; Nalepa, 1984). However, subsequent phylogenetic evidence suggests a more distant link between cockroaches and termites (Thorne & Carpenter, 1992; DeSalle *et al.*, 1992; Kambhampati, 1995). On the other hand, the ancestor of termites may still have resembled *Cryptocercus* behaviourally (Noirot, 1989), since occupying and eating logs is the hallmark of many termites to this day.

Previously, in the flush of excitement caused by the haplodiploidy hypothesis, biologists sought genetic effects in the diploid termites that might mimic the influence of haplodiploidy on Hymenopteran relatedness levels. Two types of haplodiploid analogy were proposed. The first was prompted by the occurrence of chromosome rings incorporating the sex chromosomes in some species (Luykx & Syren, 1979; Lacy, 1980), the second by alternations of inbreeding and outbreeding in the typical termite life cycle (Hamilton, 1972; Bartz, 1979). Both these traits could in some circumstances make potential workers more closely related with siblings than with offspring. However, further investigation failed to confirm an important role in the origin of termite eusociality for either phenomenon. For example, their assumptions are not universally met (chromosome rings seem to be of recent origin, and the required type of cyclic inbreeding is absent from many termite species), and neither are their predictions always fulfilled (workers in species with chromosome rings do not favour same-sex siblings as expected) (Crozier & Luykx, 1985; Luykx *et al.*, 1986; Hahn & Stuart, 1987; Myles & Nutting, 1988).

In any case, the search for a haplodiploidy analogy in termites was unnecessary. Termites live in extended families that are typically founded by a single, monogamous pair. Furthermore, termite workers most probably evolved as sibling altruists in monogamous ancestors (Nalepa & Jones, 1991). Typical relatedness levels between early termite workers and the brood they reared would therefore have been at least 0.5, the level for diploid siblings (Hamilton, 1972; Nalepa & Jones, 1991). In fact, in one of the few allozyme analyses of termite kin structure, Reilly (1987) found a relatedness among nestmates of 0.57. So, the condition in Hamilton's rule would have been satisfied in a proto-termite simply by a benefit-to-cost ratio greater than unity.

This seems an easy condition to fulfil when considered alongside other aspects of termite biology. For one thing, the work in the so-called 'lower' termites is performed by immatures (unlike the Hymenoptera, termites have a continuous metamorphosis), and these typically retain the ability to moult to a reproductive form (Myles & Nutting, 1988). Sterile, morphologically distinct workers are a later development (Noirot & Pasteels, 1987; Higashi *et al.*, 1991, 1992; Roisin, 1994). Therefore, the reproductive sacrifice made by early termite workers may not have been very great. On top of this, logs represent a patchily distributed habitat, suggesting that attempting to disperse and breed alone (and risk failing to find a suitable log) was costly in early termites (Andersson, 1984; Nalepa & Jones, 1991; Alexander *et al.*, 1991; Nalepa, 1994). Both helper and defensive behaviour (as found in termite soldiers) are similarly favoured in other organisms that live and feed in resource-rich patches. Prime examples are gall-dwelling aphids and thrips, and the famous naked mole-rat (Alexander *et al.*, 1991; Crespi, 1992, 1994; Stern & Foster, 1996). In summary, the unusual ecology of termites, combined with their ordinary diploid relatedness levels, seems entirely capable of tipping the balance in favour of eusociality.

9.3 The evolution of a stable reproductive skew

9.3.1 Ecological constraints and skew theory

Social insects vary greatly in the number of queens within their colonies (see

Table 9.1). They are typically classified as having one queen per colony (*monogyny*) or several (*polygyny*) (e.g. Wilson, 1971). In polygynous species, variation also exists in the level of reproductive skew (the degree of sharing of total reproduction among individuals). For example, in some multiple-queen ants, only a single queen lays eggs at any one time (*functional monogyny*: Heinze & Buschinger, 1988). In others, all queens lay eggs. Respectively, these represent cases of high and low reproductive skew. The question is, what accounts for such differences?

Skew theory is a recently revived set of ideas, based on kin-selection theory, that attempts to address this problem (Emlen, 1982a, 1982b; Vehrencamp, 1983b; Reeve & Ratnieks, 1993). It is important because it integrates in a single explanatory framework ecological, genetic and social factors. It also potentially applies to many kinds of social organism (e.g. Keller & Reeve, 1994; see Chapters 10 and 11). In the previous section, the analysis (using Hamilton's rule) of Stark's (1992a) data on the bee *X. sulcatipes* lead to the conclusion that helping behaviour by the non-reproductive subordinate within pairs was favoured (in the first year of Stark's study). Another way of viewing this finding would be to conclude that, in the conditions of that year, the stable skew among pairs was high (indeed maximal, since the dominant bee produced all of a pair's adult offspring). Skew theory allows investigators also to analyse cases less extreme, in which several individuals in a group share reproduction (but not necessarily evenly). So, another merit of the theory is helping to unify in a single conceptual framework the evolution of eusociality and communal breeding (Vehrencamp, 1983b; Keller & Reeve, 1994; Sherman *et al.*, 1995).

Imagine a newly adult queen. Should she set up a colony of her own, so leading to monogyny, or join an existing colony, leading to polygyny? Clearly, joining behaviour (including remaining to breed in the natal nest) would be favoured if ecological circumstances made attempting to nest alone very difficult. In other words, joining should occur when there are high *ecological constraints* on solitary breeding (e.g. Emlen, 1982b; see Chapter 10). This is why, for example, the readoption of daughters and polygyny should evolve in ants when unoccupied nest sites are scarce (e.g. Herbers, 1993; Bourke & Heinze, 1994; Keller, 1995).

A corollary of this reasoning is that societies with multiple breeders are unstable if joiners could receive greater fitness payoffs (assessed via Hamilton's rule) from dispersing to breed alone. This forms the basis of skew theory. Consider a queen whose colony is joined by a newcomer. A critical assumption is that the resident queen benefits from the joiner's presence (for example, because the joiner boosts the resident's productivity). Another key assumption is that the resident can control the joiner's offspring output (for example, by physical domination). Given these conditions, one would expect the resident to skew reproduction in her own favour only up to the threshold beyond which the subordinate would be selected to abandon her and nest alone. This is because, if the dominant queen left too small a share of the colony's reproduction for the subordinate, the subordinate's departure would deprive the dominant of the benefit of her presence. Therefore, the stable level of skew is determined by those factors that affect the relative attraction of remaining in the colony to subordinates (Vehrencamp, 1983b; Reeve & Ratnieks, 1993; Keller & Reeve 1994).

Skew theory suggests several factors of this type (Fig. 9.2). To start with, a rise in either the joiner's contribution (via aid) to colony productivity, or its relatedness to the dominant, or the level of ecological constraint, permits a rise in the stable level of skew (Vehrencamp, 1983b). The reason is that these changes either make staying in an association more profitable to a joiner (rising productivity and relatedness), or make leaving it less profitable (rising ecological constraints). Therefore, the dominant can take a greater share of the colony's reproduction without risking the loss of the subordinate. Relatedness also affects skew through the generational structure of the group. *Subsocial* societies (mother–daughter associations) are expected to show greater skew

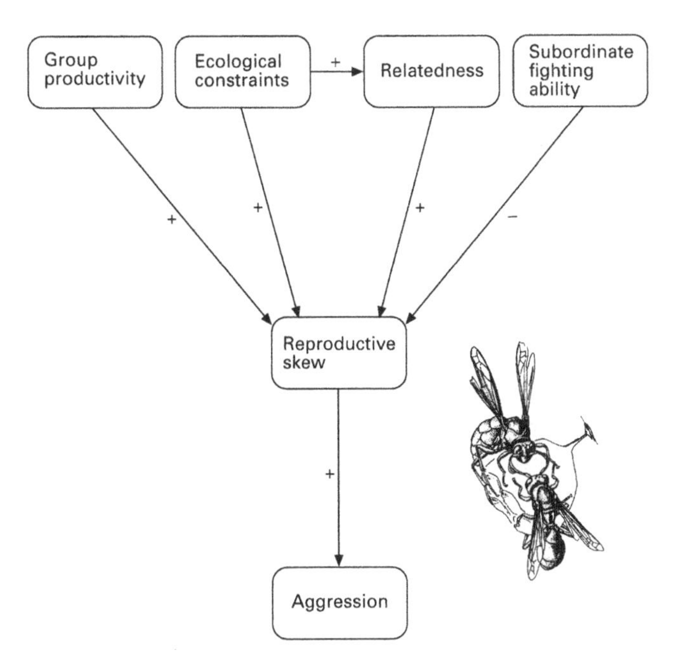

Fig. 9.2 Several factors affect the stable level of reproductive skew, which itself influences the expected level of within-group aggression. Each + sign means that a rise in that factor causes a rise in the one at which the arrow is pointing; the – sign means that a rise in one factor causes a decrease in the other. (From the models of Vehrencamp, 1983b; Reeve and Ratnieks, 1993.) The level of ecological constraints can influence relatedness (top arrow) if some dispersing individuals enter unrelated groups. Then, as ecological constraints rise and dispersal becomes less frequent, average relatedness within groups rises (Bourke & Heinze, 1994). Inset: dominance battle between *Polistes* wasp foundresses. (From Ross & Matthews, 1991.)

than *semisocial* ones (sister–sister associations). This is because daughters are (on average) equally related to their own and their mother's progeny, and so are essentially indifferent between rearing siblings or producing offspring. But, females in sister–sister groups are more closely related to offspring than to nephews and nieces, and so are less willing to let another individual reproduce instead of themselves (Reeve & Keller, 1995).

A final factor affecting skew is the subordinate's fighting ability; as this rises, skew should decrease (Reeve & Ratnieks, 1993). This happens because a relatively strong subordinate is better placed to challenge the dominant for the right to reproduce and, to avoid this, the dominant must offer the subordinate a greater share of reproduction. Reeve and Ratnieks (1993) also argued that the degree of skew itself affects the expected level of aggression. When skew is high, the prize from usurping the dominant is also relatively high, leading to aggressive probing of the dominant by the subordinate. When skew is low, becoming the dominant yields little extra reproduction, so aggression should be muted. Similarly, inside human societies one might expect the level of social discontent to depend not on average earnings but on the variation in incomes — the gap between rich and poor.

9.3.2 Skew evolution in the polistine wasps

The polistine wasps form one of the main groups with which social insect biologists have attempted to test skew theory. However, all tests to date have been indirect. This is because to measure skew accurately one needs to allocate offspring to their parents, which requires molecular genetic methods of high resolution. Happily, the development of techniques for analysing microsatellite DNA variation promises to provide such a method (Evans, 1993, 1995; Hamaguchi *et al.*, 1993; Queller *et al.*, 1993; Peters *et al.*, 1995).

Reeve and Nonacs (1992) conducted a simple experimental test of one of the assumptions of skew theory. Say a dominant attempted to skew reproduction beyond the 'agreed' level by eating some of the subordinate's eggs. The theory assumes that the subordinate can detect such cheating and retaliate. Reeve and Nonacs (1992) removed eggs from field nests of *Polistes fuscatus* wasps. Although the experimenters did not know the maternity of the removed eggs, they argued that subordinates would perceive any loss of eggs as potentially due to egg-eating by the dominant. As expected, alpha queens showed no consistent response to egg removal, but beta ones became more aggressive on average. Furthermore, this occurred only when eggs destined to be reproductive were taken. When worker-destined eggs (which are of far lower fitness value) were removed, betas showed no significant rise in aggression.

Reeve and Keller (1995) performed a comparative test of the idea that the generational structure of colonies affects the stable skew. They compiled skew estimates from the literature for a set of social insect species (mostly polistine wasps) in which some colonies were semisocial and others were mother–daughter groupings. In line with the theory, skew in the second type of colony exceeded that in the first in 13 of 17 cases.

Finally, in the North American species *P. dominulus*, Nonacs and Reeve (1995) found that multiple-foundress associations with a small size difference among their members were less productive than ones with a large size difference. This matched the idea that, as subordinates become closer in size to dominants (and therefore more similar in fighting ability), they should challenge the dominants more vigorously for supremacy. This would lead to higher aggression which, in turn, could account for the observed loss of productivity.

9.3.3 Skew evolution in the leptothoracine ants

Skew theory predicts an association between low ecological constraints, low relatedness, low skew and low aggression inside colonies on the one hand, and high ecological constraints, high relatedness, high skew and high aggression on the other (Reeve & Ratnieks, 1993; Fig. 9.2). Bourke and Heinze (1994) found evidence in the leptothoracine ants for these associations. Facultatively polygynous *Leptothorax* live in extended uniform habitats such as pinewoods. Costs of dispersal are therefore relatively low, because dispersal in any direction still keeps a young queen within the habitat. In addition, all queens lay eggs (so skew is assumed to be low), and queens live together peaceably. By contrast, functionally monogynous leptothoracines live in patchy habitats such as rocky outcrops. This raises dispersal costs, because it means that departing queens risk failing to find another suitable patch. In fact, dispersal is so costly in these ants that some have evolved a morph of queens that is permanently wingless and therefore incapable of dispersing far. Moreover, skew is high by definition (only one queen of the several present lays eggs), and aggression is also high (the egg-laying alpha queen physically dominates the others) (Bourke & Heinze, 1994). However, genetic studies have revealed that, in some populations, cyclical changes in queen numbers over time may lead to a decoupling between the social structure of leptothoracine colonies and their genetic structure and level of skew (Heinze, 1995; Heinze *et al.*, 1995). Such phenomena complicate the testing of skew theory. On the other hand, as expected, between-queen relatedness in the single functionally monogynous species for which a measure exists is higher than the average value for the facultatively polygynous species (Heinze, 1995).

In conclusion, skew theory helps explain a large amount of diversity in the social behaviour and colony structure of wasps and ants. Nonetheless, owing to its scope and power, additional experimental tests of the assumptions and predictions of the theory are undoubtedly required.

9.4 Sex ratio evolution and kin conflict in social insects

9.4.1 Sex ratio theory for the social Hymenoptera

In a very influential paper, Trivers and Hare (1976) applied Fisher's (1930) sex ratio theory to the social Hymenoptera. This work opened up one of the most intriguing areas in social insect research: the study of sex allocation and conflicts of interest within colonies. Many important insights into kin selection and sex ratio theory have flowed from this work (Charnov, 1982; Trivers, 1985).

Fisher's (1930) theory states that the sex ratio is evolutionarily stable if whoever controls the sex ratio derives as much fitness, per unit of effort, from producing a female as from producing a male (e.g. Trivers & Hare, 1976; Benford, 1978; Trivers, 1985). The reason is as follows. If the equal returns condition holds, there is no selection for systematic overproduction of either sex (females and males yield the same payoff). Therefore, the sex ratio remains unchanged, and is by definition stable. If the condition does not hold, one sex yields greater fitness and so will be produced in comparative excess. This will be the sex that is initially rare relative to its stable frequency. The reason is that members of the rarer sex have a greater mating success. For example, if there are two females for every male in a population, then the average male has twice the number of mates of the average female. Therefore, any deviation from the stable sex ratio is self-correcting; when the equal returns condition does not hold, the comparatively rare sex is overproduced, so returning the sex ratio to its stable level.

The fitness payoff from producing a member of either sex is defined as the product, regression relatedness (r) × relative sex-specific reproductive value (V) × mating success (MS) (e.g. Grafen, 1986; see Table 9.2). Therefore, algebraically, the sex ratio is stable when:

$$(r_F V_F MS_F)/c = r_M V_M MS_M$$

where the subscripts F and M denote female and male, respectively, and c equals the cost ratio, defined as the average energetic cost of producing a female divided by that of producing a male (Boomsma, 1989; Boomsma *et al.*, 1995). Costs need considering because the sexes may differ in their energy demands, for example if there is sexual size dimorphism. Let the stable numerical sex ratio be $X : 1$ females : males. Then, if a female has one unit of mating success, a male must have X units. Therefore, $MS_F/MS_M = 1/X$, so $Xc = r_F V_F / r_M V_M$. This last quantity (the ratio of life-for-life relatednesses; see Table 9.2) is termed the *relatedness asymmetry* (Boomsma & Grafen, 1990). So, Fisher's theory can be rephrased as the statement that the stable female : male sex investment ratio (Xc, the ratio of energy spent on females and energy spent on males) equals the relatedness asymmetry. Note that this refers to the sex investment ratio for the population, and not necessarily for individual broods. This is because, assuming random mating, it is the relative abundance of the sexes in the population as a whole that determines their mating success.

This reasoning can be applied to yield many familiar results in sex ratio theory. Take the standard case of diploids with parental control of sex allocation, random mating and no sexual size dimorphism ($c = 1$). Then, X equals (life-for-life relatedness to daughters)/(life-for-life relatedness to sons), which is simply 0.5/0.5 (see Table 9.2). This is Fisher's (1930) finding that the stable sex ratio in this case is 1 : 1. Now, consider social haplodiploids. For queens, the relatedness asymmetry is still 1 : 1 (queens are equally related to daughters and sons), so the stable sex ratio must also be 1 : 1. But, for workers, the relatedness asymmetry equals (life-for-life relatedness to sisters)/(life-for-life relatedness to brothers), which is 0.75/0.25 (see Table 9.2), or 3 : 1. This is Trivers and Hare's (1976) classic result; assuming a simple kin structure (monogyny, singly mated queens, sterile workers and random mating), workers in eusocial Hymenopteran colonies favour a population sex investment ratio (among the sexual forms) biased 3 : 1 in favour of females. Since queens favour a 1 : 1 ratio, there is a queen–worker conflict over sex allocation.

Fisher's (1930) theory, with Trivers and Hare's (1976) modifications, is very general. Using the approach above, it has been extended to cases either where the kin structure is not simple, so the relatedness or reproductive value terms change (for example, if there is multiple mating, worker reproduction or polygyny), or where some of the underlying assumptions are altered (for example, mating is not random) (e.g. Pamilo, 1990, 1991b; Bourke & Franks, 1995; Crozier & Pamilo, 1996).

9.4.2 Tests of sex ratio theory in social Hymenoptera

Population sex ratio data

Trivers and Hare (1976) argued that the workers should usually win the conflict with queens over sex allocation, because they both outnumber the queen and rear the larvae, and so have the opportunity to bias the composition of the reproductive brood in their interests. In other words, there should be *worker control* of sex allocation. Consequently, in monogynous ants one would expect a 3 : 1 population sex investment ratio. In polygynous ants, the expected ratio should be close to 1 : 1 even under worker control (assuming queens within colonies are related). This is because adding extra queens to a colony reduces the workers' relatedness asymmetry, as proportionately fewer and fewer females will be full sisters (Pamilo, 1990). However, in monogynous slave-making ants, one should also expect 1 : 1 investment. Slave-makers are social parasites whose workers steal pupae from nests of other species, which after maturing raise all the slave-makers' brood. So, Trivers and Hare (1976) argued, slave-maker workers lack the practical power needed to control sex allocation. This leaves the slave-maker queen free to influence brood composition unimpeded by her workers (for example, by laying equal numbers of male and queen-destined eggs). Therefore, in slave-makers one expects unbiased, queen-controlled sex allocation.

These predictions are borne out by the data (Trivers & Hare, 1976; Nonacs, 1986; Boomsma, 1989; Pamilo, 1990; Bourke & Franks, 1995). The average sex investment ratio across 40 species of monogynous ants is significantly female-biased (0.63, expressed as the proportion of investment in females), whereas polygynous ants (25 species) and slave-makers (three species) have sex investment ratios of 0.44 and 0.48, respectively, which are indistinguishable from 1 : 1 or 0.5 (Pamilo, 1990; Bourke & Franks, 1995). These findings are especially striking in that the sex ratios are true investment ratios; numerical sex ratios in ants tend to be male-biased (Nonacs, 1986; Bourke & Franks, 1995). Overall investment can nonetheless be female-biased in monogynous ants because queens receive large fat reserves to function as an energy supply during colony foundation (in ants, fat is a feminine tissue). One might ask why in monogynous ants the average ratio is not closer to the expected 0.75 (3 : 1). The likely explanation is that few of the species in the dataset exactly meet all Trivers and Hare's (1976) assumptions. Instead, there is probably a low level of multiple mating, worker reproduction or polygyny within their populations, all of which cause the workers' stable sex ratio to decline (Pamilo, 1990, 1991b). Conceivably, there could also be a degree of queen control in some species (Bourke & Franks, 1995).

A few researchers have tested Trivers and Hare's (1976) theory by measuring relatedness asymmetry directly within populations (using allozyme methods), and then checking the predicted sex investment ratio against the observed one. This approach has been used in four monogynous ants, with results that largely confirm Trivers and Hare's (1976) idea of worker control given the errors likely to be involved in measuring both relatedness values and sex ratios (Table 9.4). However, in at least one population of *Lasius niger*, sex investment was unexpectedly male-biased (Table 9.4). A possible reason is that this population was poor in resources (Van der Have *et al.*, 1988). Nonacs (1986) suggested that resource levels have a proximate effect on sex investment ratios. Underfed colonies might redirect investment from queens into workers (for example, to enhance the colony's chance of surviving to better times), so reducing female bias. Field experiments have confirmed such an effect in one case (*Formica podzolica*: Deslippe & Savolainen, 1995) but not in another (*Leptothorax longispinosus*: Backus & Herbers, 1992).

Alexander and Sherman (1977) argued that, contrary to Trivers and Hare (1976), female bias in the sex ratios of monogynous ants arises because of *local mate competition*. This occurs when related males compete for mates (so violating the usual assumption of random mating). Female bias is expected because there are then diminishing returns on producing males but not females (there is no point producing many males bearing similar genes who will compete with one another, whereas females compete with non-relatives for new nest

Table 9.4 Tests of sex ratio theory within monogynous, non-parasitic ant species.

Species	Population sex investment ratio (fraction of investment in females		Number of colonies
	Expected	Observed	
Colobopsis nipponicus	0.75	0.75	—
Formica truncorum	0.63	0.65	22
Leptothorax tuberum	0.70	0.75	47
Lasius niger			
Population 1	0.72	0.68	125
Population 2	0.79	0.65	26
Population 3	0.59	0.36	50

The source references are Hasegawa (1994) for *C. nipponicus*, Sundström (1994) for *F. truncorum*, Pearson *et al.* (1995) for *L. tuberum* and Van der Have *et al.* (1988) for *L. niger*. The observed sex ratios for *F. truncorum* and *L. niger* were recalculated by Bourke and Franks (1995). The expected sex ratios under worker control were derived by the original authors from the measured workers' relatedness asymmetries (expressed as fractions). The relatedness asymmetries fell below 0.75 in some cases because of partial multiple mating or worker reproduction. The expected sex ratio in *F. truncorum* is also calculated on the basis of extensive sex ratio splitting in this population (see text). Under queen control, the expected sex ratios were 0.5 in every case, because the queen's stable sex ratio is unaffected by multiple mating and worker reproduction in her presence (Pamilo, 1991b; Bourke & Franks, 1995).

sites) (Hamilton, 1967). However, ants typically mate in large mating swarms drawn from many colonies (Hölldobler & Wilson, 1990), and so lack a mating system involving an appreciable degree of local mate competition. Therefore, female bias cannot usually arise from this cause. On the other hand, in a few cases where conditions are appropriate, local mate competition leads, as predicted, to a sex ratio more female-biased than that expected under random mating. One example is the socially parasitic ant *Epimyrma kraussei*, whose sexuals mate exclusively in the nest (Winter & Buschinger, 1983; Bourke, 1989). Another is the non-parasitic harvester ant *Messor aciculatus* in which, although mating occurs outside the nest, genetic evidence suggested that each mating swarm is composed of sexuals from just a few colonies (Hasegawa & Yamaguchi, 1995).

The honey bee (*Apis mellifera*) exemplifies another type of violation of a Fisherian assumption. Colonies reproduce by splitting into two (colony fission), with the young queens of a hive competing among themselves to head one of the daughter colonies. These queens therefore experience *local resource competition* (competition among relatives for resources). Males, by contrast, compete for mates with all other males in the population, as in the standard case. So, there are now diminishing returns on investment in new queens, but not on producing males. This makes the predicted numerical sex ratio among honey bee sexuals extremely male-biased, and this is what is found (Bulmer, 1983; Pamilo, 1991b).

Split sex ratios in eusocial Hymenoptera

Fisher's (1930) theory specifies only a stable sex ratio at the level of the population. In a very large population at sex ratio equilibrium, the sex ratio of individual colonies can vary randomly (because, by definition, females and males yield equal returns). However, in nature, between-colony sex ratio variation is frequently greater than random, with many colonies specializing on a single sex (e.g. Nonacs, 1986).

Boomsma and Grafen (1990, 1991) proposed an explanation for such split sex ratios based on Trivers–Hare theory. Suppose that the workers' relatedness asymmetry varies among colonies in a population (due to, say, queens heading some colonies being singly mated and queens heading other colonies being multiply mated). Then, according to Boomsma and Grafen (1990, 1991), workers in colonies in which their relatedness asymmetry is high compared to the female : male population sex ratio should concentrate on female production, and workers in colonies in which their relatedness asymmetry is comparatively low should produce mainly males. The following example shows why. Say a colony exists in which the workers' relatedness asymmetry is 3 : 1, while the population sex ratio is 1 : 1. The level of the population sex ratio means that the mating success of females equals that of males. But, the level of the workers' relatedness asymmetry means that workers are three times more closely related to females than to males. Therefore, workers must derive three times more fitness from producing a female than from producing a male (recall that fitness equals the product of life-for-life relatedness and mating success), and so should rear females alone. Boomsma and Grafen (1990, 1991) also showed that, when the workers' relatedness asymmetry varies between classes of colonies, the stable population sex ratio either equals one of the relatedness asymmetries, or lies between them. Therefore, at least one colony class (the one whose relatedness asymmetry is unequal to the population sex ratio) should always concentrate on producing a single sex.

In populations of the halictine bee *Augochlorella striata*, some colonies are mother–daughter associations (eusocial colonies), but others have lost the foundress and consist of sister–sister groups (parasocial colonies). So, the eusocial colonies should produce female-biased broods (the workers' relatedness asymmetry is relatively high because workers are rearing full-siblings), whereas the parasocial colonies should produce male-biased broods (relatedness asymmetry is relatively low because workers are rearing nieces and nephews). Mueller (1991) tested this with a simple but clever experiment. He removed the foundress from one set of colonies (so creating parasocial colonies), and a random worker from another set (so leaving them eusocial but controlling for the reduction in colony size). The result was as expected, with the fraction of investment in females being 0.67 among the eusocial colonies and significantly lower (0.31) among the parasocial ones. Mueller *et al.* (1994) later confirmed the genetic structure of each type of colony using DNA fingerprinting.

Another test of split sex ratio theory was carried out by Sundström (1994) in a monogynous population of the wood ant *Formica truncorum*. She established, using allozyme analysis, that some colonies had a multiply mated queen (so reducing the workers' relatedness asymmetry), and others had a singly mated queen (leaving relatedness asymmetry unaffected). As predicted, the first type of colony produced mostly males and the second type produced mainly females. Among polygynous ants, where workers' relatedness asymmetry varies because of variations in queen number, split sex ratios also occur (e.g. Herbers, 1990; Chan & Bourke, 1994; Evans, 1995).

An assumption of Boomsma and Grafen's (1990, 1991) theory is that workers in a colony can assess their relatedness asymmetry. Note that this would not require the ability to discriminate among different kin within the colony. Instead, workers have to judge whether, for example, their queen is singly or multiply mated, or whether the colony is monogynous or polygynous. This could plausibly occur if workers assessed the genetic variability of the brood. To test this idea, Evans (1995) experimentally added unrelated larvae to colonies of the polygynous species *Myrmica tahoensis*. As expected from the idea of worker assessment, the result was that the colonies reared more male-biased broods.

Although other explanations for split sex ratios have been proposed (e.g. Frank, 1987), none predicts an association between a colony's sex ratio and the workers' relatedness asymmetry. Such an association is not predicted if queens control sex allocation either, because queens are symmetrically related to daughters and sons. Therefore, the overall conclusion from studies of split sex ratios is that workers largely control sex allocation as Trivers and Hare (1976) predicted. In ants, studies of population sex ratios also support this conclusion (e.g. Table 9.4). The confirmation of worker control is important because, by establishing that workers bias investment according to their relatedness to the sexes, it supports both Fisher's sex ratio theory and Hamilton's kin-selection theory.

A final question concerns *how* workers manipulate sex allocation. Using histological methods to count the chromosomes inside individual ant eggs (e.g. Aron *et al.*, 1994, 1995), Sundström *et al.* (1996) have recently shown that all queens in a monogynous population of *Formica exsecta* contributed a similar fraction of haploid eggs to their colony's egg pool. However, workers in colonies headed by singly mated queens raised a female bias of adult sexuals, whereas workers under multiply mated queens raised a male bias, so achieving a split sex ratio among adults identical to that found in *F. truncorum*. These findings suggest that, in colonies where their fitness interests dictate (ones with a singly mated queen), workers selectively destroy male eggs or larvae. They therefore provide excellent evidence of one way in which workers can facultatively manipulate brood composition to achieve their sex ratio preferences.

9.4.3 Kin conflict over worker reproduction

Contrary to what is often thought, workers in many social Hymenoptera can produce male offspring from unfertilized eggs (Bourke, 1988; Choe, 1988). Another important type of kin conflict occurs over worker reproduction (Trivers & Hare, 1976; Ratnieks & Reeve, 1992). In a colony with a single, once-mated queen, the workers are more closely related to sons (life-for-life relatedness = 0.5) than to nephews (relatedness 0.375) or brothers (relatedness 0.25). The queen, however, is more closely related to her sons (relatedness 0.5) than the workers' sons (relatedness 0.25) (see Table 9.2). Therefore, each party favours the production of its own male offspring, leading to queen–worker conflict (Trivers & Hare, 1976). This arises even if no single worker can monopolize male production, since in this case any given worker will still be more closely related to the average worker-produced male (a nephew) than to a brother.

Conflict over male production could account for the otherwise puzzlingly high level of friction inside bumble bee (*Bombus*) colonies, which typically have singly mated queens (e.g. Page, 1986). In these, the queen attacks laying workers and eats their eggs, while the workers try to eat some of the queen's eggs (e.g. Van Honk *et al.*, 1981; Owen & Plowright, 1982). This happens most often towards the end of the colony's life, when the queen is more likely to be laying male eggs. In fact, the escalating violence may even lead to the workers killing their own queen (Bourke, 1994).

Workers of monogynous ants with singly mated queens are usually non-reproductive in the queen's presence (Bourke, 1988; Choe, 1988). This is despite the prediction of the simple relatedness arguments that these workers should favour rearing worker-produced males, and despite the queen's lack of aggression to workers. A previous explanation for this finding was that the queens suppressed worker reproduction with chemical secretions (Wilson, 1971; Bourke, 1988). However, Seeley (1985) and Keller and Nonacs (1993) proposed that, instead, workers do not benefit from attempting to reproduce in the queen's presence. Worker reproduction could reduce colony productivity, or workers might sometimes be incapable of singling out the queen's male eggs for replacement (Cole, 1986; Nonacs, 1993). If so, the workers' interests would be served best by rearing the queen's offspring alone, but only so long as the queen remained fecund and healthy. So, the queen's pheromones could be an honest signal of her vigour, rather than an instrument of coercion. This *signalling hypothesis* for the function of queen pheromones is an open issue at present (Bourke & Franks, 1995), and deserves testing.

Multiple mating by queens alters the reproductive behaviour expected among the workers (Starr, 1984; Woyciechowski & Lomnicki, 1987; Ratnieks, 1988). Specifically, when the number of mates per queen (k) exceeds two, the workers' relatedness to the average worker-produced male (nephews, relatedness equals $0.25[0.5 + 1/k]$), falls below worker–brother relatedness (0.25). Therefore, under a multiply mating queen, workers are expected to favour rearing the queen's males (assuming that no single worker can dominate worker reproduction). To achieve this, they should actively prevent each other's reproduction, a phenomenon that Ratnieks (1988) termed *worker policing*.

The honey bee provides striking confirmation of the worker policing hypothesis. Honey bee queens mate up to 17 times (Winston, 1987). Genetic studies also indicate that only about one in 1000 adult males is worker-produced (Visscher, 1989). This is because although a few workers lay eggs in colonies with a queen, other workers selectively eat these eggs (Ratnieks & Visscher, 1989; Ratnieks, 1993). Furthermore, workers with developed ovaries are physically attacked by their nestmates (Visscher & Dukas, 1995). In short, worker policing both occurs in honey bees and leads to the almost exclusive rearing of queen-laid males. On the other hand, additional genetic evidence suggests that, on rare occasions, workers from one patriline (paternal lineage) can somehow evade policing and undergo a dramatic burst of worker reproduction (Oldroyd *et al.*, 1994). Most remarkably, honey bee queens apparently chemically label their male eggs, so allowing 'police' workers to discriminate between these eggs and those of the laying workers (Ratnieks, 1995). This arrangement is evolutionarily stable because it benefits both queens and the average worker. Honey bees therefore show how mutual inhibition through 'policing' can lead to harmony inside the colony despite the potential for kin conflict (Ratnieks & Reeve, 1992).

9.5 Conclusion

The study of social evolution in insects is a dynamic area of research. The kin-selection approach has clearly been uniquely fruitful in improving our understanding of the evolution of both cooperation and conflict in insect societies. It also provides the best basis for integrating the study of social evolution in vertebrates and invertebrates (e.g. see Chapter 10). Yet, in every area covered by this chapter, unanswered questions remain. For example, how well do the genetics, demography and sex ratios of facultatively social bees and wasps meet the assumptions and predictions of models for the origin of eusociality invoking split sex ratio and life-history effects? Are the quantitative predictions of skew theory met? What mechanisms apart from selective egg-eating do workers use in kin conflicts over sex allocation and male parentage? How widespread is worker policing? Other unsolved issues include two not dealt with in this chapter: (i) the adaptive significance of within-colony kin discrimination (Carlin, 1989; Grafen, 1990a; see Chapter 4) and (ii) the evolution of multiple mating (Crozier & Page, 1985; Bourke & Franks, 1995). Social insects therefore promise to remain a rich source of ideas and discovery in behaviour, ecology and evolution.

Chapter 10

Predicting Family Dynamics in Social Vertebrates

Stephen T. Emlen

10.1 The changing scope of cooperative breeding research

The focus of research efforts on cooperatively breeding birds and mammals has changed dramatically during the past two decades. Cooperative breeding refers to breeding systems in which adults provide significant care to young that are not their own genetic offspring. Such systems are now known to occur in roughly 3% of bird and mammal species. That number will undoubtedly increase as additional field studies are conducted.

At the time of publication of the first edition of this volume in 1978, interest centred on verifying the existence of 'helping' behaviours, on describing the diversity of types of helping associations found in nature and on explaining the paradox that these seemingly altruistic behaviours presented. Did 'helpers' really contribute significant assistance to the breeders that they attended? If they did, how could helping others (rather than breeding oneself) be reconciled with the tenets of natural selection?

By the time the second edition was published in 1984, a sufficient number of long-term field studies had been conducted to provide a generally affirmative answer to the first question, and to develop a framework for addressing the second. The most common form of group found among cooperative vertebrates was that of grown offspring helping their parents to rear younger siblings. The second question thus became partitioned into two. Why did offspring remain with their parents rather than disperse and attempt to breed independently on their own? And why did such grown offspring help, rather than ignore (or even hinder), the breeding efforts of other adults in their group?

Ecological constraints theory provided the most plausible answer to the first. Offspring stayed home when opportunities for successful dispersal and independent breeding were 'constrained' — when such opportunities were temporarily non-existent or of poor quality relative to the situation at home. Answers to the second were more diverse. It turned out that helpers themselves benefited in several additive ways by their helping actions. They increased their own probability of becoming breeders in the future (by expanding or inheriting the parental territory itself). They were better parents when they did breed (having benefited from the experience of helping). Finally, they increased their inclusive fitness by helping to rear close genetic relatives.

The early 1980s also bore witness to a growing realization that conflict goes hand in hand with cooperation, and that even in the most highly cooperative of societies, genetic conflicts of interest are inevitable. In particular, competition will exist between group members over who breeds and who does not. A general theory of such conflict and its resolution, now termed reproductive skew theory, was in its early stages of development. These models (Emlen, 1982a, 1984; Vehrencamp, 1983a,b; Emlen & Vehrencamp, 1983) attempted to predict: (i) when dominant members in a group would monopolize reproduction rather than share it; and (ii) when shared, with whom and how equitably it would be shared.

By 1991, the year of the third edition of this volume, numerous long-term studies had provided rigorous data on the actual costs and benefits of helping, both to breeding recipients and to helpers themselves. The original paradox of cooperative breeding largely disappeared with the widespread confirmation that: (i) helpers frequently *do* improve their chances of becoming breeders by staying at home and helping temporarily; and (ii) they frequently *do* obtain large indirect genetic benefits by helping to rear collateral kin. For a summary of such data, the reader is referred to Emlen (1991).

With the original set of questions largely answered, I wish to shift my attention in this current edition to the topic of social dynamics within cooperative societies, and particularly to *family* dynamics. There are several reasons for an emphasis on the family. First, the vast majority of birds and mammals that exhibit cooperative breeding do, in fact, live in multigenerational family groups (Box 10.1). Second, families provide an excellent arena for developing

Box 10.1 Family definitions

In this chapter, the definition of families is restricted to cases where offspring continue to interact, into adulthood, with their parents. I categorize families as *simple* or *extended* depending on whether reproduction is totally monopolized (skew = 1) or is shared (skew < 1). In simple families only a single male and female group member breed, while in extended families, two or more group members of one or both sexes reproduce.

The presence of a breeding male is not essential to the definition of a family. Rather, the presence or absence of reproductive males forms the basis of a second partitioning into *biparental* (also called *nuclear* or *conjugal*), versus *matrilineal*, families.

It is useful to further differentiate between *intact* families, those where the original breeders are still the reproductives, and *replacement* (or step-) families where, because of death, divorce or departure, a breeder has been replaced.

Despite these differences, most vertebrate families form in the same way when offspring delay dispersal and continue to reside with their parent(s).

and testing evolutionary social theory because families are comprised of genetic relatives that vary both in their degrees of relatedness and in their social dominance ranks relative to one another. Relatedness and dominance are expected to be major predictors of the social dynamics of any group.

By focusing on families one seeks similarities, rather than differences, among cooperative species. In most cooperatively breeding birds, adults form long-term pairbonds and males contribute considerably to the care of dependent young. Similar groupings occur in some canids, but pairbonds among mammalian cooperative breeders more typically are lacking or are of short duration. Fathers can even be absent from the group during the time of offspring dependency. Despite these differences, social groupings in both avian and mammalian cooperative breeders typically form via the retention of grown young with their parent(s). The result is a tight kin group, as well as a group with a built-in generational dominance asymmetry. These common characteristics feature prominently in the predictions developed later in this chapter.

Finally, focusing on the family structure of many animal societies highlights parallels with the early social organization of our own species. It raises the question of whether findings emerging from animal studies can usefully be extrapolated to better understand both the past and the current human condition.

Many of the ideas expressed here have been developed elsewhere (Emlen, 1994, 1995a,b). A list of specific predictions concerning the social dynamics of kin groups is reproduced in Appendix 10.1 at the end of this chapter.

10.2 Ecological constraints and the formation of family groups

In most organisms, young disperse from their area of birth well before or at least by the age of sexual maturity. In seeking independent breeding opportunities, they cease interacting with their parents. If offspring and parent come into contact later in life, they show no signs of recognition or preferential interaction with one another.

What sets species that form multigenerational families apart is the tendency for offspring to remain in association with their parent(s) beyond the age of sexual maturity and, commonly, throughout their lifetimes. The key to understanding the evolution of families is understanding delayed dispersal.

Why should a growing offspring postpone its dispersal? Imagine you are a Darwinian accountant. Your task is to maintain a ledger sheet on which you record the probable costs and benefits associated with the alternatives of: (i) remaining at home and continuing to associate with parent(s); or (ii) leaving home and attempting to reproduce independently. Families are expected to form only when the expected lifetime fitness associated with the option of staying home exceeds that associated with early dispersal. What factors might tip the balance in favour of staying?

Most mature birds or mammals that remain in their natal group do not reproduce; their breeding is suppressed by their more dominant parents. There is thus an almost automatic cost to staying home — the forfeiture of fitness associated with missed reproductive opportunities (Emlen, 1994). All else being equal, offspring would be expected to leave home.

But, all else is not always equal. Suppose there are very few 'vacancies' for dispersing individuals, or that the vacancies are of poor quality. Competition for the few good vacancies will be intense, lowering the probability of successful establishment by a dispersing youngster. Add to this a heightened mortality risk associated with dispersal itself, and the benefits on the 'leave home' side of the ledger sheet begin to pale. As ecological constraints on the dispersal option intensify, it takes fewer benefits in the 'stay home' ledger column to tip the balance in favour of remaining with one's parents.

Some researchers have emphasized the importance of the constraints on leaving; others the benefits of staying home. I view the two as complementary. (For further discussion of this issue, see Woolfenden & Fitzpatrick, 1984; Brown, 1987; Stacey & Ligon, 1987, 1991; Zack, 1990; Emlen, 1991, 1994; Koenig *et al.*, 1992; Mumme, 1992a; Zack & Stutchbury, 1993).

Together these arguments form the basis of an economic model of family formation (Emlen, 1994, 1995a). Stated simply:

1 delayed breeding occurs when the production of mature offspring exceeds the availability of acceptable opportunities for their independent reproduction;

2 under such circumstances, some offspring must postpone personal reproduction until acceptable breeding opportunities arise and they are able to successfully compete to obtain them;

3 families will form when such waiting is best done at home, when remaining on the natal territory and/or associating with one's family members somehow augments the offspring's inclusive fitness.

Family groupings are thus expected to be inherently unstable. They should form and expand when independent breeding opportunities are constrained, but decrease in size and break up as outside reproductive opportunities improve. In essence, family social organizations can be viewed as 'solutions' to the often temporary problem of a shortage of acceptable breeding opportunities.

If offspring are assessing the relative fitness profitabilities of staying home versus dispersing, then offspring residing on territories of high quality should require higher quality outside reproductive opportunities to induce them to leave home. The result will be greater stability of families that control high-quality resources.

The following studies illustrate the predictability of family formation and dissolution. Acorn woodpeckers, *Melanerpes formicivorus*, live in groups of two to 12 individuals and typically occupy permanent, year-round territories in the American Southwest. High-quality territories are in short supply, and most offspring become reproductives by waiting for an established breeder on an existing territory to die, and then successfully competing to fill the breeding vacancy (Koenig & Mumme, 1987). With assistance from numerous researchers, I (Emlen, 1984) compiled field data from several locations across many years on: (i) the proportion of offspring that remained on their natal territories; and (ii) the local availability of breeding vacancies. As predicted, the data show a clear, increasing tendency for yearlings to delay dispersal and remain home with decreasing local availability of breeding vacancies (Fig. 10.1a).

All family members in acorn woodpeckers help build and defend large granaries in which thousands of acorns are stored. These acorns provide an essential resource during times of food shortage (Koenig & Mumme, 1987). By constructing granaries, the birds increase the real estate value of their territory. The quality of territories can vary greatly within a local area. Both nesting success and adult survival are positively correlated with granary size. Stacey and Ligon (1987) found that offspring from high-quality territories were unlikely to disperse to fill vacancies in lower quality areas. Specifically, a higher percentage of yearlings remained (eventually ascending to breeder status) on territories with large granaries. Thus, family stability was greatest on the best quality territories (Fig. 10.1b).

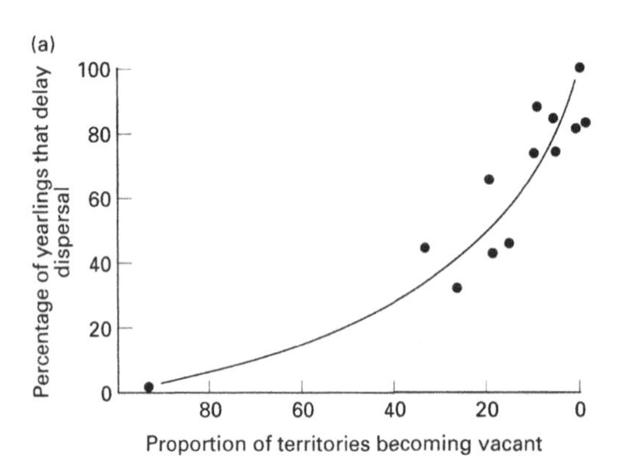

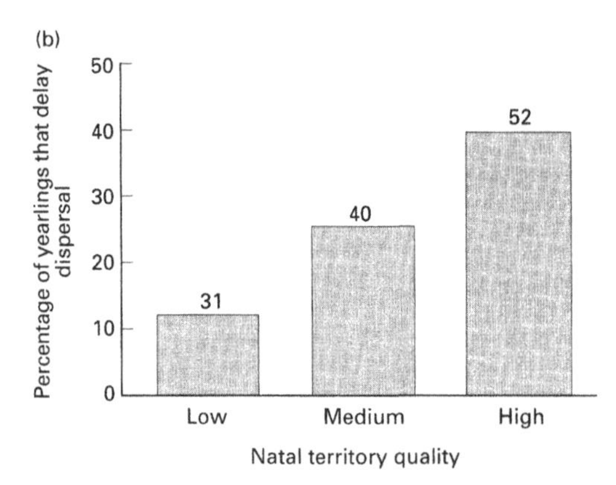

Fig. 10.1 Ecological constraints and family formation in acorn woodpeckers. (a) The proportion of yearlings that remain at home, plotted as a function of the severity of territorial shortages. (From Emlen, 1984.) (b) The likelihood that yearlings stay at home, plotted as a function of the quality of their natal territories. Numbers above histograms are sample sizes of yearlings in each territory quality category. (Data from Stacey & Ligon, 1987, from a population experiencing moderate territorial shortage.)

Superb fairy-wrens, *Malurus cyaneus*, live in southeastern Australia in families consisting predominantly of parents and grown sons. Shortages of both territories and mates (females) have been suggested as possible constraints to independent breeding (Rowley, 1965; Emlen, 1984; but see Rowley & Russell, 1990). Pruett-Jones and Lewis (1990) tested these ideas experimentally. By removing breeding males from nearby territories, they created breeding vacancies where none previously had existed. The result was the dissolution of virtually all family groups under observation. Thirty-one of 33 mature sons left home to fill the newly created breeding vacancies. Analogous results have recently been reported for red-cockaded woodpeckers, *Picoides borealis*, by Walters *et al.* (1992).

The Seychelles warbler, *Acrocephalus sechellensis*, is a formerly endangered species whose range is restricted to a few small islands north of Madagascar. In 1960, when the population consisted of just 26 individuals, habitat restoration programmes were implemented. Over the following 30 years, the population grew impressively. No family groups were reported among Seychelles warblers until 1973, roughly the time at which all suitable breeding habitat became occupied (Fig. 10.2a). In essence, families formed when acceptable breeding opportunities first became constrained (Komdeur, 1992). In subsequent years, the population of mature birds has consistently exceeded the number of occupied territories. Family groups have been the norm.

To enhance the numbers of this endangered species, birds were introduced onto two nearby, previously unoccupied, islands. To obtain birds for these

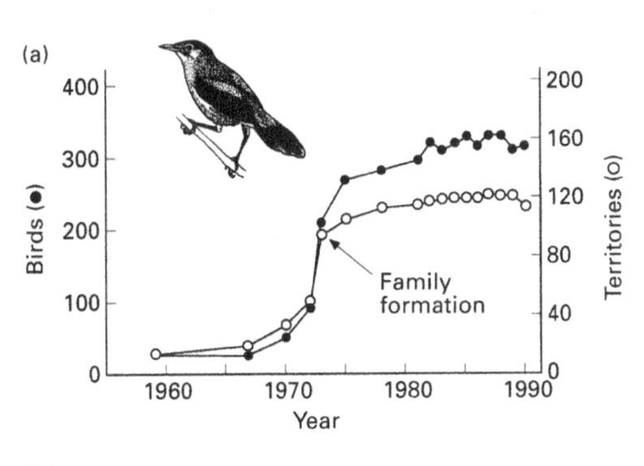

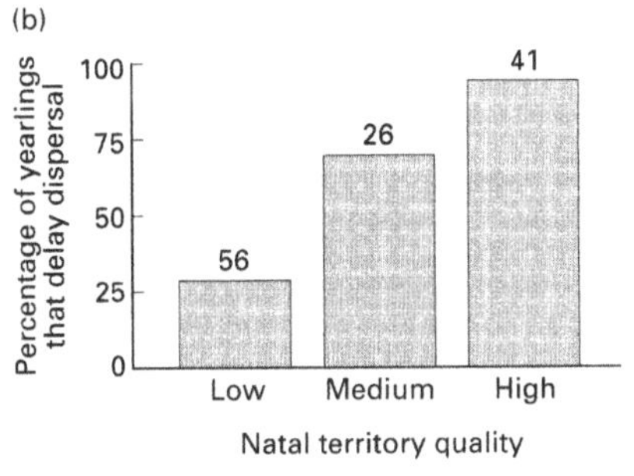

Fig. 10.2 Ecological constraints and family formation in Seychelles warblers. (a) The number of individuals and occupied territories on Cousin Island between 1959 and 1990. Family formation was first observed when all available territories became filled. (Modified from Komdeur, 1992.) (b) The likelihood that yearlings stay at home, plotted as a function of the quality of their natal territories. Numbers above histograms are sample sizes of yearlings in each territory quality category. (Data from Komdeur, 1992, from a population experiencing severe territorial shortage (Cousin Island, 1986–90).)

transfers, Komdeur removed breeding adults from occupied territories on the original island. He thus experimentally created breeding opportunities in a manner analogous to Pruett-Jones and Lewis (1990). As with the fairy-wrens, the manipulations caused widespread family dissolution. All experimental vacancies were rapidly filled by mature offspring that previously had been residing with their parents.

Komdeur had also independently evaluated the quality of all territories both in terms of adult survival and nesting success. He discovered that dispersals to his newly created vacancies followed the same pattern as in acorn woodpeckers — offspring only filled vacancies that were of equal or higher quality than their natal situations. Offspring from families residing on high-quality areas thus had fewer 'acceptable' outside opportunities; consequently they remained home for longer periods of time (Fig. 10.2b). If this result proves general, families controlling the highest quality resources will be those with the greatest temporal stability.

One result of prolonged residency at home is an enhanced probability of ascending to the parental breeding position itself, in effect inheriting the family resources. Such inheritance of the family territory has been reported in a large number of species and represents a major route to achieving breeding status among family dwelling species of both birds and mammals (see Macdonald & Moehlman, 1982; Emlen, 1984; Brown, 1987; Stacey & Koenig, 1990; Solomon & French 1996). When offspring from successive generations stay home to inherit the family holdings, the result is the continuous occupancy of the same area by a genetic lineage — the formation of a family dynasty.

As long-term monitoring studies of familial species continue, I expect the discovery of dynasties to become commonplace. It will then be possible to test the prediction that dynastical inheritance will occur preferentially in those family groups that control the highest quality resources.

10.3 Kinship and the tendency to cooperate

Families are fundamentally different from other forms of social groupings because they form by the retention of grown young with their parents. As a result, families are comprised primarily of close genetic relatives. Inclusive fitness theory explains how an individual may enhance its fitness in two ways, directly through the production of its own offspring, and indirectly through its positive effects on the reproduction of relatives (Hamilton, 1964).

Kin selection has long been hypothesized to be a selective factor favouring the evolution of cooperative breeding (e.g. Hamilton, 1964; Brown, 1978; Emlen, 1978, 1984; Vehrencamp, 1979). It predicts that assistance in caretaking should be more prevalent in family groups than in groups of less related individuals. More specifically, within families, such assistance should be expressed to the greatest degree between those individuals that are the closest genetic relatives.

With help from N.J. Demong, I searched the literature for all cases of reported family social structure in birds and mammals. We found 112 species of birds and 63 species of mammals (excluding primates) reported to live in multigenerational family groups. Of these, fully 96% of the birds and 90% of the mammals exhibit cooperative breeding (Emlen, 1995a). Because of the difficulty in generating a complete list of family-living species these figures are undoubtedly overestimates, but their true values will remain high. Assistance in rearing young appears to be the norm within vertebrate family groups.

We next examined all species for which cooperative breeding has been reported and asked what percentage are family-dwelling. Fully 88% of the birds and 95% of the mammals that breed cooperatively live in multigenerational family groups (Emlen, 1995a). Cooperative breeding as a reproductive system in social vertebrates is thus largely *restricted* to familial societies.

To test the prediction that individuals preferentially assist their closest relatives, I examined data from species that live in extended families. Multiple females (or pairs) can breed simultaneously in such families, thereby providing potential alloparents with choices in whom to aid.

Data available to date provide strong support for the prediction. White-fronted bee-eaters, *Merops bullockoides* (Emlen & Wrege, 1988), Galapagos mockingbirds, *Nesomimus parvulus* (Curry, 1988), bell miners, *Manorina melanophrys* (Clarke, 1984, 1989), noisy miners, *Manorina melanocephala* (Poldmaa, 1995), pinyon jays, *Gymnorhinus cyanocephalus* (Marzluff & Balda, 1990) among birds, and lions, *Panthera leo* (Pusey & Packer, 1994), brown hyenas, *Hyaena brunnea* (Owen & Owen, 1984), and dwarf mongooses, *Helogale parvula* (Creel *et al.*, 1991) among mammals, all show the predicted preferential allocation of aid. Only one species, the Mexican jay, *Aphelocoma ultramarina*, is reported to show no apparent kin favouritism; breeders provision fledglings from other nests in their family as much as their own (Brown & Brown, 1990).

My own work on white-fronted bee-eaters (Emlen & Wrege, 1988) provides a clear example of a species whose behaviour supports these predictions. These birds live in extended family groups in which up to four pairs may breed simultaneously. Helpers are non-breeding individuals that join one of the active nests and aid in incubation as well as nestling and fledgling care. Only 50% of the non-breeders become helpers; the rest sit out the season as non-participants.

Kinship proved to be a strong predictor of both: (i) whether a given individual becomes a helper; and (ii) to whom it provides aid. Non-breeders are most likely to become helpers when the breeding pairs in their family are close genetic relatives. When faced with a choice of potential recipient nests, they preferentially help the breeding pair to whom they are most closely related (94% of 115 cases; Fig. 10.3).

There is an additional genetic reason for increased amicable behaviour within intact families: sexual competition is predicted to be largely absent. One of the major recent discoveries in the area of mating system research has been the high frequency of extrapair fertilizations in species previously thought

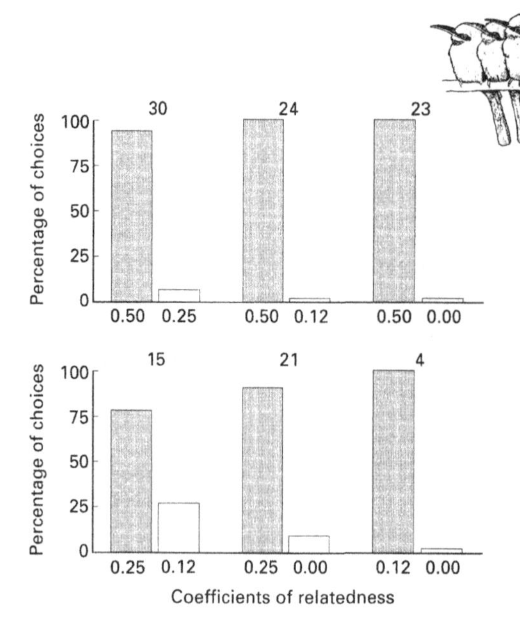

Fig. 10.3 The importance of kinship to helping decisions in white-fronted bee-eaters. Graphs show choices of nests actually assisted by helpers that had multiple recipient nests available within their family groups. Data are presented as dyadic comparisons of nest choices plotted according to the helper's relatedness to the recipient nestlings. Numbers above histograms are sample sizes of helpers in each choice comparison. (Data from Emlen & Wrege, 1988.)

to be monogamous (e.g. Birkhead & Møller, 1992). This risk of cuckoldry has selected for intense mate-guarding and other forms of aggressive defensive behaviour by males during their females' fertile periods.

Within family groups, however, most potential extrapair sexual partners are close genetic relatives. It is well documented that incestuous matings between close kin have deleterious genetic consequences in most normally outbred species (Ralls *et al.*, 1986, 1988; Thornhill, 1993; Jimenez *et al.*, 1994; Keller *et al.*, 1994). For this reason, natural selection is expected to have fostered the development of inbreeding avoidance mechanisms. Thus, sons are rarely expected to compete with their fathers, or daughters with their mothers, for sexual access to the parent of the opposite sex. Neither will siblings compete for sexual access with one another. Instead, mating partners will be selected from outside the family group.

Family-dwelling species provide an excellent testing ground for incest avoidance predictions because mature offspring remain in close social contact with their parents and siblings throughout subsequent reproductive episodes. They thus have unparalleled opportunities to interact sexually with other family members.

A review of the literature indicates that, despite such opportunities, incestuous matings within families (parent–offspring or sibling–sibling) are statistically rare. Fully 18 of 19 avian, and 17 of 20 mammalian (non-primate), familial species for which relevant data are available show strong tendencies to pair exogamously (Emlen, 1995a). Mate-guarding, courtship disruption and other forms of sexually-related aggression are consequently expected to be reduced within multigenerational families, promoting more harmonious interactions within the group.

Incest avoidance, however, is not universal among vertebrates. As ecological constraints on the option of independent breeding become increasingly severe, a point may be reached where it is better to breed incestuously than to risk not breeding at all. These conditions have been modelled by Bengtsson (1978) and Waser *et al.* (1986).

10.4 Conflicts with changing family composition

The death, divorce or departure of a breeding parent, and its replacement from outside the group, will alter the basic genetic and dominance structure of the family unit. As a result, many aspects of the resulting social dynamics of replacement families (the equivalent of stepfamilies in the case of socially monogamous species) are predicted to be different from those of biologically intact families.

A replacement mate will typically be unrelated to extant family members. As such, it will be exempt from incest restrictions. The arrival of such an individual creates potential reproductive opportunities for subordinates that were closely related to the deceased breeder. For example, a son could obtain a share of personal reproduction by mating with its stepmother, or a daughter could do likewise by reproducing with its stepfather.

While such shared mating might be advantageous to the replacement mate as well as to the subordinate, it generally will be *disadvantageous* to the surviving parent. Males will incur lost paternity; females will incur reduced male contributions to parental care. The surviving breeder is thus expected to strongly oppose extrapair mating attempts both by subordinates and by its new mate. The result is a predicted increase in sexually-related aggression.

Stripe-backed wrens, *Campylorhynchus nuchalis*, provide an example of this change in sexual dynamics following a parental repairing. These birds live in nuclear families of two to seven individuals in the savannas of Venezuela. Piper and Slater (1993) contrasted the behaviour of sons in intact versus replacement families. Sons displayed no sexual interest in their mothers, but they frequently engaged in courtship activities with their stepmothers. Paternity data showed that sons never sired offspring with their biological mothers, but they did with their stepmothers. This increase in sexual-behaviour was accompanied by an intensification of sexual competition between fathers and sons. When sons attempted to consort sexually with their stepmothers, fathers upped their mate-guarding behaviours and became increasingly aggressive toward their sons (Fig. 10.4).

Analogous changes occur when a breeder takes a new mate in socially monogamous, extended family situations. Among white-fronted bee-eaters, male offspring pair with unrelated females but bring their mates home to breed

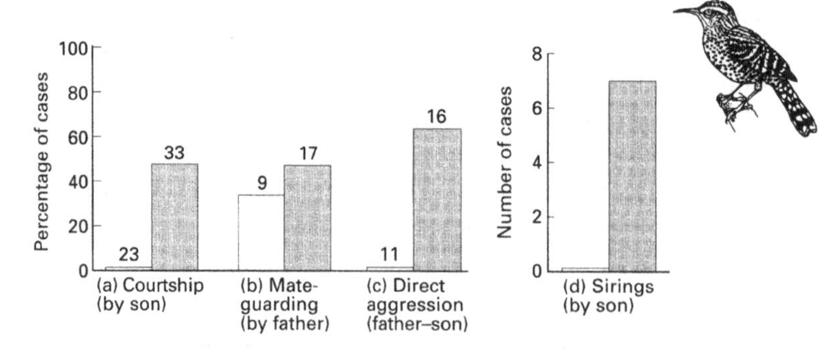

Fig. 10.4 Differences in sexually related behaviours within intact (open bars) and replacement (shaded bars) families of stripe-backed wrens. Histograms plot the likelihood that: (a) sons engage in courtship activities with their mothers or stepmothers; (b) fathers guard their original or new mate; and (c) fathers engage in direct aggression against their sons. Numbers above histograms are sample sizes of cases. (d) The actual number of detected cases of sons siring offspring with their mothers or stepmothers. (Data from Piper & Slater, 1993.)

within their natal family groups. Sons show no sexual interest in their mothers, but brothers do attempt sexual activity with their sisters-in-law and fathers occasionally copulate with their daughters-in-law. Males mated to such females closely monitor allofeedings to their mates and physically disrupt mounting attempts by any extrapair males. These behaviours are in stark contrast to the dynamic between males when the breeding female is a close genetic relative of the extrapair males. In this context, the female is not mate-guarded (i.e. fathers do not monitor feedings by sons to their mothers).

Replacement mates, for their part, are not expected to invest substantially in dependent young remaining from previous breedings because they are genetically unrelated to such offspring. Consequently, they will gain little fitness benefit by investing in the continued care of such individuals. In fact, continued care for extant young may be costly to a new, incoming mate if such care significantly delays its own reproduction or decreases the survival probability of its own, future offspring. When these costs are sufficiently large, the replacement mate may benefit by permanently evicting, or infanticidally killing, dependent young remaining from a previous breeding.

A dependent offspring whose parent takes a new mate is thus predicted to be at increased risk for neglect, abandonment and/or even death caused by the step-parent. This risk will be greatest when the step-parent is of the physically dominant (typically male) sex. This point has been repeatedly confirmed in studies of family-dwelling rodents, carnivores and primates (reviewed in Hausfater & Hrdy, 1984), as well as in many species of birds (e.g. Stacey, 1979; Stacey & Edwards, 1983; Rowher, 1986; Emlen *et al.*, 1989; Koenig, 1990).

Offspring in replacement families also suffer a reduction in the indirect fitness benefit available to them from helping to rear future young, should the new pair (parent plus step-parent) breed. The offspring will share only 25% of their genes with future half-siblings, in contrast to the 50% shared with full-siblings (produced by both biological parents) or with offspring of their own, should they leave and attempt to breed independently. All else being equal, this reduction in indirect benefits is predicted to lead to decreased family cooperation: offspring will exhibit a reduced tendency to provide assistance in the rearing of half-siblings.

The available data pertinent to this prediction are mixed. White-fronted bee-eaters, Florida scrub jays and Seychelles warblers exhibit the predicted adjustment in helping behaviour (S.T. Emlen, N.J. Demong and P.H. Wrege unpublished data; Mumme, 1992b; Komdeur, 1994, respectively). These studies contrasted the likelihood that non-breeding family members would feed young at nests where both parents were breeding with those where one, or both parents had been replaced. Each study found that the proportion of non-breeders which helped decreased when step-parents became breeders (Fig. 10.5). In the Seychelles warblers, those individuals that did help also decreased both their level of provisioning and the number of days they provisioned in families with replacement mates.

Red-cockaded woodpeckers and stripe-backed wrens, in contrast, did not alter their feeding rates when provisioning full- versus half-siblings (Rabenold, 1985; Walters, 1990). In the wrens, the prediction of reduced helping is confounded by the recent finding that offspring engage in copulations with their stepmothers (see Fig. 10.4d; Rabenold *et al.*, 1990; Piper & Slater, 1993; W.H. Piper, personal communication). Such shared breeding may turn out to be common in replacement families. When this occurs the amount of provisioning

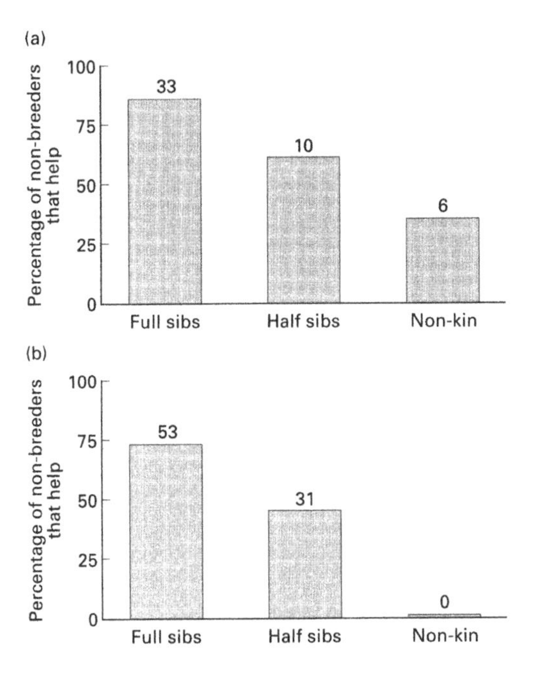

Fig. 10.5 The likelihood that non-breeding family members will serve as helpers, plotted as a function of family type (full-siblings = intact families; half-siblings and non-kin = replacement families with one, and no biological parent remaining, respectively). (a) Florida scrub jay. (Data from Mumme, 1992b.) (b) Seychelles warbler. (Data from Komdeur, 1994.)

expected may depend on the subordinate male's probability of parentage (since the dependent young would now consist of a mix of half-siblings and offspring of the 'helper').

Additional data on changing patterns of helping behaviour following mate replacement, as well as information on the likelihood of helper reproduction with a step-parent, are greatly needed.

Finally, replacement families are predicted to be less stable than intact, biological families. This will be true whether instability is defined in terms of increased dispersal tendencies of mature offspring (family dissolution) or in terms of increased separation rates of re-paired breeders (parental divorce). When reproduction occurs in replacement families, the result is the coexistence of offspring from different sets of biological parents (Fig. 10.6a). A similar mixing occurs when the replacement mate brings offspring of its own into the new pairing (creating what sociologists call 'blended' families; Fig. 10.6b). Such mixing is predicted to intensify conflicts between different family members.

Trivers (1974) was the first to stress how asymmetrical levels of kinship inevitably produce evolutionary conflicts of interest between parents and their offspring. Because offspring, on average, share only 50% of their genes with their siblings, they will have been selected to devalue the welfare of siblings, relative to themselves. Parents, because they are equally related to all of their offspring, will have been selected to counter this devaluation by encouraging offspring to behave more altruistically toward siblings than is in the genetic interests of the offspring.

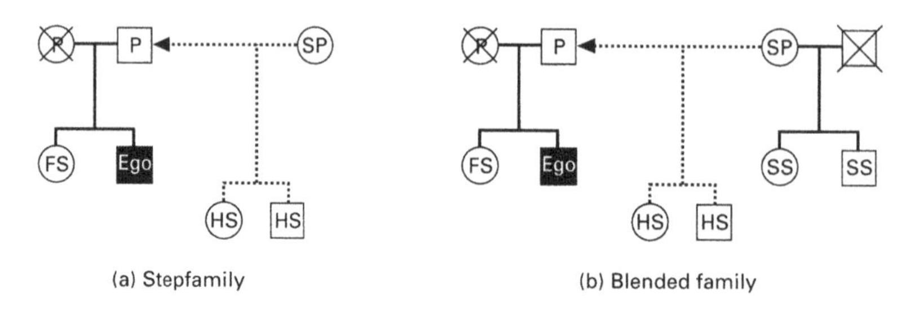

Fig. 10.6 Generalized genealogical diagrams of two forms of replacement family. (a) A stepfamily formed from a formerly intact nuclear family when, following the death of the original breeding female, the male parent takes a new mate and reproduces again. (b) A 'blended' family in which two formerly nuclear families, each with offspring, join when the widowed male parent from one pairs with the widowed female of the other. Diagrams show breeders (top line) and their offspring (symbols connected by lines to the breeders). Individuals with an X through them are deceased. Different rows represent offspring from successive breedings. Arrows denote immigration into the family unit. Symbol shape denotes sex (squares, males; circles, females). A representative offspring, Ego, is shown as a filled square. Letters within symbols denote relationships of different family members to Ego, as follows: FS, full-sibling, $r = 0.5$; HS, half-sibling, $r = 0.25$; P, biological parent, $r = 0.5$; SP, step-parent, $r = 0$; SS, stepsibling, $r = 0$. Families containing members with many high relatedness asymmetries are predicted to have high levels of conflict and social stress.

Consider the magnitudes of these conflicts when young from two or more different pairings are pooled. Each surviving parent continues to be equally related to all of its offspring, but each offspring now shares only 25% of its genes with its half-siblings, and none at all with its stepsiblings. Offspring are predicted to behave progressively less benevolently toward half-siblings and stepsiblings, respectively, resulting in greater levels of conflict both between the sets of offspring and between the parents and young (see Briskie *et al.*'s 1994 study, discussed in Chapter 7). The replacement parent is also expected to 'disagree' with its mate over the allocation of its investment in its own, versus its stepoffspring. If these darwinian predictions are valid, manifestations of this conflict should be detectable in higher rates of aggression between predictable dyads of family members, leading to higher divorce and dissolution rates in replacement, in comparison with intact, families. To my knowledge, such data have not yet been reported systematically for any familial species of bird or non-human mammal.

10.5 Conflict over who reproduces

The preceding section dealt with situations where the dominant pair monopolizes breeding (or the dominant female does, in the case of matrilineal families). But, subordinates would often increase their fitness if they could become reproductives themselves. A dynamic tension thus is expected in many family systems, a tension resulting from conflict over who will and who will not reproduce.

There are two ways a subordinate may become a breeder within its natal group. First, it can simply replace the dominant breeder either by successfully challenging and displacing the dominant, or by waiting to inherit the position following the latter's death or disappearance. Alternatively, it can share reproduction alongside the dominant. This latter produces a more egalitarian, extended family structure.

Consider the first option of simple breeder replacement. When a parent in a socially monogamous nuclear family dies, divorces or disappears, a breeding vacancy is created. Recall that families form when independent breeding options are constrained, and that a major route for becoming a breeder is to inherit the breeding position. The loss of a parent creates the opportunity for a same-sex offspring to ascend to breeding status on its natal territory.

There remains, however, a major obstacle — the surviving parent. The surviving parent usually attempts to retain the breeding position itself. The seemingly obvious solution of the offspring incestuously pairing with its parent virtually never occurs (Emlen, 1995a), presumably because of the high costs of close inbreeding. Instead, whichever individual succeeds in filling the breeding position takes a new replacement mate from outside the family group.

The loss of a parent is thus predicted to result in overt competition between the surviving parent and its opposite-sex mature offspring over the former's retention, versus the latter's assumption, of breeder status. The conflict may be severe, and can extend to interactions between the challenging offspring and any potential replacement mates that are courting the surviving parent.

The outcome of such conflict will depend, in part, on the relative dominance and fighting abilities of the participants. In most birds and mammals, dominance is influenced by gender and age. Males are typically dominant over females, and older individuals are dominant over younger ones. Daughters in socially monogamous families thus will seldom be able to challenge successfully and displace their widowed fathers for breeder status. Sons, however, frequently will be able to challenge and displace their widowed mothers.

Vivid accounts exist of within-family power struggles for breeding status after the loss of a parent (e.g. Hannon *et al.*, 1985; Zack & Rabenold, 1989; Zahavi, 1990). In avian nuclear family species, sons replace fathers as breeders much more often than daughters replace mothers (see Brown, 1987; Stacey & Koenig, 1990). In red-cockaded woodpeckers, whenever a father died, the mother dispersed and the son assumed breeder status (23 instances; Walters, 1990). Active, forceful eviction of mothers by sons following the death of fathers has been observed in Seychelles warblers (J. Komdeur, personal communication), Arabian babblers, *Turdoides squamiceps* (Zahavi, 1990) and white-breasted robins, *Eopsaltria georgiana* (M.N. Brown & R.J. Brown, personal communication).

Analogous conflicts are expected in most matrilineal families. The death of a breeding female creates an opportunity for a subordinate female to assume the breeding position, and an increase in aggressive challenges is predicted at this time. Such power struggles are widespread in matrilineal family-dwelling mammals following the death of a dominant breeder (e.g. MacDonald & Moehlman, 1982; Solomon & French, 1996).

The second way in which a subordinate can become a breeder at home is to share reproduction with the dominant. This can occur if a subordinate either obtains a mate of its own and continues to live and reproduce within the original group, or if it shares sexual access to the mate of the dominant, leading to the production of broods or litters of mixed parentage. Each increases the direct fitness of the subordinate. Each, however, typically *reduces* the fitness of the dominant. When, then, will a dominant share reproduction with a subordinate? What determines how extensive such sharing will be, and among whom the sharing will occur?

These questions are the domain of reproductive skew theory. (For a more formal treatment of this theory, the reader is referred to Emlen, 1982a, 1984, 1995a, 1996; Vehrencamp, 1983a,b; Emlen & Vehrencamp, 1983; Reeve & Ratnieks, 1993; Keller & Reeve, 1994; Reeve & Keller, 1995; see also Chapter 11.) Skew refers to the distribution of direct reproduction among same-sex individuals within a group, and can vary from zero to one. Societies with high skew are those where a few dominant individuals monopolize breeding, whereas societies typified by low skew have more egalitarian breeding. My categorization of families into simple and extended is based on the distinction of whether reproduction is totally monopolized (skew = 1) or is shared (skew < 1). This categorization, however, belies the fact that skew will often be a flexible attribute, with family structure changing predictably as ecological, demographic and social conditions change.

To understand skew theory, we must return to the basic idea that families form when grown offspring delay their dispersal and remain with their parents. Such dispersal decisions may be reversed at any time, however, and it is likely that offspring reassess their options frequently. The profitability of dispersing (relative to staying home) is expected to change with changing conditions. The stability of the family unit, and the distribution of reproduction among its various members, are expected to change as well.

The presence of helpers often increases the fitness of the breeders they assist (their parents, in the case of nuclear families). Helpers may enhance the ability of the familial group to hold or enlarge its territorial holdings. By their alloparental assistance they may reduce the work load and thereby increase the survivorship of the breeders. Finally, their direct helping contributions may result in the production of increased numbers of offspring.

Given the above circumstances, parents, not surprisingly, may be expected to engage in behaviours that lead to the prolonged retention of their offspring in the group.

One way a parent could influence the dispersal decision of its offspring would be to increase the payoff associated with the offspring's staying home. One way of doing this would be to relax its monopolization on breeding and allow the offspring to reproduce concurrently within the group. In the initial paper outlining the ideas of reproductive skew, I stated that '...breeders might be selected who yield a portion of their fitness to auxiliaries...to the point where the fitness gained by the breeder via the assistance of the helper...equaled the fitness forfeited to insure the retention of that helper in the group' (Emlen, 1982a, p. 46). More recently, Reeve and Ratnieks (1993) speak of the dominant offering a 'staying incentive' to keep the subordinate. The key point is that if a parent (or other dominant) can still realize a higher inclusive fitness while sharing reproduction than it would if it monopolized reproduction but the offspring (or other subordinate) left the group altogether, the result can be a 'win–win' situation for both participants (Emlen, 1995a).

Skew models identify four parameters which, together, specify both the conditions under which reproductive sharing should occur and the amount of sharing expected. These are:

1 the magnitude of any benefit realized by the dominant if the subordinate should stay;

2 the expected success of the subordinate if it should leave;

3 the genetic relatedness between potential cobreeders (the dominant and subordinate);

4 the relative asymmetry in dominance between them.

Each parameter influences the relative payoffs of the staying versus the leaving option for both the dominant and subordinate. Collectively, they determine the leverage that the dominant has in 'withholding', and the subordinate has in 'demanding', (anthropomorphically speaking) a share of reproduction.

Predicting the outcome of conflicts over reproductive sharing requires the integration of all four parameters. However, for simplicity, I discuss the influence of each one separately.

10.5.1 The benefit of group-living

This provides the adaptive explanation for why dominant individuals share breeding at all. In the absence of such benefit a dominant is expected to be indifferent to the dispersal decisions of subordinates. Family dissolution will occur before reproductive sharing is expected. However, if the dominant realizes a sufficiently large benefit from the presence of the subordinate, it can forfeit some of its own direct fitness through shared reproduction and still do better than if it retained its monopoly on reproduction but lost the subordinate from the group. Provided that dispersal is a viable option for a subordinate, the greater the magnitude of the group-living benefit realized by the dominant, the greater the potential leverage wielded by the subordinate.

10.5.2 The probable success of attempted independent reproduction

This specifies the profitability of the dispersal option available to a subordinate family member. The relative magnitude of this payoff, in comparison to that of staying at home, is the second major determinant of a subordinate's leverage.

Consider the case where the chance of successful independent reproduction is near zero. (This might be the case when there are no nearby available breeding vacancies.) In such instances, parents have extreme leverage because an offspring has little option but to remain at home. Assuming parents are physically dominant over their young (see Section 10.5.4), strong monopolization of breeding is expected.

The situation changes when opportunities for independent reproduction improve. As ecological constraints become relaxed, a threshold is reached where the fitness associated with dispersal exceeds that of remaining at home as a non-reproductive. The offspring's leverage is enhanced: unless it can gain some direct reproduction within its natal group, it should leave. Two additional conditions typically must be met, however, before reproductive sharing is expected.

1 The parent(s) must benefit sufficiently from the continued retention of the offspring.

2 A non-incestuous mating opportunity must exist for the subordinate.

Interestingly, if ecological constraints become too benign, it will often be in the interests of all parties for the subordinate to disperse, and families will again dissolve. There are two reasons for this:

1 the benefits of independent reproduction for the subordinate may now exceed those of staying, even with shared reproduction;

2 the group productivity benefit to the dominant may decrease to the point where there is no incentive for further retention of the subordinate.

Reproductive sharing thus is expected primarily at intermediate levels of severity of ecological constraints, when conditions afford viable independent reproductive opportunities for subordinates, but are not so benign that the benefits of continued group-living for dominants disappear altogether.

10.5.3 Kinship

Genetic relatedness between group members is the third critical parameter in skew models. One might expect that when a dominant breeder shares reproduction, it will do so preferentially with its closest genetic kin. In fact, exactly the opposite is predicted. This seemingly counter-intuitive result becomes understandable if we contrast the indirect fitness benefits of remaining home as a helper for two subordinate individuals that differ in their degrees of genetic relatedness to the dominant breeder(s). Let us assume that in the absence of personal reproduction, the direct benefits of dispersing slightly exceed those of staying home. The close genetic relative gains a larger indirect benefit through its helping activities such that it requires a smaller 'staying incentive' (if any) to keep it in the family group. Conversely, the more distant or unrelated individual gains little or no indirect benefit. It therefore requires a larger amount of personal reproduction before it will pay it to remain.

This prediction has been confirmed in numerous species, ranging from lions (Packer *et al.*, 1991) and dwarf mongooses (Creel & Waser, 1991; Keane *et al.*, 1994), to Pukeko, *Porphyrio porphyrio*, (Jamieson *et al.*, 1994) and bee-eaters (Emlen & Wrege, 1992a,b). In fact, the most egalitarian of all cooperatively breeding species are non-familial; cooperative groups of Galapagos hawks, *Buteo galapagoensis*, and groove-billed anis, *Crotophaga sulcirostris*, are comprised largely of unrelated individuals (Faaborg *et al.*, 1995; Koford *et al.*, 1990, respectively).

Kinship considerations also predict that reproduction will be shared more equitably in sibling–sibling associations than in parent–offspring groupings (Reeve & Keller, 1995). (Same-sex siblings become potential cobreeders in social vertebrates when they disperse as a coalition to fill a breeding vacancy, or when they compete to inherit a breeding slot following the death of a parent.) This is because the genetic relatedness (r) between each sibling and the offspring of the other will be symmetrical ($r = 0.25$ for each). All else being equal, under conditions where reproductive sharing is favoured by one sibling, it will also be favoured by the other.

In contrast, reproductive sharing generally will not be favoured between mothers and daughters (or fathers and sons). Assuming mate fidelity on the part of the parents, an offspring will be more closely related to its parents' future offspring (its full-siblings, $r = 0.5$) than the parents will be to the offspring's offspring (their grandoffspring, $r = 0.25$). This reduces the offspring's leverage by creating a situation where parents have more to gain from withholding shared reproduction than the offspring do from demanding it (Reeve & Keller, 1995).

Two additional factors reinforce the prediction of reduced reproductive sharing in parent–offspring associations (Emlen, 1996).

1 The age asymmetry between parents and their young assures a dominance asymmetry as well, decreasing the likelihood that an offspring can directly challenge its parent for a share of reproduction.

2 In intact biparental families (where both biological parents are still a mated pair), the option of gaining sexual access to the dominant opposite-sex breeder is unavailable because of selection to avoid incestuous matings.

Because these effects are additive, the contribution of each to the maintenance of high skew in parent–offspring groupings will be difficult to determine (Emlen, 1996).

10.5.4 Social dominance and fighting ability

Dominance hierarchies exist in most vertebrate families. Dominant individuals enjoy certain privileges (e.g. breeding status), but they are always at risk of losing their top position. When a challenger is successful it typically ascends to breeder status. The costs of such challenges can be high for both participants, however, because fights may lead to injury, eviction and even death.

All else being equal, the benefits of such challenges for a subordinate, as well as the risks for a dominant, will be greatest when the disparity in their fighting abilities is least. When the risk is sufficiently great, it will become advantageous for the dominant to share reproduction as a 'peace incentive' with its potential challenger (Reeve & Ratnicks, 1993). Such sharing increases the profitability to the subordinate of staying and continuing to cooperate within the group.

10.6 The myth of the stable family?

It is the dynamic interaction of all four variables of reproductive skew theory that forms the basis of an integrated theory of family social dynamics (Emlen, 1995a). Such theory predicts that both family structure and social interactions among family members will change as conditions vary. Specifically, predictable changes should occur:

1 as the benefits of large group size wax and wane;

2 as ecological opportunities for independent breeding increase and decrease;

3 as breeder deaths and replacements, as well as immigrations, alter family composition;

4 as the social dominance of individuals changes with age and experience.

The idea of the stable family may thus be largely a myth. Even in dynastic situations, where inheritance of the high-quality parental territory is favoured as the primary route to becoming a breeder, severe conflict is expected among potential inheritees. Here too, reproduction may become increasingly or decreasingly shared as outside opportunities, kinship and dominance factors change.

Thus, my initial categorization of family types (see Box 10.1) into simple versus extended, or intact versus replacement, should not be taken to imply fixed entities, but rather to describe various 'family states' that are expected to change in predictable ways with varying circumstances.

The explanatory potential of reproductive skew theory in understanding these changes is illustrated below in three case studies.

10.6.1 Babblers (Zahavi, 1990)

Arabian babblers live in family groups on year-round territories in the Israeli desert. Ecological constraints are severe, and independent breeding opportunities are exceedingly scarce. Males often live their entire lives within their parental territory, and successful groups may pass their territories through the male line over many generations (dynasties). Groups range from two to 22 individuals, with larger groups benefiting by greater stability and growth of their territories as well as by reduced nesting losses caused by intruding bands of floaters. Roughly 50% of groups are simple, nuclear families comprised of parents plus their various aged offspring. The remaining groups are extended families, most commonly consisting of two or more males sharing reproductive access to a single breeding female.

Incest is strictly avoided. Thus, transitions in breeding structure typically occur following the death of a parent. At such times severe fights may break out, during which the loser is either killed or permanently evicted from the group. Zahavi witnessed 'five cases in which an offspring killed its parent of the same sex, following the demise of its other parent' (1990, p. 123). More typically, the loss of a breeding female is followed by acceptance of one or more new females into the group. Subsequently all adult male babblers have the option of breeding (because incest restrictions disappear) and several may copulate with the same female(s). The amount of mate-sharing depends upon rank. Old birds (typically fathers) do not share reproduction with younger males (typically sons), but two males close in age and dominance (often siblings) do frequently engage in mate-sharing.

These flexible changes in family structure, from nuclear to extended and from monogamy to mate-sharing, conform to the predictions of skew theory. Breeding individuals reap advantages from group-living, while the options of subordinates for independent reproduction are severely constrained. As expected, dominant pairs typically monopolize breeding within the family group. Challenges for breeding status occur:

1 following the death of a parent;

2 when the incest restriction is lifted;

3 when the indirect benefits of continued residence as a non-breeder would otherwise decrease.

Furthermore, the challengers (the individuals that often become cobreeders) are more often of similar age (equal dominance) than of different ages. Specifically, father and son rarely mate-share, whereas same age brothers frequently do.

10.6.2 Bee-eaters (Emlen, 1982b, 1990; Emlen & Wrege, 1992a,b)

White-fronted bee-eaters live in socially monogamous, extended family groups of from two to 17 individuals. There is a strong group advantage in that helpers have a large effect on increasing the number of young produced. In our study area in central Kenya, the severity of ecological constraints varies unpredictably across years. In benign years, a large proportion of pairs breed. When conditions are harsh, breeding is more skewed and a larger proportion of family members act as helpers for the breeding few.

Interestingly, not all such helping is voluntary. Dominant breeders harass subordinates, disrupting the latters' breeding efforts. Natal members of such disrupted pairs typically then become helpers for their harassers. Skew theory predicts that dominant breeders have their greatest leverage over those individuals for whom the indirect benefits of staying are greatest (their closest genetic kin), and over those with whom the disparity in fighting ability is greatest (their youngest subordinates). Both are true. The most common category of harassers are fathers disrupting the breedings of their youngest sons. Harassment of distant kin, and of similar aged individuals, is practically non-existent. As a result of changing social and ecological conditions, family dynamics are highly fluid with the degree of reproductive skew changing in a predictable manner given the ecological constraints operating in each breeding season.

10.6.3 Mongooses (Creel & Waser, 1991; Creel *et al.*, 1991; Keane *et al.*, 1994)

Dwarf mongooses live in family groups of variable composition comprising three to 18 individuals. Dominant individuals benefit from living in large groups because non-breeding adults assist in various caretaking activities. The dominant female typically suppresses reproduction in subordinates. Occasionally, however, another female becomes 'pseudopregnant', undergoing hormonal changes resulting in the production of milk which is then used to nurse the young of the dominant female. These 'superhelper' females are virtually always full siblings of the young they nurse; as such they are the family members that stand to gain the largest amount of indirect fitness benefit from engaging in such costly help.

At the other end of the spectrum are females that co-reproduce along with the dominant breeder. Only 13% of subordinate females produce litters, but when they do they den their young with those of the dominant and all young are reared communally. Skew theory predicts that dominant females should share reproduction preferentially with those females that either pose the greatest threat to their own status, or are at greatest risk of dispersing. Again, both are true. The few subordinate females that breed are among the oldest (i.e. the most physically dominant) and the least closely related to the dominant breeders (i.e. the least likely to gain indirectly from staying in the group).

DNA fingerprinting data has shown that reproductive sharing also occurs among male mongooses. Subordinate males sire 24% of offspring, mostly by copulating with the dominant female who then produces a litter of mixed paternity. Again, as predicted by skew theory, 'those subordinates that reproduced were of high social rank and tended to be distantly related to the same-sex dominant' (Keane *et al.*, 1994, p. 65).

10.7 Toward a unified evolutionary social theory

Wilson, writing over two decades ago (1975, p. 4) stated '...when the same parameters and quantitative theory are used to analyze both termite colonies and troops of rhesus macaques, we will have a unified science of sociobiology.' In my opinion, we are very close to reaching that goal.

The building blocks of such a unified science, as described in this chapter, consist of ecological constraints theory, kin-selection theory, social dominance theory and reproductive skew theory. Together they specify the conditions leading to family formation and describe the factors that influence family dynamics, structure and stability.

Cooperatively breeding species have played, and will continue to play, a pivotal role in the development of evolutionary social theory. In the past two decades, studies of such species have largely answered our initial questions concerning the evolution of altruism. In the decades ahead, they will provide answers to a wider array of questions concerning social dynamics. I believe that the intellectual excitement of studying cooperative species no longer lies in their hallmark behaviour of helping, per se, but rather in the opportunities they provide for understanding the complex workings of kin-structured societies.

As the search for general principles continues, there will be a need for greater cross-taxonomic comparison. Too often researchers have partitioned themselves as studying *either* social insects, social birds and lower mammals, or social primates. This artificial distinction must be broken down so that social insect perspectives are incorporated into social vertebrate studies and vice versa. Chapter 9, on social insects, provides grounds for optimism that similar theory applies to all social taxa (see also Reeve & Ratnieks, 1993; Bourke & Heinze, 1994; Keller, 1995).

There is also a need for workers to re-examine existing data for evidence either supporting or refuting the types of predictions outlined here. When searching the literature for such evidence, I was disappointed by the surprising paucity of such analyses (Emlen, 1995a). Knowing that many long-term studies of familial species are underway or already completed, I believe that considerable unpublished data already exist that could be used to test and improve existing theory. I urge my colleagues with comprehensive databases to re-examine them with this goal in mind.

Finally, I suggest to researchers planning future vertebrate studies that they focus on species exhibiting shared reproduction (species living in extended family groupings). Since most of the currently available data come from nuclear family situations, information describing other family systems will be particularly valuable.

Studies of complex, kin-structured societies are important for an additional reason. They provide the best available models for understanding any heritable social predispositions that humans may possess.

Not all persons will agree with this position. It is based on three assumptions (developed more fully in Emlen, 1995b).

1 The expression of many social behaviours is governed, at least in part, by heritable assessment algorithms and decision rules that have been shaped by natural selection.

2 Some of the variance in the expression of human social behaviours is influenced by decision rules that were selected during our long evolutionary history of living in family kin groups. If one is willing to accept assumption number two, where does one look for animal models to provide hints of the types of decision rules which we humans might be predisposed to employ?

3 My third assumption is that organisms living in similarly structured societies are those most likely to have evolved similar sets of algorithms and rules. Most anthropologists and evolutionary biologists believe that during our recent evolutionary history, humans have lived in multigenerational family-based societies (Lee & DeVore, 1968; Lovejoy, 1981; Foley & Lee, 1989; Smith & Winterhalder, 1992). The closest vertebrate parallels, in terms of societal structure, are not most primates, but rather are family-dwelling, cooperatively breeding, birds and mammals. To the degree that early humans were socially monogamous with males providing significant parental care, complex avian societies (where shared care is the norm) and the societies of social canids will provide our best models.

As anthropologists learn more about the 'environment of evolutionary adaptation' of early humans (e.g. Alexander, 1979; Irons, 1990; Wright, 1994), we will be better able to select the optimal animal analogues. But, unless our concept of our ancestral social environment changes drastically, the study of complex, multigenerational, kin-structured societies will provide the looking glass by which we will come to see and understand better the human social condition.

Appendix 10.1 Evolutionary predictions of living within family kin groups

(Emlen, 1995a)

Prediction 1

Family groupings will be inherently unstable. They will form and expand when there is a shortage of acceptable reproductive opportunities for mature offspring, and they will diminish in size or dissolve (break up) as acceptable opportunities become available.

Prediction 2

Families that control high-quality resources will be more stable than those with lower quality resources. Some resource-rich areas will support dynasties in which one genetic lineage continuously occupies the same area over many successive generations.

Prediction 3

Assistance in rearing offspring (cooperative breeding) will be more prevalent in family groups than in otherwise comparable groups comprised of non-relatives.

Prediction 4

Assistance in rearing offspring (cooperative breeding) will be expressed to the greatest extent between those family members that are the closest genetic relatives.

Prediction 5

Sexually-related aggression will be less prevalent in family groups than in otherwise comparable groups comprised of non-relatives. This is because opposite-sex close genetic relatives will avoid incestuously mating with one another. Mating partners will be selected from outside the natal family group (i.e. pairings will be exogamous).

Prediction 6

Breeding males will invest less in offspring as their certainty of paternity decreases.

Prediction 7

The loss of a breeder will result in family conflict over the filling of the resulting reproductive vacancy. In the specific case of simple conjugal families, the surviving parent and its mature opposite-sex offspring will now compete for breeder status. The conflict will be especially severe when offspring are of the dominant sex, and when resources controlled by the family are of high quality.

Prediction 8

Sexually-related aggression will increase after the re-pairing of a parent. In the specific case of simple conjugal families, the surviving parent and its mature same-sex offspring will now compete for sexual access to the replacement mate (step-parent). This conflict will be especially severe when the asymmetry in dominance between the surviving breeder and its same-sex offspring is small.

Prediction 9

Replacement breeders (step-parents) will invest less in existing offspring than will biological parents. They may infanticidally kill current young when such action speeds the occurrence, or otherwise increases the success, of their own reproduction. This will be more likely when the replacement mate is of the dominant sex.

Prediction 10

Non-reproductive family members will reduce their investment in future offspring following the replacement of a closely related breeder by a more distantly or unrelated individual.

Prediction 11

Replacement (step-) families will be inherently less stable than biologically intact families. This will be especially true when offspring from the originally intact family are of the same sex as the step-parent.

Prediction 12

Reproduction within a family will become increasingly shared as the severity of ecological constraints decreases, that is, as the expected profitability of the subordinate's option of dispersal and independent reproduction increases.

Prediction 13

Reproduction within a family will become increasingly shared as the asymmetry in social dominance between potential cobreeders decreases.

Prediction 14

Reproduction within a family will be shared more equitably when the potential cobreeders consist of siblings than when they consist of parent(s) and grown offspring.

Prediction 15

Reproduction will be shared most with those family members to whom the dominant breeders are least closely related. In species in which dominants actively suppress reproduction by subordinates, such suppression will be greatest in those subordinates to whom the dominant is most closely related.

Chapter 11

The Ecology of Relationships

Anne E. Pusey & Craig Packer

11.1 Introduction

Whenever two individuals meet, their behavioural interaction may have consequences that will influence all of their subsequent interactions. Each animal may obtain information about its companion's competitive ability or its propensity to cooperate, or one animal may behave in a manner that raises or lowers its companion's chances of survival. As a result, behavioural ecologists have long realized that social behaviour must often be interpreted in the context of long-term social relationships (Hinde, 1983), and that these relationships are likely to be complex.

By living in groups, companions may benefit from reduced predation risk, improved defence of resources or communal rearing, but they also suffer from increased competition over critical resources. Relationships will therefore be structured both by cooperation and by competition, and, because self-interests will never entirely coincide, social relationships are expected to reflect a certain degree of coercion or compromise. Indeed, individuals can benefit by manipulating not only their own relationships, but also the relationships between their companions (de Waal, 1982; Cheney & Seyfarth, 1990).

What theory helps specify the general principles governing the form of social relationships? Most models of social behaviour have been restricted to assessing the net effect of isolated behavioural interactions. However, treatments of repeated interactions are becoming increasingly common. In this chapter, we first consider how competition structures relationships and then examine the long-term consequences of cooperation.

11.2 Competitive relationships

Animals generally compete for resources on an individual basis. However, the summation of competitive relationships within any given group can lead to complex patterns, and these have often been the subject of research in their own right. We start with the simplest pairwise interactions, discuss how pairwise relationships may lead to hierarchies, then discuss how these hierarchies vary in form and intensity across species.

11.2.1 Competition in pairwise contests

Single encounters

If opponents meet for only a single encounter, game-theoretical models predict that contestants will be more likely to engage in escalated fighting when the costs of injury are low relative to the value of the resource (Maynard Smith & Price, 1973). However, when fighting has significant costs, disputes are more likely to be settled according to certain asymmetries between the contestants (Parker, 1974a; Maynard Smith & Parker, 1976). These asymmetries may be based on differences in resource holding power (RHP) such as size, strength or fighting ability, or they may be uncorrelated with RHP (Hammerstein, 1981). Detectable differences in RHP are expected to lead to some form of assessment, where individuals 'size each other up' before committing themselves to an escalated fight, and the inferior competitor retreats before losing a costly battle. But, when assessment cannot provide a reliable prediction of the outcome and the costs of fighting are especially high, animals may settle their disputes according to an arbitrary cue. For example, individuals following the 'Bourgeois strategy' escalate if they are the first to reach the resource and retreat if they are a latecomer. Finally, if the value of the resource varies between individuals, the hungrier animal will be expected to tolerate greater costs and thus fight longer or harder (Parker, 1984; Houston & MacNamara, 1988).

Empirical studies have shown that all these factors influence fighting behaviour (assessment: Davies & Halliday, 1978; Clutton-Brock & Albon, 1979; Austad, 1983; bourgeois: Davies, 1978; Packer & Pusey, 1982; Waage, 1988; differences in resource value: Krebs, 1982; Rodríguez-Gironés *et al.*, 1996), and Maynard Smith and Riechert (1984) have demonstrated how these different asymmetries interact (e.g. large rival versus small owner, etc.).

Repeated encounters: dominance relationships and winner–loser effects

In stable social groups, the same two individuals are likely to compete repeatedly. If the difference in RHP between each opponent is consistent and detectable, the superior competitor should consistently win each contest and the inferior should defer to its opponent. The pair may then be said to have a dominance relationship. Dominance relationships pervade animal societies (reviewed in Wilson, 1975; Smuts *et al.*, 1987; Langen & Rabenold, 1994; Fournier & Festa-Bianchet, 1995), and they have been measured by the outcome of fights in dyadic encounters, the direction of approach–retreat interactions and the direction of threats and/or submissive gestures. These measures are generally correlated.

However, certain features of dominance relationships suggest that they involve more complicated processes than the simple summation of repeated contests over resources. First, although dominance relationships are frequently correlated with some measure of competitive ability (Wilson, 1975; Fournier & Festa-Bianchet, 1995) (Fig. 11.1a), they often appear to be more clear-cut than expected from the individuals' relative RHP, and some are even based on traits that are uncorrelated with RHP (e.g. seniority or age). Second, the pattern of interactions between pairs often changes through time. Initially, fighting may occur even in the absence of a resource, and the subordinate subsequently defers to the dominant without a fight.

These observations suggest that individuals may compete for dominance per se and learn from the first few encounters with their opponent (see Huntingford & Turner, 1987; Clutton-Brock & Parker, 1995).

Both processes are well illustrated by the development of dominance relationships in blue-footed boobies (Drummond & Osorno, 1992). Two chicks hatch asynchronously in each nest. The older chick frequently pecks and jostles the younger for the first few weeks even in the absence of food and then maintains dominance by less frequent but daily aggression. The younger chick responds to aggression with submissive gestures and rarely challenges the older. Thus, the older chick establishes dominance and subsequently gains greater access to food. The younger chick usually behaves submissively even if it grows larger than the older (females grow more quickly and are eventually larger than males).

The importance of early experience is illustrated by a series of experiments that paired chicks with different backgrounds. When chicks that had been raised alone (and thus lacked any experience of dominance or subordinacy) were introduced to other singletons, the pairs established dominance relationships solely on the basis of size. But, when birds that had been raised in pairs were given new companions, dominant chicks maintained their dominance even when they were smaller than the new subordinate. Although large subordinates challenged the dominant chicks more often, they did not fight as tenaciously as the dominants and were more likely to adopt submissive postures.

Such winner–loser effects are common in species as diverse as crickets and mice (Huntingford & Turner, 1987) and may explain why dominance is sometimes determined solely by age or seniority rather than by differences in RHP. In some species, dominance changes with age according to a bell-shaped function which reflects the greater RHP of prime-aged animals (Fig. 11.1a). However, in other cases, dominance appears to increase continuously with age (e.g. dwarf mongooses: Creel *et al.*, 1992; cooperatively breeding birds: see Chapter 10), although this may still reflect rising RHP if very old individuals are not able to survive in the population. Among male hyenas (Smale *et al.*, 1996), dominance is determined by the length of time an individual has resided in the group (Fig. 11.1b).

Older or senior individuals presumably enjoy an initial competitive advantage so that the younger or newer individuals learn to defer and then continue

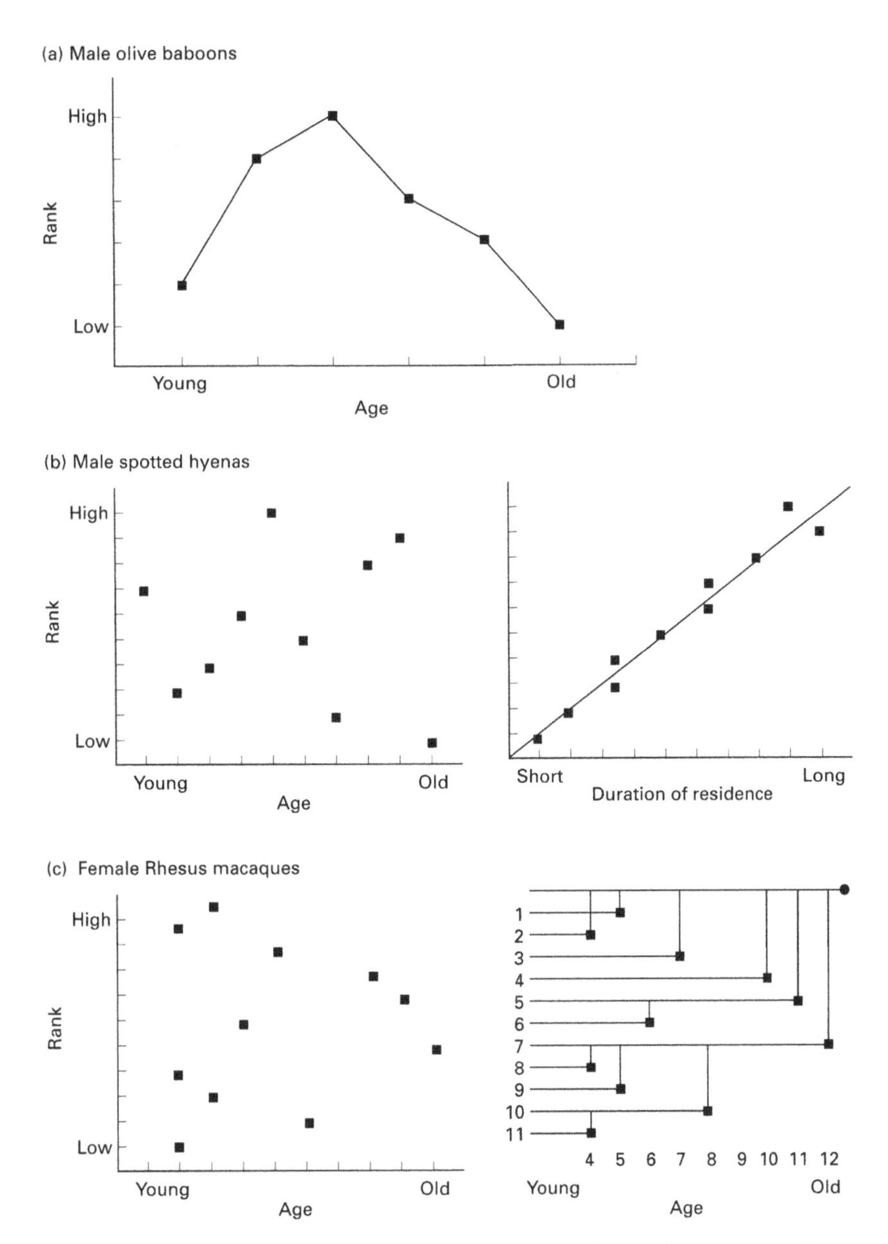

Fig. 11.1 The relationship between age and dominance in three different mammalian species. (a) In immigrant male baboons, dominance shows a bell-shaped relationship because prime-aged males have the highest RHP. (Modified from Packer, 1979.) (b) In immigrant male hyenas, dominance is not correlated with age, but instead increases with the length of residence in the clan. (Left-hand graph provided by K. Holekamp & L. Smale, personal communication; right-hand graph is modified from Smale *et al.*, 1997.) (c) In female rhesus macaques, dominance is unrelated to age except as revealed by birth order. In the right-hand graph, each female's age is indicated by the length of a horizontal line, and her maternity is indicated by each vertical line. Six females were daughters of a deceased female that is marked with a circle in the top-right corner. Note that each daughter outranks all of her older sisters with the exception of the second-ranking female. Also note that each daughter ranks beneath her mother but above every adult that ranks lower than her mother. (Both graphs modified from Hinde, 1983.)

to do so. However, there are currently no theoretical models to explain why some species show such persistent deference and thereby form breeding 'queues'. Specifically, we need to know why RHP provides the basis of pairwise dominance relationships in some species whereas other species form long-term queues that are uncorrelated with RHP (see Ens *et al.*, 1995). Indeed, persistent respect of an arbitrary asymmetry is unexpected unless latecomers gain some incentive from waiting their turn (see Grafen, 1987; Schwagmeyer & Parker, 1987), although there may be some insights into this pattern from the study of alliances (see Section 11.2.2).

11.2.2 Dominance hierarchies

Hierarchies as the outcome of pairwise interactions

Dyadic dominance relationships can often be arranged in a linear hierarchy in which individual A is dominant to the rest of the group, B dominates all but A, C dominates all but A and B, and so on (Schjelderup-Ebbe, 1935; Walters & Seyfarth, 1987). This strict linearity is difficult to explain on the basis of pairwise contests. Unless individual differences in RHP are extreme, the outcome of a dyadic fight will sometimes differ from the direction of RHP differences, especially between animals with near-average abilities. Consequently, the correlation between dominance and RHP will be imperfect, and, since the probability that a hierarchy will be perfectly linear depends on the product of all these dyadic probabilities, linearity should decline rapidly with increasing group size (Landau, 1951; Chase, 1974).

Using a game-theoretical model, Mesterton-Gibbons and Dugatkin (1995) show that the probability of linearity is increased where individuals assess their relative RHP in each dyad and fight only if this exceeds an evolutionarily stable threshold. This threshold increases with the magnitude of variation in RHP, the reliability with which RHP predicts the outcome of a fight and the ratio of the cost of fighting to the value of winning. Such assessment reduces the number of fights and their associated uncertainty of outcome. Nevertheless, their model still predicts that linearity should decrease with group size, becoming highly unlikely in groups larger than about nine individuals. When the observed degree of linearity is greater than predicted by these models, dyadic dominance relationships are likely to be ordered by additional factors besides differences in RHP. These may include psychological reinforcement of 'losing status' through winner–loser effects (see Section 11.2.1) as well as the ordering of relationships through alliances.

Alliances

Some animals (e.g. female Old World monkeys and spotted hyenas) form stable linear hierarchies in groups much larger than 10 (reviewed in Chapais, 1992;

Frank *et al.*, 1995a). In most cases, these hierarchies are not based on individual RHP. Instead, daughters rank directly below their mother in reverse order of age, and the whole matriline ranks above the next most dominant matriline (Fig. 11.1c). Rank inheritance is achieved by mothers supporting their daughters against females from lower-ranking families and by supporting their youngest daughter against her older sisters (Kawai, 1958; Cheney, 1977; Datta, 1983). Females also intervene in disputes between non-matriline members, allying against individuals that rank lower than themselves (Chapais, 1983; Hunte & Horrocks, 1986; Netto & van Hooff, 1986). The consequences of female alliances on rank inheritance and maintenance is demonstrated by a recent series of experiments with Japanese macaques (Chapais, 1988a,b; Chapais *et al.*, 1991)(Box 11.1).

Box 11.1 Experiments on rank determination in Japanese macaques

Chapais and his colleagues performed experiments with a captive group of Japanese macaques consisting of three unrelated matrilines.

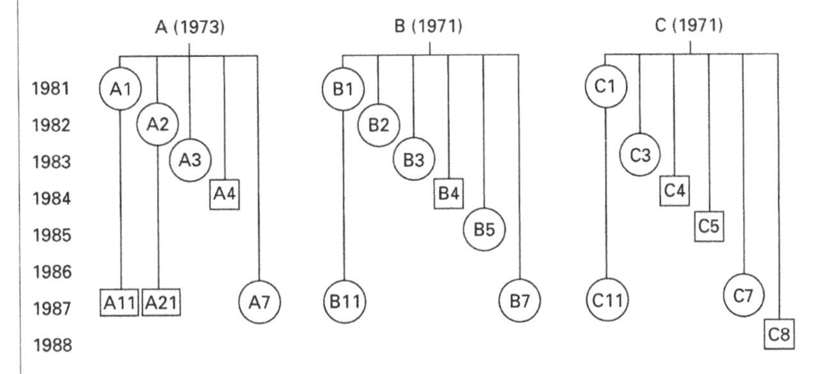

Fig. B.11.1 Composition of the group at the end of the study period (summer of 1989). Circles, females; squares, males. The alpha male is not represented. (From Chapais *et al.*, 1991.)

In the intact group, members of the A matriline ranked above the B and C matrilines, and B ranked above C.

Chapais created a variety of experimental groups to examine how alliances influenced the inheritance and maintenance of dominance rank.

Inheritance of maternal rank

Lone females from high-ranking matrilines were placed with several females from lower-ranking matrilines, and an ally of the high-born female was later added. For example, female A1 was placed with the C matriline. She dropped

Continued on p. 260

Box 11.1 *Cont'd.*

in rank until she ranked below all the members of the C matriline. When her mother (A) was added to the group, A maintained her rank above the C matriline, and defended her daughter against members of the C matriline, so that A1 eventually regained her rank above the C matriline. In all other cases, mothers or adult daughters were also able to help their daughters or younger sisters to regain their rank (Chapais, 1988b).

Maintenance of rank

In a series of 58 experiments, a female from a high-ranking matriline was deprived of her allies and placed with:

1 a single but older/larger low-born female;

2 two or three low-born sisters;

3 a complete subordinate matriline.

Out of 148 dyads of high- and low-ranking females, dominance reversals occurred in 56%. The likelihood that the female lost her rank depended on two factors: (i) her absolute age; and (ii) the number of lower-ranking females. Females were most likely to maintain their rank when they were older and were placed with few rivals.

For example, when a 3-year-old member of the A matriline (A3) was placed with a B female that was 1 year older, A3 maintained her rank, and she was also able to maintain her rank above a pair of B sisters. But, when she was placed with two C sisters, one of whom was 2 years older, she did not maintain her rank. When a female aged 4 years or older from the A or B matriline was placed with an entire lower-ranking matriline, the female only managed to maintain her rank in five of 12 cases. For example, the matriarch A, maintained her rank over the entire B matriline, but B was not able to maintain her rank above the C matriline (Chapais, 1988a).

Chapais concludes the following.

1 Competition for rank is ubiquitous in matrilineal hierarchies, with low-ranking females constituting a potential and constant threat for any high-ranking female.

2 Kin can form revolutionary alliances.

3 Rank maintenance is conditional on the high-ranking female having enough alliance power to counter revolutionary alliances.

4 Competition for rank is somewhat constrained. In cases where the high-ranking female was outnumbered, rank reversals only occurred when the power asymmetry (relative size/age or relative alliance power) was pronounced. Chapais refers to this as a 'minimal risk strategy' of competition for dominance.

Continued on p. 261

Box 11.1 *Cont'd.*

Non-kin alliances

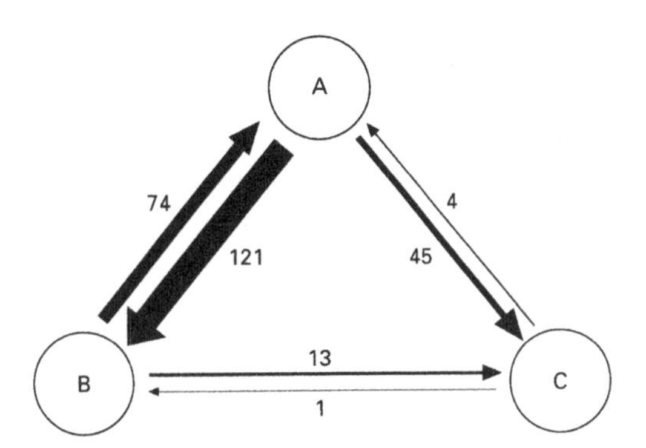

Fig. B. 11.2 Distribution of non-kin interventions among the three matrilines. The thickness of the arrows is proportional to the frequencies of interventions (numbers). Support given to a matriline (direction of arrow) is necessarily against the third matriline. (From Chapais *et al.*, 1991.)

Figure B.11.2 shows the pattern of support between matrilines in the intact group. Members of A and B matrilines consistently supported each other against C, and rarely supported members of the C matriline against the other matriline.

In a series of experiments, a single female from the A or B matrilines was placed with the two other matrilines. In each case, the higher-ranking matriline supported the lone female against the C matriline rather than vice versa. For example, B was placed with the A and C matrilines. When B was alone with the C matriline she lost her rank (see p. 260). However, when the A matriline was also present, they supported her against the C matriline so that she maintained her rank above the Cs.

When A1 was placed with the B and C matrilines, the B matriline outranked A1, but supported her against the C matriline, rather than supporting the C matriline against her. Thus, she maintained her rank above the C matriline (Chapais *et al.*, 1991).

Chapais (1992) suggests that matrilineal hierarchies evolved from an initial state in which females were ranked by age or individual RHP. Females protected their daughters when they were most vulnerable and helped them side-step the lowest rungs of an age-graded hierarchy. Because mothers would be expected to support their daughters against more distant relatives, females eventually banded together with their daughters and granddaughters to overpower any high-ranking female that attempted to act on her own.

A linear hierarchy, however, would be unstable if females only supported members of their own matriline because a large low-ranking family could overthrow a small high-ranking family. However, upheavals in rank (e.g. Erhardt & Bernstein, 1986; Samuels *et al.*, 1987) are rare, and stability apparently results from the pattern of non-kin support. Consider a descending hierarchy of three matrilines: A, B, C. If females supported members of any non-matriline, matriline B could be threatened by 'bridging' alliances between A and C. Also, matriline A could be threatened by 'revolutionary alliances' between B and C. However, non-kin generally only aid each other against matrilines that rank lower than either of them (i.e. A and B against C) (Box 11.1). By assisting higher-ranking matrilines against lower-ranking families, midranking females may prevent the formation of bridging alliances, and, by supporting the most dominant of lower-ranking opponents, the highest-ranking females create a state of dependency that forestalls the formation of revolutionary alliances and maintains the status quo (Chapais, 1992).

In other cases, alliances can destabilize hierarchies and prevent them from being linear. Male chimpanzees live in a fission–fusion society in which males spend considerable periods apart. Male rank depends heavily on alliances, and if male A relies on the support of male C to dominate male B, he may be defeated by B when C is absent (Bygott, 1979). In addition, subordinate males are sometimes fickle in their alliances, with C sometimes supporting A against B and sometimes supporting B against A, thereby inducing a state of dependence in each and gaining greater access to resources (de Waal, 1982; Nishida, 1983). Again, however, there are no satisfactory explanations for why alliances act to stabilize the status quo in some species whereas they are destabilizing in others.

11.2.3 Benefits and costs of dominance

Stable dominance relationships may benefit both dominants and subordinates by minimizing the incidence of serious fighting. A classic study in chickens showed that hens in groups with stable hierarchies fought less and laid more eggs than those with unstable hierarchies (Guhl *et al.*, 1945). Also, damaging fights in a colony of chimpanzees were five times more frequent when male ranks were unstable (de Waal, 1982). Nevertheless, dominance relationships are generally expected to result in inequitable access to resources, and thus high-ranking individuals should enjoy greater reproductive success than subordinates. Although there is considerable evidence that dominant individuals do gain greater access to scarce or monopolizable resources, such as mates, food or safe refuges (reviewed in Huntingford & Turner, 1987; Langen & Rabenold, 1994; Fournier & Festa-Bianchet, 1995), high rank is not always associated with higher reproductive success. This sometimes results from the existence of equally successful alternative strategies, such as satellites or sneaks (Gross, 1996), which opt out of the competition altogether, or because there are also costs associated with maintaining high rank or high RHP. We consider two examples where high rank confers costs as well as benefits.

Spotted hyenas live in clans of up to 80 individuals in which all adult females rank above the adult males, and females form a stable matrilineal hierarchy (Frank *et al.*, 1995a). Hyenas feed on medium-sized antelope, and clan members compete intensely for carcasses with the result that high-ranking animals gain greater access to food (Frank, 1986). High-ranking females reach sexual maturity earlier, have slightly shorter interbirth intervals and recruit more offspring of each sex (Frank *et al.*, 1995a). However, female aggressiveness is associated with a syndrome that includes large body size, elevated levels of androgens and dramatic masculinization of the external genitalia. Female masculinization is suggested to have evolved because of the advantages of aggressive competition (Frank, 1986), but the syndrome carries considerable costs. Females give birth through a penis, and primiparous mothers suffer a high incidence of birth complications that may reduce their lifetime reproductive success by as much as 25% (Frank *et al.*, 1995b).

Female savanna baboons also form matrilineal dominance hierarchies (Cheney, 1977; Hausfater *et al.*, 1982; Johnson, 1987), although a small number of females are able to rise in rank above their matrilines (Samuels *et al.*, 1987; Packer *et al.*, 1995). High-ranking females enjoy higher infant survival, shorter interbirth intervals and younger ages at sexual maturity (Packer *et al.*, 1995). However, rank is not significantly correlated with lifetime reproductive success in some populations because high-ranking females suffer more miscarriages and occasionally suffer from reduced fertility (Wasser, 1995; Packer *et al.*, 1995; but, see Altmann *et al.*, 1995). The source of these costs is unknown, but high-ranking female baboons show an imbalance in their ratio of oestrogen and progesterone (Wasser, 1995), high-ranking juvenile females have higher levels of androgens (Altmann *et al.*, 1995) and, in social carnivores, high-ranking animals show elevated levels of stress hormones (Creel *et al.*, 1996).

Packer *et al.* (1995) suggest that traits conferring high rank in females are subject to strong stabilizing selection because of their potentially negative effects on reproductive physiology. Although individuals that invest too much in aggressive competition suffer these costs, those that opt out of direct competition would suffer reduced feeding success. Thus, dominance behaviour will be expected to persist even if high rank does not confer significantly higher reproductive success.

11.2.4 Alternatives to dominance: ownership

When contestants are evenly matched and the costs of fighting are high compared to the value of the resource, animals may settle contests on the basis of asymmetries that are not correlated with RHP, including prior ownership of the resource (the bourgeois strategy: Hammerstein, 1981; Maynard Smith, 1982), provided that the contest does not approximate a war of attrition (Hammerstein & Parker, 1982). In social groups, this convention would translate into a state-dependent dominance that lasted only as long as one animal retained ownership of the resource, and no persistent dominance relationships would be expected. African lions provide an apparent example.

Female lions compete with their pridemates for access to meat but respect each others' ownership of specific feeding sites at each carcass (Packer & Pusey, 1985). Their prey have tough hides, and meat can be extracted most readily from certain parts of the carcass. The first female to feed from each site refuses to yield to any other female in her pride, lunging and snarling if another ventures too close. Females 'respect' each pridemate's ownership, and roles are reversed as active feeders become temporarily sated and move away from the kill. Breeding appears to be egalitarian, and dominance hierarchies have never been reported in female lions (Packer *et al.*, 1988) — in marked contrast to every other species of social carnivore (e.g. hyenas: Frank *et al.*, 1995a; mongooses: Creel *et al.*, 1992; wild dogs: Malcolm & Marten, 1982).

Male lions form coalitions that compete intensively against other coalitions for access to female prides, but coalition partners also compete for access to individual oestrous females (Packer & Pusey, 1982). When one male forms a consortship with an oestrous female, his coalition partners respect his 'ownership' of the female, avoiding the pair and retreating from the threats of the consorting male. Again, dominance is temporary and state dependent, with an owner on one occasion becoming a rival on another. Individual mating success is unequal in coalitions where companions vary in age or size but is equivalent in coalitions where companions are evenly matched. However, paternity tests revealed a considerable skew in reproductive success in larger coalitions (Packer *et al.*, 1991), even in groups that showed no obvious dominance hierarchy. We do not know whether ownership rules are ignored by those few males who do enjoy high reproductive success, or if reproductive skew results from some other form of male–male competition or from the preferences of the females.

We have argued that lions respect ownership both because contestants are usually evenly matched in age and size and because of the high costs of fighting. Besides having sharp teeth and strong jaws, lions can easily blind an opponent with their claws. However, hyenas are also formidable fighters yet show strict dominance (see Section 11.2.3). An additional cost of fighting to lions is that companions are vital alliance partners for territorial defence and damaging or killing a companion would be highly detrimental (see Section 11.3.3).

11.2.5 Dominance and models of reproductive skew

Models of reproductive skew (see Chapter 10) may help explain the occurrence, strength and intensity of dominance hierarchies. Assuming that the dominant can control the subordinate's reproduction, these models predict that four parameters will affect the extent to which reproduction will be skewed:

1 the probability a subordinate can breed successfully on its own elsewhere;
2 the extent to which the subordinate can increase the dominant's productivity;
3 the genetic relatedness of the pair;
4 their relative fighting abilities.

Under certain circumstances, dominants are expected to cede reproductive opportunities to subordinates to induce them to stay (staying incentives), or to induce them to cooperate without fighting (peace incentives). In general, these incentives will increase with increasing prospects for breeding alone (since leaving becomes more advantageous for the subordinate), increasing productivity by the dominant (since the dominant can afford to surrender more to the subordinate), and decreasing kinship (since the subordinate gains fewer inclusive fitness effects from assisting the dominant). Peace incentives will increase as fighting ability becomes more similar.

How do these incentives influence dominance behaviour? If dominants cede more reproduction to subordinates, do they merely permit subordinates more access to food or mates, while continuing to enforce strict dominance, or do dominance relationships themselves become less clear-cut? Keller and Reeve (1994) suggest that as the incentives increase, the frequency and intensity of dominance interactions should decrease, not only because subordinates gain a lower payoff from 'testing' the fighting ability of the dominant, but also because dominants have less to lose from such challenges. Dominance should therefore be most pronounced when skew is high, but this does not appear to be the case in several mammalian species. Male lions generally exhibit high skew but inconspicous dominance, while female baboons exhibit relatively low skew but pronounced dominance. However, much more research is necessary to quantify the severity of dominance interactions across species and to measure their associated skew.

11.3 Cooperation

Although kinship plays an essential role in the evolution of cooperative behaviour (as reviewed in Chapters 9 and 10), the formation of cooperative relationships between non-relatives has also generated considerable theoretical interest and extensive empirical research. Cooperative behaviour is often assumed to involve a short-term cost (if only through an 'opportunity cost' from failing to behave selfishly), but the long-term consequences of repeated cooperation may be fundamental to understanding social relationships. In fact, most theoretical work explicitly recognizes that cooperation involves repeated interactions between the same pair of individuals.

11.3.1 Reciprocity

The basic paradigm for most evolutionary models has been the 'Prisoner's dilemma' (Box 11.2) where individuals have only two alternatives: to cooperate or to defect. By definition, mutual cooperation gives a higher payoff than mutual defection. However, in the prisoner's dilemma, a defector gains an even higher payoff when paired with a cooperator. Thus, although everyone would do best if everyone cooperated, a selfish mutant could always invade a cooperative population.

Box 11.2 Reciprocity: the basic problem

Considerable theory has been developed to specify the conditions when individuals might form long-term cooperative relationships despite short-term advantages from selfish behaviour. Most models specifically deal with interactions between two unrelated individuals where each individual only has two options, either to cooperate or to defect.

	Against	
Payoff to	Cooperate	Defect
Cooperate	R	S
Defect	T	P

Models of reciprocity are based on a specific set of payoffs called the 'prisoner's dilemma'. If both players cooperate, they both receive the reward (R) from mutual cooperation, but if both defect they receive the punishment (P) from mutual defection. If the opponent cooperates, there is a temptation (T) to cheat which causes the lone cooperator to receive the sucker's payoff (S). Because $T > R > P > S$, defection confers the highest short-term benefits. However, repeated interactions theoretically permit the evolution of cooperation through some form of reciprocity. Such strategies cooperate in the first encounter but only continue to cooperate if their opponent also cooperated and to defect if their opponent defected.

The simplest strategy involving reciprocity is called tit-for-tat (TFT). TFT initially cooperates then behaves according to its opponent's behaviour during the prior move. In an iterated game between two individuals playing TFT, both contestants start out cooperatively and both continue to cooperate in all subsequent interactions:

```
   1's payoff   RRRRRRRRRRRRRR...
Player 1: TFT   CCCCCCCCCCCCCC...
Player 2: TFT   CCCCCCCCCCCCCC...
   2's payoff   RRRRRRRRRRRRRR...
```

Continued on p. 267

Box 11.2 ***Cont'd.***

In an iterated game of N encounters, each individual therefore receives a cumulative payoff of RN.

If TFT meets a pure defector (ALL-D), TFT suffers in the initial encounter, but then gains the same payoff as ALL-D in all subsequent interactions:

```
   1's payoff      SPPPPPPPPPPPPP...
Player 1: TFT      CDDDDDDDDDDDDD...
Player 2: ALL-D    DDDDDDDDDDDDDD...
   2's payoff      TPPPPPPPPPPPPP...
```

An iterated prisoner's dilemma gives a cumulative payoff matrix of:

	Against	
Payoff to	TFT	ALL-D
TFT	RN	$S + P(N-1)$
ALL-D	$T + P(N-1)$	PN

In a population of TFT, ALL-D cannot invade if $RN > T + P(N-1)$, which is most likely when N is very large. Although TFT can never invade ALL-D, as $N \to \infty$, $PN \approx S + P(N-1)$, and TFT may persist long enough to gain a foothold in the population, then spread if it can primarily interact with other TFT strategists (i.e. if the population is spatially structured).

Although selfish behaviour is expected to dominate whenever individuals meet for only a single encounter, repeated interactions allow the evolution of cooperation through some form of reciprocity (Trivers, 1971). By basing its own behaviour on the prior behaviour of its opponent, an individual can benefit from mutual cooperation while preventing cheats from prospering.

Following Axelrod and Hamilton (1981), numerous authors have investigated the conditions where reciprocity can evolve. Their original analysis suggested that animals might follow a simple strategy called 'tit-for-tat' (TFT) which shows an initial bias towards cooperation then copies each of its opponent's moves. Box 11.2 presents a simplified version of the iterated prisoner's dilemma and outlines the major features of the game. We seek a strategy that is not only able to withstand invasion by a selfish strategy ('all defect' or ALL-D) but that can also invade a purely selfish population. This latter point is important since selfishness is generally assumed to be the ancestral condition.

Although TFT has often been considered an evolutionarily stable strategy (ESS: *sensu* Maynard Smith, 1982), there is no ESS in the iterated prisoner's dilemma. This partly results from the mathematical pathologies of the repeated game: in a population of TFT, an unconditionally cooperative strategist (ALL-C) would gain the same overall payoff and thus be able to persist by drift. Once enough ALL-C have appeared, the population could be invaded by ALL-D. In addition, certain strategy combinations can replace TFT. A population playing TFT can be invaded by a pair of mutant strategies called tit-for-two-tats (TF2T), which waits for its opponent to defect twice in a row before retaliating, and suspicious TFT (STFT), which has an initial bias to defect before playing TFT (Boyd & Lorberbaum, 1987).

The most important weakness of TFT, however, is that any mistakes can lead to a permanent breakdown in cooperation (Box 11.3). Thus, if both

Box 11.3 The problem of mistakes

As originally formulated, the iterated prisoner's dilemma assumed that animals never made any mistakes in judging their opponent's behaviour, but TFT is so responsive that a single mistake leads to a long series of mutual retaliation:

```
                       Mistake
                          ↓
   1's payoff     RRRRRRRRTSTSTS...
Player 1: TFT     CCCCCCCCDCDCDC...
Player 2: TFT     CCCCCCCCCDCDCD...
   2's payoff     RRRRRRRRSTSTST...
```

By the convention of these models $(T + S)/2 < R$, so these over-reactive responses confer lower fitness than a strategy that can recover from such mistakes and resume mutual cooperation. The simplest of these is TF2T which only stops cooperating after two successive defections by its opponent:

```
   1's payoff       SSPPPPPPPPPPPP...
Player 1: TF2T      CCDDDDDDDDDDDD...
Player 2: ALL-D     DDDDDDDDDDDDDD...
   2's payoff       TTPPPPPPPPPPPP...
```

TF2T is not provoked into defecting by a single mistake:

```
                       Mistake
                          ↓
   1's payoff      RRRRRRRRTRRRRR...
Player 1: TF2T     CCCCCCCCDCCCCC...
Player 2: TF2T     CCCCCCCCCCCCCC...
   2's payoff      RRRRRRRRSRRRRR...
```

Continued on p. 269

Box 11.3 *Cont'd.*

The most robust of these mistake-correcting strategies is called 'Pavlov' because it changes its behaviour after receiving a poor payoff (*P* or *S*), but repeats its behaviour after receiving *T* or *R*. Pavlov also has an initial bias to cooperate:

```
                               Mistake
                                  ↓
   1's payoff          RRRRRRRRTPRRRR...
Player 1: Pavlov       CCCCCCCCDDCCCC...
Player 2: Pavlov       CCCCCCCCCDCCCC...
   2's payoff          RRRRRRRRSPRRRR...
```

Note that the individual receiving *T* defected again in the following move whereas the individual receiving *S* changed its behaviour to defect. But, once both received *P*, both changed their behaviour to cooperate, and mutual cooperation was thereby restored.

Pavlov can also exploit an unconditional cooperator (ALL-C). Whereas TFT or TF2T would quickly return to mutual cooperation after a single round of exploiting ALL-C, Pavlov continues to defect and therefore 'never gives a sucker an even break':

```
                               Mistake
                                  ↓
   1's payoff          RRRRRRRRTTTTTT...
Player 1: Pavlov       CCCCCCCCDDDDDD...
Player 2: ALL-C        CCCCCCCCCCCCCC...
   2's payoff          RRRRRRRRSSSSSS...
```

Although Pavlov performs better than TFT in an imperfect world, it fares poorly in a population of ALL-D (only receiving *SPSPSPSP*...).

	Against	
Payoff to	Pavlov	ALL-D
Pavlov	RN	$N(S+P)/2$
ALL-D	$N(T+P)/2$	PN

Thus, Nowak and Sigmund's (1993) analysis suggests that Pavlov is most likely to appear in a population after TFT has replaced ALL-D and can only persist if $R > (T+P)/2$.

opponents play TFT, a single mistake sets up an alternating pattern of retaliation with each animal changing its behaviour in response to its partner until a second mistake either restores mutual cooperation or leads to mutual defection. Box 11.3 outlines the various strategies that might permit reciprocity in the face of mistakes. These include TF2T which 'forgives' a single mistake and 'Pavlov' which repeats its prior move after receiving a good payoff (*R* or *T*), but changes its behaviour after a poor payoff (*S* or *P*). If cooperation breaks down, Pavlov results in mutual defection for just one move before cooperation is restored. Nowak and Sigmund's (1993) analysis suggests that while TFT may often be the first form of cooperation to appear in a selfish population, it will generally be replaced by Pavlov.

It is important to note, however, that all these analyses are based on quite restrictive assumptions. Pairs of animals are assumed to make simultaneous decisions without any information about the opponent's current move (for a rigorous examination of alternating decisions see Nowak & Sigmund, 1994). Contests are always between pairs of individuals; reciprocity based on communal resources is unlikely to evolve in groups larger than two (Boyd & Richerson, 1988). Payoffs are assumed to be perfectly symmetrical and to remain unchanged through evolutionary time. Asymmetrical payoffs may mean that one partner faces a prisoner's dilemma while the other does not (Packer & Ruttan, 1988), or that a dominant individual might have the option of manipulating its subordinate partner (Clutton-Brock & Parker, 1995; see also Section 11.3.2).

Even in games that allow mistakes (e.g. Nowak & Sigmund, 1993), the mistake rate is low, the game is assumed to be infinitely iterated and players can only remember the opponent's behaviour in the prior move (thus excluding tolerant strategies such as TF2T). The assumption of an infinite iteration allows the outcome of the initial encounter to be ignored. Thus, TFT behaves the same as ALL-D in a population of ALL-D and so can persist until drift enables a critical mass of TFT cooperators to interact and reap the rewards of mutual cooperation; TFT then permits the coexistence of ALL-C or Pavlov, but ALL-C is prone to invasion by ALL-D, whereas Pavlov is less so. Infinite iteration also assumes that animals will not 'discount' future payoffs, and thus that they will forego immediate payoffs for higher future rewards, whereas empirical studies show that animals have strong preferences for immediate gain (Stephens *et al.*, 1995).

The theoretical literature on the iterated prisoner's dilemma may therefore be summarized as supporting a general tendency toward some form of cooperation in very long-term relationships. The conditions where cooperation is actually expected, however, are extremely limited — provided that animals are only allowed two options in each interaction: either to cooperate or to behave selfishly.

11.3.2 Other routes to cooperation

Given the substantial difficulties facing the evolution of reciprocity, it is important to consider alternative explanations for cooperation, especially when evaluating empirical data. We outline three of these in Box 11.4.

Box 11.4 Other routes to cooperation

A: Short-term mutualism

Cooperative interactions do not necessarily involve a prisoner's dilemma. Two heads may be better than one (with both individuals receiving the largest award from mutual cooperation), and there may be no advantage from exploiting a companion's behaviour.

	Against	
Payoff to	Cooperate	Defect
Cooperate	$R = 6$	$S = 3$
Defect	$T = 2$	P

In this case, cooperation is expected even when opponents meet for only a single encounter. By definition, $R > P$, but if $P > S$, defection is also an ESS, and cooperation is most likely to evolve in a structured population (where families of cooperators typically receive R, while families of defectors mostly receive P).

B: Long-term models of mutualism

Payoffs may conform to the prisoner's dilemma in the short term but become mutualistic in the long term. This transition, however, does not result from following complex behavioural strategies (as is the case for reciprocity), but from the rebounding consequences of selfishness.

Consider a daily game where the payoffs depend not only on the behaviour of each contestant, but on the number of animals playing the game each day, and each animal receives a higher payoff as a member of a pair than as a solitary. Let the pairwise payoffs follow the prisoner's dilemma, but assume that defection increases the partner's risk of mortality. The survivor subsequently becomes a solitary who receives a daily payoff of X, where $X < S$.

In the extreme case, a single defection kills the partner, and the partner cannot be replaced. Thus, the payoffs are accumulated as follows:

```
   1's payoff       S
Player 1: ALL-C     C
Player 2: ALL-D     DDDDDDDDDDDDDD...
   2's payoff       TXXXXXXXXXXXXX...
```

When both defect, one or both is killed. Payoffs are either:

```
   1's payoff       P
Player 1: ALL-D     D
Player 2: ALL-D     DDDDDDDDDDDDDD...
   2's payoff       PXXXXXXXXXXXXX...
```

Continued on p. 272

Box 11.4 ***Cont'd.***

Or:

```
   1's payoff      P
Player 1: ALL-D    D
Player 2: ALL-D    D
   2's payoff      P
```

But when both cooperate:

```
   1's payoff      RRRRRRRRRRRRRR...
Player 1: ALL-C    CCCCCCCCCCCCCC...
Player 2: ALL-C    CCCCCCCCCCCCCC...
   2's payoff      RRRRRRRRRRRRRR...
```

Thus, the cumulative payoffs are:

	Against	
Payoff to	ALL-C	ALL-D
ALL-C	RN	S
ALL-D	$T + X(N-1)$	P *or* $P + X(N-1)/2$

If $N > 1$ and $R > (T + X)/2$, the most likely outcome would be unconditional cooperation. However, the population would have to be structured for cooperation to invade a population of pure defectors. Note that in this case, pure cooperation clearly fares better than any conditional strategy (such as TFT) if animals risk mistaking their companion's behaviour.

Group-living often confers advantages strong enough to counter the short-term advantages of selfish behaviour. A good example is Lima's (1989) model of cooperative vigilance. Each individual must trade-off vigilance for foraging. A more vigilant individual protects itself and its companion from predation, but spends less time foraging. The short-term selfish optimum should therefore involve relatively little vigilance. However, because the predator is assumed to capture only one prey at a time, an animal's survival greatly reduces its companion's risk of predation (due to the dilution effect). Thus, in a repeated-encounter game, animals will have to improve their companion's chances of survival so as to minimize their own long-term risk of predation.

C: Producers versus scroungers

Obtaining resources can often entail an inherent cost, but no resources will be

Continued on p. 273

Box 11.4 ***Cont'd.***

available unless someone works to extract them. In this case, there may be a temptation to defect ($T > R$), but cooperators are at an advantage when they are rare.

	Against	
Payoff to	Cooperate	Defect
Cooperate	$V/2 - C/2$	$V/2 - C$
Defect	$V/2$	0

Two individuals compete for a resource of value, V; extracting the resource incurs some cost, C. As long as someone works to extract the resource, it is divided between both contestants. If both work together, each pays 50% of the costs. As long as $V/2 - C > 0$ cooperators (producers) can invade defectors (scroungers), but if defectors gain equal access to the resource without paying any costs, scroungers can always invade producers.

In this case cooperation will be frequency dependent, and in large groups subsets of cooperators may coexist with a proportion of defectors.

D. Coerced cooperation

Most models of cooperation assume that animals have only two options, either to cooperate or to defect. In addition, these models assume that both players possess exactly the same skills and RHP. However, most animals live in groups with definite dominance hierarchies and their relationships will generally be asymmetrical (see Section 11.2). Thus, a difference in RHP or dominance rank may permit one animal to coerce its companion into behaving cooperatively (Clutton-Brock & Parker, 1995). Note that the following analysis now describes phenotype-limited strategies.

Consider a two-person game with the same basic structure as the prisoner's dilemma. However, now allow a dominant animal to punish any subordinate that fails to behave cooperatively. Let d be the costs to the dominant of punishing the subordinate, and i the costs to the subordinate of being punished. If $S > P - i$, then a defecting subordinate will do better to change its behaviour to cooperate than to risk further punishment. If the subordinate subsequently starts to cooperate and $T - d > P$, the dominant benefits by punishing the subordinate every time it defects:

Continued on p. 274

Box 11.4 *Cont'd.*

```
                          Punishment
                              ↓
   1's payoff               P-i SSSSSSSSSSSS...
Player 1: subordinate       D   CCCCCCCCCCCC...
Player 2: dominant          D   DDDDDDDDDDDD...
   2's payoff               P-d TTTTTTTTTTTT...
```

However, if $S < P-i$, the subordinate would be better off incurring repeated punishment rather than submitting to the demands of a tyrannical despot. In which case, the dominant might be expected to behave as a benevolent despot, habitually cooperating and forcing the subordinate to cooperate as well:

```
                          Punishment
                              ↓
   1's payoff               T-i RRRRRRRRRRRR...
Player 1: subordinate       D   CCCCCCCCCCCC...
Player 2: dominant          C   CCCCCCCCCCCC...
   2's payoff               S-d RRRRRRRRRRRR...
```

Here the subordinate will cooperate if $R > T-i$, and the dominant will be expected to punish each defection if $R-d > S$.

The most obvious explanation for cooperation is that when an individual meets a cooperative partner, it gains a higher payoff from cooperating than from defecting. In the absence of any temptation to cheat, cooperation is clearly an ESS even in a single-encounter game and no elaborate safeguards against cheating will be expected. The set of payoffs described in Box 11.4A are often described as 'byproduct' mutualism (West-Eberhard, 1989) because cooperation is the best strategy regardless of the opponent's behaviour, and thus cooperation is merely a byproduct of following the optimal strategy.

Mutualistic advantages can also arise in long-term interactions, even if the payoffs from any single encounter are consistent with the prisoner's dilemma. In Box 11.4B, we present a simplified example in which individuals gain a short-term advantage from cheating, but this selfish behaviour results in the loss of their companion with the consequence that they must subsequently live alone. With strong enough advantages of grouping, unconditional cooperation will be favoured in an iterated game (Lima, 1989). Again, in such situations, no elaborate cooperative strategies are necessary: the well-being of one's companions has an important effect on personal fitness. However, the outcome of more realistic games will depend on the ease with which partners can be replaced.

Cooperation may not always be two-sided or ubiquitous. 'Producers' may extract more resources per capita than 'scroungers', yet a population of producers will often be prone to invasion by scrounging and vice versa (see Barnard, 1984). In pairwise interactions, this may result in one individual cooperating while the other defects (Box 11.4C), but in larger groups a subset of individuals would be expected to cooperate while the remainder defect (see Section 11.3.3). Thus, a majority of individuals may cooperate unconditionally.

Finally, several recent models have examined the possibility that cooperation may be imposed by coercion (e.g. Boyd & Richerson, 1992; Clutton-Brock & Parker, 1995). This approach is especially relevant here, since dominance relationships are explicitly incorporated into the structure of the payoffs — and dominance relationships are nearly universal in animal societies. The important addition of these models, however, is that individuals can physically punish defectors whereas earlier models only permit TFT or Pavlov to withhold further cooperation.

As outlined in Box 11.4D, dominant animals may often be expected to coerce subordinates to behave cooperatively. For simplicity, we have assumed that a single punishment induces the subordinate into cooperating in all subsequent encounters. In reality, of course, the evolution of 'enforced cooperation' will depend on the speed with which the subordinate 'learns' to cooperate in response to the dominant's behaviour (Clutton-Brock & Parker, 1995) as well as on the rate of recidivism. If dominants can impose sufficiently heavy costs they may be able to exploit the subordinate and maintain a highly one-sided relationship. However, in other circumstances the dominant would be forced to settle for a mutually cooperative relationship (also see Section 11.2.5 and Chapter 10).

In summary, none of these routes to cooperation requires strategies as elaborate as TFT or Pavlov, and in many cases such 'reactive' strategies would be highly disadvantageous compared to a simple strategy of 'all cooperate'. Coercion may be reactive (i.e. the subordinate is punished if it does not cooperate), but only leads to mutual cooperation when the dominant cannot maintain a purely exploitative relationship with the subordinate.

11.3.3 Empirical studies

Several recent reviews (e.g. Packer & Ruttan, 1988; Emlen, 1991; Clements & Stephens, 1995) have concluded that reciprocity is rare to non-existent in nature, while others suggest that reciprocity is widespread (e.g. Dugatkin *et al.*, 1992). Reciprocity remains one of the most beguiling concepts in behavioural ecology, but the most formidable obstacle to proving its existence stems from the difficulty of measuring short-term payoffs. Does a particular example of cooperation really follow a prisoner's dilemma or is it mutualistic?

Here we outline three different studies of cooperative behaviour that highlight the difficulties in disentangling the alternatives.

Cooperation in African lions

Field studies confirm that African lions are highly cooperative. Lions hunt in groups (Scheel & Packer, 1991; Stander, 1992), females raise their young in a communal crèche (Pusey & Packer, 1994) and both sexes defend joint territories against like-sexed individuals (McComb *et al.*, 1994; Heinsohn & Packer, 1995; Grinnell *et al.*, 1995). Lions of each sex gain higher reproductive success by living in groups (Packer *et al.*, 1988): males in larger coalitions gain higher per capita reproductive success than those in smaller coalitions, and females in moderate-sized prides have higher fitness than those in very small prides. These advantages, however, are the summed outcome of a variety of behaviours, and we discuss the evolutionary basis of each in turn.

Group hunting. By hunting together, individuals can increase their joint hunting success so that $R > P$. However, an individual joining a hunt is likely to incur costs, whether from the energetic effort required to capture the prey or from the risk of injury while subduing the prey. When the prey is large enough to feed an entire group, the decision to join a group hunt may follow a prisoner's dilemma (Packer & Ruttan, 1988). If the success rate of a lone hunter is already high, a second hunter may not be able to improve the group's success rate sufficiently to overcome its own costs of hunting, in which case $T > R$. However, if individual success rate is very low, two heads may be so much better than one that the improved chances of prey capture can overcome the personal costs of joining the hunt, in which case $R > T$.

Lions, like most other group hunters (Packer & Ruttan, 1988), show the clearest evidence of cooperation when R is likely to be greater than T, and thus when cooperation is mutualistic. Scheel and Packer (1991) found that individual behaviour during group hunts could be classified into several discrete strategies, one of which, 'refraining', involves almost no participation in the hunt. 'Refrainers' remain stationary or move only a few paces towards a distant prey animal and do not join in the final charge. Other lions clearly cooperate, actively stalking the prey and participating in the final charge. Consistent with predictions of mutualism, the probability that Serengeti lions cooperated during a group hunt depended on two factors relating to hunting success.

1 Lions were most likely to cooperate when the prey was difficult to capture, and showed higher levels of 'refraining' when their companions pursued prey that was more easily captured.

2 Since male lions are less skillful hunters than females, they are less able to improve the females' success rate, and they 'refrained' more than females.

Perhaps the most impressive data on cooperative hunting in lions comes from Stander's (1992) studies in Namibia. These animals demonstrated a clear division of labour, with certain individuals habitually stalking to the left, others habitually stalking to the right and the remainder moving directly towards the prey. However, they hunted fleet-footed prey in open habitat, and individual hunting success was close to zero. Thus, their cooperative hunting was highly likely to have been mutualistic.

Across a variety of species, cooperative hunting appears to be restricted to circumstances of simple mutualism (see Caro, 1994; Creel & Creel, 1995), even though cooperation might theoretically evolve by reciprocity. However, it might often be impossible to detect whether a hunt has failed due to the elusiveness of the prey or to the defection of another hunter (Packer & Ruttan, 1988). Strategies such as TFT and Pavlov are sensitive to the mistake rate (see Box 11.3), and thus reciprocity may not evolve in situations where failure is the most likely outcome of a cooperative interaction.

Although cooperative hunting appears to be largely mutualistic, this conclusion must be considered provisional until experimental tests of reciprocity can be devised. All published studies are strictly correlative; no field study has ever measured T, R, S and P.

Communal cub-rearing. Female lions pool their cubs in a 'crèche' and raise them communally, even nursing each other's cubs. In the short term, communal nursing may involve a prisoner's dilemma (see Caraco & Brown, 1986). A defector directs all her investment to her own cubs while a cooperator allows all cubs to nurse equally. Thus, the defector pays some small costs of vigilance to ensure that only her cubs nurse from her, but her cubs gain exclusive access to her milk. A pair of cooperative females avoid the costs of excluding other cubs ($R > P$), but when a defector is paired with a cooperator, the defector's cubs gain all their mother's milk plus a portion of the cooperator's (T), while the cooperator's cubs only gain a portion of their mother's milk (S). With trivial costs of enforcement, $T > R > P > S$.

However, females do not form crèches in order to engage in communal nursing. Crèches are protective coalitions against infanticidal males, and groups of mothers are more effective than solitaries (Packer *et al.*, 1990). Once together, though, the mothers have to balance a number of different activities, and cubs often try to nurse when their mothers are resting or sleeping. Thus, mothers are constrained to be gregarious and communal nursing is largely a consequence of their cubs' parasitic behaviour (Pusey & Packer, 1994). It is therefore misleading to view communal nursing in isolation. Instead, losing milk to non-offspring appears to be a cost of forming the crèche, with individual females adjusting their milk distribution according to their personal costs.

Field observations revealed that female lions nursed indiscriminately only when paired with their closest female kin. When their companions were distant relatives, mothers directed most milk to their own offspring. Lion litter size varies from one to four, and milk production appears to be independent of litter size. Females with small litters were more generous than mothers of larger litters, presumably because of the lower costs of milk loss (Pusey & Packer, 1994). Although the precise payoffs are impossible to measure, there is no evidence that communal nursing in lions involves any form of reciprocity; after kinship and litter size were controlled for, females did not nurse their companions' cubs according to the extent to which their own cubs were nursed by their companions.

Group territoriality. Lions maintain joint territories that persist over generations, and successful reproduction requires a high-quality territory. The lion's roar is a territorial display, and both sexes respond cooperatively to the pre-recorded roar of a like-sexed stranger, approaching the speaker and even attacking a stuffed lion hidden nearby. Playback experiments reveal that lions are most likely to respond when they outnumber their opponents and that they often monitor each other's behaviour while approaching the speaker (McComb *et al.*, 1994; Grinnell *et al.*, 1995). Territorial defence is likely to involve a short-term prisoner's dilemma, with joint defence being more effective than non-defence ($R > P$) and a defector suffering fewer risks of injury than a cooperator during an intergroup encounter ($T > S$). A single defender can repel a lone intruder, so $T > R$.

However, several studies suggest that lions cooperate unconditionally, rather than basing their responses on their companions' behaviour. Grinnell *et al.* (1995) found that males would approach the speaker when their companions were absent, thus cooperating even when their companions' response could not be monitored. Even more strikingly, Heinsohn and Packer (1995) found that certain females habitually lagged behind their companions during the approach to the speaker. 'Leaders' could recognize whether their companions were also leaders or if they were 'laggards'. However, when paired with a laggard, a leader would continue toward the speaker, arriving considerably earlier than the laggard. Thus, the leaders cooperated rather than 'defected' in response to their partner's defection, nor did they physically punish a laggard for failing to cooperate.

As in communal nursing, a focus on short-term payoffs is probably inadequate: male lions enjoy only a brief tenure within a pride and they need their companions for future interactions (Grinnell *et al.*, 1995). Males must therefore behave cooperatively at every opportunity to enjoy long-term mutualistic advantages (see Box 11.4B). For females, the territory is a long-term resource that must be defended habitually and any failure to defend the territory today will result in fewer resources tomorrow (Heinsohn & Packer, 1995). However, as long as enough females maintain effective defence of the territory, a proportion of laggards can take advantage of their companions' behaviour, resulting in a mixture of producers and scroungers (see Box 11.4C).

In summary, the precise evolutionary basis of the lions' cooperative behaviour is difficult to determine because of their complex social system and the long-term consequences of their short-term decisions. Furthermore, behaviour in one context may well rebound on another, with foraging success influencing territorial defence, and so on. Nevertheless, it is striking that one of the most cooperative of all mammalian species shows no obvious adherence to the rules of reciprocity.

Predator approach behaviour

Many species of fish approach and orientate towards a novel predator, and this behaviour has been dubbed 'predator inspection' whereby individuals obtain information about the predator's hunger and aggressiveness (Pitcher *et al.*, 1986; Milinski, 1987). Repeated tests have shown that experimental groupings of sticklebacks, guppies and mosquito fish will move into close proximity of the predator, and they often do so in pairs (Milinski, 1992; Turner & Robinson, 1992). Milinski (1987) was the first to suggest that predator inspection involved a prisoner's dilemma, defining any move towards the predator as 'cooperation' and any movement away from the predator as 'defection'.

Cooperative pairs gain R because of the advantages of inspecting the predator from the comparative safety of a group. Defecting pairs only gain P because they learn nothing about the predator's state. Single defectors gain T by remaining at a safe distance and watching the predator's response to the cooperative companion. Finally, the lone cooperator gain S from incurring the risk of being selected by the predator. Thus, $T > R > P > S$. Milinski (1987, 1990) and others (Dugatkin, 1988, 1991; Masters & Waite, 1990; Huntingford *et al.*, 1994) have shown that sticklebacks typically copy each other's behaviour while approaching a predator, and most have interpreted this pattern as evidence of TFT.

However, several other authors have viewed the same behaviour quite differently, interpreting it instead as 'pursuit deterrence' (e.g. Lazarus & Metcalfe, 1990), and one study has shown that non-approaching fish suffer higher risk of predation than the approaching partner (Godin & Davis, 1995a,b). This result is consistent with a broad literature suggesting that cryptic predators are less likely to attack once they have been detected (e.g. FitzGibbon & Fanshawe, 1988). If this is also generally true in predatory fish, the short-term payoffs from approaching the predator would be mutualistic, with $R > S > P > T$. Copying the partner's behaviour in this case can simply be explained by shoaling: individuals benefit from the dilution effect by staying as close together as possible, whether moving toward or away from the predator.

Despite nearly 10-years research in a highly tractable system, where straightforward manipulations can be performed with mirrors, models and trained fish, it is still not clear which of these alternative viewpoints is more nearly correct. While it seems intuitive that $R > P$ (the fish do typically approach the predator in pairs, so the behaviour must confer some benefit), it has proven remarkably difficult to measure S and T to everyone's satisfaction.

Godin and Davis's (1995a,b) data suggest that $S > T$, but their methods have been criticized by Milinski and Boltshauser (1995), and data collected by Dugatkin (1992) suggest that, in mixed groups, individuals that had previously been classified as inspectors suffered greater predation risk than non-inspectors.

In the absence of any consensus concerning the payoffs, it is still possible to test hypotheses that distinguish between TFT and shoaling. Assuming that the fish are in a prisoner's dilemma and that they are playing some form of reciprocity, they should copy each other's behaviour according to certain rules (Lazarus & Metcalfe, 1990; Reboreda & Kacelnik, 1990; Huntingford *et al.*, 1994; Stephens *et al.*, 1996). If both animals cooperate on one move they should continue to cooperate on the following move; if both defect, both should either defect again (if playing TFT) or cooperate (if playing Pavlov). If one cooperates and the other defects, then the behaviour of one or both will be expected to change depending on whether the animals are playing TFT or Pavlov.

The predicted transition matrices for these two strategies are given in Table 11.1, as well as empirical data collected by Stephens *et al.* (1996) who took as separate 'moves' the animals' net movement every 1.09 s. These data show that while mutual cooperation is followed by further cooperation 57% of the time, and that mutual defection is typically followed by further defection (consistent with TFT), the most important predictions of TFT and Pavlov are not met. Whenever a non-cooperator is paired with a cooperator, the fish were not reactive but instead tended to repeat their prior move (the most common response to CD was CD, etc.).

Consistent with shoaling, however, fish were most likely to copy each other's movements when they were in close proximity and to move towards each other when farther apart (even when this meant moving in opposite directions with respect to the predator). Thus, a major goal of their movements seems to be to maintain close proximity with their companion, rather than to maintain a close check on each other's cooperative tendencies.

Critics of the shoaling hypothesis counter that if there was an inherent, mutualistic advantage of grouping, then every fish in the tank should approach the predator together whereas they typically approach in pairs (Milinski *et al.*, 1990; Dugatkin, 1996). Pairwise approaches are also considered evidence of reciprocity since TFT is most likely to evolve in small groups (Boyd & Richerson, 1988). Proponents of shoaling counter that the advantages of predator deterrence are likely to be frequency dependent, so that the optimal group size for approaching the predator may only be one or two (Godin & Davis, 1995b; Stephens *et al.*, 1996).

The chief attraction of this system compared to most field studies has been that short-term payoffs are likely to be the sole determinant of the fishes' behaviour. However, it is disturbing to realize how inconclusive the predator inspection/deterrence controversy has been. Perhaps it is futile to study the

Table 11.1 Transition matrices for predator-approach behaviour predicted by TFT, Pavlov and empirical data collected by Stephens *et al.* (1996). The unique predictions of TFT are printed in **bold**, and of Pavlov in *italics*. By TFT, if both defect (DD) in the first move then both should continue to defect (DD) in the next move; if one defects when its partner cooperates in the first move (DC), the two should reverse their behaviour (CD) in the second, and vice versa (CD → DC), while a cooperative pair (CC) should continue to cooperate (CC). With Pavlov, mutual defection (DD) should provoke mutual cooperation (DD → CC), as should mutual cooperation, whereas defection by a single partner should lead to mutual defection (CD → DD and DC → DD).

	To			
From	DD	DC	CD	CC
Tit-for-tat				
DD	**1**	0	0	0
DC	0	0	**1**	0
CD	0	**1**	0	0
CC	0	0	0	1
Pavlov				
DD	0	0	0	*1*
DC	*1*	0	0	0
CD	*1*	0	0	0
CC	0	0	0	1
Observed				
DD	**0.609**	0.156	0.147	*0.088*
DC	*0.266*	0.362	**0.117**	0.256
CD	*0.240*	**0.167**	0.359	0.234
CC	0.090	0.135	0.204	0.571

evolutionary mechanism of cooperation when short-term payoffs cannot be experimentally controlled and manipulated.

Cooperative key-pecking in blue jays

In order to clarify the role of short-term payoffs in eliciting cooperative behaviour, Clements and Stephens (1995) conducted a series of operant feeding experiments where pairs of blue jays were each allowed to press one of two coloured keys. One key had the effect of being cooperative because if both cooperated, both received R, whereas if both pressed the opposite key, both received P, with $R > P$. The payoffs for these experiments (measured in numbers of food pellets) either conformed to a prisoner's dilemma or to simple mutualism (Fig. 11.2) and were held constant until the birds showed the same response more than 90% of the time for 4 consecutive days.

The birds typically completed 200 trials per day, and all three pairs showed the same result (Fig. 11.2): after a brief period of random pecking, they reacted to the prisoner's dilemma by settling into a pattern of mutual defection. Once the payoffs were changed to mutualism, the birds showed mutual cooperation.

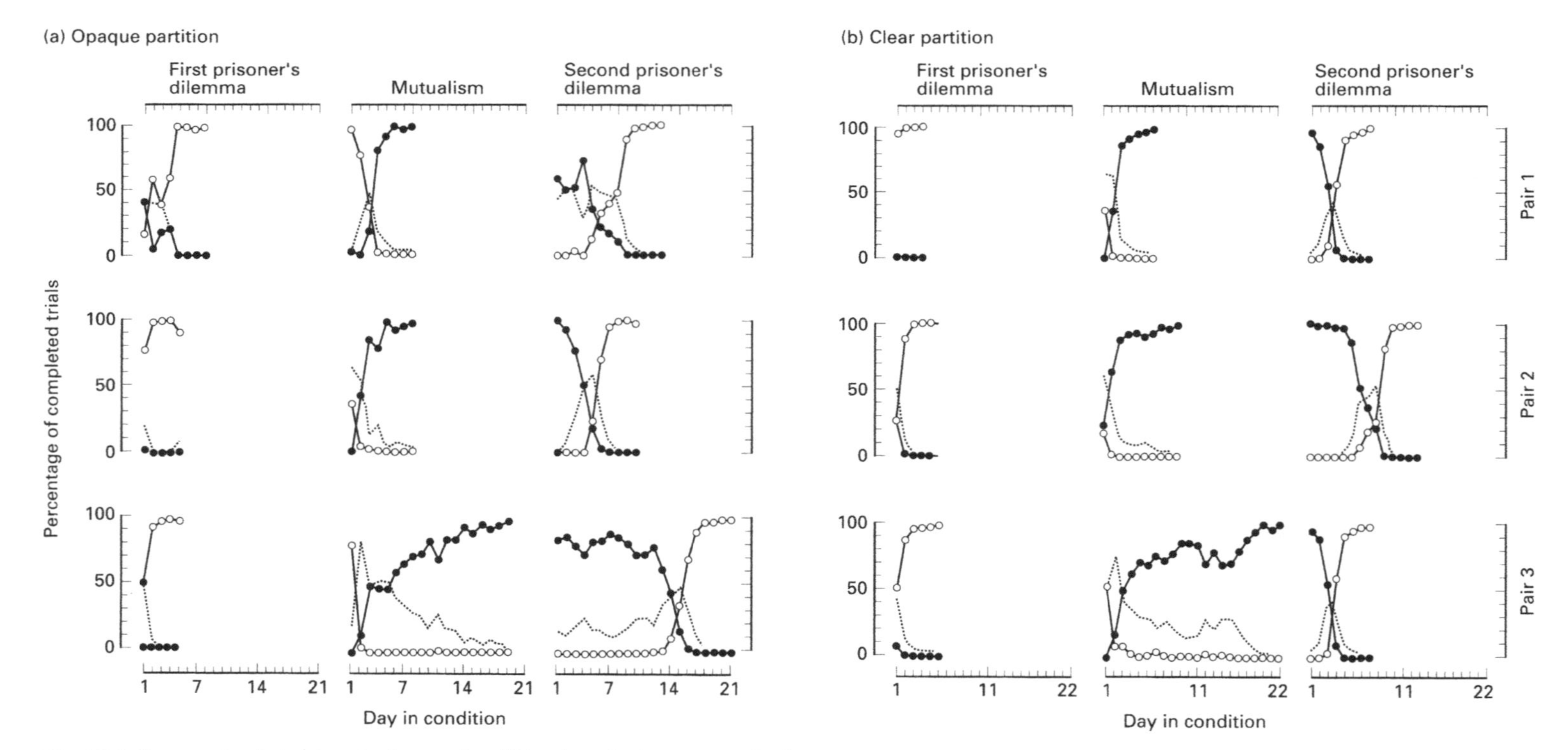

Fig. 11.2 Cooperative behaviour in three pairs of blue jays during operant feeding tests (Clements & Stephens, 1995). Black circles plot the percentage of trials each day in which the birds showed mutual cooperation (CC), open circles show mutual defection (DD), dotted lines show CD plus DC. When the payoffs followed a prisoner's dilemma, the number of food pellets received by each bird was $T = 5$, $R = 3$, $P = 1$, $S = 0$; when mutualistic, $R = 4$, $T = S = 1$, $P = 0$. The birds were separated by a glass partition in one treatment and by an opaque partition in the other, but the nature of the partition had no effect on the results.

Then, most intriguingly, when once again faced with the prisoner's dilemma, the birds quickly switched back to mutual defection, even though they entered the final phase playing mutual cooperation.

Although these results derive from a highly artificial experimental design, the study clearly shows that animals may not be as predisposed toward reciprocity as has often been claimed. Because the birds were separated by a partition, there was no scope for punishment (see Box 11.4C), and it would be interesting to test whether birds might enforce cooperation if they could physically punish non-cooperation by their partner. Future studies may seek to find a more naturalistic system but the first concern should be to have absolute control over the four possible payoffs from a pairwise game.

In summary, these studies emphasize the importance of carefully controlled experiments in determining whether cooperation results from mutualism or reciprocity. In practice, it may be impossible to provide definitive tests in naturalistic settings: animals do not behave in a vacuum and behaviour often has long-term as well as short-term consequences. There are many putative examples of reciprocity, but almost all suffer from the same general problem: does the behavioural exchange truly involve a prisoner's dilemma? For example, because of their greater cognitive abilities, non-human primates are widely believed to exchange altruistic acts (de Waal & Luttrell, 1988) and yet claims of reciprocity in these species have also been contested (see Bercovitch, 1988; Noë, 1990; Hemelrijk, 1996). One commonly cited context has been the exchange of altruistic behaviour for social grooming, yet there is no convincing evidence that grooming is itself costly (Dunbar, 1988; Hemelrijk, 1994). Without a clearly designed test where R, S, T and P are all known with certainty, any claims of reciprocity seem increasingly tenuous.

Chapter 12

The Social Gene

David Haig

12.1 Introduction

The complex behaviours and structures that have evolved by the process of natural selection can be viewed as adaptations for the good of the relevant genes ('replicators') rather than for the good of individual organisms ('vehicles'). A common criticism of this view is that organisms are integrated wholes, in which no gene can replicate without the assistance of many others. The implicit metaphor is of the organism as a machine, with the genes as instructions for the assembly of its component parts. But, an alternative metaphor is possible: genes as members of social groups. Societies, like machines, can display intricate mutual dependence and elaborate divisions of labour; but, unlike machines, societies are not designed. Cooperation and coordination cannot be assumed; when present, they must be explained. Social theories vary in the causal relationships they posit between individual and society: some emphasize the power of individual actions to shape society whereas others emphasize the social constraints on individual freedom. This chapter views properties of organisms as social phenomena that arise from the actions of individual genes, and explores the internal conflicts that can disrupt genetic societies and the social contracts that have evolved to mitigate these conflicts.

Two disclaimers are necessary. First, the chapter emphasizes the current state of genetical systems, and asks why conserved features of organisms are evolutionarily stable relative to conceivable alternatives. Phylogenetic questions, although important, are not my principal concern. Equilibria can often be reached by multiple paths and, in this sense, are independent of history. Thus, too narrow a phylogenetic approach runs the risk of becoming Whig history, telling the story of the winners, and not the innumerable losers, those ephemeral less functional genes that strut and fret their hour upon the stage and then are heard no more. Second, gene-centred theories are often reviled because of their perceived implications for human societies. But, even though genes may cajole, deceive, cheat, swindle or steal, all in pursuit of their own replication, this does not mean that people must be similarly self-interested. Organisms are collective entities (like firms, communes, unions, charities, teams) and the behaviours and decisions of collective bodies need not mirror those of their individual members. As I write this paragraph, my replicators — my genes and my memes — are in constant debate, even dissension, yet somehow I muddle through. I am glad I am not a unit of selection.

12.2 Genes as strategists

Genes are catalysts. They facilitate chemical reactions but are not themselves consumed. A gene influences its own probability of replication by the reactions it catalyses, usually indirectly via transcripts and translated products. These effects can be likened to the gene's strategy in an evolutionary game. When, where and in what quantity the gene is expressed is part of that gene's strategy to the extent that changes in the gene's sequence (mutations) could produce a different pattern of expression. The evolutionary theory of games has illuminated many issues in behavioural ecology (Maynard Smith, 1982), but has usually been phrased in terms of payoffs to individuals rather than their genes. This sleight of hand is possible because outcomes that enhance an individual's reproductive success also enhance the transmission of most (if not all) of the individual's genes.

Individual and genetic payoffs are no longer in such close harmony when an individual's actions affect the reproductive success of relatives or when conflicts occur within an individual's genome. Interactions among relatives can be reconciled with an individualistic perspective by recourse to the concept of *inclusive fitness* (Hamilton, 1964; see Chapter 9), but intragenomic conflicts pose a more intractable problem because an individual's fitness, inclusive or otherwise, is ill-defined when different genes have different fitnesses. Such conceptual difficulties do not arise if genes, rather than individuals, are treated as the strategists. (See Hurst *et al.* (1996) for a recent review.)

Why use strategic thinking, which anthropomorphizes genes, instead of the well-developed infrastructure of population genetics? My reasons are pragmatic. Molecular biology reveals genes that are much more sophisticated than the stolid dominant or recessive caricatures of classical genetics. A gene may be expressed in some tissues, and some environments, but not in others; may have multiple alternative transcripts; may respond to signals from other genes; may have a history (be expressed when maternally derived but silent when paternally derived); and so on. Such complexities are difficult to model by traditional genetic methods. Game theory, however, allows evolutionarily stable strategies to be selected from among a large range of alternative patterns of gene expression. The realism of a strategic analysis depends on the realism of the set of alternatives from which candidate strategies are chosen. Some conceivable strategies may be unavailable in the real world, but too restricted a set of alternatives can also mislead.

12.3 Variant genes

There are at least two defensible answers to the question 'How many words in this chapter?' The first is the number tallied by the word processor of my computer. In this answer, a *word* is a string of characters terminated by a space or punctuation mark. Each time 'vehicle' appears it is an extra word added to the tally. The second is the size of my vocabulary. In this answer, 'replicator' counts as a single word no matter how many times it appears. *Gene* has a similar ambiguity. It can refer to the group of atoms that is organized into a particular deoxyribonucleic acid (DNA) sequence — each time the double helix replicates, the gene is replaced by two new genes — or it can refer to the abstract sequence that remains the same gene no matter how many times the sequence is replicated. The *material gene* (first sense) can be considered to be a vehicle of the *informational gene* (second sense).

Debates about the 'units of selection' are interminable, partly because different meanings of 'gene' are conflated. When hierarchical selectionists (e.g. Wilson & Sober, 1994) describe the gene as the lowest level in a nested hierarchy of units (species, populations, individuals, cells, genes), their sense is closer to the material gene, whereas, when gene selectionists (e.g. Dawkins, 1982) refer to the gene as the unit of selection, their sense is closer to the informational gene. The latter may be materially represented at multiple levels of the vehicular hierarchy, but is not itself a level of the hierarchy. The informational gene, however, is not precisely the meaning of gene selectionists. I will call their gene the *strategic gene* because their sense corresponds to the gene that is a strategist in an evolutionary game.

Every genetic novelty (new informational gene) originates as a modification of an existing gene and is initially restricted to a few vehicles at lower levels of the material hierarchy, solely because it is rare. Therefore, the gene's material copies will interact with each other only when they are present in different cells of the same body or in the bodies of closely related individuals. If such a gene is ever to become established, it must be able to increase in frequency under these circumstances. As the gene's frequency increases, its fate may be influenced by selection at higher levels of the hierarchy, but it will still retain the features that ensured its success when rare. Thus, the gene can be said to commit itself to a strategy when rare that it must maintain at all frequencies. The phenotypic effects of successful genes will consequently appear to be adaptations for the good of groups of material genes that interact because of recent common descent. A *strategic gene* corresponds to such a set of material genes and can be considered the unit of adaptive innovation. In philosophical jargon, the strategic gene is a concept intermediate between the *type* (informational gene) and its *tokens* (material genes).

The meanings of words (like genes) evolve and it would be futile to legislate a single meaning of 'gene', just as it would be futile to legislate a single meaning of 'word'. Semantic flexibility can even be useful when precise distinctions are unimportant, because it allows subtle shifts of sense without becoming embroiled in long terminological explanations. Occasional inconsistency is sometimes the price of brevity.

12.4 The reach of the strategic gene

A strategic gene is defined by the nature of the interactions among the copies of an informational gene that influence transmission of its sequence when copies are rare (Fig. 12.1). If all copies of the informational gene acted in isolation, the only phenotypic effects that would promote its transmission would be effects that directly promoted the replication of its individual copies. A strategic gene would then be coextensive with a material gene. Material genes need not act in isolation. For example, material genes that are expressed in the soma of multicellular organisms do not leave direct descendants but promote the transmission of their replicas in the organism's germ line. The strategic gene now corresponds to an organism-sized cluster of material genes. Similarly, a gene in the soma of one individual may promote the transmission of its copies in the germ lines of relatives. In this case, the strategic gene becomes

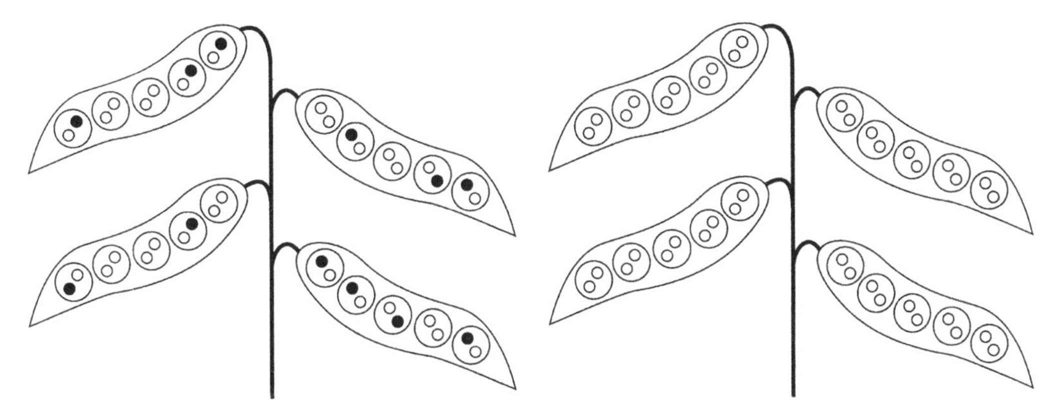

Fig. 12.1 An informational gene is an abstract sequence embodied by material genes (its physical copies). Material genes can be arranged in nested hierarchies: multiple copies within each multicellular pea; multiple peas within a pod; multiple pods on a plant; multiple plants within the local population; and so on. In the diagram, filled circles represent the copies of a gene that are identical by recent common descent (IBRCD); open circles represent other alleles (the multiple copies of a gene within each pea are not represented). The definition of a strategic gene depends on how the phenotypic effects of material genes influence the transmission of their IBRCD copies. The effects of a material gene could affect the transmission of its IBRCD copies solely via flowers of the seedling that develops from its own pea; or could influence the transmission of its IBRCD copies via other peas in the same pod; or could influence the transmission of its IBRCD copies via peas in other pods; and so on. These possibilities are arranged in order of increasing numbers of material genes being lumped together in the concept of the strategic gene. Gene selectionists and hierarchical selectionists use different terminology to explain the same reality. Protagonists of both schools want the other side to admit that their terminology is wrong.

a cluster of material genes distributed among some, but not all, of the members of a family. Such a gene's strategy could be 'treat all offspring equally', not because all carry its copies, but because the gene has no way of directing benefits preferentially to the offspring who do.

If a gene copy confers a benefit B on another vehicle at cost C to its own vehicle, its costly action is strategically beneficial if $pB > C$, where p is the probability that a copy of the gene is present in the vehicle that benefits (see Chapter 9). Actions with substantial costs therefore require significant values of p. Two kinds of factors ensure high values of p: relatedness (kinship) and recognition (green beards). The action of kin selection is distinguished from a green beard effect because, under kin selection, a gene's strategy is blind to the outcome of each toss of the meiotic coin. Thus, the treatment of the individual members of a class of relatives does not depend on which genes they actually inherit, and p corresponds to a conventional coefficient of relatedness. By contrast, green beard effects will discriminate between brothers with and without the relevant gene. Genes recognize kinship by historical continuity: a mammalian mother learns to identify her own offspring in the act of giving birth; a male preferentially directs resources to the offspring of mothers with whom he has copulated; the other chicks in a nest are siblings; and so on. A green beard effect is in operation if genes are recognized directly by their phenotypic effects.

Green beard effects gained their name from a thought-experiment of Dawkins (1976), who considered the possibility of a gene that caused its possessors to develop a green beard and to be nice to other green-bearded individuals. Since then, a 'green beard effect' has come to refer to forms of genetic self-recognition in which a gene in one individual directs benefits to other individuals that possess the gene. The recognition of self and the recognition of non-self are two sides of the same coin (see Chapter 4). Thus, the rejection of individuals that do not possess a label can also be considered a green beard effect, if the absence of the label is correlated with the absence of the genes responsible for rejection. Green beard effects have often been dismissed as implausible because a single gene has been considered unlikely to specify a label, the ability to recognize the label and the response to the label. However, these functions could also be performed by two or more closely linked genes. If X and Y are in linkage disequilibrium (Box 12.1) and X causes its vehicle to treat vehicles with Y differently from vehicles without Y, then (on average) X causes its vehicle to treat vehicles with X differently from vehicles without X.

These concepts are applicable to many kinds of genetic interaction. When homologous centromeres segregate at anaphase I of meiosis, their orderly behaviour is made possible by the prior recognition of some degree of sequence identity between homologous chromosomes (a green beard effect), whereas when sister centromeres segregate at anaphase II, recognition is not necessary because the centromeres have been physically associated since their joint origin from an ancestral sequence (kinship). Similarly, the physical cohesion

Box 12.1 Linkage disequilibrium and recombination

Consider two informational genes X and Y. If $p(X)$ and $p(Y)$ are the probabilities that X and Y are present in a vehicle and $p(XY)$ is the probability that they are jointly present, then the distributions of X and Y are statistically non-independent if $p(XY) \neq p(X)p(Y)$. In other words, knowing whether Y is present provides information about X. If X and Y are alleles at different loci within a species, this non-independence is known as linkage disequilibrium, but the definition of linkage disequilibrium can be generalized to refer to all cases of non-independence whether these occur within or between species, and whether or not the concept of alleles at a locus (positions in a team) is well defined.

Species boundaries are a major cause of linkage disequilibrium (using the broad definition above). For example, the genes of the introduced grey squirrel have displaced the genes of the indigenous red squirrel from many English forests. Some genes from red squirrels might do just fine in grey squirrel bodies, but they never get the chance. Thus, alleles at two loci may be in linkage equilibrium within a species but in linkage disequilibrium between species. Chromosomal inversions can thus have similar genetic consequences to speciation events if they restrict recombination between inverted and uninverted chromosomes.

Linkage disequilibrium results from one of three causes.

1 A new mutation initially occurs in a single vehicle, on a single genetic background. Recombination will spread the gene to new backgrounds, but the approach to equilibrium will be slow for closely linked genes, and cannot occur where there are absolute barriers to recombination.

2 Epistatic selection generates linkage disequilibrium because it causes a gene to leave more descendants when it is present in some combinations than in others. The strength of disequilibrium will be determined by the balance between the selective elimination of less favoured combinations and their regeneration by recombination.

3 Sampling effects can cause linkage disequilibrium in small populations.

of the body is made possible because sister cells have remained in intimate contact since their origin from a common zygote (kinship), but this rich source of nutrients is defended against interlopers by an immune system that distinguishes self from non-self (a green beard effect).

12.5 The prokaryotic firm: managing a cytoplasmic commons

Genetic replication makes use of energy and substrates that are supplied by the metabolic economy in much greater quantities than would be possible without a genetic division of labour. These materials are common goods, available to every gene in the cytoplasm. Thus, genetic communities are potentially vulnerable to free-riders, genes that take more than they contribute, and the gains of trade from biochemical specialization would not have been possible without the evolution of institutions and of procedures that limit the opportunities for social exploitation. In particular, strict controls are expected on access to the machinery of replication.

DNA-based replicators are believed to have evolved from ribonucleic acid (RNA)-based replicators, possibly because DNA is copied with greater fidelity than RNA (Lazcano *et al.*, 1988). The change also had implications for cellular security. Communities in which RNA polymerases were responsible for both replication and transcription would have been less easily policed than communities in which replication ('self-aggrandizement') was performed by DNA polymerases and transcription ('communal labour') by RNA polymerases. As a bonus, a community that modified its own genes to DNA, and periodically cleansed its cytoplasm with ribonucleases, would in the process eliminate most RNA-based parasites.

An effective way to manage the cytoplasmic commons is to link genes to a single origin of replication, and to exclude non-members from the cytoplasm. The chromosome becomes a team whose members' interests coincide. The solution is egalitarian, at least within the group. Each gene that joins the chromosome is given equity and replicates once per cycle, no matter what its contribution during that cycle. Efficiency might conceivably be improved if genes that contributed more to productivity in the local environment were rewarded with increased copy number, but this argument ignores the costs of negotiating fair shares and of policing complex rules. The suppression of internal conflict by replication from a single origin has a price because genes can be copied more quickly from multiple origins than from one (Maynard Smith & Szathmáry, 1993).

12.6 Dangerous liaisons

Cosmides and Tooby (1981) called a set of genes that replicated together, and whose fitness was maximized in the same way, a *coreplicon*. They argued that intragenomic conflicts are likely if an organism contains more than one coreplicon, because the members of a coreplicon will sometimes be selected to maximize their own propagation in ways that interfere with the propagation of other coreplicons. Selection for short-term and long-term replication may be opposed. A coreplicon that replicated faster than other coreplicons within its cell lineage would increase in frequency. However, if differential replication were costly for cell survival, the cell lineage would eventually be eliminated in competition with other lineages. Thus, the long-term interests of coreplicons that share a cell lineage will coincide if they never have opportunities to form new combinations with other genes in other lineages. *Recombination* decouples genes' fates and is therefore essential for the indefinite persistence of intragenomic conflict (Hickey, 1982).

Many bacteria contain multiple circular genomes. By convention, one of these circles is designated the bacterial chromosome, and the extra circles are called plasmids. Plasmid replication consumes energy and substrates. Whether a plasmid pays for its keep — from the perspective of chromosomal genes — depends: (i) on the metabolic skills that its genes bring to the cell; (ii) on whether these skills are required in the current environment; and (iii) on the degree of coadaptation between plasmid and chromosome. Many of the genes for antibiotic resistance that are the scourge of modern hospitals are carried on plasmids. Such a plasmid may be essential in the presence of antibiotics but a burden in their absence (Eberhard, 1980).

Most plasmids promote conjugation between their host and other bacteria, although some smaller plasmids rely on larger plasmids for these functions. In the process, a copy of the plasmid is retained by the donor cell and an uninfected bacterium acquires a plasmid. Thus, conjugation allows plasmids to colonize new cytoplasms. Chromosomal genes, by contrast, are usually not transferred. Therefore, the chromosome bears the costs of replicating the donated plasmid, and the costs of increased exposure to viruses during conjugation, but seemingly gains little in return. If the plasmid encodes beneficial functions these are transferred to a potential competitor. Perhaps, if a plasmid is a burden to chromosomal genes, the chromosome benefits from sharing its cold with rivals.

Plasmids cannot be categorized simply as parasites or mutualists. For example, a plasmid that initially reduced its host's competitiveness enhanced fitness after plasmid and chromosome were propagated together for 500 generations (Lenski *et al.*, 1994). A plasmid has two modes of transmission, vertical and horizontal, and selection can favour its propagation by either path. Selection for greater horizontal transmission, at the expense of vertical transmission, will generally increase the costs of the plasmid to chromosomal genes, whereas selection for increased vertical transmission will generally benefit the chromosome. In the limit, when there is no horizontal transmission of the plasmid, the long-term fates of chromosome and plasmid are inexorably linked and they effectively become a single coreplicon. Similar arguments apply to the viruses, transposons and other coreplicons that populate bacterial cytoplasms.

12.7 Plasmid protection rackets

Plasmids, once acquired, are difficult to discard. Plasmid genes encode multiple functions that ensure their stable transmission within an infected lineage (Nordström & Austin, 1989). Many plasmids encode a persistent 'poison' and its short-lived 'antidote'. Thus, if a cell segregates without the plasmid, it is cut off from its supply of antidote and succumbs to the poison. The gene for the poison can be said to recognize the presence or absence of the gene for the antidote. Because poison and antidote are inherited as a unit, the plasmid can be said to recognize itself (a green beard effect). Such protection rackets take many forms (Lehnherr *et al.*, 1993; Salmon *et al.*, 1994; Thisted *et al.*, 1994). Some plasmids, for example, encode a methylase and its matching restriction enzyme. The methylase modifies bacterial DNA so that it is protected from the restriction enzyme, which cleaves unmodified DNA. Bacteria that lose the plasmid die, because methylation must be restored each time the chromosome replicates. The restriction enzyme simultaneously defends its cytoplasm against viruses and rival plasmids that lack the appropriate methylase, just as a gangster defends his turf (Kusano *et al.*, 1995).

Mitochondria of trypanosomes contain a large DNA maxicircle and many small minicircles. The maxicircle encodes essential genes in garbled form, whereas the minicircles encode guide RNAs that edit the otherwise unreadable transcripts to yield translatable messenger RNAs (Benne, 1994). Could RNA editing have evolved as a minicircle-maintenance system? If so, one would predict that minicircles can also edit DNA and encrypt maxicircle genes in ways that only they can decipher. Minicircles have been observed to spread from one mitochondrial lineage to another (Gibson & Garside, 1990), strengthening their analogy with bacterial plasmids.

12.8 Team substitutions

A non-recombining bacterial chromosome is a team that does not change its members (except by mutation). Its social contract is 'all for one, and one for all' not 'every gene for itself.' Chromosomal recombination occurs on rare occasions as a coincidental side-effect of the horizontal transfer of plasmids (conjugation) or viruses (transduction). Some bacteria, however, have evolved mechanisms by which DNA is taken from the environment and used to replace homologous sequences of the chromosome (natural transformation). Transformation, unlike conjugation and transduction, is controlled by chromosomal genes (Stewart & Carlson, 1986). Uptake of DNA is induced under conditions of nutritional stress and may have evolved primarily as a means of gaining nutrients (Redfield, 1993). Nevertheless, the expression of DNA-binding proteins that prevent the degradation of the donor sequence and the induction of the enzymatic machinery of recombination (Stewart & Carlson, 1986; Lorenz & Wackerknagel, 1994) suggest that recombination is not a mere side-effect but has been positively selected.

Why should a team replace one of its members? The repair hypothesis views transformation as a means of replacing injured team members (damaged DNA). However, repair is unlikely to be the principal function of transformation because uptake of DNA is not induced by damage to the chromosome (Redfield, 1993). The recombinant-progeny hypothesis views transformation as a means of trying out new players. Replacement of one gene by another occurs in a single cell of a clone and does not commit other team members to the new combination, because the old combination survives in other cells of the clone. A team's chances of remaining successful in a changing environment will presumably be improved by some degree of experimentation with new combinations. The problem is that, for each member of the team, the advantages arise only from changes at positions other than its own. Genes would be selected to increase their chances of replacing an established gene on other chromosomes, but to decrease their chances of being so replaced. The important social question becomes whether some positions are privileged and not subject to replacement; in particular, whether the genes responsible for transformation are themselves transformed.

12.9 Multicellular corporations

The development of resistant spores by *Bacillus subtilis* illustrates the differentiation of soma and germ line in simple form. A bacterium undergoes an unequal cell division to produce a mother cell (soma) and a prespore (germ line). The mother cell engulfs the prespore, assists in formation of the spore coat and is then discarded. The process is coordinated by an exchange of signals between mother cell and prespore (Errington, 1996). The genes of the mother cell sacrifice themselves for their replicas in the spore. Some bacterial somas are more complex. *Myxococcus xanthum* is a motile, predatory prokaryote that forms a multicellular fruiting body. Individual myxobacteria forage and divide in the soil but, when nutrients become scarce, they aggregate to form a structure in which sacrificial stalk cells (soma) raise myxospores (germ line) above the substrate (Shimkets, 1990).

Organisms develop somas to gain the benefits of a cellular division of labour. A soma, however, is a rich resource that can be exploited by genes of other germ lines. Therefore, the advantages of somatic specialization can be realized only if the genes of the soma have some degree of confidence that their copies are represented among the beneficiaries of their labour. The simplest means by which genes in somatic cells ensure that their efforts are well directed is physical cohesion between soma and germ line. The genes of the *Bacillus* mother cell can be assured that their copies are present in the prespore because cell division and sporulation take place within an enclosed sporangium that excludes outsiders. However, as somas become larger and more complex, interactions between somatic cells and germ cells become less direct. This creates additional opportunities for parasites to misappropriate somatic effort, and necessitates more elaborate security systems to protect the soma from exploitation. The genes of my liver are almost certain to have copies in my testes (because of my body's physical cohesion), but my lump of lard and mass of meat would not last long without a sophisticated system of immune surveillance.

Uninterrupted physical cohesion cannot protect the genes of somatic cells from exploitation if sister cells become detached to forage (as in *Myxococcus*) or to form complex organs (as in multicellular animals). Some form of cellular recognition is required. When two cells meet, their responses can be influenced by what they learn about each other. Molecules on their surface can provide clues about which genes are present in a cell and whether the cell is friend, foe or indifferent. Two categories of molecular interactions can be distinguished. Homotypic interactions occur between identical molecules on the two cells and are a particularly direct means for a gene to recognize itself in other cells. Heterotypic interactions occur between molecules encoded by different genes and can also provide a gene with information about its presence or absence in another cell if there is linkage disequilibrium between the interacting genes (Haig, 1996). Thus, green beard effects may play an important role in the somatic security systems of multicellular organisms (particularly making use of the complete linkage disequilibrium between genes of different species).

The origin of molecules that were able to discriminate between themselves and closely similar molecules greatly expanded the strategies available to genes and made possible the evolution of large multicellular bodies. The ancestor of the immunoglobulin superfamily probably interacted with itself in homotypic adhesion or signalling, but the family now includes many heterophilic adhesion molecules as well as the T-cell receptors, major histocompatibility complex (MHC) antigens and immunoglobulins of the vertebrate immune system (Williams & Barclay, 1988). The cadherins, to take another example, are a family of cell-surface proteins that bind to copies of themselves on other cells. Nose *et al.* (1988) introduced the genes for P-cadherin and E-cadherin into a cell line which lacked cadherin activity, creating two sublines that were identical except for this single gene difference. When the cells were mixed, they spontaneously segregated into discrete populations, like oil mixed with water. Cadherins play a pivotal role in organogenesis, but similar mechanisms could clearly be used to distinguish self from non-self.

12.10 A chimeric menagerie

Slime moulds are eukaryotes with a life cycle remarkably similar to *Myxococcus* (Kaiser, 1986). They feed as unicellular amoebae, but aggregate when starved to form a fruiting body with a simple division of labour between spores and somatic stalk. Slime moulds are thus particularly vulnerable to somatic exploitation because there is no guarantee that the amoebae who respond to an aggregation signal are members of the same clone, or that a predator will not use the signal to lure amoebae to their doom. The dangers are real, although somewhat mitigated by the mechanisms of cell-surface recognition discussed in the previous section. *Dictyostelium caveatum* is a predator that responds to the aggregation signals of other species and devours their amoebae before forming its own fruiting body (Waddell, 1982). Zygotes of *D. discoideum* produce the aggregation signal and devour haploid amoebae of their own species as they respond to the signal (O'Day, 1979). Some strains of *D. discoideum* form chimeric fruiting bodies with amoebae of other strains without contributing to the stalk (Buss, 1982).

Chimerism between members of a single species has also been described in animals. Vascular fusion frequently occurs between neighbouring genotypes of the colonial urochordate *Botryllus schlosseri*. The progenitors of germ cells circulate in the blood and will colonize, and in some cases totally replace, the gonads of the neighbouring soma (Pancer *et al.*, 1995). As another example, 'hermaphrodite' females of the haplodiploid scale insect *Icerya purchasi* are host to spermatogenic cells derived from sperm that entered the cytoplasm of an egg, but which failed to fertilize the egg nucleus because they were pre-empted by another sperm (Royer, 1975). Thus, a sperm that fails to fertilize the eggs of the mother can try again with those of the daughter or granddaughter, or persist as a permanent haploid inhabitant of female somas. Occasionally winged males are produced from unfertilized eggs (Hughes-Schrader, 1948), but must compete for fertilization with the 'reduced males' resident in female gonads.

Marmosets and tamarins regularly produce dizygotic twins in a uterus ancestrally designed for singletons (i.e. in a simplex uterus which lacks long uterine horns to keep squabbling offspring apart). The placental circulations of the twins fuse, with the result that each adult marmoset carries blood cells derived from its twin (Benirschke *et al.*, 1962). If germ cells were also transferred, and equally mixed between twin brothers, the genes of their respective somas would be indifferent about which brother copulated, although competition within the brothers' testes and ejaculates could be intense. Chimerism between dizygotic twins is the rule for marmosets but is the exception for human twins (van Dijk *et al.*, 1996). Chimerism is common, however, between human mothers and their offspring. Fetal cells circulate in a mother's blood from the early weeks of pregnancy and descendants of these cells may persist in a mother's body for decades after the child's birth (Bianchi *et al.*, 1996). Are these cells simply lost, or do they manipulate the maternal soma for the offspring's benefit?

Exploitation of host somas by pathogens and parasites remains a major problem for multicellular organisms. This section's collection of intraspecific chimeras emphasizes that the risk of somatic exploitation is not restricted to members of different species. Of course, somatic exploitation within species usually involves the everyday strategies of coercion and deceit.

12.11 The nuclear citadel

The speed of replication limits the amount of DNA that can be copied efficiently from a single origin of replication. The chromosome of *Escherichia coli* takes about 40 min to replicate (Zyskind & Smith, 1992). If the 1000-fold larger genome of *Homo sapiens* were similarly organized as a circular chromosome with a single bidirectional origin of replication, it would take almost 1 month to replicate (comparative genome sizes: Morton, 1991; Fonstein & Haselkorn, 1995). Humans and other eukaryotes avoid this problem by using multiple origins of replication. The attendant risk that some parts of the genome will replicate faster than others is exacerbated because the alternation of gametic fusion and meiotic segregation creates ample opportunities for rogue elements to colonize new genomes (Hickey, 1982). For these reasons, eukaryotes are expected to have evolved sophisticated systems for controlling unauthorized replication.

Two characteristic features of eukaryotic cells probably contribute to replicative security. The first is the separation of the machinery of protein synthesis (in the cytoplasm) from the genetic material (in the nucleus). Passage of large molecules to and from the nucleus is controlled at the nuclear pore complex. Before a protein can dock with this complex, it must possess nuclear localization signals that are recognized by docking molecules in the cytoplasm (Davis, 1995; Hicks & Raikhel, 1995). The second is the eukaryotic cell cycle. Replication is confined to a specific S phase. Before DNA can replicate, it must acquire a 'replication licensing factor' that authorizes it to replicate once, but once only, per cycle (Rowley *et al.*, 1994; Su *et al.*, 1995). The origin recognition complex (ORC) that marks a site for future initiation of replication causes transcription to be silenced in its vicinity (Rivier & Pillus, 1994). Thus, genes near an ORC are prevented from producing locally acting RNAs that could tamper with the genes' own replication.

The bread mould *Neurospora crassa* has evolved a highly effective defence against genetic elements that replicate more than once in a cell cycle. If a sequence is repeated within a haploid nucleus, both copies are inactivated by methylation and subject to a process of repeat-induced point (RIP) mutation until their sequences have diverged sufficiently to be no longer recognized as similar (Selker, 1990). Thus, if a DNA sequence replicates faster than other members of its collective, both the additional copies *and* the master sequence are corrupted by a process of programmed mutation.

Vertebrates compartmentalize their DNA into active regions and methylated regions that are maintained in a compact transcriptionally inactive state (Bestor, 1990; Bird, 1993). The inactive portion of the genome often contains large amounts of simple repetitive sequences that do not encode proteins and which are subject to high rates of sequence turnover because of replication slippage and unequal crossing over (Dover, 1993). This arrangement may function, in part, as a system of defences against intragenomic parasites. First, a higher proportion of insertions will occur in non-critical sequences. Second, foreign DNA (once inserted) is transcriptionally inactivated by a methylation process which may specifically recognize structural features of parasitic DNA (Bestor & Tycko, 1996). Third, inserted DNA is subject to sequence degradation by the processes of genomic turnover.

It would be misleading to argue that the sole function of the organizational changes of the eukaryotic nucleus has been internal security. Even when agents have identical interests there is still a problem of coordination. The genome of *E. coli* contains about 4000 genes whereas the genomes of humans, mice and pufferfish contain about 80 000 different genes. Bird (1995) has argued that the nuclear envelope and histone proteins of eukaryotic cells, and the extensive methylation of vertebrate chromosomes, are adaptations for the reduction of the transcriptional noise associated with larger genomes. New security measures would have evolved hand in hand with new mechanisms of control.

12.12 The sexual revolution

Bacterial recombination involves the formation and dissolution of partnerships between coreplicons or the substitution of one gene for another in a process that has clear winners and losers. By contrast, meiotic recombination involves a symmetrical relationship in which two temporary teams come together, swap members and form new temporary teams. The members of successful teams get to play more often in the next generation than members of unsuccessful teams. Thus, a successful player is one who performs well as a member of many different teams, and the system favours teams of champions rather than champion teams. Team members pursue the same goals, not because their long-term destinies are indissolubly linked, but because the rules of meiosis ensure that all receive the same opportunities if only their team can make it through to the next lottery.

If there were completely independent assortment of genes at meiosis, players could not form long-term partnerships because any two players present in a haploid team before gametic fusion would have an even chance of parting at meiosis. This 50% probability of recombination per generation applies to almost all randomly chosen pairs of genes in organisms with multiple chromosomes. Genes that are linked on the same chromosome can expect to remain associated for longer periods. If some combinations of linked genes work more effectively together than others, these combinations will tend to occur in successful teams and leave more descendants than less favoured combinations. By this process, selection generates non-random associations of players (i.e. linkage disequilibrium), but these associations are constantly being disrupted by recombination.

One of the major preoccupations of evolutionary genetics has been the question why so many genetic collectives regularly break up successful teams to take a chance on untried combinations. Zhivotovsky *et al.* (1994) summarized numerous models that reached the conclusion:

> in a random mating population, if a pair of loci is under constant viability selection (the same in both sexes), with recombination between them controlled by a modifying gene, and if this system

attains an equilibrium at which the major genes are in linkage disequilibrium, then new alleles at the modifying locus can invade only if they reduce the rate of recombination between the major loci.

A similar principle applies for an arbitrary number of loci (Zhivotovsky *et al.*, 1994). The intuitive explanation of this 'reduction principle' is that new teams generated by recombination will, on average, be less successful than existing teams that have survived a generation of selection. Therefore, individual players are more likely to be successful in the next generation if there is less recombination of their current team.

Recombination is widespread in nature, and one or more of the assumptions of models that predict selection for reduced recombination must be violated. Genes that increase recombination can be favoured if a population has not reached selective equilibrium because recombination increases the efficiency with which currently favoured players are brought together in the same team. Technically, the advantage a team gains from having both gene A and gene B must be less than the sum of their individual contributions to team success (Barton, 1995). A similar process favours increased recombination if the cost of injury to A and B (i.e. mutation) is greater than the sum of the costs if A and B were damaged individually (Charlesworth, 1990). In both these examples, increased recombination improves the efficiency of selection because it reduces the risk that inferior players will 'hitch-hike' on the success of their team mates or, what amounts to the same thing, that superior players will be dragged down by lesser players. Theories that ascribe the adaptive advantage of recombination to increased resistance to parasites are explanations of this sort because recombination rates evolve in a constantly deteriorating environment in which the most favoured allelic combination is always in flux (Hamilton *et al.*, 1990).

12.13 The open society and its enemies

The reduction principle also breaks down in the presence of multilocus green beard effects. Green beard effects allow genes to direct benefits to teams in which they have a high probability of being present. As we have seen, a gene (or coalition of genes) can profit from conferring a benefit B on another team at cost C to its own team, if $pB > C$ where p is the probability that the gene (or coalition) is present in the team that benefits. If this probability were the same for all genes in the donor team, all members would gain equally from the transaction. But, if the probabilities differ — as they do when benefits are directed to green-bearded relatives at the expense of other relatives (Ridley & Grafen, 1981) — some team members will lose while others gain. Linkage disequilibrium can enable small coalitions of genes to conspire against the common good, but high levels of recombination will disrupt the persistent non-random associations on which multilocus green beard effects depend.

Other team members would suspect the motivation of a group of players who were simultaneously members of a rival team, and can benefit from disrupting cliques before they form.

The best-studied conspiracies are systems of meiotic drive. A haplotype in a heterozygous diploid causes the failure of gametes that do not carry its copies, usually by means of a two-locus poison–antidote mechanism. If the haplotype does not go to fixation, it must be associated with countervailing fitness costs that will be experienced in full by team members that are unlinked to the haplotype. Therefore, selection at unlinked loci favours increased recombination to disrupt the conspiracy and separate the poison from its antidote (Haig & Grafen, 1991). Leigh (1971; see also Eshel, 1985) has compared the genome to 'a parliament of genes: each acts in its own self-interest, but if its acts hurt the others, they will combine together to suppress it.' Segregation distortion and related phenomena are departures from fairness. Leigh (1971) suggested:

> The transmission rules of meiosis evolve as increasingly inviolable rules of fair play, a constitution designed to protect the parliament against the harmful acts of one or a few.... Just as too small a parliament may be perverted by the cabals of a few, a species with only one, tightly linked chromosome is an easy prey to distorters.

12.14 The eukaryotic alliance

Most internal conflicts within the nucleus are defused by the procedures of fair segregation and allelic recombination. However, eukaryotes also contain genes, in mitochondria and plastids, that are not part of the meiotic compact. The eukaryotic cell originated as an alliance between nuclear genes and the genes of symbiotic bacteria. Many of the latter eventually joined the nuclear firm, but some retained a limited independence as the mitochondrial and plastid genes of today. We do not fully understand why some genes have accepted (or been granted) nuclear equity whereas others have maintained a separate contractual arrangement, nor why these shifts of allegiance have been predominantly one-way, from organelle to nucleus. Nuclear and organellar genes are mutually dependent, yet their different rules of transmission can be a source of conflict in their partnership.

If different organellar lineages occupied the same cytoplasm after gametic fusion, the lineages would be expected to compete for occupation of the cytoplasm, with concomitant costs to nuclear genes. Cosmides and Tooby (1981) suggested that nuclear genes have been selected to minimize conflicts among organellar genes by causing the destruction of the organelles of one gamete, either before or after syngamy. For this reason, they proposed, the nuclear genes of one kind of gamete (sperm) discard their organellar partners before fertilizing a different kind of gamete (eggs) which retain their organelles (see Hastings, 1992; Hurst & Hamilton, 1992; Law & Hutson, 1992 for related arguments). Nuclear-enforced suppression of cytoplasmic conflict may thus have been the key factor in the evolution of eggs and sperm, with all other differences between the sexes arising from this initial dichotomy. In support of this conjecture, Hurst and Hamilton (1992) have noted that morphologically distinct sexes are absent in taxa that exchange nuclear genes without cytoplasmic fusion.

Uniparental inheritance of mitochondria and plastids resolves one conflict but creates another. Nuclear genes are transmitted by sperm and eggs whereas organellar genes are transmitted by eggs alone. Organellar genes would therefore benefit from preventing reproduction by male function, if this increased the resources available for female function. Cytoplasmic male sterility has evolved many times in flowering plants. In all well-studied cases, male sterility is caused by mitochondrial genes but their effects are often countered by nuclear genes which restore male fertility. Chloroplasts also have predominantly maternal inheritance, but chloroplast genes are not known to cause male sterility (Saumitou-Laprade *et al.*, 1994). The plastid genome may lack mechanisms to abort male function, or, if such mechanisms exist, they may be circumvented easily by nuclear genes.

Despite its internal conflicts, the eukaryotic alliance has been an outstanding success. The philosopher Daniel Dennett (1995, pp. 340–1) considers humans to be a radically new kind of entity, comparable in importance to the eukaryotic cell. In his view, we are a symbiosis between genetic replicators and cultural replicators (memes). Just as eukaryote cells cannot survive without both nucleus and organelles, we cannot survive without both genes and memes; neither genes nor memes are dispensable; and neither genes nor memes can claim priority as representing our true selves. Genes and memes have very different rules of transmission and a meme cannot simply be incorporated into a chromosome where it follows the rules of meiosis. Conflicts are therefore expected. Some people will die for an idea. Others will abandon their faith for a brief sexual encounter.

12.15 Sex chromosomes

An average gene from a species with two sexes spends equal time in male and female bodies because every individual has a mother and a father. However, at a selective equilibrium some genes (or combinations of genes) may be more successful than average in one sex and less successful than average in the other. Such sexually antagonistic genes will benefit from being associated with other genes that bias sex determination towards the sex in which they have a relative advantage (Rice, 1987), whether this is a conventional fitness advantage (viability) or a segregational advantage (meiotic drive). The process is self-reinforcing because genes that influence sex determination spend more time in one sex than the other and can thus persist in linkage disequilibrium with genes whose disadvantage in the less frequent sex is greater than their advantage in the more frequent sex. Genomes thus have a tendency to split into factions that spend equal time in the two sexes (autosomes) and factions that specialize in one sex or the other (sex chromosomes).

Meiotic drive in spermatogenesis or oogenesis (but not both) can favour the evolution of sex chromosomes, because a distorter has an advantage in one sex that is absent in the other. Segregation distorters will also be favoured if they arise on existing sex chromosomes. Associations between agents of meiotic drive and the genetic determiners of sex result in biased sex ratios, but these biases will be opposed by the parliament of genes (or at least by its autosomal majority). Half of the genes in the next generation will come from males and half from females. This means that members of a minority sex will leave more descendants, on average, than members of the majority sex, and autosomal genes will benefit from being present in the minority sex (assuming that the sexes are equally costly). Autosomal genes are expected to enforce fair segregation of sex chromosomes in the heterogametic sex because neither sex will then be in a majority.

Hamilton (1967) recognized that autosomal genes will sometimes favour biased sex ratios. He considered a model in which small numbers of unrelated females founded local populations, their offspring mated among themselves, and the newly mated females dispersed to found new local populations. If males were heterogametic and segregation was strictly Mendelian, the sex ratio in the global population would be very close to unity, with some variation among local populations because of random fertilization by X- and Y-bearing sperm. The expected fitness of a female offspring would be the same as the average female fitness for the global population, regardless of the local sex ratio, but the expected fitness of a male offspring would increase with the local proportion of females. Therefore, an autosomal gene that caused itself to be present in female-biased local populations would have higher than average fitness, and a balanced sex ratio would no longer be the unbeatable strategy. In this example, the parliament of genes contains a number of parties with different policies concerning the sex ratio. The X party, the autosomal party and the mitochondrial party would enter into coalition against the Y party to force a female-biased sex ratio among offspring, but the coalition partners would lack unanimity about precisely which ratio (Hamilton, 1979). Sexual politics can profoundly destabilize the 'parliamentary rules' of meiosis (Haig, 1993).

12.16 Genomic imprinting and the altercation of generations

Relatives are genetic collectives that share some, but not all, of their members. A gene can benefit from employing a contingent strategy that treats collectives differently depending on information about the probability r that a collective includes one of its copies. This section will assume that a gene's only information about r comes from the family tree (pedigree of collectives) and the Mendelian probabilities associated with the pedigree, in some cases supplemented with knowledge of parental origin and the uncertainty of paternity. Green beard effects will not be considered.

The simplest relationship between a diploid mother and her sexually produced diploid offspring is one in which the mother produces a series of eggs that are provisioned, fertilized and scattered, without subsequent maternal care. Each gene in the mother has an equal probability of being present in each offspring — determined by a flip of the meiotic coin — and the quantity of yolk received cannot be influenced by genes expressed in offspring because provisioning is completed before meiosis. Genes of the mother can do no better than produce the size and number of eggs that maximize the mother's expected lifetime reproductive success. In a simple model in which eggs are produced sequentially on a production line from a fixed quantity of resources, maternal fitness is maximized when each egg is supplied with an amount of resources such that the marginal benefit (δB) that an offspring would gain from a little bit extra committed to its egg would equal the marginal cost of these resources (δC) to another offspring that develops from an egg at the end of the line. Marginal costs and marginal benefits are given equal weight because a gene in the mother has the same chance of being present in either offspring.

The relationship becomes more complex if offspring receive post-zygotic maternal care, because the amount of care received can be influenced by genes expressed in offspring. A gene expressed in the current offspring gains the full marginal benefit of extra resources received from the mother but has only a probability r of being present in the offspring that experiences the marginal cost. Therefore, genes expressed in offspring will favour receiving extra resources as long as $\delta B > r\delta C$, whereas genes expressed in the mother will favour terminating investment once $\delta C > \delta B$. Thus, parent–offspring conflict exists whenever $\delta C > \delta B > r\delta C$ (Trivers, 1974; Haig, 1992). This conflict arises from the difficulty of making binding agreements. Even though genes in a parent would agree among themselves to terminate investment in each offspring when $\delta B = \delta C$, the agreement is generally unenforceable once genes find themselves in offspring. All genes would do better if offspring demanded less, but unilateral restraint will be exploited.

The probability r is 50% for maternal genes expressed in offspring, whereas r will generally be less than 50% for paternal genes. This is because the offspring that gains a marginal benefit from extramaternal resources may have a different father from the offspring who suffers the marginal cost. Therefore, paternal genes in offspring are predicted to make greater demands on mothers than are maternal genes in the same offspring. Such conditional strategies are made possible by genomic imprinting which causes genes to have different patterns of expression depending on whether they spent the previous generation in a male or female germ line (Moore & Haig, 1991). For example, during murine development, *insulin-like growth factor 2* (*Igf2*) is expressed from the paternal allele and the maternal allele is silent, whereas the *insulin-like growth factor 2 receptor* (*Igf2r*) has the opposite pattern of expression. Mice with an inactivated paternal copy of *Igf2* are 60% normal size at birth, whereas mice with an inactivated maternal copy of *Igf2r* are 20% larger than normal at birth. Birth weight is normal in mice that inherit the inactivated genes from the opposite parent (DeChiara *et al.*, 1991; Lau *et al.*, 1994). *Igf2r* has been proposed to function by degrading the products of *Igf2* (Haig & Graham, 1991). Thus, these genes employ a conditional strategy: 'make greater demands when paternally-derived than when maternally-derived.'

The relatedness asymmetry between maternal and paternal genes is maximal for half-siblings, but most kinds of relatives will have different degrees of maternal and paternal relatedness. These patterns can be quite complex. Consider a hypothetical social system in which males disperse and females remain in their natal group. If all offspring within a group are fathered by a single male who maintains a monopoly on sexual access to the females until he is displaced by a new unrelated male, female group members of different ages will be closer relatives on the maternal side than on the paternal side because of female philopatry, whereas offspring of the same age will be either full siblings or paternal half-siblings. It is not known whether genes have evolved conditional strategies that take account of such patterns of relatedness.

All the well-studied examples of genomic imprinting so far appear to be simple conditional strategies of the form 'do one thing when maternally derived and something else when paternally derived.' More complex conditional strategies are logically possible — for example, 'do one thing when derived from an egg; something else when derived from a sperm of a resident male; and something else when derived from a sperm of a cuckolding male' — but, whether such logical possibilities are actually realized depends on costs, benefits and the existence of appropriate mechanisms. In the social system of the previous paragraph, a gene's probability of being present in other female group members increases as it passes through successive female germ lines. One could imagine a gene subject to a cumulative imprinting effect that was reset to zero every time the gene passed through a male germ line.

12.17 Reprise

Our intuitive concept of the genetic boundaries of an organism approximates the membership of a coreplicon (i.e. of a set of genes that are transmitted by the same rules). A coreplicon evolves as a unit with common goals because its members benefit from the same outcomes, whereas the genes of different coreplicons can have conflicting interests. Thus, prophage genes inserted into a bacterial chromosome are distinguished from 'true' bacterial genes because of their alternative mode of transmission. They can be mobilized to outreplicate their companions; package themselves into resistant phage particles; and lyse their host. The relationship between coreplicons need not, however, be strictly adversarial. One coreplicon can obtain a good from another by trade, as well as by theft, but with room for haggling over the price. A coreplicon, however, functions as a commonwealth, without an internal market. Its members thus avoid the transaction costs of finding buyers; of learning the prices and quality of the goods on offer; and are protected from hucksters and frauds in the marketplace (cf. Coase, 1993).

Bacterial cells usually contain a small number of coreplicons (sometimes only one). Recombination between bacterial chromosomes is rare and, when it occurs, is a substitutional process in which one gene (a winner) is substituted for another (a loser). By contrast, eukaryotic recombination is much more frequent, and is a segregational process in which genes are swapped between chromosomes, without winners and losers. The coreplicon is no longer a group of non-recombining genes who cooperate because their long-term fates are intimately bound, rather it becomes a group of temporary associates who obey the same rules and who gain an equitable division of resources when their ephemeral partnership is dissolved. High-frequency recombination creates a market (of a sort) for currently favoured team players.

Recombination creates *relatives* — genetic collectives that share some, but not all, of their genes. Interactions with relatives are a potential source of internal dissension within the collective, because some members of the collective can gain at the expense of others, for example, by sabotaging other members' gametes or by favouring some offspring over others. High levels of recombination can be a partial solution to the conflicts created by lower levels of recombination, because randomizing devices disrupt the 'cabals of the few'.

The reasons for the eukaryotic sexual cycle of gametic fusion, recombination and meiotic segregation remain somewhat unclear, but the process probably enhances individual genes' chances of long-term survival in a changing selective environment.

PART 4

LIFE HISTORIES, PHYLOGENIES AND POPULATIONS

Habitat quality and competition with other birds determine how an oystercatcher Haematopus ostralegus *chooses where to feed and where to breed. Can we use an understanding of individual decision-making to predict how habitat loss will influence population number?* [Photograph by Jan van de Kam.]

Part 4: Introduction

How should individuals allocate resources to their own growth and survival versus reproduction? When they reproduce, how should they divide their effort between current and future reproduction or between number and quality of offspring? These trade-offs form the basis of life-history theory, discussed by Daan and Tinbergen in Chapter 13. Much of the current research is still stimulated by David Lack's classic studies in the 1950s and 1960s on optimization of clutch size. Today's studies use more sophisticated measures of energy expended by parents in raising broods of different sizes and test how these influence lifetime reproductive success. Daan and Tinbergen emphasize that life-history trade-offs cannot be studied by using natural variation in clutch size and parental survival because this often reflects adaptive decisions made by individuals of different quality. Instead, experimental manipulations are needed where, for example, clutches of different sizes are allocated randomly across individuals to measure the effects of reproductive effort on adult survival. Some of these experiments provide good evidence for optimization of reproductive effort in relation to individual quality.

Comparative studies across species show that many aspects of life history, from longevity to heart beat rate, scale with body mass. Traditionally, these scaling relationships have been assumed to reflect basic physiological constraints within which organisms are forced to operate. Daan and Tinbergen point out that this cannot explain why species may vary by a factor of 10 from the allometric prediction and favour the view that the relationships may arise as a consequence of resource allocation between growth and reproduction to optimize life-history trade-offs. As well as providing a clear account of current theoretical issues, the chapter summarizes some neat recent experimental work on life histories in lizards and birds, revealing that individuals exhibit 'phenotypic plasticity', namely a single genotype can produce a range of phenotypes (e.g. combinations of clutch size and time of breeding) depending on environmental conditions.

Ever since John Crook's pioneering studies of weaver birds and primates, comparative studies have provided important insights into the relationships between behaviour and ecology. In Chapter 14, Harvey and Nee show how these can be improved by using phylogenies. Phylogenies describe the diversity of life. By mapping characters onto phylogenetic trees it is possible to unravel how evolutionary change in behaviour has been influenced by changes in ecology. The main advantage of this approach (the method of independent contrasts) is that it can be used to identify independent evolutionary events and so to test statistically whether various characters change in relation to each other (e.g. warning colours and gregariousness in insects or sexual dimorphism and extrapair mating in birds). However, as Harvey and Nee point out, the problem is that the conclusions may vary depending on the model assumed for evolutionary change, on the exact phylogeny used and on the statistical methods. Faced with this minefield of complexity, some workers have used the simpler technique of 'pairwise' comparisons between closely related taxa (e.g. species in the same genus) to provide independent evolutionary events. Finally, Harvey and Nee discuss some recent exciting studies of lizards in the Caribbean and warblers in the Himalayas where phylogenies can be used to infer how ecological niches are filled as speciation occurs.

The importance of evolutionary history for interpretation of current adaptation is continued as the theme in Chapter 15, where Hewitt and Butlin consider how the genetic structure of populations is influenced by both historical factors and current selective pressures. Behavioural ecologists often assume that the organisms they study are well adapted to local conditions and they then ask how various behaviour patterns might maximize an individual's fitness. However, Hewitt and Butlin point out that the distribution of many organisms in north temperate regions has been influenced dramatically by recent ice ages. Colonization patterns following the retreat of the ice sheets have been complex so that populations now close to each other in space may have distant geographical origins and diverged genomes (e.g. the grasshoppers in Fig. 15.2). Many populations are still evolving after their latest colonization and so traits may be changing under selection and, furthermore, may have different genetic bases in other parts of the species' distribution. Hewitt and Butlin provide some examples where restricted gene flow may permit local adaptations. However, in other cases dispersal from neighbouring populations may prevent this. Daan and Tinbergen discuss a possible example from a study of great tits on the island of Vlieland in the Netherlands, where clutch sizes are larger than is optimal probably because of immigration from mainland populations where larger clutches are selected for.

Environmental change, much of it caused by humans, is having profound effects on animal numbers and species diversity. Conservation efforts will be enhanced greatly if we can predict what will happen to populations when a patch of forest is destroyed or a barrage is built across an estuary, for example. In the final chapter, Goss-Custard and Sutherland show how studies of individual decision making can help us to understand how populations will respond to changed environments. Population ecologists usually take various parameters as given, such as birth rate, death rate and dispersal, and then explore the consequences for population numbers and distribution.

Behavioural ecologists, on the other hand, are interested in how these various parameters have evolved and how individuals vary their life-history decisions adaptively in relation to selective pressures. The advantage of this additional knowledge is that it may then be possible to predict what will happen to populations when conditions change and individuals vary their behaviour. Goss-Custard and Sutherland build on the idea of the 'ideal free distribution' to predict how individuals will distribute in relation to resource abundance and competition with others. They consider how the stable distribution across habitat patches will be influenced by interference and unequal competitive ability. Then they discuss the problems of measuring these parameters in a shorebird, the oystercatcher, and show how predictions can be made as to how populations will respond to loss of breeding or wintering habitat.

There are signs that some species will evolve new behaviour patterns in response to environmental change. Berthold's remarkable work on the new migratory patterns of blackcaps provides a fascinating example. Other species may be less capable of change. We hope that Goss-Custard and Sutherland's approach will inspire more behavioural ecologists to consider the population consequences of behavioural decision making by individuals and so contribute to conservation efforts.

Chapter 13

Adaptation of Life Histories

Serge Daan & Joost M. Tinbergen

13.1 Introduction

Life history is the distribution of major events over the lifetime of individuals. Life-history studies concern the timing and the intensity of reproduction, as well as the processes generating this temporal distribution. They analyse life span, age and size at maturity, the trade-offs between somatic growth, maintenance and repair versus reproduction, the decisions on number and size of the offspring, the investment in current offspring and in future reproductive attempts. Life-history research aims to reveal why this temporal organization varies among species — between fruitflies which start breeding after 12 days of life and giant tortoises which wait until they are 30 years old. Also, it aims to understand variation among individuals in a species: why does one pair of great tits lay twice per year, and another pair in the same population only once? The 'why' here is the general evolutionary question: how has natural selection led to variation in adaptive strategies, both between and within species?

Life-history theory provides the functional framework to evaluate results and generate hypotheses. It aims to explain why evolution produced particular life histories out of innumerable possibilities, and why a diversity of strategies persists in a population even if they yield different reproductive outputs. Life-history theory is concerned with the consequences of temporal patterns for the fitness of the individual. It is formulated largely independent of the genetic and physiological processes causing these patterns. Its predictions are thereby fairly general and applicable to a wide array of organisms. Life-history theory is most advanced in the analysis of reproductive decisions, where the consequences of different strategies are most readily related to fitness. For other aspects of life-history organization, such as ageing and sex ratio control, separate bodies of theory have been developed. Eventually these will all have to be integrated in life-history theory.

Two elements are crucial in the process of natural selection and in the variation of phenotypes: environmental and genetic variability. In the present context, environmental variation is the more general. Even in the absence of genetic variation, life-history strategies may vary between members of a species as phenotypic adaptation to different environments. In the absence of environmental variation, different strategies may also stably persist due to genetic variation, but only under the specific conditions defined by evolutionarily stable strategies. We therefore choose to emphasize environmental adaptation rather than the genetic basis of the strategies.

In this chapter, we summarize the main elements of life-history theory, and illustrate these with recent experimental studies designed to test the hypotheses. We further outline questions poorly addressed by current theoretical or empirical work. We start by clarifying our usage of several concepts which form the core of life-history theory.

13.2 Concepts in life-history theory

13.2.1 Traits

A trait is any quantitative property of a living organism. Life history concentrates on traits such as age of first reproduction, clutch size and sex ratio of the offspring.

13.2.2 Fitness

Life-history theory is based on the idea that the temporal distribution of life events affects the contribution of an individual animal to the gene pool in the next generation. This contribution represents the (evolutionary) fitness of that individual. At any age t, where a life-history 'decision' is taken, the expected rate of gene propagation by an individual over the rest of its life is equal to Fisher's (1930) reproductive value:

$$V_t = \sum_{x=t}^{\infty} \lambda^{-x} f_x l_x$$

where f_x is the expected fecundity (zygotes produced) at age x ($> t$), and l_x is the probability of survival until age x. In the case of sexual reproduction V_t has to be multiplied by 0.5 to take the contribution of male and female to each zygote into account. Clearly, the product $f_x l_x$, when integrated over the lifetime of an individual (from $t = 0$) represents the total production of copies of its genes. To obtain the rate of change in relative frequency which these individual copies represent, the product at each age has to be weighted by the general change in frequency for all individuals in the population. This weighting factor is λ^{-x}, and λ, the innate rate of increase of the population, is found by solving the Lotka–Euler equation:

$$1 = \sum_{x=0}^{\infty} \lambda^{-x} f_x l_x$$

In practice, f_x and l_x are often determined in population studies by measuring the number of offspring produced per age class x, and the survival from one age class to the next. The innate rate of increase is often implicitly assumed to be 1, in which case reproductive value would equal the *lifetime reproductive success* (LRS) of an individual. Since populations on which empirical demographic parameters are based may well be stable in numbers, but yet have surplus emigration ($\lambda > 1$) or immigration ($\lambda < 1$), LRS gives only an approximation of fitness.

13.2.3 Trade-offs

Since fitness is a complex measure, based on multiple components, a change in strategy may have negative consequences for one component, and positive effects on another. Such consequences determine a trade-off, e.g. between current and future reproduction, or between the number of offspring and their reproductive value. We use the concept of trade-offs in the evolutionary or functional sense. However, trade-offs have their roots in the physiology of the organism. Due to physiological limitations, variation in the value of one trait which affects fitness may have direct consequences for the value of another trait. This would appear to lead to a trade-off between two or more traits, but in fact is better considered as a constraint (see below) on the system.

13.2.4 Optimization

By identifying trade-offs, we may hope to define models predicting optimal values for particular life-history traits; optimal at least under particular conditions. These values are those which maximize fitness. A fitness maximum usually emerges as the consequence of a trade-off. We emphasize that optimization refers to the individual in its particular situation, and that the trait value which leads to maximal fitness in a population need not be the one which is optimal for every individual.

13.2.5 Decision rules and reaction norms

A reaction norm (Stearns & Koella, 1986) describes the variation in trait values as a function of environment and/or condition. Decision rules refer to the mechanism of response to these conditions. In essence the two concepts are the same. Reaction norms or decision rules are optimal if they maximize fitness for each environmental condition.

13.2.6 Constraints

The use of the constraint concept is confounded in life-history theory, and Stearns (1992, pp. 16–18) teaches us that we should not expect biologists to

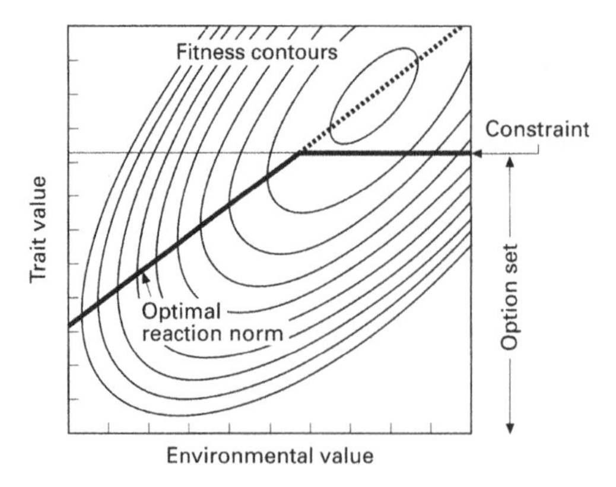

Fig. 13.1 Graphical representation of some basic concepts in life-history theory. In the presence of environmental variation (here simplified as unidimensional on the abscissa) diferent trait values lead to different fitness contours. The dashed line is the optimal reaction norm connecting the optima for different environments. The option set is the collection of trait values possible in the species. A boundary of the option set forms a constraint; when the optimal trait value is outside the option set it constrains the optimum. The constrained optimal reaction norm is given by the solid line.

agree on a certain definition. We use the word constraints in the sense of the boundaries of the option set or parameter space, i.e. the extreme values a trait can assume. These boundaries are set, e.g. by physical restrictions. When discussing individual phenotypic plasticity, we should be aware that individual decisions may in addition be constrained by the genetic make-up of the species and thus by historical limitations. The optimal solution is located either inside or outside of the option set. If it is outside, maximization (or minimization) of the trait may represent the optimal realizable solution. Figure 13.1 shows the connection between several of the concepts.

13.3 Growth and maturation

The first major life-history problem an animal faces is *when to start reproduction*? Most animals are much larger as adults than the zygote they have grown from. Some time must elapse for the organism to grow, even with the simplest form of reproduction, which is cell division. There is always some mortality, however. The longer maturation is delayed, the lower the probability of reaching the reproductive stage. The dilemma is a clear example of a trade-off; fecundity often increases with size but there is a lower probability of surviving to a larger size. The expected duration of reproductive life is thus negatively associated with age of maturity.

The theory on optimal age at maturity is basically due to Gadgil and Bossert (1970). Their fundamental idea of a trade-off between fecundity and survival was later expanded to provide quantitative predictions. Such predictions were, for instance, derived and tested for age and size at maturity in different fish species by Roff (1984). Stearns and colleagues went a step further by developing predictions for intraspecific variations, given different growth and mortality rates (Stearns & Crandall, 1981, 1984; Stearns & Koella, 1986). The resulting optimal reaction norms were based on maximization of the innate rate of increase λ as a measure of fitness. Predicted reaction norms could be tested, for instance with experimental data on age and size of pupal eclosion in *Drosophila* (Gebhardt & Stearns, 1988). Kozłowski and colleagues (Kozłowski & Wiegert, 1986; Kozłowski, 1992) have provided a formulation of the theory based on the allocation of energy to growth and reproduction. Their models yield optimal solutions by maximizing LRS. The direct link with energy allocation gives these models additional power in terms of predictions of other life-history patterns. We therefore choose to illustrate the main line of reasoning using Kozłowski's approach.

Imagine an organism at birth. It starts acquiring energy from the environment at a rate A (e.g. in watts = joules per second) and spending energy for maintenance and acquisition at a rate R. The excess energy ($P = A - R$) can be used for growth and/or reproduction. It is assumed that P initially increases with increasing body size (w). The larger an animal grows the more spare energy it will have to turn into reproduction: the potential fecundity $f(w)$ increases as a function of age (x), as depicted in the backplane of Fig. 13.2. If there were no mortality, it would always pay to keep growing. However, since there must be some mortality, life expectancy at birth is finite. In Fig. 13.2, mortality is assumed to be independent of age. Expected duration of reproductive life decreases with increasing age of first reproduction (α). This is in fact true no matter whether mortality is dependent or independent of size. The prominent result of Kozłowski's and other allocation models is that it is optimal to invest each spare joule of energy in growth as long as this increases expected future reproduction by more than 1 J, and otherwise each spare

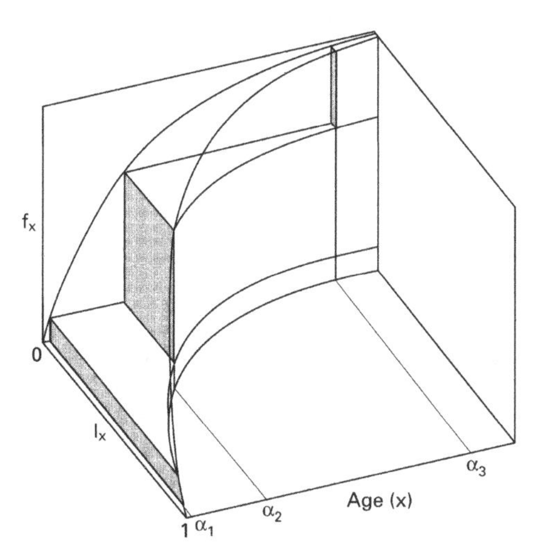

Fig. 13.2 Optimization of age and size at maturity in Kozłowski's model. The curve in the groundplane relates the probability (l_x) to survive until age x. The curve in the backplane relates the potential fecundity f_x to the age of maturity α (switch from growth to reproduction). Expected lifetime reproduction for three different ages at maturity, α_1, α_2 and α_3, is represented by the three solids. The switch from growth to reproduction at age α_2 yields a higher lifetime reproduction than at α_1 or α_3. Thus, there is an intermediate optimal α.

joule should be allocated to current reproduction. After the optimal age of maturity is reached there should be no further growth, since the condition remains fulfilled. The necessary existence of a single switchpoint from growth to reproduction for a plausible assumption of diminishing returns was derived mathematically by Ziołko and Kozłowski (1983). A constant rate of reproduction after the switch point means that total expected lifetime reproduction is represented by $f(\alpha)\int l_x$, or by a volume in the f, t, l space. Three such volumes are shown in Fig. 13.2 for different ages of maturity (α_1, α_2, α_3). Clearly, there is an intermediate age, α_2, where the volume is maximal. Maturation at this age maximizes fitness. Since there is no further growth beyond age α_2, the model simultaneously yields a prediction for the optimal size of mature animals.

The general prediction that adult size remains constant after reaching maturity is violated by *indeterminate growers*, such as many fish, molluscs and other poikilotherms. These continue to grow after their first reproduction. However, the model so far did not take into account variation in environmental conditions. In variable environments, for instance due to seasonal changes in energy expenditure (R) and energy intake (A), the fecundity curve of Fig. 13.2 is no longer monotonic. Instead, during the winter after first reproduction, fecundity is low and, moreover, expected future reproduction gradually rises as the season of high mortality passes by. Under such conditions, LRS maximization predicts that animals revert to allocating energy to growth rather than reproduction after initial reproduction. Instead of continuous reproduction beyond age α, there is an alternation of reproductive and growing seasons. Kozłowski and Uchmanski (1987) have shown that under such conditions the optimal solution is a gradual increase with age in the annual allocation of resources to reproduction and a decrease in growth rate. Indeed, they were able to predict growth and fecundity curves, e.g. for wild populations of Arctic charr (Fig. 13.3). These growth curves can be closely fitted by the well-known Bertalanffy curve, $z_x = z_\infty \{1 - \exp(-kx)\}$, where z_x is length at age x, and k is the growth coefficient. Kozłowski (1996) has recently shown that such growth curves may be the general outcome of maximizing LRS through energy allocation.

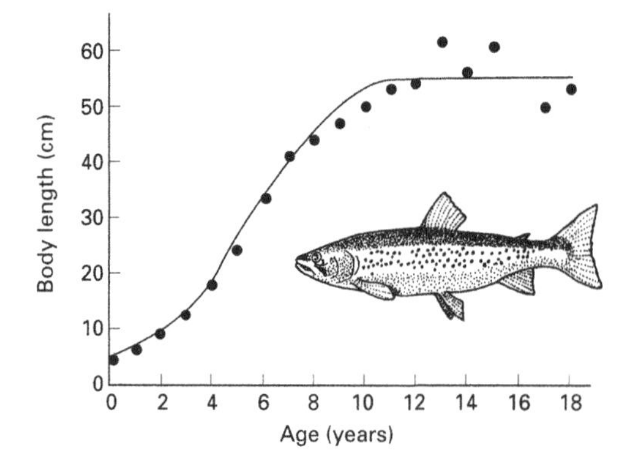

Fig. 13.3 Body mass as a function of age in female Arctic charr *Salvelinus alpinus* in Labrador. (Symbols: data from Dempson & Green, 1985; line: prediction from the resource allocation model of Kozłowski & Uchmanski, 1987.)

The resource allocation approach to body size still has significant tasks waiting ahead. It should eventually be able to predict which circumstances should lead to determinate growth, which to indeterminate growth and which to pronounced (seasonal) fluctuations in body mass such as observed in many small rodents (Heldmaier, 1989). It should further lead to precise predictions for some species where the necessary field data on size dependence of energy allocation and mortality can be obtained. Such information is now coming within reach with the development of isotope techniques for the assessment of energy turnover in nature (Nagy, 1980).

13.4 Scaling of time and energy

Many aspects of life history are known to scale in a systematic fashion with body mass between species (Fig. 13.4). Traditionally, explanations for such patterns have been sought in a physiological dependence of time and/or energy constraints on body size. Such explanations face particular problems, such as the difference of scaling factors when individuals within species or species within higher taxa are compared (Heusner, 1987), or the large variance around the regression lines. Above, we have seen that body size itself may be optimized dynamically, along with measures of time (age of maturity) and energy (fecundity). In this section we discuss to which insights the optimization approach leads with respect to problems of scaling (Kozłowski & Weiner, 1996). We address the question of how random variations in rates of mortality (m) and in rates of reproduction (f), as may occur between species, will affect optimal size.

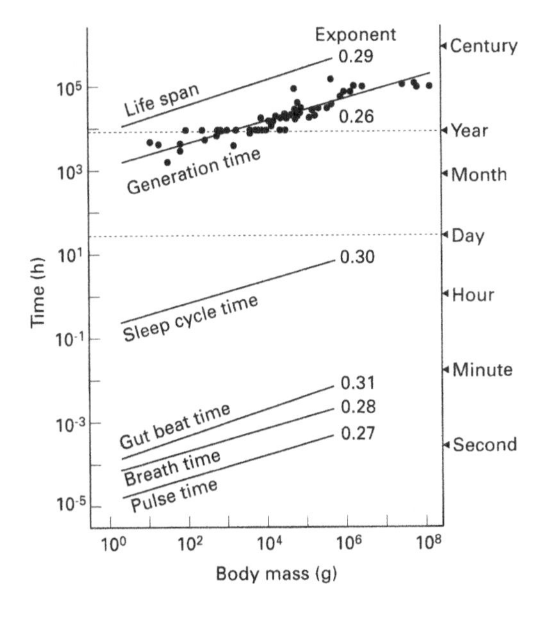

Fig. 13.4 Allometry of the 'rate of living'. Interspecific regressions of heart beat time, breath time, gut beat time, sleep cycle, age at maturity and maximum life span on body mass in mammals. For age at maturity, the data points on which the regression is based are indicated. Numbers indicate the slope of the regressions (exponent of body mass). Note: (i) that the exponents are similar, varying between 0.26 and 0.31; and (ii) that the variation around the regressions is in the order of one log unit. (Modified from Daan & Aschoff, 1982.)

In the model of Kozłowski and Weiner (1996) both respiration and assimilation rates are assumed to be power functions of body mass for individuals within a species: $A = aw^b$, and $R = \eta w^\beta$. For simplicity we assume that $b = \beta$, so that production P $(= A - R = dw/dt)$ also scales as a power function of mass: $P = cw^b$; where $c = a - \eta$. We further assume that mortality $(m = dP/dw)$ is mass independent (Kozłowski and Weiner explore consequences of different associations between mass and mortality). We keep parameter b, the exponent of the intraindividual power function, constant at an arbitrary value of 0.5. Different species may have slightly different parameters c and m, which can be interpreted as different rates of food intake or energy expenditure (c) and of mortality (m). To simulate such variations, we take for 50 species random values of c and m from normal distributions around mean $m = 0.0002$, and mean $c = 0.015$. These means were again arbitrarily chosen. Their particular value does affect the mean size and mean life-history traits. We shall see later what a change in the mean values brings about.

For each combination of c and m, the switch criterion $(w_\alpha = (bc/m)^{1/(1-b)})$ is calculated. This yields the optimal adult size w_α, as well as the age of maturity $\alpha = w_\alpha^{(1-b)}/c(1-b)$, and production $P = cw_\alpha^{b}$. Figure 13.5 shows the results of three such simulations in double logarithmic plots of P (energy) and α (time) against mass (w). In the first simulation (filled squares), the coefficient of variation (cv_m) of m was set at 0.2, while $cv_c = 0$. All the solutions are distributed on a straight line with slope 0.5, both for log P and for log α. This can easily be understood: since c does not vary between species, they all follow the same growth curve, determined by $b = 0.5$, hence the switch points must be on this curve. In the second simulation (filled circles), cv_c was set at 0.2, while $cv_m = 0$. Now the slopes of the distributions are changed: age at maturity is the same for all 'species' (slope 0), since it is solely determined by m. Production rate (log P) increases proportionally with mass (slope 1.0), i.e. twice as steep as the intraspecific curve (slope $b = 0.5$). The highest values of c (largest difference between energy intake and energy expenditure) lead to the largest optimal adult sizes. In the third simulation, c and m were both allowed to vary independently, with $cv_c = cv_m = 0.2$. In this case, the data (circles) show more scatter on all three axes of mass, production and maturation age. The regressions of log P and log α on log w are now intermediate between the previous regressions (where either c or m was constant): $\log P = -2.76 + 0.80 \log w$ $(r^2 = 0.90)$ and $\log \alpha = 3.07 + 0.20 \log w$ $(r^2 = 0.36)$.

The associations of production P and age at maturity α with mass reflect allometries of energy turnover (power) and time in general. The length of different intervals in the lives of birds (Lindstedt & Calder, 1981) and mammals (see Fig. 13.4), such as heart beats, gestation and life span, all scale with exponents similar to age of maturity. Predictions of age at maturity thus have implications also for patterns on other time scales. Keeping this in mind, we should make three general points about the results of Fig. 13.5.

1 Environmental variation in c leads to a steeper allometric relationship

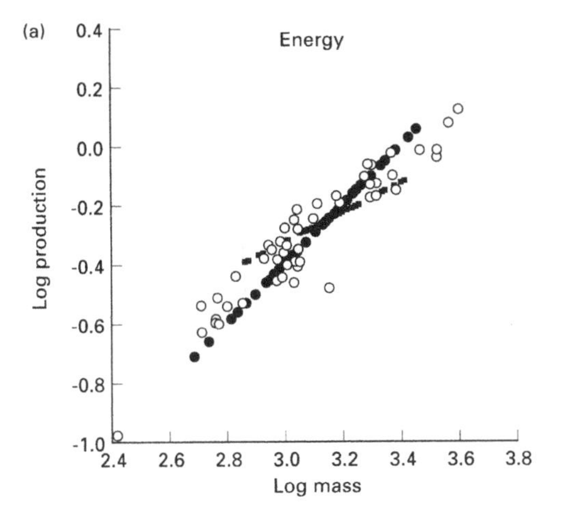

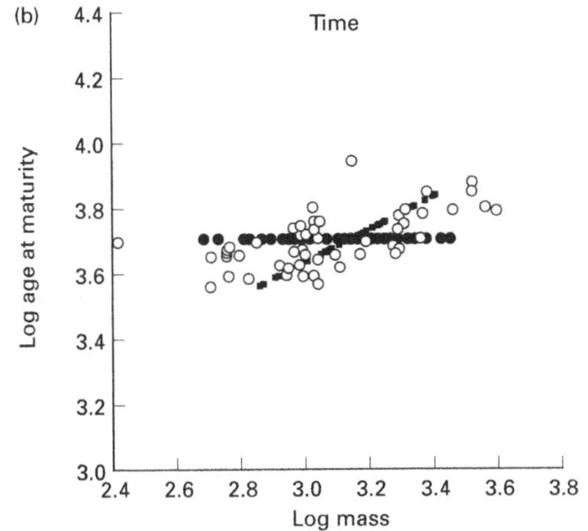

Fig. 13.5 Production (a) and age of maturity (b) plotted against body mass for 50 'species' in a resource allocation model (simplified from Kozłowski & Weiner, 1996), with c and m drawn from normal distributions. (Mean $m = 0.0002$; mean $c = 0.015$ throughout.) Squares (c constant, m varying), coefficient of variation of c (cv_c) = 0; $cv_m = 0.2$. Dots (c varying, m constant), $cv_c = 0.2$; $cv_m = 0$. Circles (c varying, m varying), $cv_c = 0.2$; $cv_m = 0.2$.

between production (energy intake – expenditure) and mass than there is along the individual growth curve. This provides a general explanation for the fact that the scaling of metabolic rates between species is typically steeper than within species (e.g. Heusner, 1987).

2 In all three simulations, the slopes of the regressions of log P and log α on log w add up to 1, which means that $P\alpha$ is proportional to w^1, or $P\alpha/w$ is a constant. This result, noted first by Kozłowski and Wiegert (1987), is contingent on the basic assumption in the model that $w_0 = 0$, or the calculation of optimal size starts at size 0. $P\alpha/w$ has been considered a 'life-history invariant' (Charnov, 1993), and has the dimensions power × time/mass, or energy/mass. Its size independence among animals, predicted by Kozłowski's theory, possibly provides a general explanation for the fact, first noted by Pearl (1928) in mammals, that the total mass-specific metabolism integrated over the lifetime (or over a heart beat) is weight independent; per unit of tissue, the same number of joules are consumed in a lifetime from mouse to elephant.

Speculations on the nature of this phenomenon have concentrated on specific mechanisms relating metabolism to the process of ageing and senescence (see references in Rose, 1991). The size independence of the relationship is, however, a consequence of evolutionary theory without specification of any particular biological mechanism.

3 Due to independent variance in both production and mortality, the slopes are intermediate between those generated by each source of variance separately. Since the maximum value of b must be 1, this means that the observed interspecific slope of energy metabolism on mass should be less than 1, or, in other words, larger animals should have lower mass-specific metabolic rates than smaller animals.

We now expand our exploration of the Kozłowski–Weiner approach by introducing two other groups of 50 imaginary 'species'. These are characterized by the same $cv_c = cv_m = 0.2$ as in the third simulation of Fig. 13.5, but one group (black circles in Fig. 13.6) has a higher mean c (= 0.03), the other (triangles in Fig. 13.6) a higher mortality (mean $m = 0.0004$). Increasing mean c causes the whole distribution to shift to larger optimal body mass. The average production for the group is increased (Fig. 13.6a, black circles). In contrast, the average age at maturity has remained the same (Fig. 13.6b). Within the group, the regressions have retained the same slopes: $\log P = -2.61 + 0.79 \log w$ ($r^2 = 0.91$) and $\log \alpha = 2.91 + 0.21 \log w$ ($r^2 = 0.44$). There is a range of overlapping body masses between the two groups with mean $c = 0.015$ and $c = 0.03$. In this mass range the 'species' from the latter group have higher production rates, and lower α. This result may be interpreted in various ways. The $c = 0.03$ group may represent, for example, a phylogenetically or ecologically coherent group of birds adapted to exploit a high-protein energy source, such as shorebirds which have typically high-energy turnover for their mass (Kersten & Piersma, 1986).

In yet another group, mortality m is drawn from a distribution around $m = 0.0004$ instead of 0.0002. In this case, we find a reduction in the mean age at maturity, as well as in the adult production rates (Fig. 13.6 triangles). Again, the same regression slopes are found: $\log P = -2.52 + 0.78 \log w$ ($r^2 = 0.92$) and $\log \alpha = 2.82 + 0.22 \log w$ ($r^2 = 0.51$). Again, there is a range of overlap in body mass with the $m = 0.0002$ group. In this range the species from the high-mortality group have increased production rates and decreased ages of maturity. In this case, we may attribute the increased mortality in the group, for instance, to the increased degree of seasonality in the temperate zone compared to tropical areas (Curio, 1989). Thus, life-history theory readily gives an explanation for increased productivity and metabolism of similarly sized animals in temperate compared to tropical zones of the globe. Such global clines are well known for mammalian litter size (Lord, 1960) and avian clutch size (Klomp, 1970) and metabolic rates (Weathers, 1980; Ellis, 1984). Increased productivity for a given body mass away from the equator is further reflected in an increase in the size of the metabolic machinery such as found in heart, liver and kidneys (Rensch

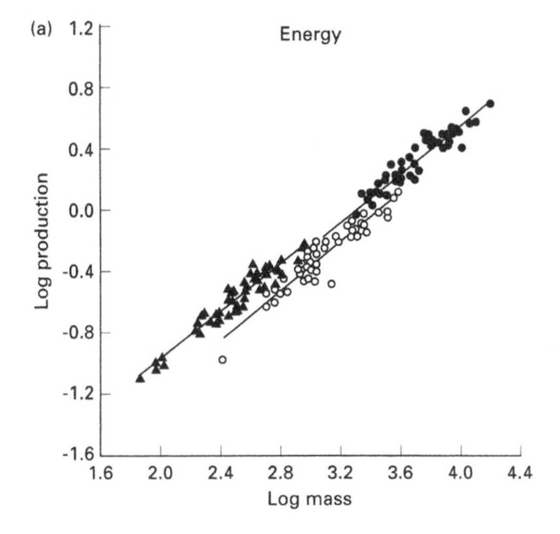

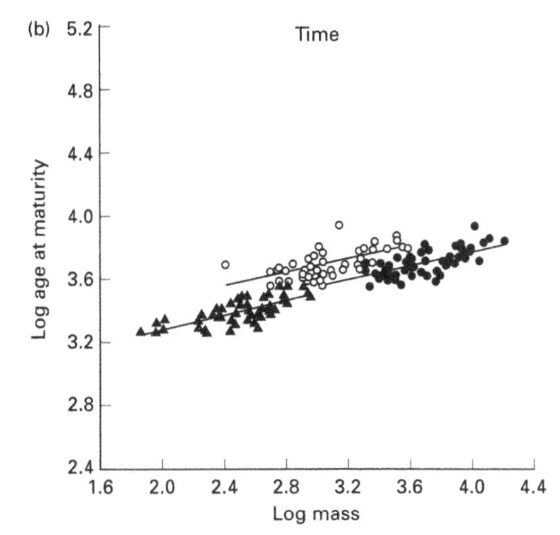

Fig. 13.6 Production (a) and age of maturity (b) plotted against body mass for 50 'species' in a resource allocation model ($cv_c = 0.2$; cv_m m = 0.2 throughout). Open symbols: same simulations as circles in Fig. 13.4 (mean $m = 0.0002$; mean $c = 0.015$). Dots: mean $m = 0.0002$; mean $c = 0.030$. Triangles: mean $m = 0.0004$; mean $c = 0.015$.

& Rensch, 1956; Daan *et al.*, 1990b). This illustrates how deeply into morphology and physiology the life history of animals eventually is reflected.

The general importance of the approach is that it removes the problem of scaling from the realm of physiological constraints to that of optimization. For over one century theoreticians have concentrated on the allometric exponents of the relationships with body mass. Much of this work was based on the intuition that these exponents provide a set of physiological 'laws of nature', of constraints within which organisms are forced to operate. Rubner (1893) postulated that energy metabolism necessarily scales with body mass to the power $^2/_3$ because of the surface : volume ratio. When more data became available, this exponent turned out to be closer to $^3/_4$ (Kleiber, 1947), and this led to other more complex theories of dimensionality (e.g. Stahl, 1962; McMahon, 1973; Heusner, 1987). These are all based on the assumption of constraints restricting the design of animal physiology. They all treat body mass as the independent variable on which all the derived variables are dependent. The resource allocation theory suggests that body mass itself may be dynamically optimized along with other biological variables. It also makes clear that interspecific allometric relationships can vary a great deal between groups of animals, depending on the variances and mean values of production and mortality rates, even with allocation models simplified to such a degree as the ones used here. Physiological 'laws' predict neither that species may deviate by a factor of 10 from allometric prediction nor the fact that the allometric exponent itself varies with the taxonomic level of analysis (Elgar & Harvey, 1987; Bennett & Harvey, 1988).

13.5 Parental effort and investment

Williams (1966) advanced the seminal idea that some of the effort animal parents dedicate to the production and growth of their offspring is at the expense of later reproduction. This *cost of reproduction* reflects a trade-off between current and future reproduction and is a key concept in life-history theory. The reduction in future reproduction which is the result of the current effort has been termed *parental investment* by Trivers (1972).

Neither Williams nor Trivers had strong evidence for a fitness cost of reproduction, let alone for particular mechanisms generating such a cost. Their hypothesis led to a new field of theoretical and experimental enquiry. One powerful strategy has been to manipulate parental effort by modifying the family size. Reduction of the effort is usually possible by taking dependent offspring away from the family. Increasing the effort has at least proved feasible in some altricial birds that accept extra young in the nest as their own. Three studies are now available in which the parental effort has been assessed in terms of the mass-specific daily energy expenditure (DEE; Fig. 13.7). In all three, DEE/mass increased with the experimental brood size raised.

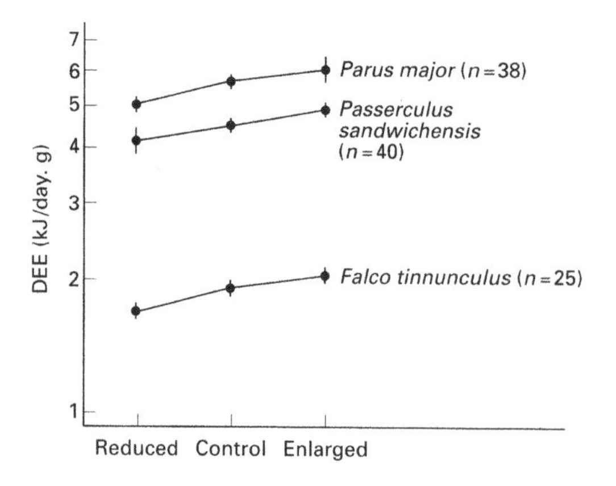

Fig. 13.7 Effects of brood size manipulations on mass-specific DEE in three species of birds: the savanna sparrow, *Passerculus sandwichensis* (Williams, 1987), kestrel, *Falco tinnunculus* (Deerenberg *et al.*, 1995) and great tit, *Parus major* (J. M. Tinbergen, in preparation).

However, the increase may be either due to mass loss (great tit: J. M. Tinbergen, in preparation) or to increased DEE (kestrel: Deerenberg *et al.*, 1995). Hence, we should not take it for granted that parents in all species necessarily respond in the same way to brood size manipulations, or that such experiments always test Williams' theory.

The measured consequences of brood size manipulations in birds for different fitness components have been summarized by Dijkstra *et al.* (1990) (see also Stearns, 1992). Of 14 studies analysing the fate of parents after brood manipulations five have reported significant reductions in local survival, and eight out of 14 reported reductions in later fecundity (e.g. Gustafsson & Sutherland, 1988). Reduced local survival may be due to increased mortality or increased emigration from the study area. To distinguish between these possibilities, we have recently studied the time of death among 39 kestrels that had raised manipulated broods and were subsequently reported dead by the general public, mostly away from the study area. The results demonstrate that mortality in winter is sharply increased in birds that had extra nestlings to raise in the previous summer (Daan *et al.*, 1996). The rates of survival and future reproduction of parent birds raising reduced, control or enlarged broods are integrated in the *reproductive value* (V_p) in Fig. 13.8. The curve, approximated by an exponential function as proposed by Kacelnik (1989), suggests that there is a non-linear reduction in V_p with increased parental effort. These data lend quantitative support to Williams' original proposition, but they do not identify the causal chain of events between parental effort and mortality.

The time of death, half a year after the experiments, is inconsistent with any theory invoking enhanced predation due to increased efforts. It rather points in the direction of physiological impairment. Recent research on the cost of reproduction focuses on enhanced parasite infection and susceptibility to disease. Indeed, in great tits, *Parus major*, raising enlarged broods increased

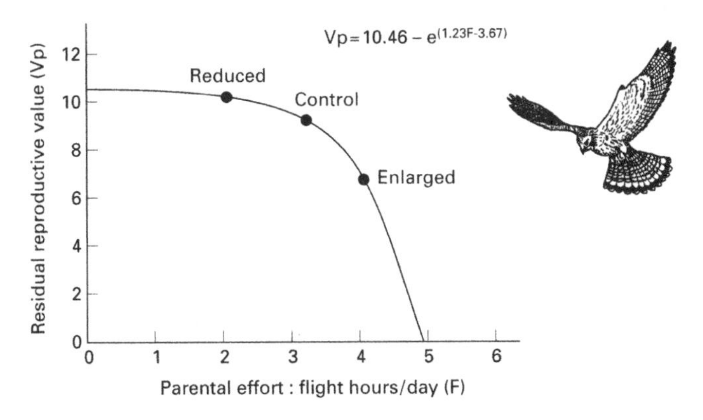

Fig. 13.8 Kestrel residual reproductive value plotted as a function of their parental effort (hours of flight per day). (From Daan *et al.*, 1990a.)

rates of infection — with malaria (Richner *et al.*, 1995) and haematozoans (Norris *et al.*, 1994) — have been found. Such effects may be caused by increased rates of contact with parasites or by reduced effectiveness of the immune system. Studies by Deerenberg (1996) measure immunocompetence by challenging the immune system with a standard antigen — sheep red blood cells — and assessing antibody titres several days later. This approach has so far been restricted to laboratory situations. Zebra finches, *Taeniopygia guttata*, show increased breeding intervals in response to brood size enlargements (Deerenberg, 1996) and reduced reproductive success if the efficiency of food acquisition is reduced (Lemon, 1993). Similar manipulations of brood size and daily work schedules reduce the formation of antibodies against the sheep red blood cells (Deerenberg, 1996). This is a promising direction of research, which may lead to the specification of (physiological) allocation of energy to reproductive effort versus repair and maintenance; and of the corresponding trade-off between current offspring and expected future reproduction.

We are dealing here more generally with plasticity of energy allocation. The allocation models assumed that assimilation A and respiration R are inflexibly determined by body size. But, there is evidence that both may be flexible. A, the total energy collected, may be increased during reproduction, as we know from the brood manipulation studies. R, or the DEE, is partly channelled into foraging activities, and this part cannot be reduced without incurring a penalty in a reduction of A. The remainder is associated with maintenance and repair processes, including the immune system, and this part may well be more flexible. Resting metabolism can, for instance, be reduced in response to low planes of nutrition (Daan *et al.*, 1989b), and to increased work rates in daytime (Deerenberg, 1996), and increased by low-temperature acclimatization (Konarzewski & Diamond, 1994). Such flexibility may be the key to the question why reproduction is not continuous after maturation as predicted by the allocation models.

13.6 Seasonal timing

David Lack was the first to discuss seasonal timing in a life-history context. Lack (1950) proposed that breeding seasons in birds have evolved in such a way that, on average, the peak demand of the nestlings coincides with the peak in food supply. In this manner the average bird would raise the largest number of offspring. Variations around the optimal time would be retained in the population as a balanced polymorphism by year-to-year variations in the timing of food supply. Population studies inspired by Lack's ideas later showed that the earliest breeders are typically the most productive, both in terms of offspring produced and recruits surviving (Klomp, 1970). Thus, the average bird in the population appeared to behave suboptimally. The pattern of seasonal clutch size variation can be reconciled by the concept of individual optimization (Hoegstedt, 1980; Drent & Daan, 1980). Declining prospects of offspring with progressive date of birth lead to optimal breeding times preceding the time when the maximum number of young can be raised. The trade-off between raising fewer superior offspring early in the year and more inferior offspring late in the season would determine the optimal time. Optimal times would vary between individuals, from early in rich territories to late in poor territories. Drent and Daan (1980) and Rowe *et al.* (1994) saw female body condition as a proximate constraint on number of eggs produced, and its vernal increase as the cause of the seasonal increase in the maximum raisable number of offspring. However, the trade-off causes a seasonal decline in optimal clutch size–date combinations regardless of the phase (laying eggs or raising nestlings) at which constraints operate (Daan *et al.*, 1989a).

This can be illustrated using the relationship between parental effort and residual reproductive value (V_p) in Fig. 13.8. Let us consider first a pair of birds raising a single nestling. Parental effort for one offspring is minimal at the time when food supply is maximal. If we represent the seasonal environment by a sine wave in foraging yield (y), then parental effort will be minimal (and hence V_p maximal) when it coincides with the time of maximum food supply (Fig. 13.9). This reflects Lack's original argument. However, as we have seen, the earliest born young often have the best prospects to become recruits. The most general reason is probably that by their earlier birth they gain an advantage over their year class in competition for food, territories, mates, etc. The reproductive value of the young (V_0) therefore declines with date of birth. V_0 has to be added to the residual reproductive value of the parent (V_p), and the sum V is maximized (circles in Fig. 13.9) at a date preceding the 'Lack date'.

In Fig. 13.10(a,b) we have further introduced differentiation between habitats by varying the amplitude of the seasonal yield (y) curve. Maximizing V_p in all cases leads, of course, to reproduction at the seasonal peak (Fig. 13.10a). Maximizing V leads to earlier optimal dates in the richer habitats (Fig. 13.10b). We have extended the same argument to broods of more than one offspring in a formal optimization model (S. Daan *et al.*, in preparation). In Fig. 13.10(c,d) the optimal clutch–date combinations are calculated, again for three levels of

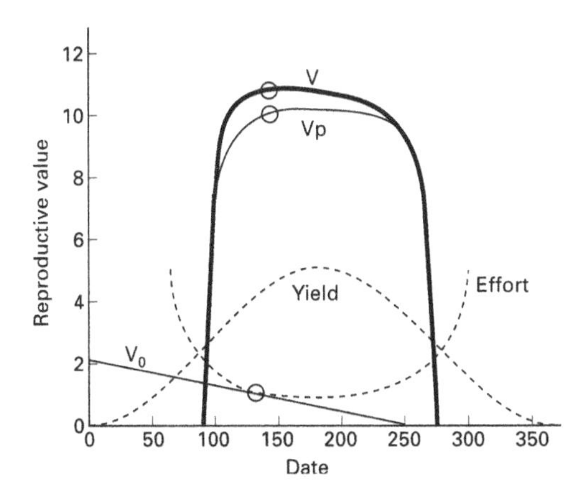

Fig. 13.9 Seasonal timing: (sinusoidal) seasonal variation in food supply (yield) leads to minimal parental effort and maximal residual reproductive value (V_p) at the yield maximum (the 'Lack date'). A seasonal decline in offspring reproductive value V_0 leads to an optimal date where the slopes of V_p and V_0 cancel each other (circles), i.e. before the yield maximum.

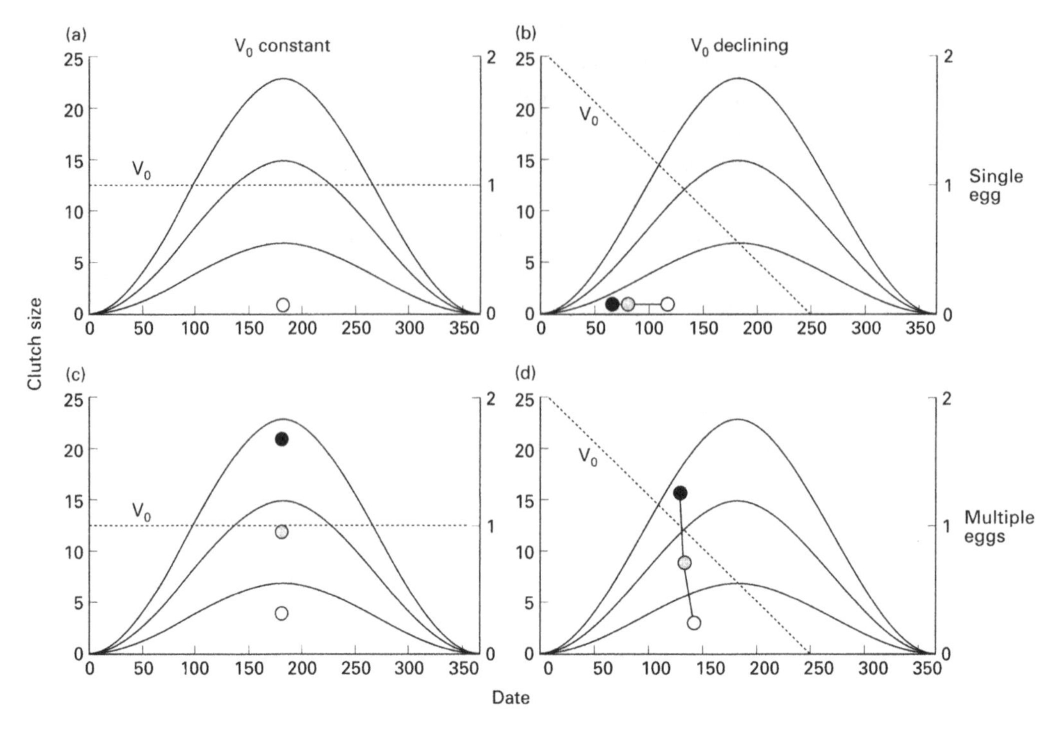

Fig. 13.10 Optimal date and clutch size combinations obtained from the model in Fig. 13.9, when there is variation between territories in environmental quality (three amplitudes of seasonal yield curves; solid lines): (a,b) For single egg clutches; (c,d) variable clutch size. (a,c) For constant V_0; (b,d) for seasonally declining V_0. The solid dots give the solution for the richest, the open symbols for the poorest territory, with shaded symbols intermediate. In (a) the three solutions are identical.

food supply (territory quality). The optimal clutch size now increases with increase in y. The optimal date is again the 'Lack date' if seasonal change in V_0 is not taken into account (Fig. 13.10c). It is advanced relative to the peak food supply when V_0 declines with progressive date of birth (Fig. 13.10d). The important point here is that the advance is greater in the richer territories. Together, these effects cause a seasonal pattern of declining clutch sizes in the population, without invoking any particular constraint such as female condition.

Thus, the variation in brood size and timing of breeding within populations can, in principle, be explained on the basis of individual optimization. The optimal combinations have been worked out for the kestrel (Daan *et al.*, 1990a; Lessells, 1991), and were found to be in reasonable harmony with observations. In this species, the seasonal decline in V_0 was based on correlative rather than experimental data, and, hence, we cannot be sure that for an individual V_0 would indeed change along this same line with date as it does between individuals. This particular weakness of the analysis was resolved in a recent study of the European coot, *Fulica atra* (Brinkhof, 1995).

Brinkhof studied the life-history consequences of a change in the date of laying by exchanging whole clutches of eggs which differed by 10 days in laying date. By this manipulation he either advanced or delayed for the parents the date at which their brood hatched. The experiment is suitable to resolve two different issues (Fig. 13.11).

1 A seasonal trend in any component of fitness may be caused by a general environmental variable associated with date and acting on all birds in the same manner (date hypotheses). Alternatively, such a trend may be caused by differences in quality of the birds or their habitat which simultaneously also affect the date of breeding (quality hypotheses). When parent birds are confronted with earlier or later broods than they chose themselves, the same trend in the fitness component as observed in the population should be found under the date hypothesis, but deviations from this trend under the quality hypothesis. This issue is important for the derivation of optimal solutions for individuals.

2 If individuals optimally tune the date of breeding to their own circumstances, advance and delay manipulations should both lead to a depression in fitness (Fig. 13.11). Note that this question of individual optimization cannot be answered for single components of fitness but only for total reproductive value.

Brinkhof followed post-hatching growth and survival of the young coots as well as survival of the female parents of his experimental broods, and calculated the offspring and parental components of reproductive value. The results are as follows: for the survival rate of hatchlings until 7 weeks of age (S_0) and the probability that parents produce a second clutch (P_2), the variation in the population is apparently fully due to *date* itself. Indeed, Brinkhof could identify the seasonal change in insect food as the cause of the variation in S_0. The seasonal variation in the survival of young coots until age 1 year

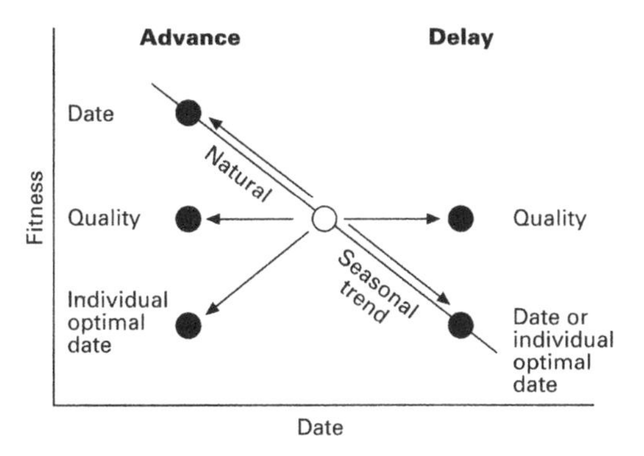

Fig. 13.11 Expected results of manipulations of date on fitness components under different hypotheses. Date hypothesis: advanced and delayed broods are expected to follow the seasonal trend observed in the population. Quality hypothesis: advanced and delayed broods are not expected to differ from controls. Individual optimization hypothesis (only for total reproductive value): advanced and delayed broods are expected to have lower fitness than controls. (From Brinkhof, 1995.)

(L_1) was attributable to quality (of the step-parents!), not to date variation. Finally, the survival of the female parent (L_{p1}), which did not vary with date in the population, remained unaffected in parents raising a delayed brood. However, it was sharply reduced compared to controls in parents raising an advanced brood. The cause of this increased mortality remained unidentified. It is possibly associated with increased rates of parental effort due to the fact that advanced dates caused more young to survive the early stage of post-hatching development. Be this as it may, on the basis of these analyses Brinkhof was able to reconstruct the final picture for total reproductive value V (Fig. 13.12). This shows that, except for the very earliest birds in the population, V was negatively affected by both advanced breeding dates (due to reduced parental survival) and delayed dates (due to reductions in hatchling survival and the probability of second broods). The study — unique in its completeness — thus strongly supports the notion that variation between

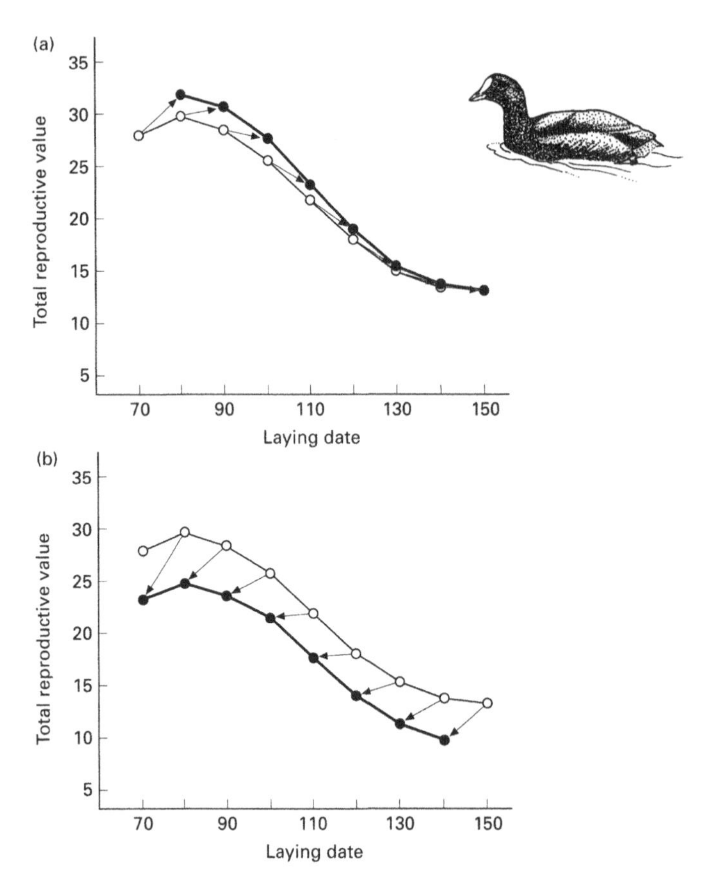

Fig. 13.12 Estimates of individual reproductive values of coot *Fulica atra* parents plus brood as a function of laying date in the natural population (open circles) and following experimental 10-day shifts (arrows) of the hatching date of the brood (filled circles). Both delay shifts (a) and advance shifts (b) reduce fitness for most laying dates, favouring the individual date optimization hypothesis. (From Brinkhof, 1995.)

individuals in the timing of breeding reflects variation between individuals in the optimal solutions. At each date, the birds starting a clutch would have done worse in terms of reproductive value if they had chosen an earlier or a later date.

The most important features of seasonal timing from a life-history perspective are thus twofold.

1 Optimal timing may often be based on the core life-history trade-off between current and future reproduction; in this case between reduced parental fitness (due to increased effort) and increased offspring fitness at earlier dates. This trade-off explains why most animals concentrate reproduction in spring, at the beginning rather than the height of the good season.

2 The general seasonal decline in clutch and litter size is best understood on the basis of differentiation between individuals in optimal timing.

13.7 Offspring size and numbers

Another fundamental trade-off in life-history theory is that between the number of independent offspring produced per attempt and their size. Generally, the prospects for each offspring will be positively associated with size, but size is negatively associated with the number of offspring over which each unit of parental investment is distributed. The life-history problem is to find the number of offspring per reproductive event which maximizes the total reproductive value for the parent. Size at independence is determined both by the energy stores allocated to the eggs, and — in species with parental care — the energy invested in raising and protecting the brood. The simplest situation is offered by species without parental care. We illustrate the experimental analysis of the size versus number trade-off with a study on eggs of the side-blotched lizard *Uta stansburiana* by Sinervo *et al.* (1992).

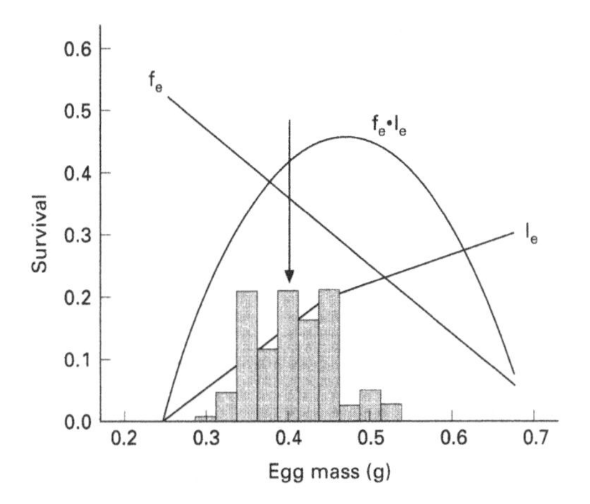

Fig. 13.13 Number of eggs per clutch (f_e), survival of hatchlings (l_e) and the fitted function through their product $f_e l_e$ as an estimator of fitness, in a population of eggs of manipulated size of the lizard *U. stansburiana*. Bars show the natural variation in egg mass. (From Sinervo *et al.*, 1992.)

Sinervo *et al.* (1992) collected near-term gravid females of this small iguanid lizard in the inner coastal range of California. They measured egg sizes and established that there is generally a negative association between the number of eggs laid by a lizard and their average size. This is represented in Fig. 13.13 by the linear regression of clutch size (f_e) on egg mass in a sample from one study site (1989, Del Puerto, late clutches). The authors further experimentally assessed the consequences of egg size for post-hatching survival. The experiments made use of two techniques:

1 egg size was reduced after laying by extracting some of the egg yolk with a syringe;

2 in some of the growing follicles in early vitellogenic females the authors stimulated extra yolk deposition by removing yolk surgically from other follicles.

After laying, eggs were incubated under standard conditions and the hatchlings released in the study area. One month after release the study area was thoroughly searched for survivors. The fraction l_e of surviving offspring, plotted in Fig. 13.13, was usually positively associated with egg mass, although in some samples a decline was observed for the largest hatchlings. The product $f_e l_e$ is an important component of the fitness of the offspring, and indeed may have an over-riding effect on total fitness. This component showed a maximum at intermediate egg mass in seven out of eight repeats (two locations, 2 years, two halves of the breeding season) of the experiment. A close match was found in most cases between observed and predicted mean egg mass.

This ingenious study did not relate the consequences of the experimental modification to the original unmanipulated egg mass. Testing whether interindividual variations in a trait reflect variations in the optimal value for this trait would require the comparison of manipulated with control groups. Only by experimental manipulation can we be sure that the fitness consequences for individual alternatives are measured. Such experiments are often difficult, and assessing the full consequences for fitness is tedious, but without them it is not possible to evaluate whether life-history strategies represent optimal solutions. A study where the difference between descriptive and experimental analysis has been worked out is that of the decision on the *number* rather than the *size* of eggs in the great tit, *Parus major* (Tinbergen & Daan, 1990; Verhulst, 1995).

The analysis is based on a population of great tits breeding in nestboxes in a woodland 'Hoge Veluwe', which has been under study for nearly 50 years. Tinbergen and Daan (1990) analysed the reproductive value of parents raising clutches of different size, as well as of their offspring. The results, summarized in Fig. 13.14(a), show that in the population both V_p and V_c are positively associated with clutch size. This is due to the fact that parent birds laying larger clutches have a higher probability of producing a second brood (hence V_p is larger), and raise more fledglings. Over 5 years, brood size was experimentally manipulated in 341 nests by either taking 50% of the nestlings out

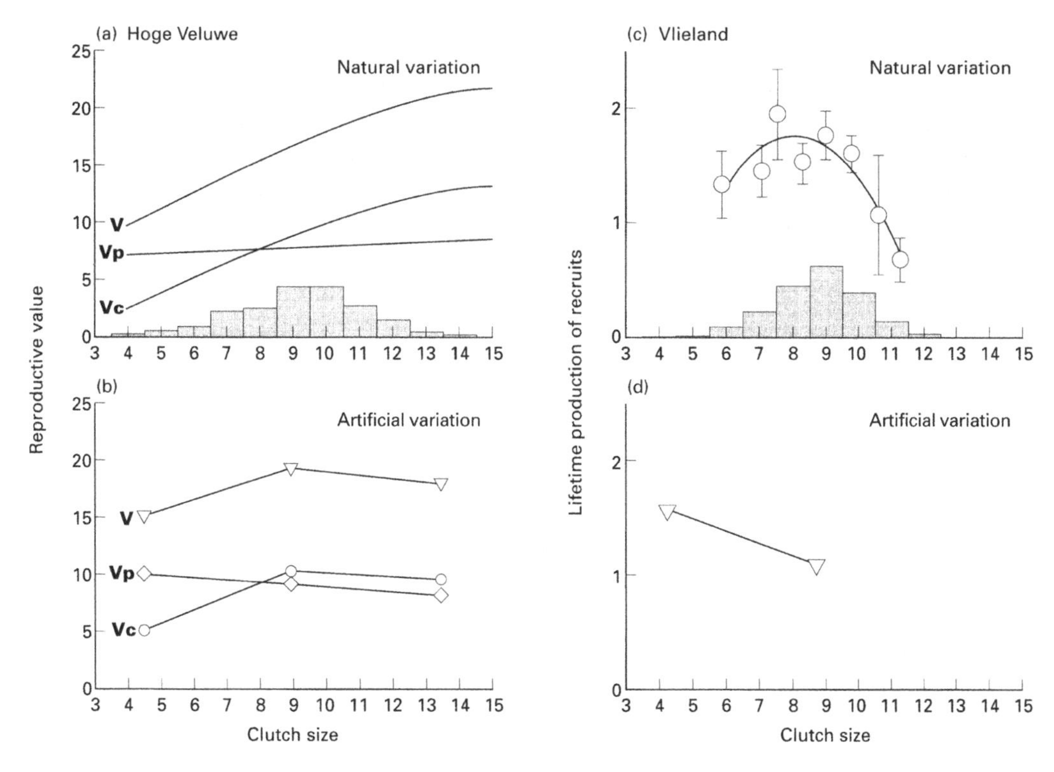

Fig. 13.14 Natural and artificial variation in clutch size in two populations of great tits *P. major*. (a) Mainland study site the Hoge Veluwe. Lines show reproductive values of parent (V_p) and clutch (V_c), and their summation V as fitted functions of natural clutch size. Bars: distribution of natural clutch sizes. (b) Artificial reduction and enlargement of clutches reduce fitness. (Modified from Tinbergen & Daan, 1990.) (c) Island population Vlieland. Open symbols: lifetime production of recruits by parents of different clutch sizes +/– one standard error of the mean. Line: fitted quadratic function. (d) Lifetime production of recruits for parents with control (average 9 eggs) and reduced clutches (average 4 eggs). Artificial reduction of clutch size enhances fitness in this population. (From Verhulst, 1995.)

or adding 50%. Such experimental variation in brood size leads to a wholly different picture (Fig. 13.14b). Now, V_p declines with increasing brood size and V_c shows a curvilinear relationship with brood size. On the basis of the descriptive analysis one might surmise that the clutch size associated with maximum fitness (V) is 15 nestlings. The experimental analysis shows that optimal brood size is in fact the control brood size (which for practical reasons was around the most common brood size), since both enlargements and reductions led to a reduction in fitness (V). Hence, clutches manipulated from nine to 13 represent lower fitness than natural clutches of 13. The implication is that natural variation in brood size must have fitted the variation in optima. This is another instance of individual optimization as we encountered earlier in the kestrel and coot.

More recently, Verhulst (1995) has applied the same approach to another population of great tits, living on the Dutch island of Vlieland. He used lifetime production of recruits rather than reproductive value as a fitness measure, and applied only brood reductions, not brood enlargements, as the experimental intervention. Interestingly, the Vlieland results deviate from those in the mainland population: in the descriptive analysis, a curvilinear relationship was found, with maximum fitness associated not with the maximal clutch size but with clutches of about eight eggs (Fig. 13.14c). In the experimental analysis, artificial reduction of brood size led to a significant enhancement of LRS (Fig. 13.14d). This means that the observed clutches on Vlieland were on average larger than is optimal. Parents in the long run would contribute more genes to the population if they would lay fewer eggs. This discrepancy shows clearly that the method does not necessarily lead to a fit between average and optimum, but that it is possible also to detect maladaptive features in life history. Verhulst (1995) has argued in this case that probably the tendency to lay too many eggs is maintained in the population by immigrants from the mainland. Such gene flow would prevent the population from becoming fully adapted to the local environmental circumstances.

13.8 Population heterogeneity in life histories

Variation of life-history parameters within animal populations has stimulated the analysis of their adaptive significance. When there is such variation, we cannot directly assess whether this variation is either distributed along the same reaction norm for all genotypes in the population (i.e. be solely based on phenotypic plasticity) or represents purely genetic diversity, or any mixture of environmental and genetic diversity. In the early days of life-history studies it was more or less implicitly assumed that most of the variation is genetic. In this vein, Lack's (1950) original propositions were that the *average* clutch size and the *average* date of laying would be the traits maximizing fitness in stable populations, with stabilizing selection weeding out the extreme genotypes. Population studies in birds, mostly inspired by Lack's ideas, have not confirmed these predictions. In general, the earliest breeders, and those with the largest clutches, contribute most to the next generation. If these trends would reflect directional selection, the genotypes leading to late and small clutches would soon disappear from the evolutionary scene. Yet, population geneticists claimed, on the basis of mother–daughter correlations in reproductive traits, that a considerable part of the variation in traits such as clutch size and laying date was of genetic origin (e.g. Van Noordwijk *et al.*, 1981). Current views on how environment interacts with genotypes leave more scope for environmental effects on variation in life-history traits via phenotypic plasticity or reaction norms.

Such variation may not be related exclusively to the environment of the adult animal but also to the conditions during early ontogeny. Grafen (1988) has coined the term 'silver spoon effect' to indicate the lifetime benefits an animal may gain from growing up in rich nutritional conditions. Examples of such effects are now slowly accumulating in the literature (e.g. Huck *et al.*, 1987; Tinbergen & Boerlijst, 1990). If proving general, they may help explain mother–daughter correlations in reproductive traits as well as the high degree of variation in total lifetime reproductive success, which is now emerging as a general phenomenon in long-term population studies (Newton, 1989). Be this as it may, a sizeable extent of environmental variation, whether acting at the time a reproductive decision is taken or early in ontogeny, implies that we cannot find quantitative solutions of any optimization problem without recourse to experimental manipulations of the traits. By their very nature such traits are concerned with rare life events, and it will not usually be possible to use individuals as their own control in experiments. This and the fact that fitness costs and benefits have to be assessed in the natural environment makes it necessary to perform large numbers of experiments and renders analytical progress in this field slow.

This reasoning should not distract from the fact that some of the variance may be of genotypic origin. From the experiments on the mainland great tits we have seen that the mean clutch size closely corresponds with the mean optimal solution. Yet, in the case of the Vlieland great tits, clutch sizes were larger than optimal, and such deviations may be attributed to genetic constraints imposed by gene flow from a *source* population (Verhulst, 1995). Life-history research faces important challenges in unravelling — both theoretically and empirically — the degree to which genetic and environmental trait variation is found. With a strong unidirectional gene flow from sources to sinks, we intuitively expect more directional selection in the sinks than in the sources. Following this reasoning, one may predict more genetic diversity in traits strongly affecting fitness to be maintained in sinks by the combination of directional selection and gene flow.

Furthermore, the degree of predictability of the temporal environment is bound to affect the contributions of genetic and environmental variance. In a perfectly predictable, but spatially heterogeneous environment, we expect that animals can individually optimize their decisions better, and the observed variation will largely reflect environmental components. In a temporally unpredictable environment, variation may be more prone to containing a strong component of balanced genetic polymorphism. The time-scale of these variations is necessarily relative to the time-scale of individual lives. Evolution must have found many different solutions for adjusting the traits to whatever environmental predictor is available. The ubiquitous photoperiodic response is a physiological mechanism suppressing reproductive activity when the harsh winter season approaches for many short-lived animals; it triggers reproduction in those large mammals where winter gestation is required to synchronize parturition with the next springtime. The interrelationships between genetic and environmental variation, between physiological control mechanisms and their fitness consequences, and between temporal characteristics of the environment and of a species' rate of living will be fascinating future foci of life-history research.

Chapter 14

The Phylogenetic Foundations of Behavioural Ecology

Paul H. Harvey & Sean Nee

14.1 Introduction

The vibrant synergism between studies of behavioural ecology and evolutionary biology is well illustrated by the results of using phylogenetic trees to help tackle problems in the two areas. Evolutionary biologists want to understand the reasons for the diversity of species. Phylogenetic trees describe that diversity, but it is becoming increasingly obvious that behavioural ecology can help explain why that diversity arose. In particular, species with particular behavioural ecologies are more likely to speciate so that each of several bushy areas of phylogenetic trees may contain species with similar life-styles. On the other hand, behavioural ecologists want to understand the intimate links between behaviour and ecology. Mapping character states on to phylogenetic trees provides critical insights into how behaviour and ecology have shown correlated evolutionary change, or when new associations between the two have come about.

This chapter is divided into four sections. The first examines how species differences in behaviour and ecology can help explain differences in the structure of evolutionary trees. The second section asks how character change can be mapped on to phylogenetic trees and used to test ideas about how particular behavioural ecologies evolved and whether some types of character evolve more rapidly than others. The third section moves from the consideration of single characters to correlations among characters, and shows how, if we are to understand why character covariation across taxa occurs, phylogenetic tree structure must be an integral part of the analysis. Phylogenetically based comparative analysis has proved a powerful tool for generating and testing ideas about the links between behaviour and ecology. The fourth and final section integrates the first three by showing how, during the course of evolution, speciation may be accompanied by predictable changes in behavioural ecology. The consequence is that similar sequences of phenotypes may evolve independently at different times and places.

This chapter is constructed around illustrative examples taken from the recent literature. We should emphasize from the start an assumption that is common to those examples, but which may not be correct in some cases and will almost certainly not be true in others. When a statistical test is performed in the various examples described, it is often assumed that the phylogenetic tree used for the analysis is perfectly correct (always in topology, and occasionally in branch length also). In fact, phylogenetic trees are working hypotheses generated from limited data. As the dataset improves, so does the likelihood that the working hypothesis will be correct (for consistent methods of tree construction see Huelsenbeck *et al.*, 1996). We are lucky to be working at a time when there has been a revolution in the reconstruction of phylogenies which show the relationships among contemporary species (Hillis *et al.*, 1996). This revolution results from technical advances (notably the polymerase chain reaction) which makes the sequencing of large sections of genetic material (deoxyribonucleic acid (DNA) and ribonucleic acid (RNA)) routine. Sequence data have added enormously to our ability to reconstruct more accurate phylogenies.

14.2 Tree structure

14.2.1 Key innovations and cichlid fishes

Occasionally, a phylogenetic lineage or clade branches at a higher rate than others. For example, cichlid fishes are extraordinarily speciose, as are phytophagous insects. If we are interested in why particular clades are very speciose we might attempt to identify some 'key innovation' common to members of the clade but which is not shared by its close relatives. 'Innovation' must be interpreted broadly as simply any biological feature of a clade which either promotes speciation or, for example, simply allows speciation to occur if the clade finds itself in a situation where competition is absent, leaving many niches empty. Liem (1973, 1980) may have identified such an innovation. He pointed out that a small shift in position of a single muscle attachment ultimately allowed the pharyngeal bones of cichlids to manipulate their prey items while still holding them. As a consequence, the pre-maxillary and mandibular jaws were freed to evolve along new routes that did not involve manipulating prey. This, he argues, allowed cichlids to evolve a whole new diversity of feeding mechanisms and there seems little doubt that the adaptive radiations of cichlids in African lakes, in the face of competition from other fish families, does result from their evolved diversity of feeding mechanisms. However, Lauder (1981) pointed out that while such explanations may well be correct, any other characteristic common to cichlid fishes but which differs from a sister group could, in principle, be the key innovation which resulted in high speciation rates. Conclusions are strengthened if several sister-group comparisons can be demonstrated where the same key innovation is associated with high species diversity (Heard & Hauser, 1995). Such comparisons control for the many differences between sister lineages because any particular difference apart from the one under test is less likely to be partitioned in the same manner among different sister taxa. We now use two examples, one from insects and the other from birds, to show how such sister comparisons are performed and how they have been used in hypothesis testing.

14.2.2 Phytophagy and insect diversification

Insect species that feed on vascular plants are termed phytophagous. Half the world's insect species are phytophagous, but they are restricted to only nine of the 30 orders of insects. Southwood (1973) imagined that there were considerable barriers to an insect group evolving phytophagy. Behavioural and morphological adaptations would be required: (i) to reduce the risk of desiccation; (ii) to remain attached to the host; and (iii) to deal with low-nutrient food. Once those problems were overcome, Strong *et al.* (1984) suggested that the diversification of phytophages would be promoted by the great diversity of plants species and diverse plant parts as well as the absence of competitors. As a test of the idea that phytophagy promotes insect diversification, Mitter *et al.* (1988) were able to perform 13 sister taxon comparisons between the diversity of a phytophagous clade and its non-phytophagous sister clade. Since sister clades originated at the same time, they have had identical times for diversification. In 11 of the 13 comparisons, there were more species in the phytophagous clade than in the sister group, and in each of those cases the difference was greater than twofold.

14.2.3 Mate choice and speciation in passerine birds

It has been suggested frequently that sexual selection by female choice might increase the rate of reproductive divergence between populations and thereby increase the speciation rate of a clade (e.g. Darwin, 1871; Lande, 1981; Dominey, 1984; Carson, 1986; Schluter & Price, 1993). Barraclough *et al.* (1995) tested this idea using passerine birds as a case study. It is generally accepted that mate choice is responsible for the evolution of sexual dichromatism in passerine birds (Andersson, 1994), and so Barraclough *et al.* used dichromatism as an indirect measure of the importance of mate choice for a species. With few exceptions, passerine species can be scored unambiguously as dichromatic or not. On significantly more than 50% of occasions, the clade with a higher proportion of sexually dichromatic species was more speciose than its sister taxon, thereby supporting the idea that mate choice can increase speciation rate.

14.3 Character change

How and why do characters evolve? We can begin to answer these questions if we have: (i) a phylogenetic tree showing the relationships among a series of contemporary species; (ii) character states for those species; and (iii) a model for character evolution.

14.3.1 Mapping character change on to the phylogenetic tree

There are various ways of mapping character change on to the phylogenetic tree, and the one we choose will depend on the model of character evolution that we have in mind. Maddison and Maddison (1992) give a clear exposition of the practicalities associated with choosing and employing a model of character change. Although their discussion is given in part with particular reference to their computer program MACCLADE, the major issues are well described. The basic tool is the 'step matrix' which specifies how easy it is to go from each character state to any other. They discuss, for example, how a step matrix can be derived from a matrix of probabilities of change between states. However, so little is generally known about probabilities of character transitions that some sort of parsimony or minimum-evolution criterion is used for ancestral character-state reconstruction. If required, probabilities of changes can then, themselves, be estimated from the tree which provides ancestral character states at the different nodes.

14.3.2 Tracing character evolution: the evolution of parental care

It is often of interest to behavioural ecologists to trace behavioural change on the phylogenetic tree. For example, some models for the evolution of parental care assume that evolutionary transitions involving a near-simultaneous change in parental behaviour from care to no care (or vice versa) by both parents is extremely unlikely, and Gittleman's (1981) analysis of the phylogeny of parental care in fishes provides support for that assumption (Fig. 14.1). Of 21 transitions, all were between biparental and maternal care, biparental and paternal care, no care and maternal care, or no care and paternal care; none was between biparental care and no care, or maternal care and paternal care. The most common transitions were from no care to male care to biparental care to female

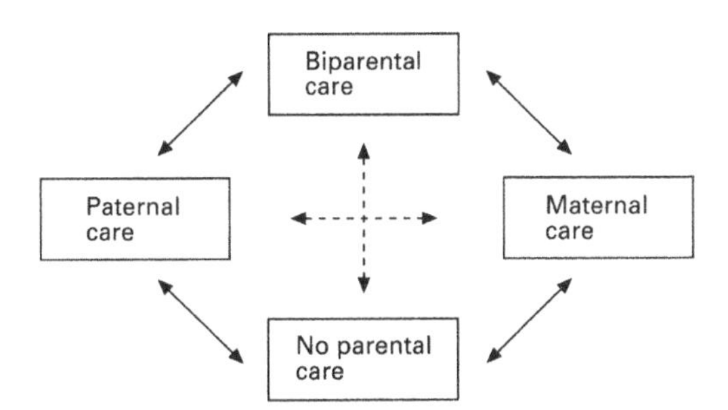

Fig. 14.1 The directions of change in parental care in fishes. Four states are possible. The solid arrowed lines are hypothesized transitional routes and the dashed lines, on the other hand, do not accord with the hypothesis that simultaneous change by both parents is unlikely. All transitions observed by Gittleman (1981) were in accord with the hypothesis.

care. Subsequently and similarly, Carpenter (1989) tested West-Eberhard's (1979) model for the evolution of patterns of social behaviour in vespid wasps; in that example some transitions did not accord with expectation, although others did.

A more recent and thorough analysis of evolutionary transitions returns to parental care, this time in shorebirds, which for birds show an astonishing diversity of incubation and brood-rearing patterns, ranging from pure maternal through pure paternal to biparental. There had been two schools of thought concerning the evolutionary transitions likely to be responsible for the diversity. The first is that biparental incubation was ancestral, and uniparental care followed depending upon the costs of care and the benefits of desertion to each parent (Jenni, 1974; Pitelka *et al.*, 1974; Emlen & Oring, 1977). The second suggestion involved the evolution of biparental care from male care, as males reduce their contribution (van Rhijn, 1985, 1990). To help distinguish between the two scenarios, Székely and Reynolds (1995) used a phylogeny based on Sibley and Ahlquist's (1990) DNA–DNA hybridization data, adjusted to account for recent criticisms and improvements. (Fortunately, their results seem robust to alternative likely phylogenies.) Parental care was classified for almost 50% of the 203 known species in two ways: (i) which parent, if either, is involved in most care; and (ii) what is the duration of parental care for each parent. Parsimony was used to determine ancestral character states, although some branches and nodes had equivocal states (e.g. biparental and maternal care equally likely). The major results to emerge (Fig. 14.2) were that most

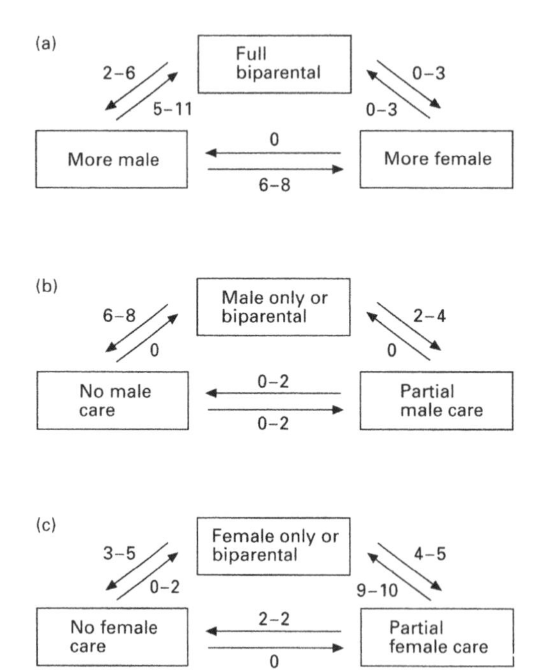

Fig. 14.2 The minimum and maximum number of evolutionary changes in patterns of parental care in a sample of shorebirds according to Székely and Reynolds (1995). (a) The number of transitions between care types. (b,c) Transitions in duration of care within each sex.

evolutionary transitions in the shorebirds had been towards a reduction in paternal care, which was sometimes but not always compensated for by an increase in maternal care. Székely and Reynolds (1995) conclude that their analysis provides more support for the idea that males have been reducing their parental support in comparison with females over evolutionary time. They argue that in shorebirds the costs of care per clutch are so high that males never guard more than one clutch at once. At the same time, there is always the possibility for the male of extrapair fertilization or multiple pair bonds. Accordingly, they reason, there has been continued selection for reduction in male care, which leaves them the problem of why there are actually more contemporary species in which males provide more care than females. Presumably, one might answer, either the rate of speciation has been higher among lineages in which males provide more care than females or the rate of extinction has been higher in lineages in which females provide more care than males.

14.3.3 Rates of character evolution: behaviour evolves rapidly

Given that we can trace character evolution on a phylogenetic tree, it should be a fairly straightforward exercise to examine rates of character change. The idea that some types of character are evolutionarily more labile than others is frequently assumed but rarely tested. In particular, Gittleman *et al.*'s (1996) review of the literature indicates a general belief 'that behavioural traits are more labile than morphological or physiological traits'. There are a variety of ways evolutionary rates can be measured (Gingerich, 1993), the more sophisticated of which incorporate the phenotypic standing variation of the character and generation time. In their attempt to test the idea that some traits have more rapid rates of evolution than others, Gittleman *et al.* (1996) could not use such sophisticated measures owing to paucity of data, and made do with Haldane's (1949) crudest measure, the 'darwin'. A rate of 1 darwin corresponds to a change of 1 logarithmic unit per million years. They examined variation in two behavioural characters (group size, home-range size), two life-history characters (gestation length, birth weight) and two morphological characters (adult brain weight, adult body weight) in each of eight mammalian taxa for which there were reasonable phylogenies. In summary, the behavioural characters seem to have been more evolutionarily labile than the morphological characters, which were more labile than the life-history characters. Although Gittleman *et al.* (1996) wisely urge caution over their results, which must be regarded as preliminary, the analysis does point the way forward to the use of phylogenetic information for determining which characters show more evolutionary inertia than others. As that information accumulates, general patterns may well emerge which will suggest why some characters evolve more rapidly than others.

The fact that some characters are usually more evolutionarily labile than others, does not necessarily mean that rates of character evolution remain the same in different lineages. Garland (1992) has developed statistical tests designed to compare rates of character evolution in different clades. As an illustration of the method, Garland uses data on metacarpal : humerus ratio (an index of cursoriality), metatarsal : femur ratio, body mass, maximal sprint running speed and home-range size from 16 species of Carnivora and 27 species of ungulates whose phylogenetic relationships are reasonably well known. As expected (Janis, 1996), the limb proportions evolved more rapidly in the ungulates than the Carnivora, but the other three variables seem to have evolved at very similar rates in the two groups. A comparison of overall rates of change of the various characters is also 'consistent with the long-standing idea that behaviour generally evolves more rapidly' (Garland, 1992, p. 516) than other characters.

14.4 Correlated character change

Behavioural ecologists are frequently interested in the reasons why some species behave differently or occupy different habitats from other species. One perfectly reasonable way to proceed is to search for correlates of the difference(s) we are interested in. However, two variables can be correlated because the first has influenced the second, because the second has influenced the first or because both have been influenced by an unknown third variable. Biologists are fortunate because many third, or confounding, variables can be eliminated effectively from their comparative analyses by careful use of phylogenetic information.

14.4.1 Why incorporate phylogeny?

We know that closely related species are similar among themselves, and they differ in many ways from other less closely related species. The similarities among closely related species are often said to arise from 'phylogenetic constraint', but that can be a misleading term because of the different processes which may be involved. For example, new species are likely to invade niches very similar to those occupied by their immediate ancestors because those niches provide environments to which the species is well adapted; here, the 'constraint' is adaptive resulting from similar selective pressures. Alternatively, there may not be genetic variance available for new characters to evolve, in which case the 'constraint' is genetic. These and other reasons for similarity to be retained by descent through common ancestry are discussed elsewhere (Harvey & Pagel, 1991). The evolutionary perspective provided by phylogenetic information allows comparative biologists to identify cases where change in one character or habit is accompanied by change in another. It is then possible to use independent evolutionary events to statistically test the null hypothesis that two or more characters change independently of each other.

14.4.2 Types of character change

There is a useful but somewhat artificial distinction to be made between continuous and categorical characters. Examples of continuously varying characters are body mass or amount of time spent foraging, while typical categorical characters might be presence versus absence of eyes or type of habitat occupied (woodland, hedgerow, pool, stream). Continuous variables are, in fact, discrete variables where the width of the discrete units is vanishingly small. Nevertheless, different comparative methods have been developed to deal with the different types of character. Whatever method is used, an important point to grasp is that it will incorporate a model of character change. For continuously varying characters, the most usual model is that of Brownian motion or a random walk in which, for example, it is supposed that the average body size of a species changes continuously and randomly over time. An alternative, punctuational, model is that characters change only at speciation events (i.e. at nodes on the tree). In any event, our null model would be that any two characters we were interested in changed independently of each other through the tree. We shall consider the most widely used method of testing for correlated change in continuously varying character below.

Before considering comparative methods per se, it is important to return to the point that character change in either direction may not be equally likely (see Sections 14.3.1 and 14.3.2). For example, Dollo's law states that complex characters are more frequently lost in evolution than they are gained. The question is, how much more frequently? If we were dealing with the presence and absence of eyes and we had considerable variation among contemporary species and a well-resolved phylogeny, we might impose the restriction that eyes are homologous organs which were present in the common ancestor of the group in question and evolved just once: all evolutionary transitions would be towards the loss of eyes. Similarly, although less extreme, there appear to be many lineages in which size increases more frequently than it decreases through time (Cope's law), and we might wish to impose some differential likelihood of size increase versus size decrease on our character-state reconstruction.

14.4.3 The method of independent contrasts

For continuously varying characters, Felsenstein (1985) developed the comparative method of independent contrasts. The key realization here is that differences in character states between each pair of sister taxa in a phylogeny evolved independently of differences between all other pairs of sister taxa. Each node in a bifurcating phylogeny subtends two sister taxa, and so there are as many comparisons possible as there are nodes in the phylogeny. Each comparison is termed a contrast. For a bifurcating phylogeny, the number of independent contrasts adds up to the number of species minus one. After estimating ancestral character states for each node, contrasts are calculated for each character being considered. The direction of comparison is arbitrary, although it is conventional to order the taxa so that all contrasts for the independent variable are positive (or zero). The contrasts for the dependent variable(s) then use the same order of taxa and can take both positive and negative values. The contrasts for the dependent variable(s) can then be compared with those for the independent variables, and conventional statistical techniques (correlation or regression) are used to test for correlated evolutionary change. The null hypothesis is that change is uncorrelated, and significant departures from the null hypothesis can indicate either positive or negatively correlated character change (Box 14.1).

Box 14.1 The independent comparisons method for two characters in a single phylogeny

Under a Brownian motion model of evolution, d1, d2 and d3 provide independent comparisons. Path length differences are ignored in this illustration.

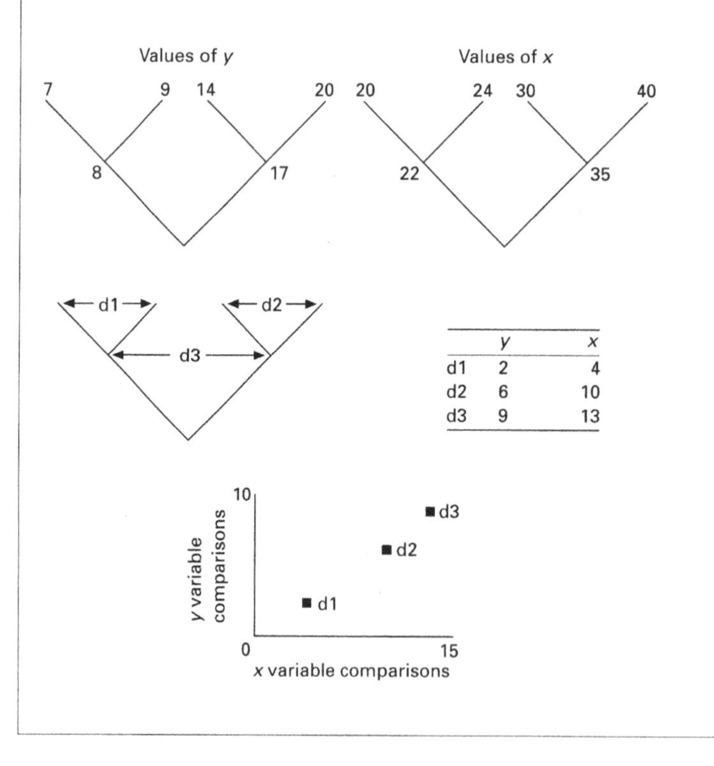

	y	*x*
d1	2	4
d2	6	10
d3	9	13

Independent contrasts have been used to test fairly sophisticated ideas. For example, in 1927 the ecologist Charles Elton argued that birds feeding on small prey items are likely to provide less food for their offspring than species that select large but rare prey. A similar suggestion was subsequently made by Sutherland and Moss (1985). Saether (1994) found that, indeed, a smaller amount of food was brought back to the nest in taxa of altricial birds feeding on small prey items. Saether reasoned that if those species feeding on larger prey items could bring back more food, then they could either provision a larger clutch or fledge their young earlier. In either event, their reproductive success would be increased. However, species that feed on larger prey tend themselves to be larger bodied, often produce smaller clutches and need to provision larger offspring. As a consequence, Saether needed to control for the correlations with adult body size when examining the relationship between prey size and clutch size or incubation period. This was done by regressing the contrasts of prey size, clutch size and incubation period on the contrasts for adult body weight and calculating the deviations from the regression line. Those deviations removed the correlations with body size and were then plotted against each other. In fact, it turned out that when altricial birds feed on relatively large prey they produce relatively large clutches (Fig. 14.3), but they do not have a more rapid nestling growth rate or earlier fledging time.

Saether's (1994) analysis, like most others in the recent literature, assumed that the different variables being considered evolved according to a Brownian motion model of evolution. Statistical conclusions can, in fact, change when different models of character change are entertained. For example, Harvey and Purvis (1991), report a study of the evolutionary relationship between a measure of running activity and relative foreleg length (foreleg length corrected for body size) in a group of *Anolis* lizards. Three models of evolutionary

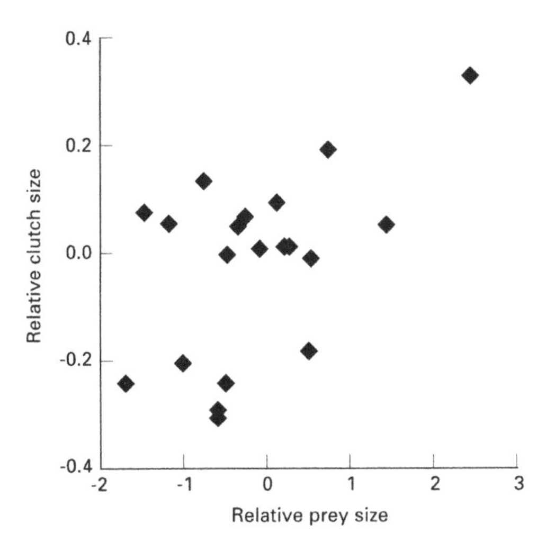

Fig. 14.3 Birds which, for their body size, feed their chicks larger prey items tend to also have relatively larger clutches. (Adapted after Saether, 1994.)

change were considered (Brownian motion, punctuation and minimum evolution), only the first of which produced a significant relationship between contrasts for the two variables being studied ($p < 0.025$ compared with $p > 0.10$ for the other two models).

While independent contrast methods are proving an extremely powerful tool for investigating correlated character change, apart from the need to consider alternative models of character evolution, it is important to bear in mind two other factors which sometimes lead to faulty or overstated conclusions. First, when comparing contrasts, linear regression models are usually used and, because the expectation of no change in one character is associated with the expectation of no change in the other character, regressions are forced through the origin. Such procedures may hide important features of the data. For example, larger bodied species typically achieve lower population densities in favoured habitats, and contrast analyses are expected to reveal a negative relationship between population size and body mass. However, when closely related bird species are being compared, the relationship can be reversed (Nee *et al.*, 1991): larger bodied species have larger populations (for reasons which will be considered below: see Section 14.4.5). If a linear regression is forced through the origin of a contrast plot, this interesting phenomenon might well be overlooked (Fig. 14.4).

The second note of caution when using independent contrast methods concerns error in the phylogeny being used. Independent contrast methods assume that the phylogeny is correct but, in fact, many phylogenies can mislead. For example, branch lengths may be incorrect, the extreme being the representation of a dichotomous phylogeny as a star phylogeny where all

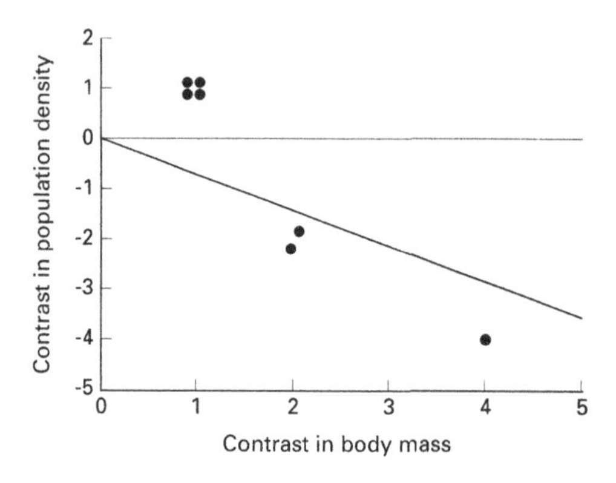

Fig. 14.4 An independent comparison (or contrast) analysis of the relationship between population density and body mass among eight hypothetical bird species (which gives seven independent contrasts). When contrasts of population density are plotted against contrasts of body mass, the relationship is non-linear and a model 1 regression line forced through the origin fails to distinguish the interesting fact that, when contrasts are small, population density actually increases with body mass. Small contrasts are between closely related species, such as congeners.

internal branch lengths are zero, while node to tip lengths are the same. Alternatively, and additionally, the wrong topology may be used so that rather than representing a coarsened variant of the true phylogeny, what we are actually analysing is the wrong phylogeny. Whatever the sources of error, as they are increased the chances of failing to reject an incorrect null hypothesis are also increased.

When designing comparative methods, it is always necessary to test for acceptable statistical properties using simulation studies. Essentially, characters are evolved along the branches of phylogenetic trees according to set levels of character covariation. The comparative method is then applied to the resulting species data, and both type 1 and type 2 statistical errors are estimated. There have been several such simulation studies which reveal that, on the whole, independent contrast methods have acceptable statistical properties (e.g. Martins & Garland, 1991), and are a vast improvement over cross-species studies which treat species values as independent items of information. Although evolutionary biologists have now largely accepted this fact, there is still considerable resistance to it from some ecologists (Harvey, 1996).

14.4.4 Methods for categorical variables

Starting with Ridley (1983), there have been several methods proposed for analysing character covariation between pairs of categorical variables. When the different methods have been applied to comparative datasets in order to test particular comparative predictions, almost invariably there has been a subsequent reply in the literature seeming to suggest that the original conclusion was incorrect because the method used had faults. Then, an 'improved' method is used to produce different conclusions. Cases relating to behavioural ecology include the relationships between warning coloration and gregariousness in insects (Harvey & Paxton, 1981; Sillén-Tullberg, 1988, 1993; Maddison, 1990), and between plumage dimorphism and lekking in birds (Höglund, 1989; Oakes, 1992; Höglund & Sillén-Tullberg, 1996). This is clearly an unsatisfactory state of affairs. Furthermore, there is a view that many popular methods for the comparative analysis of discrete traits, even though they were explicitly constructed to 'take account of phylogeny', do not solve the problem of phylogenetic non-independence at all (Read & Nee, 1995; Ridley & Grafen, 1996). Although there is a general consensus that contrast-based methods for continuous characters are basically sound, there is no hint of a consensus about any aspect of comparative methods for discrete characters.

Why is the status of the comparative method for discrete characters so different to that for continuous characters? One simple reason is as follows. With continuous characters, as we have seen, analyses begin by constructing contrasts and, almost always, such contrasts do, indeed, exist; for continuous characters, even closely related species differ to some extent. This is no longer the case for discrete characters. Typically, large areas of the tree show no variation in one or both characters of interest. If, and how, such regions of the tree are to be incorporated into the analysis is a very contentious issue. More subtle problems that arise in the study of discrete characters are discussed in Ridley and Grafen (1996).

14.4.5 Sister taxon and other pairwise comparisons

It is not at all surprising that working biologists often back away from performing a full-blown comparative analysis. The literature is peppered with reanalyses using different statistical methods and models of evolution, so that few conclusions seem robust, and evermore rococo comparative methods continue to proliferate. Yet, it is now widely acknowledged that comparative biologists seeking correlated character change must take phylogenetic information into account: statistical inferences based on cross-species comparisons are virtually worthless. Fortunately, it is possible to perform reasonably valid comparative analyses that do not require elaborate models and statistics.

So-called pairwise methods simply compare taxa whose separate evolutionary histories traced from a common ancestor are not shared with other taxa being compared. (Note that the taxa being compared could be two populations of a species, or two species, but that if intraspecific comparisons are made, those same species cannot be used in an interspecific comparison.) The advantages of pairwise methods are frequently bought at a considerable cost. The advantages are that ancestral character states need not be inferred and, consequently, precise models of character evolution need not be specified. The cost is that a lot of information is ignored; frequently, the vast majority of variation in character states is found among the most distantly related taxa, and that is the variation that tends to be ignored with the consequence that very few degrees of freedom are available for statistical analysis. In short, the methods lack statistical power and may find it difficult to reject the null hypothesis of no character covariation. However, despite having high type 2 statistical error rates, pairwise comparisons are expected to have low type 1 error rates; they are unlikely to reject a correct null hypothesis.

Møller and Birkhead (1992) have championed the use of pairwise comparative methods, and illustrate their use by testing the idea that high copulation rates among birds are a 'paternity insurance' mechanism. First, they asked whether rates of extrapair copulation increased with local population density, which they did in all eight possible comparisons, which is a highly significant result. (Note: 15 potential comparisons are possible with 16 taxa when ancestral states are considered.) Second, they asked whether birds living in colonies had higher rates of *intra*pair copulations, which they did in 12 of 13 comparisons. Unfortunately, in this the results are not as striking as they first appear because sometimes the same bird species was used in an interspecific and intraspecific comparison. For example, solitarily and colonially breeding *Hirundo rustica* were compared, and then colonially breeding *H. rustica* was compared with solitarily breeding *H. daurica*.

A fine example of the use of pairwise comparisons in behavioural ecology is provided by Robinson and Terborgh's (1995) observational and experimental study of Amazonian bird species. The territories of more than 330 bird species were mapped along a successional gradient by the side of an Amazonian river. Species pairs from more than 20 genera were found to hold non-overlapping but contiguous territories, while others held partially or totally overlapping territories. Congeneric species pairs were then chosen for reciprocal heterospecific song playback experiments. One species of a pair which held non-overlapping territories typically approached the speaker aggressively, and when this happened it was always the heavier species. Robinson and Terborgh argue that the larger congeneric species pairs thereby occupy the more productive end of habitat gradients, and that it is the marked successional gradients typical of Amazonia which thereby helps to explain the increased congeneric species richness of Amazonian bird communities. Where they occur, the larger species will be at higher densities as a consequence of contest competition for resources; in this case, territories in the higher productivity habitats. Robinson and Terborgh's study provides a test of one proposed explanation for the unexpected positive correlation between population density and body size mentioned above (see Section 14.4.3).

Finally, we will mention an intriguing pairwise behavioural analysis that has provided an insight into population genetic history. Until the Panama seaway closed about 3 million years ago, the Caribbean and eastern Pacific were connected. The final closure of the seaway split many species of snapping shrimps of the genus *Alpheus* in two (among many other species, of course). For each of seven pairs of such species, Knowlton *et al.* (1993) studied the level of intolerance (snapping and other aggressive contacts) shown between individuals, one from each species in the pair, when placed in a glass dish. They also studied the degree of mitochondrial DNA divergence and allozyme divergence between the two species in each of the seven pairs. Treating degree of intolerance as a divergence measure, they found that all three measures are highly correlated, providing confidence in their final conclusion that, for three of the pairs, gene flow was disrupted millions of years before the final closure of the seaway.

14.5 Character change and tree structure combined

Phylogenies can be used to ask a series of questions which integrate evolution, behaviour, ecology and community structure. For example, when speciation occurs, how and to what extent is the old ecological niche expanded to accommodate the two daughter species, and how and to what extent is it partitioned between them? If, by chance, a species invades a new location, to what extent is its evolutionary future predictable? These and other similar questions are now being tackled by a combination of techniques which involve: (i) phylogenetic analysis; (ii) morphological, behavioural and ecological comparisons between contemporary species; and (iii) long-term field experiments. Losos and colleagues' work on *Anolis* lizards in the Greater and Lesser Antilles islands in the Caribbean provides an excellent case-study.

There are about 150 species of *Anolis* lizards in the Caribbean, but only six 'ecomorphs': phenotypically similar species adapted to particular ecological niches. The ecomorphs are mainly described by the habitats which they exploit: twig, crown–giant, trunk–crown, trunk, trunk–ground, grass–bush. Morphological and ecological comparisons show that the ecomorphs are a valid classificatory tool because they tend to form clusters in multidimensional space when data are summarized by principal components analysis (Losos, 1996). However, the same ecomorphs have often evolved independently by parallel or convergent evolution, so that different ecomorphs from the same island may be more closely related to each other than the same ecomorph from different islands.

Phylogenetic trees for the two independent radiations on Puerto Rico and Jamaica are reasonably well established. When Losos mapped ecomorphs on to phylogenetic tree structures, the two islands revealed very similar pictures (Losos, 1992). The first speciation event resulted in a generalist arboreal lizard and a twig species. The generalist then produced daughter species, a crown–giant and a trunk–ground ecomorph. The trunk–crown ecomorph was the next to evolve (on Puerto Rico this evolved from the trunk–ground species, and on Jamaica it evolved from the crown species). The evolution of the trunk–crown species on both islands was followed by speciation into a larger and a smaller trunk–crown species. Finally, on Puerto Rico but not on Jamaica, the trunk–ground form spawned a daughter species that lives on grass and bushes independently of trees. A statistical randomization test demonstrated that the chances of two such similar sequences occurring by chance alone is very small indeed. Instead, the patterns point to a common process at work which Losos interprets as niche partitioning (the tree) with the occasional invasion of new niches (such as the ground, grass and bushes).

The supposed role of niche partitioning in the adaptive radiations implies that interspecific competition is likely to have been an important mechanism forcing species to occupy narrower niches, and the differences between the ecomorphs suggest that particular morphological and behavioural changes should occur over the generations when species invade new habitats. A number of experimental field studies have been used to test these ideas (reviewed in Losos, 1996).

1 When two species were placed either alone or in combination on unoccupied islands, each species achieved higher population densities when alone than when in combination as predicted if interspecific competition is important.

2 Almost 20 years ago T.W. Schoener introduced one species from a known habitat onto a number of unoccupied islands. The change in habitat (from trees to scrub vegetation) has been accompanied over about 40 lizard generations by expected changes in limb length which are significantly correlated with the availability of perches with different diameters among the islands.
3 Some 'natural experiments' have occurred where unoccupied small islands have been invaded by specialized trunk–ground and trunk–crown species. In some cases the species have evolved to become less specialized, but in other cases the specialization seems to have been retained even in the absence of competitors.

Another notable study that takes a phylogenetic approach to evolutionary ecology is that of Richman and Price (1992) on Old World leaf warblers. In Kashmir, there are eight sympatric warbler species of the genus *Phylloscopus* which constitute the insectivorous leaf-gleaning guild. Using Felsenstein's (1985) comparative method (see Box 14.1), Richman and Price first show that there are significant correlations between morphology, on the one hand, and behavioural differences in feeding behaviour and habitat choice, on the other. So, for example, species with relatively longer tarsi breed in deciduous, as opposed to coniferous, woodland and species which have a greater tendency to flycatch have wider beaks. They also show that these correlations are robust in the face of uncertainties in the phylogeny. An analysis of the contrasts suggests that the ecological diversification of the warblers can be characterized as an early differentiation of feeding method and prey size, with changes in habitat selection being more recent; closely related species tend to differ primarily along the niche axis of habitat choice and little on other axes.

These studies, in addition to the others reviewed in this chapter, illustrate the central role that phylogenetic analysis can play in studies of behavioural ecology. Gone are the days when phylogenies were used merely to describe the genealogical relationships of taxa. Now we know that a phylogeny is a necessary component of any scientific attempt to integrate the results of any biological study which is based on differences among populations or species.

Chapter 15

Causes and Consequences of Population Structure

Godfrey M. Hewitt & Roger K. Butlin

Genes exist within individuals and individuals make up populations which are geographically distributed in the habitats to which they become evolutionarily adapted through time. The behaviour of an organism, like other aspects of the phenotype, will be a product of evolution in this geographically and temporally structured context. It can, in turn, affect the distribution, interaction and evolution of genotypes. In this chapter we wish to explore the relationship between population genetic structure and behaviour. Both fields are currently producing and incorporating new ideas and approaches, but until recently have for the most part developed separately. We will consider how ancient and modern history has forged the present patchwork of diverse genomes within species, the effects of gene flow and population size, and how we may measure these. We will then consider some ways in which population structure and behaviour might interact. Does population history compromise current adaptation? How does dispersal behaviour evolve when there is local adaptation, and how is local adaptation influenced by dispersal? Does divergence in mating behaviour restrict gene flow? Are small, isolated populations sites for rapid evolution of behaviour and does this contribute to speciation?

15.1 What is population structure?

We may define population structure as the spatial variation in density and genetic composition of individuals in a species. It is clearly a considerable task to describe this adequately within even one species. There are a number of components to population structure that need to be considered and measured: primarily these include genotype variation, spatial hierarchy, life histories and temporal changes.

First, the genotypes of individuals need to be identified. Under some circumstances one local population may contain only one or two genotypes while another of similar size may contain hundreds. Quantifying genotypic variants in a population is a mountainous task compared with simply counting the number of individuals but, as we shall see, even a small input of genetic information can greatly increase our understanding.

Second, there is a hierarchy to the population structure of a species. Individuals and their genotypes exist in families (or clones), and one to several of these may be represented in the local deme, within which there is interbreeding and interaction. These demes are found in patches of suitable habitat: the number of demes in any patch depending on the size and structure of the patch. Patches are usually clustered in particular geographical areas, and there may be little or no movement between them. Figure 15.1 provides an example of this hierarchy for alpine *Podisma pedestris*, sometimes called the brown mountain grasshopper. There may be genetic differentiation at the level of demes or patches or clusters of patches.

Geographical areas often contain distinct races that are phenotypically and genetically different (Mayr, 1970). Furthermore, where their ranges meet they mate and hybridize, producing hybrid zones that are usually relatively narrow (Hewitt, 1988). These narrow strips typically represent considerable genetic transitions between the racial genomes, which themselves may cover a considerable area. It has become apparent in the last decade that such genetic subdivision of species ranges into races separated by hybrid zones is common; indeed, the use of genetic markers such as allozymes, chromosomes and deoxyribonucleic acid (DNA) sequences has revealed much more cryptic subdivision than was apparent from external phenotypic characters (Butlin & Hewitt, 1985a,b; Cooper *et al.*, 1995).

Third, the life history of each species determines a particular age structure, class or caste specialization and sex ratio of its populations. At one end of the scale are the annual plants and univoltine insects that form a major part of the biota of temperate and boreal regions. In keeping with the seasons, they have one generation a year, so that when sampling adults in the summer they are all roughly the same age. Synchronization is also found in some tropical organisms like army worm caterpillars and locusts where, even though continuous breeding is possible, good weather conditions instigate development and breeding. At the other end are redwoods, oaks, whales, elephants and man, that are long-lived with several overlapping generations and individuals of all ages. Differences in the age, caste, morph and sex structure of populations and species can have important effects on the genetic variation present and significant consequences for the evolution of behaviour. In sampling populations for genetic and behavioural studies, one must be fully aware of these factors.

Fourth, temporal variation must be considered explicitly in describing population structure. Some temporal variation is implicit in the life-history properties intrinsic to the organism. Temporal changes in distribution, density and genotype of individuals and populations are also caused by extrinsic factors of climate and habitat variation. These occur over time-scales from days to ages. For short-lived organisms that can complete a generation in a few weeks, seasonal conditions produce large changes in numbers and genotypes, and, in fruit flies like *Drosophila*, changes occur as the fruit rots over a few days.

While changes over a few years can be studied first hand, longer time series require historical evidence, fossils and inference. Volcanic islands, such

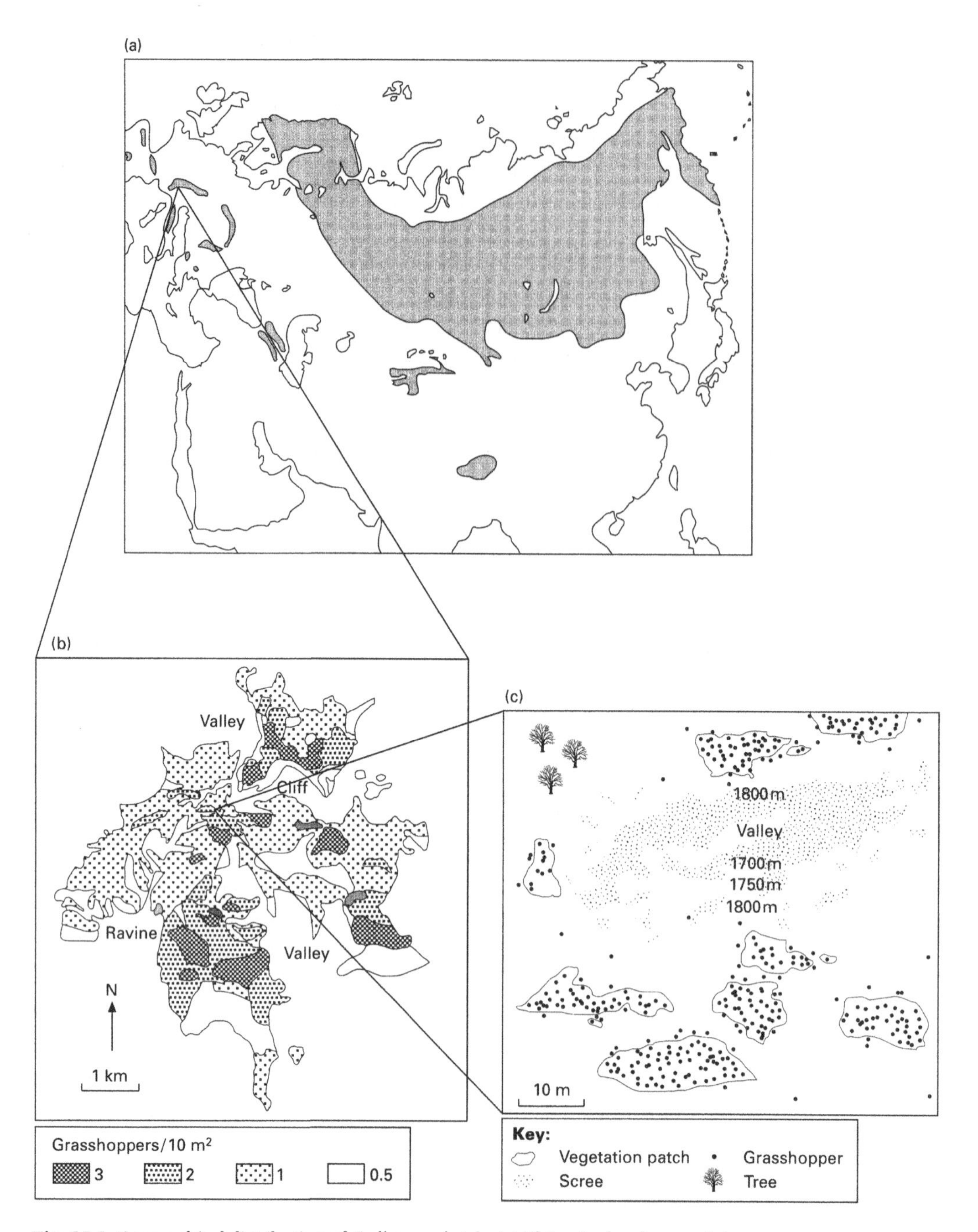

Fig. 15.1 Geographical distribution of *Podisma pedestris*. (a) This wingless boreo-alpine grasshopper is found over much of northern Europe and Asia, and in the high mountains of southern Europe and Asia. However, its distribution is very inhomogeneous. (b) Adapted to colder climates, in the Alps it lives on blocks of mountains over about 1500 m up to 2800 m, often associated with alpine shrubs like wortleberry, where it can be quite abundant. It finds hot, dry slopes and herbless scree inhospitable. This density map is around Chabanon, Seyne-les-Alpes, France. (c) This association with habitat is apparent down to the local population level, with each patch of suitable vegetation containing one or more demes. The dispersal distance of *Podisma* is 15–20 m generation^{-1}, so while occasional migrants move between neighbouring patches of suitable vegetation, crossing the valley shown is most unlikely. This location is between Tete Grosse and Les Tomples.

as Hawaii, provide a link through from recent eruptions to several million years ago. Lava flows wipe out most vegetation and leave small isolated refugia, or 'kipuka'. They also dissect the range of species. New islands are produced over a 10^5–10^6 year-scale and may be colonized progressively from neighbouring ones. Such processes must modify population structure enormously, and there are some outstanding studies of the effects of historical and ancient events on the genetic structure and behavioural adaptations of *Drosophila* (Carson *et al.*, 1990).

In summary, the population structure of a species can be seen as spatial variation in density and genetic composition of its individuals. It has a hierarchy from families to geographical races and subspecies. The life history of the organism imposes an age structure, sex ratio and polymorphism within this, and environmental fluctuations and catastrophes modify it through time. In terms of ecology and behaviour, such features have significance for local adaptation to particular conditions, the form and consequence of metapopulation dynamics, the spread of adaptations and the shifting balance theory, the persistence of varieties and species with their adaptations, and the evolution of racial and specific characters.

15.2 What shapes population structure?

In the previous section we explained that population structure changed drastically over longer periods of time, taking hundreds and thousands of generations, caused by large climatic and habitat fluctuations. In the present, over a scientific lifetime, we may study and experiment with the effects of habitat patchiness, local adaptation, population size, dispersal, extinction and colonization, and their interactions on demographic and genotypic changes. Having considered both past and present processes, it will be possible to attempt to clarify which effects are dominant in particular cases and their relative importance under different scenarios.

15.2.1 Past processes

The past 2½ million years — the Pleistocene period — is characterized by a series of major climate oscillations recognized as ice ages in temperate regions, but whose effects were global. These have a cycle of 100 000 years driven by the orbital eccentricity of the Earth round the Sun. Lesser cycles of climatic change of 41 000 and 23 000 years are overlaid on this. There were lesser fluctuations nested within these, some very rapid, and fortunately these can produce brief warm interglacials, such as the period in which we live. The physical causes and detailed effects of these processes are currently the subject of much research and debate (Hewitt, 1996). The effects on species distribution and population structure were enormous. Pollen and fossil analysis shows that organisms now in northern Scandinavia were in southern Europe, and those in northern Canada were south of the Great Lakes and the ice sheets. As the climate warmed around 15 000–8000 BP these species moved northward and up mountains, while others invaded from the south, where deserts expanded. At the level of demes, this clearly involves much colonization and extinction, dispersal and adaptation. The distinct geographical races and genotypic forms that currently inhabit these regions, and are often separated by hybrid zones (Hewitt, 1988), must have moved in during this post-glacial period, the Holocene.

Consider the example of the meadow grasshopper *Chorthippus parallelus* where DNA sequence analysis identifies five subspecific genomes in Europe (Cooper *et al.*, 1995). Most genetic variation is found in the distinct taxa in Spain, Italy, Greece and Turkey, while the DNA sequences of the fifth taxon show that it spreads from the Balkans to France, UK, Scandinavia and western Russia (Fig. 15.2). Suitable vegetation for *C. parallelus* existed in the far south of Europe during the ice age; these southern countries have kept their refugial

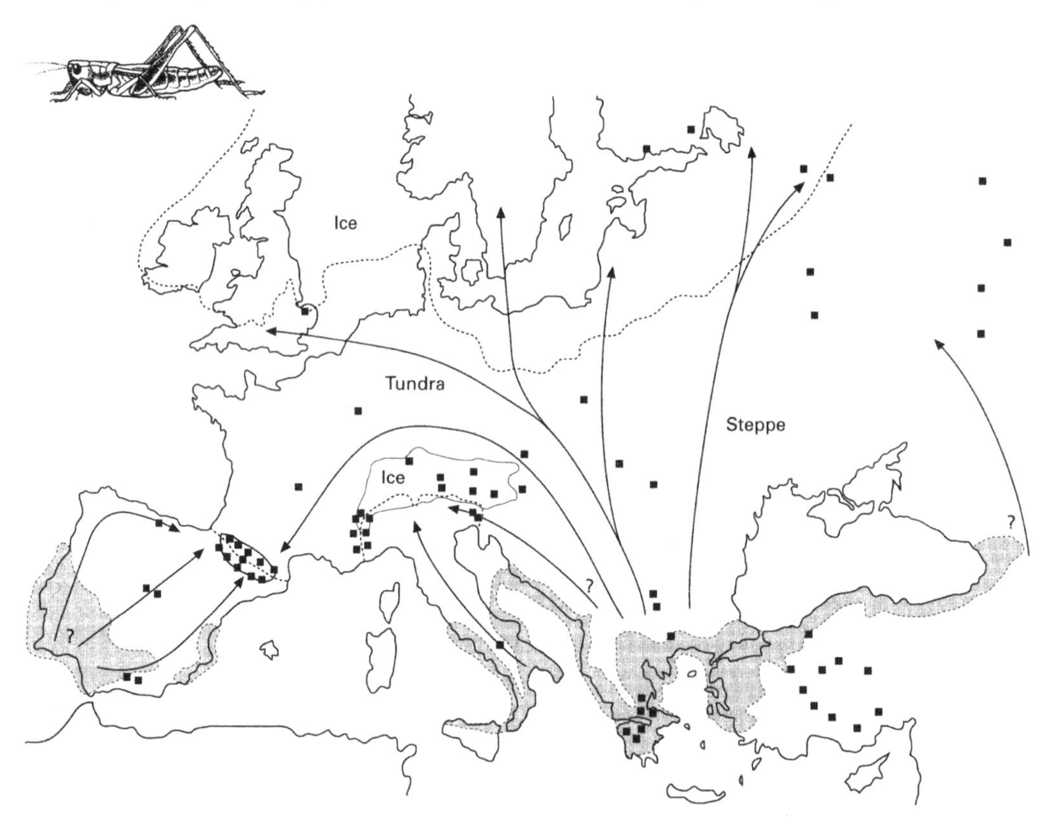

Fig. 15.2 Europe, showing the putative ice age refugia (shaded) from fossil evidence, and possible post-glacial expansion routes of the grasshopper *C. parallelus* as deduced from DNA sequences from the sample sites marked (black squares). Narrow hybrid zones occur in the Pyrenees and Alps, and possibly also in Greece, Turkey and Caucasus, where diverged genotypes meet. Thus, *C. parallelus* is a patchwork of distinct genomes with considerable diversity in the south.

genomes — now up mountains — while a small sample from the Balkans expanded north, west and east to fill the rest of Europe. The dynamics of such colonization may well have reduced and restructured the genetic variation further (Hewitt, 1996). A number of other species provide similar evidence, although their detailed geographical distributions are not identical. For example, allozyme data on the Norway Spruce indicate that different genotypes colonized different parts of Europe during its post-glacial expansion from refugia in the Dinaric Alps, Carpathians and Russia (Lagercrantz & Ryman, 1990). Several species show western and eastern forms in Europe, with hybrid zones running roughly north–south down from Scandinavia to the Alps, indicating southwest and southeast glacial refugia from which they have expanded — e.g. crow, mouse, newt, toad, snake, oak — and many others in Europe and North America have distributions which suggest similar histories (Hewitt, 1996).

Consequently, populations that are presently close to each other in space may have quite distant geographical origins and diverged genomes. If these hybridize, the offspring may be more or less unfit and form a hybrid zone, and there may be no obvious morphological sign that this is occurring. Furthermore, while separate during the ice age the two populations may have developed different adaptations which may be reflected in their behaviour. Clear differences in morphology and behaviour between such parapatric populations have been found in many cases where hybrid zones have been studied in detail; for example, Spanish and French *C. parallelus* in the Pyrenees differ in their courtship behaviour, songs and pheromones (Hewitt, 1993a; Neems & Butlin, 1994). The yellow and red fire-bellied toads *Bombina bombina/variegata* have distinct pond preference, mating calls and egg-laying strategies (Szymura, 1993), and similar examples occur in crickets, butterflies, mice, birds and flowering plants (see Harrison, 1993). Besides being excellent places to study the evolution of such characters, such cases demonstrate that the geographical history of the populations being studied can be of primary importance in understanding behavioural differences and adaptations. Methods for establishing such phylogeography are being developed using DNA sequences and offer exciting prospects (e.g. Avise, 1994).

Such range changes consequent on climatic and environmental fluctuations have some interesting population genetic consequences (Hewitt, 1993b). The colonization of new territory involves founder events by small numbers of individuals, which will reduce allelic diversity and reorganize gene interactions. Where the expansion is very rapid, this process will be repeated and accumulative, and results from an increasing number of species agree with this expectation. Classic allozyme data on lodgepole pine in northwest America coupled with radiocarbon dated fossils show how its post-glacial expansion northwards caused loss of genetic diversity and change in dispersal adaptation (Cwynar & MacDonald, 1987). The same pattern is seen for DNA sequence in *C. parallelus* in northern Europe (Hewitt, 1996).

In recent times, we have seen a number of range expansions, some caused by man, and indeed of man — out of Africa (200 000 years), across Europe (5000 years) and around the world (300 years). The expansion of the collared dove out of Turkey/Balkans to progressively fill most of Europe this century is well documented (Hengeveld, 1989). The similar spread of the oak gall wasp (Stone & Sunnocks, 1993) also showed a reduction in allelic diversity as predicted with rapid colonization. The spread of the rabbit by Romans, Normans and British all over the world, and the house mouse with agriculture across Europe and then in ships to colonies, involved founder events, so that the populations in Australia and North America carry only a small proportion of the Old World genomes. Insect pest species and diseases of plants, animals and humans show great plagues and epidemics which have changed the ecology and genetics of many if not all species, and will continue to do so.

One corollary of such Holocene expansions is that species have very different genotype composition and variation across their range. In Europe, southern populations in areas of glacial refugia may well contain greater diversity, which will affect their adaptive response to natural selection, and is of relevance to conservation strategies. Another corollary is that equilibria, both numerical and genetical, may be hard to find. Many populations are probably still evolving after their latest colonization or catastrophe, and the adaptations we study may be ongoing and genetically various. A life-history component measured in one population, such as number of eggs laid by a grasshopper, may still be changing under selection, and have a different genetic basis in another part of the species' distribution.

15.2.2 Present interactions

The habitats in which species live are more or less patchily distributed. Even before urbanization the population of generalist man was concentrated in more suitable environments. This localization is all the more apparent in flightless insects or small plants living in discrete areas on high mountains, such as the alpine biota of Europe. Such patchiness may be equally marked for isopods living in Scandinavian lakes, and at an even finer scale for insects specialized to a particular plant host that is scattered occasionally through a tropical forest.

Populations on small isolated patches of suitable habitat will be subject to genetic drift. Over the generations their allelic frequencies will vary stochastically and alleles will be lost, leading in principle to increased homozygosity. This neutral divergence will occur in the absence of selection for different alleles, but it may also accelerate any differentiation due to natural selection. While small isolated populations will continue to diverge, in general the closer they are the more likely they will receive dispersing individuals from each other. These migrants will tend to delay or prevent divergence by exchanging genes among the patches and demes. The interplay of population size, gene flow and selection, and its complexities, has received much theoretical attention with the production of many population genetic models (Wright, 1978; Slatkin, 1985). For a good basic text see Hartl and Clark (1989).

The 'island models' are most widely used, where a series of finite populations with N individuals each exchange migrants equally at a rate of m each generation. Drift will increase homozygosity (identity by descent) in each population and divergence among them. This divergence is measured by F_{st}, which can be estimated from the difference between the heterozygosity observed in the subpopulations and that expected from random mating across the whole population. Its value ranges between 0 and 1. At equilibrium between drift and migration $F_{st} = 1/(1 + 4Nm)$, and so Nm — the number of migrants — can be estimated. It is surprising how little gene flow can prevent significant genetic divergence among the island populations — e.g. even when there is only one migrant ($Nm = 1$) each generation, $F_{st} = 0.20$ at equilibrium. Another commonly used approach is the 'stepping stone model' where migration is allowed between neighbouring populations only. Under these assumptions too, the subpopulations become well differentiated when $Nm < 1$, and when $Nm > 4$ they behave essentially as one population (Tables 15.1 & 15.2). Many organisms are not distributed simply as if on small islands, but more

Table 15.1 Estimates of the number of migrants per generation between subpopulations (Nm) from genetic estimates of inbreeding in subpopulations (F_{st}) calculated from observed and expected heterozygosities. (From Slatkin, 1985.)

Species	Organism	Nm	F_{st}
Mytilus edulis	Mussel	42.0	0.006
Drosophila willistoni	Fly	9.9	0.025
Peromyscus californicus	Mouse	2.2	0.102
Plethodon ouachitae	Salamander	2.1	0.106
Drosophila pseudoobscura	Fly	1.0	0.200
Podisma pedestris	Grasshopper	0.91	0.212
Thomomys bottae	Gopher	0.86	0.225
Plethodon cinereus	Salamander	0.22	0.532

Table 15.2 Estimates of neighbourhood size (Nb) from measures of dispersal (s) and population density (d), where $Nb = 4\pi s^2 d$, and s is the standard deviation in distance between parent and offspring in metres.

Species	Organism	s (m)	d (individuals m^{-2})	Nb (individuals)
Podisma pedestris	Grasshopper	20	0.1	503
Bombina bombina	Toad	100	0.01	1257
Mus musculus	Mouse	430	0.05	116 176
Dendroica coronata	Warbler	900	0.005	50 894
Bembicium vittatum	Intertidal snail	2	26	144 105

continuously for at least some of their ranges, and models of genetic differentiation due to limited dispersal, or 'isolation by distance' have been employed. Genetic studies using allozymes and, lately, DNA markers generally show that subpopulation differentiation (F_{st}) is higher in subdivided habitats as compared with continuous ones. It is also higher in poorly dispersing species than in similar good dispersers. However, a number of cases where values of F_{st} are not apparently compatible with the observed gene flow remind us of the simplicity of these models and the complexity of factors affecting population structure.

In recent years, models of local extinction and recolonization have been examined — often called metapopulation dynamics — and have generated much debate concerning their effects on genetic variation and evolution (McCauley, 1991; Harrison & Hastings, 1996). Most theory has revolved around the classic Levins metapopulation model, where each of the set of demes has equal likelihood of extinction and recolonization, and the question is whether such population turnover generates more or less variation among the demes. If the recolonizers are few in number and from essentially one deme, then the differentiation among demes can increase, otherwise if the recolonizers are many and varied in origin then extinction/recolonization promotes gene flow and homogeneity among demes. The ecological reality of these metapopulation models is questionable, and other types of metapopulation may be more common in nature. Further insight may come from computer simulation, integrating genetic and ecological features.

Most population genetic models concerning gene flow and drift assume no selective difference among the alleles, and for many mitochondrial and nuclear non-coding regions this seems reasonable. However, there is increasing concern that occasional mutations causing increased fitness may allow an allele and genes linked to it to spread through the population and even the range of the species. There is good evidence for such selective sweeps in *Drosophila melanogaster*, where a number of loci show reduced polymorphism in their region of the chromosome (Kreitman & Akashi, 1995). In reverse, when deleterious mutations arise they will cause selective loss of their regions of the chromosome, and so reduce polymorphism. Consequently, as a population adapts to local conditions, alleles will be selected and polymorphism around them reduced; how much of the genome will be affected will depend on how many genes are selected and the tightness of the linkage. This should not affect direct measures of population structure but due caution is needed in deductions from indirect models based on neutral theory. Phylogeographical conclusions should also be robust to such effects.

The great changes in distribution and population structure caused by major climatic oscillations have been emphasized, and such extinction and recolonization occurs right down to the local patch level over a few generations. It is therefore difficult to see that the equilibrium conditions assumed for most of these models can pertain in almost any species. Furthermore, the approach to equilibrium is very slow in many models, particularly where population size is large; the clines in blood group alleles in man across Europe remain from the Neolithic agricultural colonization (Ammerman & Cavalli-Sforza, 1984). Indeed, it has been known for some time that local areas of distinct allele frequency that arose by drift can persist for hundreds of generations, as Endler (1977) showed through simulation. The spatial pattern of neutral allele frequencies set up during the colonization of an area can often include large patches of distinct genomes, which then remain for a long time under stable population dynamics (Ibrahim *et al.*, 1996). Consequently, a population genome is not only affected by current changes, but also carries residual effects of history.

15.2.3 Current versus historical effects

Until recently the study of population structure has concentrated on current distribution and determinants, i.e. deme sizes, gene flow and selection, but historical events like range expansion and colonization can produce major lasting effects on the distribution of genotypes. For example, populations scattered over a wide range may be similar genetically, indicating high gene flow among them; on the other hand, they may have been put in place by a recent expansion and colonization from a source population, which could produce the same features but not involve current gene flow. How can we determine the importance of past and present events? As Nichols and Beaumont (1996) put it while considering data from man, *Drosophila* and *Podisma* — 'Is it ancient or modern history that we can read in our genes?'

Differences in allele frequency across a geographical distribution have been analysed by principal components and spatial autocorrelation to measure the genetic similarities and relate these to known historical events, notably in man's colonization of Europe (Ammerman & Cavalli-Sforza, 1984; Sokal *et al.*, 1989). Typically, blood groups and allozymes provide the data for such analyses. The availability of DNA sequences now provides not only allelic information in the form of haplotypes, but also the genealogy of these haplotypes, which can be expressed as a phylogenetic tree.

When the tree is combined with the geographical distribution of the haplotypes it gives a phylogeography (Avise, 1989). For example, the related sequences on major branches of the tree can occur in separate parts of the species range, and the subset from a younger nested branch may be confined to a small area within a major one. Such relationships between geographical distributions and phylogenetic position allow one to distinguish among hypotheses about the historic events that could have produced such a distribution; in this case, it suggests an older geographical separation and divergence, with a younger mutation undergoing local range expansion (Avise, 1994).

These phylogeographies contain further information which can be mined by the development of suitable analyses combined with good datasets, and recently Templeton *et al.* (1995) have produced a major methodological advance in this endeavour. They developed a nested cladistic analysis of geographical distances among haplotypes, and applied this to mitochondrial DNA (mtDNA) data from the tiger salamander *Ambystoma tigrinum* from the midwest US. This method introduces statistical rigour to procedures for discriminating restricted gene flow, past population fragmentation, colonization and range expansion. In the tiger salamander very restricted gene flow is evidenced by the tips of the mtDNA cladogram being geographically localized within the interior clades, and this fits well with its biology — it is a strict pond breeder, often paedomorphic and returns to its home pond. Range expansion is also apparent where a younger clade of haplotypes is more widespread than its ancestral clade, particularly where it is found in areas that were affected by glaciation. This is the pattern noted in *C. parallelus* in Europe (Cooper *et al.*, 1995), and the *Ambystoma* data reveal two clear examples of such Holocene range expansions. Good background knowledge of the species biology and distribution, adequate sampling of the range, and information on the Pleistocene history allow the best use of such phylogeographical data and deductions.

15.3 Testing predictions about population structure

Another important research endeavour is to test the measures and predictions in natural populations of particular species, where the various life history and ecological parameters can be specified and quantified to some extent. We will demonstrate this through a series of cases which cover many, but not all, of the major approaches.

15.3.1 Turkey oaks: neighbourhood size

Several studies have attempted to discern and measure neighbourhood size and structure, particularly in plants, but with somewhat variable success, and it seems that adequate sampling in the field is critical. Studies in a natural forest of Turkey oak (*Quercus laevis*) in South Carolina, assessing its population genetic structure for the effects of isolation by distance, provide an instructive example (Berg & Hamrick, 1995). Nine polymorphic allozyme loci were typed in 3400 trees in a 160-m^2 plot in an old stand. This impressive dataset comprising allele and genotype frequencies distributed spatially was analysed by a battery of techniques, including principally number of alleles in common (NAC), hierarchical estimation of differentiation (C_{st}, a variant of F_{st}) and inbreeding (F_{is}), spatial autocorrelation of allele frequencies and of genotypes. Control comparisons were provided by computer simulations of a 10 000 tree stand (100 × 100 grid) run for 10 000 years (10 years minimum generation time) with different sets of values for pollen and seed dispersal (which define neighbourhood size). This was iterated 99 times for each set of dispersal values, and the simulated datasets analysed by the same methods. Individuals within 10 m have similar genotypes as shown by a positive correlation coefficient, but not further away indicating a limit on gene flow above this distance. In the case of these oaks, this is probably due to heavy acorns falling near the parent tree. However, no larger patch structure of genotypes was revealed, indicating that gene flow over larger scales was sufficient to overcome any significant isolation by distance. The simulations gave the best fit with the observed statistics when the seed and pollen dispersal levels were high, encompassing a neighbourhood size of 440 individuals. Most acorns may fall near the tree but there is also clearly effective long-distance dispersal due to pollen and probably some seed. Such leptokurtic dispersal is commonly described in field work with many organisms. The application of detailed ecological knowledge and complementary genetic analysis in this species has provided a much clearer understanding of its population structure.

15.3.2 Intertidal snails: gene flow

The neighbourhood size in oak was deduced by comparing genetic evidence with simulations, and the dispersal was inferred indirectly. Dispersal could also be estimated directly by mark recapture experiments, although such ecological measures can be very difficult or impossible in some species, as for example those with highly pelagic larvae. Such direct and indirect estimates (Slatkin, 1985) were compared in the intertidal snail *Bembicium vittatum* in the Houtman Abrolhos islands off western Australia (Johnson & Black, 1995). These islands provided a comparison between a linear continuous 11-km shore and a linear series of islands over similar distance and north–south disposition. Along the continuous shore, based on mark–multiple-recapture experiments, the variance in dispersal distance (s^2) of snails was calculated over 15 months. Dispersal was leptokurtic and provided estimates of neighbourhood size of 7000 adults in one location and 1600 in another. Neighbourhood size was also calculated from genetic data in the form of 13 polymorphic allozyme loci in samples collected from three nested transects of different length; several thousand individuals comprise this large dataset. This analysis showed that subdivision began to rise between 150 and 300-m distance between samples, and therefore estimates neighbourhood size at 16 650–33 300 in the dense location and 3900–7800 in the sparser one. More indirect estimates of neighbourhood size (Slatkin, 1993) from the allozyme data based on a stepping stone model gave values in the range 22–38. These are very low and cause concern about the model's suitability in this case. In comparison the isolation by distance measured by G_{st} in the series of separate islands was twice as high as the continuous population for those 1-km apart, and did not increase with distance. Furthermore, allozyme frequencies varied considerably. Clearly, small water channels between islands are effectively reducing gene flow in this snail. This is understandable as the species does not have pelagic larvae.

These experiments and genetic analyses define the amount of gene flow, neighbourhood size and isolation by distance in continuous and discontinuous habitats and they provide a firm basis for understanding population structure and its consequences.

15.3.3 House mice: DNA markers, breeding units and colonization

In recent years, with the invention of polymerase chain reaction and many new techniques for analysing DNA efficiently, a much more powerful range of genetic markers is becoming available for population and ecological studies (see Avise, 1994). The millions of base pairs of DNA sequence in an individual genome comprise an enormous reservoir of information, some parts of which are very variable and ideal for population studies (e.g. non-coding regions and microsatellites), while others are highly conserved over time. The raw data from DNA-based approaches often require new statistical and analytical methods (Slatkin, 1993, 1995).

A recent study of population subdivision and gene flow in house mice in Denmark utilizes both mtDNA control region sequences and microsatellite loci to good effect (Dallas *et al.*, 1995). Previous studies have caused some debate as to whether gene flow is very restricted due to mice living in closed breeding units, or whether it occurs more widely as suggested by the spread of introduced alleles. There is certainly major geographical structuring, with hybrid zones, and isolation by distance over several kilometres from allozyme and mtDNA studies. Dallas *et al.* (1995) sampled 11 farm populations in two clusters with the farms generally just a few kilometres apart. They used five microsatellite loci with 135 alleles altogether, and 888 base pairs of mtDNA sequence. Both Hardy–Weinberg and F statistics showed that within these populations mating was essentially at random, but θ (an estimator of F_{st}) was significant indicating that the farms are to some extent genetically separated. A plot of θ against distance between farms suggested some isolation by distance, with some gene flow among neighbouring sites. Assuming an equilibrium between migration and drift the level of θ observed would be produced by Nm = 1–5 migrants each generation for each farm. Thus, gene flow appears limited and probably occurs by active migration between neighbouring farms.

A further interesting inference is possible with such data. The effective population size (N_e) can be calculated from populations collected over short periods and compared with the long-term N_e i.e. the N_e necessary on coalescence theory to explain the nucleotide divergence (*dx*) in the control region mtDNA within a population that has been separate since its founding, possibly thousands of generations ago. These do not agree in this mouse example; the current population size is much smaller than that needed to explain the survival of the current genetic diversity unaugmented by migration. It strongly supports the proposal that recent gene flow or colonization by genetically varied mice has occurred. Such examples demonstrate how DNA data with suitable analysis can provide real insights into population structure, from family behaviour through to historical range expansions.

15.4 Interactions between behaviour and population structure

The models of population structure that we have described so far, and which are typically used to make inferences from genetic data, assume that dispersal is random with respect to both habitat and genotype, that all individuals have equal dispersal tendencies and that immigrants have the same mating success as residents. Clearly, there is much scope for the real behaviour of animals and plants to cause departures from these assumptions, and for selection to operate on behaviour in ways that might depend on population structure. We will consider some of these issues but, at present, the number of studies that explicitly deal with both behaviour and population structure is limited.

15.4.1 Non-random gene flow

In the consideration of population structure, migrants are typically assumed to be a random subset of the available individuals in each generation and the effect of migration is to reduce differentiation among subpopulations. However, it is easy to see that selective migration can have the opposite effect of accentuating divergence, for some loci at least. Habitat preference is the simplest example: alleles that increase movement out of one habitat type and/or into another will quickly become differentiated among habitats. The resulting tendency for reduced movement among habitat types, and for mating encounters among individuals originating from the same habitat type, will favour local adaptation (Diehl & Bush, 1989).

The apple maggot fly, *Rhagoletis pomonella*, provides an excellent example of genetic divergence favoured by habitat, in this case host–plant, fidelity. The apple maggot fly infests both apple and hawthorn fruits in the eastern US. Populations on the two hosts exhibit consistent allele frequency differences at electrophoretic loci (Feder & Bush, 1989) and genetically based differences in emergence time that represent adaptations to the different phenologies of the two hosts (Feder *et al.*, 1993). The contribution of host fidelity to the maintenance of this differentiation has been studied at a site in Michigan where the two hosts grow together, well within the dispersal range of the adult flies (Feder *et al.*, 1994). Active host preference, a tendency to move to the nearest tree following adult emergence and matching of emergence time to host fruiting time all contribute to restriction of interhost movement. However, these experimental results still predict about 6% gene exchange between host races per generation, enough to remove the observed allozyme differentiation rapidly in the absence of other effects, most probably host-associated selection (Feder *et al.*, 1994). The restriction of gene exchange facilitates further adaptation and thus greater isolation. However, it is not yet clear whether this process, which has come a long way in the short time since apples were introduced to the US in the mid-19th century, will continue to completion and thus produce speciation. An interesting possibility is that the initial divergence was promoted by non-random gene exchange without genetic divergence (Butlin, 1990). If the population contains heritable variation in emergence time, and development is slowed on the new, presumably suboptimal host, then matings on the new host tend to involve late developing individuals from the original host and vice-versa. Thus, alleles for late development flow preferentially to the new host, and for early development to the original host, reinforcing the initial divergence without the involvement of selection.

The structure of a population in a patchy habitat will be influenced by behavioural characteristics that influence the probability of an individual leaving a patch or being attracted to another patch. There are few empirical data on these effects but a new study by Kuusaari *et al.* (1996) on the Glanville fritillary butterfly, *Melitaea cinxia*, provides valuable pointers. Butterflies introduced into an empty network of patches on a Finnish island were less likely to leave large patches, patches with many conspecifics, with many nectar-bearing flowers or with a high proportion of the boundary forested (i.e. very distinct from the habitat within the patch). They were more likely to enter new patches that were large and had many nectar sources. Females were more likely to emigrate and moved further than males, perhaps because females were searching for oviposition sites while males were searching for mates.

15.4.3 Variation in dispersal

Dispersal behaviour is a major determinant of population structure but it need not be a static characteristic of a species: it can evolve. Inbreeding depression will favour increased dispersal away from natal groups, while local adaptation will increase the costs of dispersal. Patchy environments, where there is a risk of extinction in any one patch, will favour colonizing ability. Adaptations for dispersal, such as wings and flight muscles, may be costly in themselves, movement may expose an individual to greater predation risk and there may be a significant chance of failing to find suitable habitat after dispersal. Compromises among these various factors may be found in many possible ways, for example in mammals it is common for males to disperse and females to remain in their natal group, although there are several alternative explanations for such a pattern (Greenwood, 1980; Johnson & Gaines, 1990). More powerful tests of hypotheses about the costs and benefits of dispersal can be made where there is marked dispersal polymorphism within species.

Denno (1994) has used the widespread wing dimorphism of delphacid planthoppers to test several hypotheses. Delphacids are small sap-sucking insects in which many species have fully winged macropterous forms capable of long-distance migration that appear in otherwise brachypterous populations when population densities are high or hostplant quality is low. Short-winged flightless brachypterous females reproduce earlier and are more fecund that macropters. Brachypterous males are often more successful in competition for mates than macropters and may live longer. Thus, macroptery must have benefits in some circumstances to outweigh these costs.

Denno (1994) predicted that the frequency of macroptery would decrease with increasing habitat persistence. This prediction is very strongly supported by across species comparisons (Fig. 15.3) and the association remains significant when only phylogenetically independent contrasts are used (see Chapter 14). In species occupying persistent habitats, both males and females show marked density dependence in the frequency of macropters, whereas in temporary habitats males are macropterous at both high and low densities. This effect may be due to the need for mobility in mate location. One would clearly predict that the genetic structure of planthopper populations would also vary with the frequency of macroptery, and that local adaptation would be stronger in sedentary species, but these predictions have yet to be tested.

In some insect species, especially holometabolous insects, wing dimorphism has a simple mode of inheritance, one locus with two alleles and brachyptery dominant, but in other well-studied examples, such as crickets, planthoppers and gerrid pondskaters, a polygenic threshold inheritance pattern has been documented (Roff, 1994). There appears to be an underlying distribution of 'liability', with both genetic and environmental contributions, and a threshold liability above which development follows the macropterous pathway. Where it has been measured, the additive genetic contribution to variation is substantial, with heritabilities from 0.30 to 0.98. Thus, there is ample opportunity for wing dimorphism to evolve in response to local selection pressures.

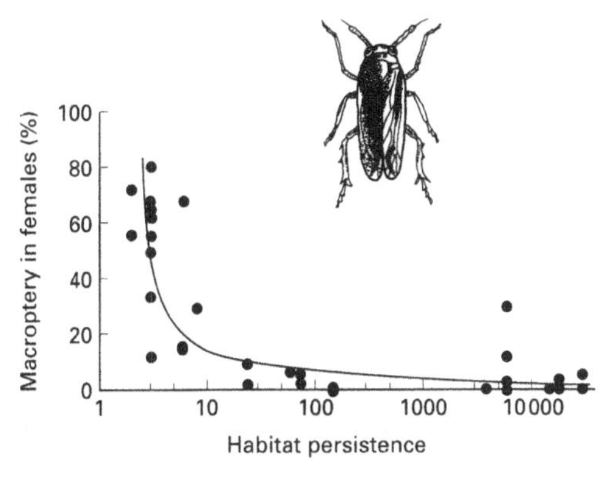

Fig. 15.3 The relationship between female macroptery in natural populations and the persistence of the habitat among species of delphacid planthoppers. Habitat persistence is estimated by multiplying habitat age in years by the number of generations per year for each species. Data from 41 populations of 35 species are included. The relationship remains significant when corrected for phylogenetic independence. (From Denno, 1994.)

In isolated patches of suitable habitat, dispersal is effectively selective death, and dispersing individuals are unlikely to be replaced by immigrants. One therefore expects to find that species living on islands, or in isolated habitats like mountain tops, have lost flight ability, even in the absence of flight polymorphism in other populations. This effect has been widely documented, although it can be difficult to separate direct selection against dispersal from selection on other life-history traits in the absence of predators, for example (McArthur & Wilson, 1967). Butterflies in isolated patches have been observed to evolve toward smaller size and there is some evidence that larger individuals are more likely to leave patches (Kuusaari *et al.*, 1996).

At the other end of the scale, patchy and ephemeral habitats will select for high dispersal ability. Given a trade-off between dispersal and fecundity, one should find a correlation between habitat stability and dispersal ability within species analogous to the correlation across species in planthoppers. Such a correlation has been observed in the African armyworm moth, *Spodoptera exempta*, by Wilson and Gatehouse (1993). This moth is a serious pest on pasture grasses and crops that depend on sporadic rainfall. Migratory potential, as measured by pre-reproductive period (PRP), shows heritable variation within populations and significant differences among populations which correlate with the frequency of rainfall in the predicted direction: short PRP where rainfall is frequent, long where it is infrequent.

A further prediction is that longer established populations will have lower frequencies of dispersive morphs or lower dispersal tendencies (Olivieri *et al.*, 1995). This has been documented in thistles, *Carduus* spp., which produce two types of seed: with or without the pappus that enables them to be dispersed by wind. Older populations produce a lower proportion of seeds with the pappus. As in wing-dimorphic insects, selection within a population would result in zero dispersive seeds but selection at the level of the metapopulation maintains dispersal since new sites can only be colonized by migrants.

15.4.4 Mating success of migrants

Dispersal is not equivalent to gene flow: it has no genetic consequences unless immigrants leave offspring and to do so they have to achieve matings. However, selection may favour either avoidance of, or preference for, immigrants as mates. This can best be illustrated with plant populations.

Electrophoretic analysis of plant populations frequently demonstrates strong spatial genetic structure even where distributions are fairly continuous. The spatial scale may be small, only a few metres in some species (Waser, 1993). Thus, there is potential for outbreeding depression either due to genetic incompatibility or because offspring are poorly adapted to either of their parents' local environments. On the other hand, normally outbreeding plants typically suffer strong inbreeding depression if selfed or crossed to close relatives (Barrett & Harder, 1996).

In the slow-growing, perennial larkspur, *Delphinium nelsonii*, long-term experiments by Waser and colleagues (Waser, 1993; Waser & Price, 1994) have explored the fitness consequences of crosses to male parents at varying distances from the maternal plant, when the offspring are planted close to the maternal parent. Crossing distances were chosen on the basis of the observed movement patterns of hummingbird and bumble bee pollinators. Seed dispersal is less than 1 m. The results are summarized in Table 15.3. There is evidence for both inbreeding and outbreeding depression, and selection is strong in both cases. The typical pollen and seed dispersal distances are such that a randomly chosen mate at 1-m distance is roughly equivalent to a first cousin, a result consistent with electrophoretic data on population structure, and yet fitness reduction is still about 0.5 relative to the 10-m crosses. Outbreeding depression is even greater. It is very unlikely to be due to genetic incompatibility at this scale but there is evidence, from reciprocal transplant experiments (Waser & Price, 1985), for adaptation to local environmental conditions on a comparable scale.

The genetic structure of the population will clearly be influenced by the reduced fitness and thus reduced genetic contribution of progeny from both short- and long-distance migrants. Further local adaptation will tend to be favoured. One might also expect the evolution of strategies on the part of the female parent to avoid outcrossing. In *D. nelsonii* there is, indeed, evidence for such an effect: pollen tube growth is slower for pollen from flowers at 100 m than from those at 10 m from the maternal parent, and this has the expected effect on fertilization success (Waser & Price, 1993).

This is analogous to the controversial process of 'speciation by reinforcement', selection for assortative mating in response to the production of unfit 'hybrid' progeny (Butlin, 1995a). In this case, it forms a step in a feedback loop: local adaptation produces outbreeding depression which favours mechanisms to reduce wide outcrosses and thus facilitates further local adaptation. How commonly this loop operates is not clear. In principle, it can certainly apply to animal populations as well and there are some data to support the idea of

Table 15.3 Effect of crossing distance on fitness of offspring in *D. nelsonii*. (From Waser & Price, 1994.)

Crossing distance (m)	Progeny size*	Progeny lifespan (years)	Flower production†	Overall fitness‡
1	78.0 ± 19.9	3.33 ± 0.36	4	0.30 ± 0.086
3	122.9 ± 23.2	3.77 ± 0.32	6	0.42 ± 0.086
10	131.1 ± 19.9	4.27 ± 0.32	8	0.65 ± 0.086
30	64.4 ± 24.2	3.11 ± 0.32	2	0.08 ± 0.086

* Leaf area in mm^2 5 years after planting.

† Number of sibships flowering within 7 years, out of 13 in each case.

‡ λ of the Lotka–Euler equation.

optimal outbreeding (e.g. in crickets: Simmons, 1991; see also Chapter 4). A similar feedback loop can be produced by the interaction between population structure and cooperative breeding (Breden & Wade, 1991; see also Chapter 10). Detecting the process is, however, critically dependent on studies at the correct scale and with thorough documentation of fitness in the appropriate habitat (Waser, 1993).

An indirect way of assessing the likely genetic contribution of immigrants is to examine geographical variation in mating signals. Both mating signals and, to a lesser extent, associated preferences have been shown to vary among populations within species (Butlin, 1995b; Bakker & Pomiankowski, 1995). However, the geographical scale of this variation is much greater than the dispersal distance so that its impact on gene flow must be negligible. Even in hybrid zones between races that differ in mating signals and preferences, where the steepest transitions would be expected, clines for these characters are wide. For example, in the fire-bellied toads, *Bombina bombina* and *B. variegata*, clines in song characters are in the region of 10-km wide, wider than the clines for allozyme markers and much wider than the dispersal distance of about 1 km (Sanderson *et al.*, 1992). Similarly, in the grasshopper, *C. parallelus*, clines in song characters and female preference are wide relative to the dispersal distance of about 30 m (Butlin & Ritchie, 1991). Assortative mating in a steep cline for direction of coil in the land snail, *Partula suturalis*, on the Pacific island of Moorea has little, if any, impact on gene exchange as measured by allozymes (Johnson *et al.*, 1993).

15.5 Evolution of behaviour in small isolated populations

A long-standing debate in evolutionary biology concerns the role of founder events, where small numbers of individuals initiate isolated populations, in divergence and speciation. The alternative points of view were crystallized in a seminal pair of papers in 1984 (Barton & Charlesworth, 1984; Carson & Templeton, 1984). The debate concerns the likelihood that the extreme sampling effects in the foundation of such a new population will cause sufficient genetic change for it to come under the domain of attraction of a different adaptive peak, in the framework of Wright's 'adaptive landscape'. This debate continues; see Whitlock (1995) for example. 'Peak shifts' in general are beyond the scope of this chapter, but specific effects of founder events and evolution in small populations have been proposed for mating behaviour and the consequent origin of pre-zygotic reproductive isolation.

15.5.1 Additive genetic variance in courtship

Characters that are closely correlated with fitness are unlikely to maintain substantial additive genetic variation within populations and, furthermore, founder events are likely to involve loss of additive variation. As a result, the potential for evolutionary response in such characters in recently founded populations is expected to be low. However, non-additive variation, due to dominance and epistasis (interactions between alleles at different loci), can be maintained under selection and may be 'converted' to additive genetic variance during founder events as a result of random changes in allele frequencies (Goodnight, 1988). The simplest example would be a recessive allele at low frequency in a source population: an increase in frequency in a founder event would cause the allele to contribute much more to the additive genetic variance.

Courtship characters are clearly fitness-related traits. There is evidence for low additive genetic variance, but significant dominance variance in natural populations for such traits as male courtship vigour or female mating propensity (Meffert, 1995; see also Butlin, 1995b; Bakker & Pomiankowski, 1995, with respect to mating signals and preferences). Epistatic effects are harder to detect. If significant additive variance is released following founder events, this could contribute to the evolution of mate discrimination and ultimately speciation. Some assortative mating has been observed in experimentally bottlenecked laboratory populations, mainly in *Drosophila*, although the effects are not strong (Rice & Hostert, 1994). Direct attempts to measure the expected release of additive genetic variation have been successful for other characters, but a study of these effects for courtship repertoire gave equivocal results (Meffert, 1995).

Meffert (1995) compared estimates of additive genetic variance, from parent–offspring covariances, in six bottlenecked and two control lines of house fly, *Musca domestica*, for 11 components of courtship behaviour, both male and female. The six bottlenecked lines showed concerted phenotypic shifts over 10 generations in seven of the 11 characters and significantly increased additive genetic variance in some cases. This could be explained by directional dominance but there was no close link between increased variance and phenotypic shifts. Thus, some other mechanism, potentially epistasis, must be involved. The measurement of these courtship characters is, however, seriously complicated by the influence of the behaviour of the mating partner. For example, male courtship may appear more vigorous and more complex when a male encounters an unreceptive female than when the same male encounters a receptive one. Since effects of bottlenecks are not likely to be coordinated across courtship elements, the most common effect is likely to be a reduced success rate of courtship. Typically, this will be analogous to inbreeding depression: it will increase the risk of extinction of the newly founded population. However, it may be that, rarely, a resolution of this incompatibility can lead to a realignment of the courtship elements in a novel way that creates incompatibility with the source population and thus contributes to speciation.

15.5.2 The 'Kaneshiro' hypothesis

The evolutionary scenario originally envisaged by Kaneshiro (1976) had much in common with the above argument. Kaneshiro observed that pre-mating isolation was asymmetrical between pairs of species in the *planitibia* subgroup of Hawaiian picture-winged *Drosophila*. Females of *D. planitibia* (P) and *D. differens* (D) discriminate against males of *D. silvestris* (S) and *D. heteroneura* (H) more strongly than females of S or H do against males of P and D in two male : one female laboratory tests. S and H occur on the youngest Hawaiian island, the 'Big Island' of Hawaii itself, whereas P and D occur on the older islands of the Maui complex to the northwest. Chromosomal phylogeny suggests that S and H are derived from P or D by a process that certainly involved colonization of the Big Island and may have been driven by founder events. Kaneshiro, therefore, proposed that the founder event may have lead to simplification of the courtship repertoire of the derived species through loss of behavioural elements in the founder event and, furthermore, that this pattern could be used to infer the direction of evolution where asymmetrical isolation was observed. If derived males lack courtship elements typical of the ancestral species, ancestral females will discriminate strongly against them while ancestral males will more than satisfy the requirements of the derived species. The pattern of asymmetry is supported by observations on natural and laboratory variation within species (Kaneshiro, 1989), although scenarios that do not involve founder events may predict different asymmetries. Further study of courtship in P and S (Hoikkala & Kaneshiro, 1993) has begun to elucidate the behavioural basis of the asymmetrical isolation. The presence of the same set of courtship elements in each species initially appears to contradict Kaneshiro's original hypothesis, but Hoikkala and Kaneshiro argue that critical transitions between male and female behavioural elements have been lost.

The connection between founder events and loss of behavioural elements has never been very clear. It is a less critical part of a more recent formulation of the same underlying idea (Kaneshiro, 1989) which concentrates attention on the changes in selection pressures associated with colonization events, rather than genetic changes. Source populations are supposed to exist at relatively high population densities and in more complex communities than are newly founded populations. Kaneshiro argues that high densities produce high rates of encounters between potential mates, while complex communities may contain related species so that the risk of interspecific courtship is also high. In these conditions, selection will favour discriminating females with stringent courtship requirements since the risks associated with rejecting a male are small and an encounter with another male can be expected quickly, while the risks of indiscriminate mating are high; a randomly selected male may well be of the wrong species. If females are very discriminating in mate choice, males will be selected for high courtship vigour and elaboration of mating signals. The opposite conditions apply in small newly founded populations. Population densities are low, at least initially, and the new habitat may not have been colonized by related species. Relatively indiscriminate mating by females is favoured by selection. Male courtship vigour will decline and signals may degenerate.

These differences alone are sufficient to generate asymmetrical isolation. The vigorous males of the ancestral population easily satisfy the weak conditions imposed by the derived females, whereas the derived males cannot provide sufficient stimulation for the ancestral females. Kaneshiro further argues that relaxed selection will allow greater genetic variation in the derived population. In time, the new habitat is likely to fill, and the pattern of selection to return to the conditions of the source population, but it is possible that a novel system of signals and responses will crystallize from the variable and loosely coordinated courtship of the derived species, eventually leading to symmetrical isolation.

Some of the necessary elements of this process have been demonstrated. For example, where female discrimination has been separated from female preference it has been found to be more variable, and with a larger heritable component, than preference (Butlin, 1993). *Drosophila melanogaster* strains from Africa, the ancestral range of the species and its close relatives, have higher courtship vigour in males and lower mating propensity in females than French strains (Cohet & David, 1980). However, it remains important that the mechanism is modelled explicitly and further investigated empirically.

15.5.3 Parallel speciation

These ideas view mating preferences and the traits on which they operate as if they evolve independent of the environment and of other adaptations of the organisms concerned. By contrast, Paterson (1985) has argued that mating signals and responses are closely adapted to the environment in which they operate and that colonization of novel environments is the primary driving force for their evolutionary divergence. Rice and Hostert (1994), from a review of laboratory experiments on speciation, conclude that reproductive isolation is most likely to evolve as a pleiotropic effect of adaptation to divergent habitats. Small isolated populations will invariably be in habitats that are distinct from those of the source populations so it is dangerous to conclude that an association between speciation and founder events is necessarily due to bottlenecking effects rather than selection.

Schluter and Nagel (1995) have recently drawn attention to the fact that a large ancestral population may often give rise to many new populations in similar environments, especially in periods of population expansion like the phase of global warming at the start of the Holocene. Such repeated colonizations can provide powerful evidence for the role of selection in both adaptation and speciation if the same patterns appear in independent replicates. Consideration of one of the possible examples offered by Schluter and Nagel will make the mechanism clear.

At the end of the Pleistocene, new freshwater habitats became available in many areas of the northern hemisphere and were colonized from a large ancestral marine population of the three-spine stickleback, *Gasterosteus aculeatus*. Adaptation to the freshwater habitat in most cases involved a substantial reduction in adult body size, although in some lakes two forms evolved: a small-bodied limnetic form and a large-bodied benthic form. Since mating is strongly size-assortative, these adaptive changes in body size incidentally produced pre-mating reproductive isolation both between the small-bodied freshwater populations and their large-bodied marine ancestors, and between the pairs of benthic and limnetic populations within lakes. The stochastic effects of founder events are most unlikely to have produced these parallel patterns, whereas they are readily explicable in terms of adaptation. Indeed, the pairs of forms within some lakes call into question the need for geographic separation, let alone bottlenecking, in the origin of reproductive isolation.

15.6 Future prospects

The great majority of examples presented in this chapter have been either examples of genetic analysis of population structure or studies of behaviour. Very few organisms have been considered from both points of view and yet there are clearly good reasons why they should be. The history of many species may compromise the optimality approach if populations are still evolving local adaptations; the current structure of populations will influence the precision of local adaptation while gene flow determines the scale of variation in behaviour. Behavioural variability and evolution in response to the costs and benefits of dispersal similarly compromise the simplifying assumptions of population genetics. Key questions that remain to be answered, and where the combination of modern genetic techniques with progress in theoretical modelling and computer simulation may permit progress, are to determine the scale of local adaptation in behaviour in relation to the genetic structure of populations (determined from putatively neutral loci), and the influence of history and contemporary dispersal and gene flow on local adaptation. Of particular interest are processes that might involve feedback loops, such as the interaction between optimal outbreeding and local adaptation, and those that involve conflicts between selection pressures at different levels, such as dispersal polymorphisms. There are great opportunities for collaboration between population geneticists and behavioural ecologists, but the magnitude of the undertaking should not be underestimated.

Chapter 16

Individual Behaviour, Populations and Conservation

John D. Goss-Custard & William J. Sutherland

16.1 Introduction

Throughout the world, habitats are being lost and degraded and natural populations are in danger of being overexploited by the ever increasing demands placed upon them by our expanding and resource-hungry population. There is widespread concern about the effect of this on species diversity and on the abundance of particular organisms. One of the many factors that will influence how energetically such concerns are translated by decision-makers into conservation-friendly action is the credibility of the predictions made by scientists as to the consequences of allowing present trends to continue; few will take seriously doom-laden forecasts that are scientifically suspect. In many of the cases where natural populations need to be managed rather than their habitats merely secured, quite sophisticated quantitative predictions may be required. For example, when some people harvest a species that other people like to watch, a conflict of interests arises that may require policy-makers to make trade-offs and scientists to provide quantitative predictions for a number of policy scenarios. The need for a strong scientific underpinning of decision making in such circumstances remains as strong as ever.

The scientific challenge arises because ecologists are increasingly being asked to predict outside their direct experience. As scientists, we are trained to treat with suspicion extrapolations beyond the empirical range. Yet, this is precisely what we must do if we are to forecast to the novel circumstances brought about by changes in such fundamentally important global systems as the weather. In the case of vertebrate populations, the subject of this chapter, this means that we need to forecast for the new environments the form and parameter values of the important demographical functions, such as density dependent recruitment and mortality rates (Goss-Custard, 1980, 1993). It is difficult enough to measure such functions in present-day circumstances, yet we are required to understand their basis sufficiently well to predict their properties in novel ones. But, unless we can do so, conservation and resource-management strategies may fail because key components of our population models do not apply in the new environments for which predictions are needed.

Our contention is that behavioural ecology has an important role to play in making predictions at the population level, even though few behavioural ecologists so far have attempted to do so. We argue that it is possible to derive demographical functions in new environments through the study of competing individuals (Goss-Custard, 1985, 1993). We focus on predicting demographical functions because quantitative population prediction requires us to use population models that include demographical functions appropriate to the new environments. The many other ways in which behaviour may have important consequences for population and conservation issues are therefore outside the scope of the chapter. As examples: the social system has considerable consequences for effective population size (Primack, 1993) and thus the amount of genetic variation that will be retained; the dispersal pattern will influence both the extinction and recolonization rate of populations; foraging behaviour can be used to examine the behaviour of hunters (Winterhalder, 1981), fishers (Abrahams & Healey, 1990) and whalers (Whitehead & Hope, 1991). It is not our aim to discuss how discoveries at the behavioural level have implications for understanding processes at the population level. Rather, it is to illustrate how behavioural studies of individuals enable us to derive demographical functions that can be inserted into population models in order to make quantitative predictions of population responses to new environments. Our fundamental point is that this approach provides a reliable basis for prediction because the choices made by the individual animals are based on decision principles, such as optimization, that are unlikely to change in the new environments, even if the exact choices made by animals, and thus their chances of surviving and of reproducing, do so.

16.2 Importance of density dependence

When populations are at high levels, individuals are more likely to starve, fail to breed or raise fewer offspring if they do breed. This simple concept of density dependence is of fundamental importance in population biology and has been known since Malthus's (1798) suggestion that increasing human populations will lead to increasing 'misery and vice'. It is, however, surprising how many behavioural ecologists, conservationists and even some ecologists have yet to grasp the widespread implications of this simple idea. If conservationists wish to predict the consequences of exploitation, changing the mortality rate caused by predators, disease, pollution or habitat loss, then it is usually necessary to understand the role of density dependence.

We concentrate in this chapter on predicting the consequences of habitat loss and change. For sedentary species, the consequences of large-scale habitat loss are relatively easy to predict. Destroying a large block of habitat containing 50% of a species of snails is likely to halve the population because the demographical functions in the area that remains have not been affected by the adjacent habitat loss. Note, however, that this is only likely to be true if the scale of habitat loss is much greater than the individual animal's home range. If the habitat loss occurs on a small scale, the results can be hard to predict without in-depth studies of species requirements. For example, with small-scale forest clearing, some species may be indifferent to, or even benefit from, the creation of open patches. But, others may show greater declines than the proportion of the habitat that is lost because of increased predation along forest edges (Wilcove, 1985), changes in microclimate or as a result of the dispersal pattern being affected (Henderson *et al.*, 1985).

For migratory species, predicting the consequences of habitat loss is much more complex and we devote much of the rest of the chapter to this problem. An understanding of density dependence is now crucial to understanding the consequences of habitat loss (Goss-Custard, 1977). If there is no density dependence, habitat loss does not matter providing that displaced individuals can locate the remaining habitat and move to it. The population can then simply crowd together in the areas that remain and fare just as well. Although the local density is increased, total population size is unaffected. At the other extreme, there is perfectly compensating density dependence, such that the addition of one individual leads to another individual either dying, emigrating or failing to breed. In this case, a loss of habitat leaves population density the same as it was before in the areas that remain, but the total population size is reduced. The extreme predictions are thus that habitat

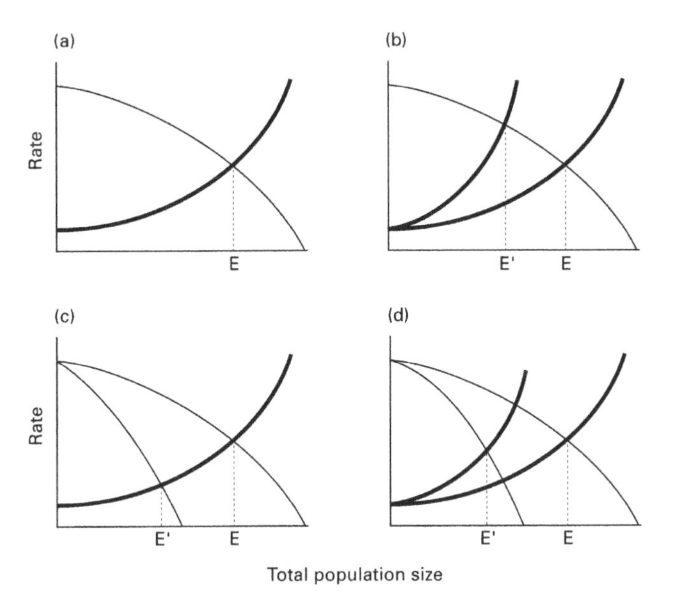

Fig. 16.1 (a) The equilibrium population size E is that at which the net birth rate (thin line) equals the winter death rate (thick line). In this case, both net birth rate and winter death rate are assumed to change with density. (b) A loss of winter habitat shifts the winter mortality curve to the left and results in a lower population. (c) The consequences of a loss of breeding habitat. (d) The consequences of a loss of both.

loss has no impact on the population (as some developers assert) or that the decline in population size is in direct proportion to the amount of habitat that is lost (as some conservationists assert).

One general way of considering the issue of habitat loss in migratory species is to consider the relationship between total population size and *per capita* population change on both the breeding grounds and the wintering grounds (Fig. 16.1). Habitat loss can then be expressed in terms of a shift in these density dependent relationships and Box 16.1 shows how the consequences

Box 16.1 The relationship between habitat loss and population size for migrating animals. (After Sutherland, 1996b.)

Consider a small loss of winter habitat so that the winter mortality curve is shifted very slightly by an amount L_w. This results in a reduced equilibrium population size (see Fig. B.16.1).
From simple algebra:

$$\frac{x_w}{L_w} = \frac{d'}{d' + b'} \qquad \text{(B.16.1)}$$

where x_w is the decline in population size and L_w is the loss of habitat, so that x_w/L_w is the relationship between population decline and habitat loss. As shown in Fig. B.16.1, b' is the slope of the relationship between net birth rate and density and d' is the slope of the relationship between winter mortality rate and density.

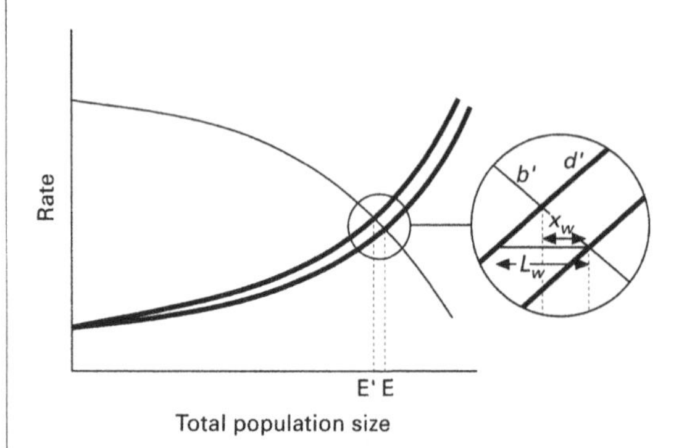

Fig. B.16.1 How the consequences of a small loss of winter habitat can be predicted. As in Fig. 16.1, habitat loss shifts the winter mortality curve (thick line). After habitat loss the population declines from E to E'. The circle to the right shows the critical part of the figure on a larger scale. Note that it is assumed the slopes are linear over such a small scale. x_w, decline in population; L_w, the change in the total population size required for the population to remain the same in the remaining patches (hence, L_w/E equals proportional habitat loss). (From Sutherland, 1996b.)

Continued on p. 377

Box 16.1 *Cont'd.*

A value x_w/L_w of 1 means that the population declines in direct proportion to the extent of habitat loss: a 2% loss of habitat then results in a 2% population decline. A low value of x_w/L_w means that the population decline will be much less than the habitat loss.

Almost exactly the same approach can be used to predict the population decline x_s as a consequences of the loss of breeding habitat, L_s:

$$\frac{x_s}{L_s} = \frac{b'}{d' + b'}. \tag{B.16.2}$$

Thus, if the density dependence can be measured in both the winter and breeding grounds, the consequences of habitat loss of average quality can be considered. The value of d' derived from studies of oystercatchers on the Exe estuary (see Section 16.4) is 0.00011 and the value of b' for studies in Schiermonikoog (see Section 16.3) for the same-sized population is 0.00005. Incorporating these values into Equation B.16.1 shows that a 1% change in winter habitat will result in a population change of 0.69%, while incorporating them in Equation B.16.2 shows that a 1% change in breeding habitat will result in a population change of 0.31%.

Resident species can be considered as a special case of this in which the wintering and breeding habitats are the same. It thus follows by combining Equations B.16.1 and B.16.2 that $x_{w,s}/L_{w,s} = 1$. The population decline will be in direct proportion to the extent of habitat loss.

of the loss of habitat of average quality can then be determined. The impact of habitat loss is more severe for the season in which the density dependence is stronger. That is, if density dependence is weak in one season but strong in the other, the loss of the habitat where the density dependence is stronger has greater impact.

At first, it may seem surprising that the impact of habitat loss is only affected by the ratio of the strengths of the density dependence in the two seasons (Fig. 16.2). The consequences are as great in a species with strong density dependance in both the breeding and wintering grounds as in one with weak dependence in both. Most importantly, Box 16.1 shows that the consequences of habitat loss on population size can only be predicted if the strengths of the density dependence in both the breeding and wintering areas are known. Much of the rest of this chapter describes how to estimate the strength of density dependence and to predict any changes in function form that might arise as a result of habitat loss and change.

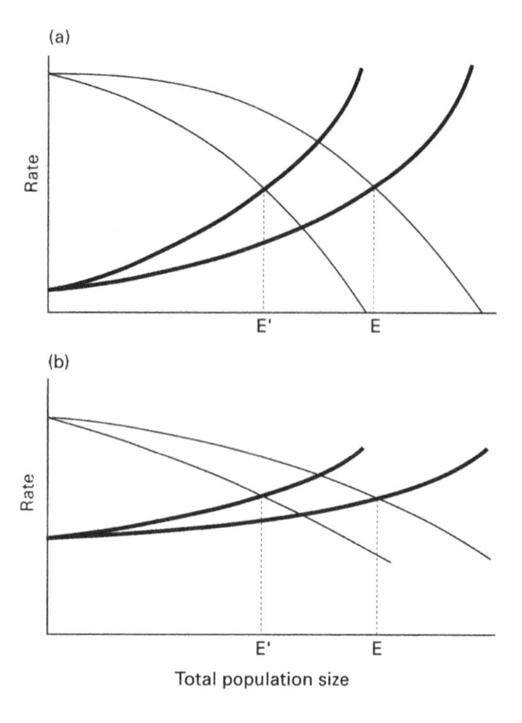

Fig. 16.2 Two examples to illustrate that it is the relative slopes of the winter (thick lines) and summer (thin lines) density dependence that is critical, and not the absolute values. In (a), the density dependence in both the wintering and breeding areas is strong, while in (b) it is weak in both. The consequences of losing 25% of the wintering habitat leads to the same proportional decline in each case. After habitat loss the population declines from *E* to *E′*.

16.3 Determining the strength of density dependence

Occasionally, it is practicable to determine density dependance by manipulating densities experimentally (Alatalo & Lundberg, 1984). But, the usual approach is to estimate the life-history parameters of mortality rate or birth rate over a naturally occurring range of population densities. Although this method has been used successfully (Sinclair, 1989), there can often be problems with this direct approach. In most vertebrates, population size is unlikely to vary enough for the density dependent functions to be defined, especially if there are large annual variations in survival or birth rates due to weather and other factors. A more fundamental problem is that a population is likely to be temporarily high in the first place for reasons which affect life-history parameters; for example, a reduction in parasites which affect mortality or breeding output. The observed density dependence may then differ from the response to density that would apply under more typical conditions. Furthermore, it may be necessary following habitat loss to extrapolate the density dependent relationships to densities that are considerably higher or lower than those which occurred during the study itself. Finally, empirical estimates of density dependent relationships will only apply to the conditions under which the study took place. If habitat of above or below average quality is lost, for example, new density dependent relationships will apply, thus making present-day functions inappropriate (Goss-Custard *et al.*, 1995c). Removing the best or worst patches changes both their average quality and spatial arrangement. An approach is needed to show how individuals would respond to this and how it would affect their chances of survival at different population sizes.

16.3.1 A behavioural approach to density dependence

An alternative approach to determining density dependence which allows this to be done is to understand the decisions individuals make, specially in their choice of feeding or breeding location, and then to create game-theory models based on these decisions to determine the probability of starving or breeding successfully at different population sizes (Goss-Custard, 1985, 1993; Sutherland & Parker, 1985). Game theory is necessary because the decisions individuals make in response to the decisions made by other individuals' have important fitness consequences. For example, more individuals may decide to defer breeding as population size increases and competition intensifies, thus reducing *per capita* reproductive performance across the population as a whole. By running game-theory models that incorporate such processes at different population sizes, it is possible to predict how mortality or breeding output changes with total population density.

These behavioural models have the advantages that a change in the environment, such as in the average quality of the habitat, can also be incorporated; thus, the effect of both habitat loss and habitat change can be explored. Two main classes of models are available. Ideal free-distribution models describe the distribution of individuals between patches within a site (Fig. 16.3). Assume that patches differ in a quality that affects their 'suitability'; for example, they

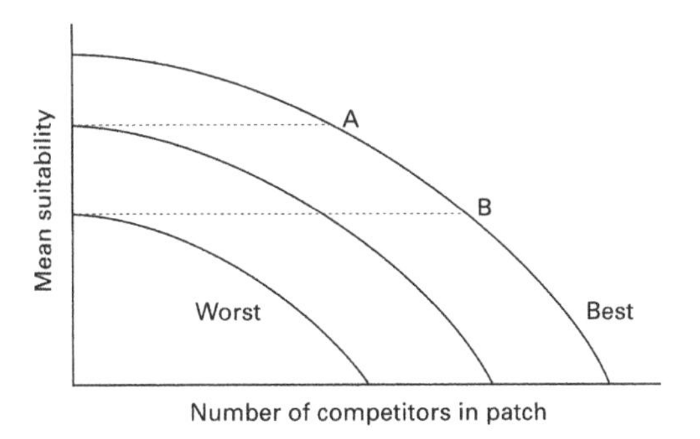

Fig. 16.3 The ideal free distribution. This shows three patches which differ in 'suitability'; this could, for example, be intake rate. As the density of competitors increases the suitability declines. The best patch is occupied until a density 'A' is reached where the suitability is the same in the best patch with many competitors as in the intermediate patch with none. At this point the intermediate patch will start being used. As the density in these patches increases further, 'B', the suitability, declines until the point is reached when it is worth individuals occupying the worst patch. (Modified from Fretwell, 1972.)

may differ in food density such that average intake rate differs between patches. Assume also that there is some form of negative feedback so that the suitability declines as the number of competitors increases. At low population sizes, all individuals occur in the best patch. As the population increases, the suitability in the best patch, where there are many competitors, becomes equal to that in the next patch, where there are none. From then on as total numbers increase still further, individuals occupy both patches and the mean rewards at any given point are equal in both.

The ideal despotic distribution (Fig. 16.4) is similar but subtly different. In this case, the order in which individuals occupy a patch is important because of, for example, territorial defence or dominance hierarchies. In this case, the mean rewards may differ but the suitabilities for the best territory of the next position in the hierarchy should be equal in all patches.

Negative feedback is an essential component of these two models and can have a number of origins, such as territorial exclusion, resource depletion and interference. Both models assume that individuals possess perfect knowledge of the quality of each patch. In reality, animals never possess perfect knowledge, but for some species this approximation is probably much better than for others, depending both on their memory and their ability to sample the environment. For example, wading birds may return to the same estuary year after year and, by capturing numerous prey, build up a reasonable picture of the habitat. By contrast, some invertebrates may live only for a few hours or days and catch or parasitize a few victims. This is important, as the ideal free and despotic distributions assume that individuals are able to respond to both patch quality and competitor density. If they are unable to sample or evaluate the alternatives available, the models are not appropriate.

It is probably much easier for individuals to evaluate feeding areas in the non-breeding season than breeding areas because the quality of breeding habitat

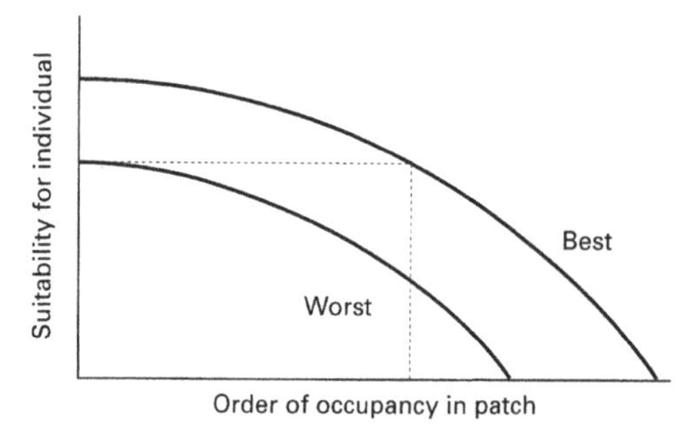

Fig. 16.4 The ideal despotic distribution. This is similar to the ideal free distribution except that individuals defend areas within a patch and exclude others. Hence, as further individuals settle in a patch, their suitability will decline even though the suitability of the first to settle will be unaffected. The dashed line shows when individuals would gain as much from settling in the best area within the worst patch as from a poor area in the best patch. (Modified from Fretwell, 1972.)

may only emerge a long time after the decision is taken. Breeding locations are often chosen early in spring, yet the most important factor determining breeding success may be the abundance of food or predators or diseases later in the summer when the young are being raised. Although yellow-headed blackbirds, *Xanthocephalus xanthocephalus*, feed their young largely on odonates, the location of males and the choice of female territories are unrelated to prey abundance, perhaps because the territories are chosen before the odonates emerge (Orians & Wittenberger, 1991). Similarly, the risk of eggs and young being taken by predators is probably difficult for individuals to determine when they are choosing where to breed because the predators may be nocturnal and scarce and predation risk may depend upon factors, such as the structure of the environment and the abundance of alternative prey, some weeks later. Kentish plovers, *Charadrius alexandrinus*, in Hungary switch back and forth between alkaline grasslands and drained fishponds even though the breeding success is twice as high in the grasslands due to the lower predation risk (Székely, 1992). The suggested explanation for this is that the food availability is much greater in the fishponds and the birds are unable to assess the higher nest predation risk. Wheareas feeding areas are sampled continuously, nest predation risk, being the probability of a single catastrophic event occurring, is hard to estimate, especially a few weeks hence. This differs from the non-breeding season where decisions tend to have immediate consequences.

Such problems in evaluating breeding season habitat quality not only present difficulties for the animals themselves but also make breeding season studies more difficult for the behavioural ecologist interested in deriving density dependent functions. Most progress has therefore been made for the non-breeding season where the payoffs seem more immediate and easier to evaluate. We therefore illustrate the approach to measuring density dependence from empirical game-theory modelling by reference to work done in that season.

16.3.2 Measuring density dependence in the non-breeding season

Introduction

When the breeding season ends and young are either completely independent or sufficiently mobile to keep up with the group, many species abandon their breeding territories. This section is concerned with such animals; for those that remain territorial, we assume that the ideas developed later in this chapter apply. Although the parents in many free-ranging species continue to care for young, we assume for simplicity that the main goal of individuals is just to survive the non-breeding season even though, in reality, they must do this in sufficiently good condition to return to the breeding areas. For the same reason, we discuss food acquisition with little reference to the trade-offs animals must often make between foraging and reducing the risk from enemies.

If the population of either herbivores or carnivores is large enough, depletion will reduce food density by the end of the non-breeding season to the threshold level at which foragers will be unable to feed fast enough to survive; in Fig. 16.5, this is where the initial population is $3n$. If all individuals are equally efficient at foraging, all will starve simultaneously when the threshold is reached. With smaller initial populations, the threshold will not be reached and all individuals will survive. With larger initial populations, all will again starve but, as depletion is more rapid, they will die sooner. Such scramble competition through depletion gives the all-or-nothing density dependent mortality function, with a slope of infinity, shown in the inset to Fig. 16.5(a).

In reality, all individuals will seldom be identical. If some animals are able to secure food at the expense of others, so that individuals vary in competitive ability, contest competition will arise. In foraging, competitive ability comprises

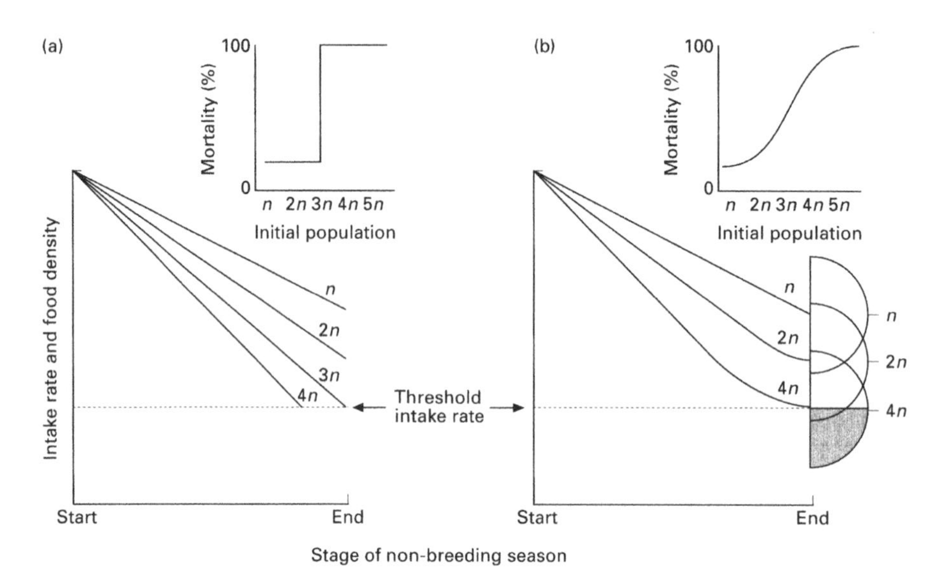

Fig. 16.5 How introducing variation between individual animals reduces the slope of the density dependence. In (a), all individuals are equally efficient at foraging and contesting food during the non-breeding season. The result is scramble competition in which all animals starve once the population at the start of the non-breeding season is large enough ($3n$) to deplete the food to the level at which the animals cannot feed fast enough to meet their energy requirements. Once the initial population size reaches this point, all individuals die and the slope of the density dependence is infinity (inset). In (b), there is variation between individuals, as shown by the hemispherical frequency distributions of individual intake rates. As the food is depleted and the average intake falls, some poorly performing individuals starve (shown by shading), thus relieving the pressure on the food supply. With such contest competition, some good performers survive the non-breeding season even though greater than $3n$ individuals occupied the habitat at the start. The slope of the density dependence is thus reduced (inset).

two characteristics. Foraging efficiency is the intake rate an individual achieves in the absence of competitors (Fig. 16.6a). Inefficient foragers will reach the threshold first and die. The subsequent rate of depletion and/or interference will then slow down, enabling the more efficient individuals to survive; case $4n$ in Fig. 16.5(b). Mortality is still density dependent, but the slope is no longer vertical (inset to Fig. 16.5b). Individuals also vary in their susceptibility to interference, the rate at which intake rate changes as competitor density increases (Fig. 16.6b). Dominants may not be affected by interference because other animals avoid them; only the subdominants thus bear the time-cost of increased interaction rates as forager density increases. Where there is food stealing (kleptoparasitism), dominants may even increase their intake rate as forager density increases because there are more opportunities to rob subdominants. This again increases the variation in performance and reduces the slope of the density dependent function. By allowing subdominants to escape

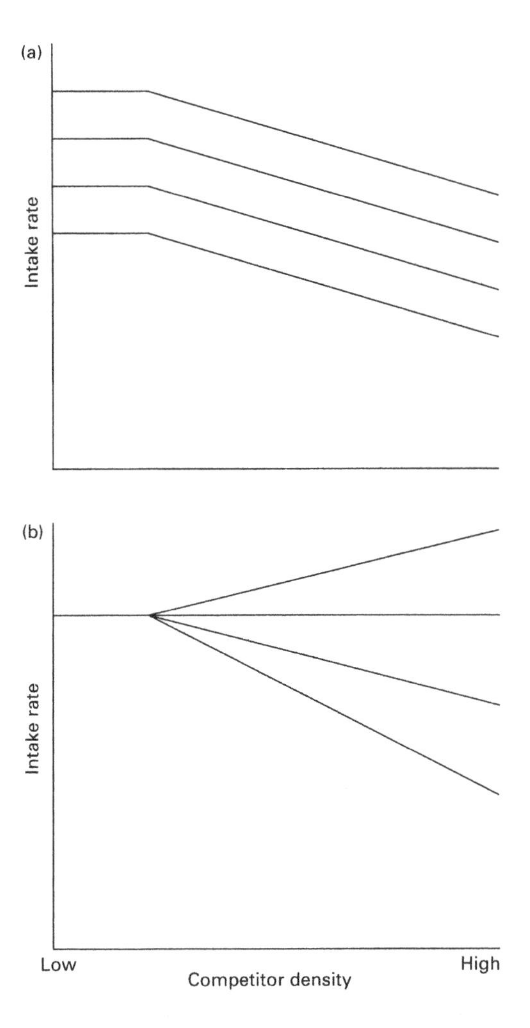

Fig. 16.6 How individual foragers may vary. Each line refers to the intake rate of an individual at different densities of competitors. In (a), all individuals are equally susceptible to interference once the density of competitors has reached the density at which intake rate is reduced. Individuals differ in foraging efficiency because, before the threshold is reached, their intake rates differ. In (b), all individuals have the same foraging efficiency but their response to increasing competitor density above the threshold differs so their individual susceptibilities to interference differ. The positive slope refers to an individual which is able to steal food increasingly often as density, and thus the opportunity to steal, increases.

areas of high competitor density by moving to poorer feeding areas with few competitors, spatial variation in food density increases the variation in individual performance and further reduces the slope of the density dependent function.

An individual's foraging efficiency determines how successful it is in responding to a reduction in the density of available prey. By definition, its susceptibility to interference will determine its response to interference. Behavioural ecology provides a framework for exploring the costs and benefits of such individual variation in behaviour through the principle of individual optimization. The optimal solution for one individual in a given environment differs from that of another because its own particular combination of efficiency and susceptibility to interference subjects it to different constraints. The following section illustrates individual variations in foraging efficiency and susceptibility to interference and their fitness consequences.

Individual variation in foraging efficiency

This arises from differences between classes of animals within the population and from differences between individuals within a class. Class differences have been widely studied and have been related in birds, for example, to age (e.g. Sutherland *et al.*, 1986) and sex (Durell *et al.*, 1993) and are often connected with morphological factors (e.g. Gosler, 1987) that affect efficiency (Marchetti & Price, 1989). Individual variation in foraging efficiency has been studied in birds (Partridge, 1976), but not frequently. An example from other vertebrates is provided by bluegill sunfish, *Lepomis macrochirus*, a generalist freshwater predator. In an experiment, fish learned to increase their search speed in open water but to reduce it in vegetated areas so as to match the requirements of the particular food supply present. In open water, the prey (*Daphnia*) are quite conspicuous and unable to avoid being attacked by the fish, which learned to pursue *Daphnia* quickly. But, in vegetation the *Daphnia* are cryptic and become motionless or hide when fish are nearby and can also escape when detected by fish which therefore learned to hunt more stealthly. Individual fish were not equally flexible in adapting to the change in behaviour required and thus differed in the searching tactics they adopted which, in turn, affected the efficiency with which individuals foraged in the two habitats (Ehlinger, 1989). In an experimental study of zebra finches, *Taeniopygia guttata*, individuals varied considerably in how successfully they selected between patches that provided quite different rates of energy gain. This between-individual variance was partly accounted by a heritable component (Lemon, 1993). In vew of their scarcity, more experimental studies of the causes of such individual variation would be worthwhile.

Two field studies on mammals have stressed the fitness consequences of foraging efficiency. Free-living Columbian ground squirrels, *Spermophilus columbianus*, varied in their ability to select the diet that maximized daily energy consumption. Body size and the feeding time available constrained what was the optimal diet for each individual, 63% of whom consumed the diet that maximized their daily intake. The remaining 37% approached an energy maximizing diet but made some incorrect foraging decisions, and so failed to reach it. These individual differences, which were consistent across seasons and were not related to social factors, seemed large enough to have fitness consequences, although these could not be studied in detail (Ritchie, 1988). However, individual differences in foraging efficiency in Soay sheep, *Ovis aries*, during an overwinter population crash caused by overgrazing, did have clear fitness consequences (Illius *et al.*, 1995). Individual variations in survival were related to variations in the breadth of the incisor arcade; a big mouth is advantageous when food density is low. In accordance with the increasing number of studies pointing to the importance of parasites to host fitness (Nelson, 1984; Møller *et al.*, 1993; Loehle, 1995), survival during the crash was also related to the gut parasite burden. Illius *et al.* (1995) argue that narrow incisor arcades nonetheless persist in the population because, when food is more plentiful at low population density, they can increase their fitness by feeding more precisely, and thus more selectively, than individuals with a broad incisor arcade. These two studies illustrate that further work on the foraging differences between free-living individuals, and thus on their variable response to increased rates of depletion and interference as population size increases, would be worthwhile.

Individual variation in susceptibility to interference

In kleptoparasitic systems, the success of individuals stealing food will, by definition, be related to dominance, since dominance is measured as the percentage of encounters won. Pusey and Packer (see Chapter 11) discuss the many correlates and the costs and benefits of dominance. Suffice to say that fitness benefits in the non-breeding season have often been identified. For example, the more dominant families in wintering geese feed in the more profitable feeding areas and parts of the flock and, apparently as a consequence, leave for the breeding grounds in better condition, thereby increasing their subsequent chances of survival and of breeding successfully (Ebbinge, 1989; Black & Owen, 1989; Prop & Deerenberg, 1991; Black *et al.*, 1991). However, the possibility that dominant individuals are also particularly efficient at foraging has not been investigated in such studies.

Although many studies have related intake rate to dominance, few have described the interference function and estimated susceptibility to interference. In seed-eating snow buntings, *Plectrophenax nivalis*, the relationship appears dome-shaped. At low densities, birds feed rather slowly, perhaps because of the need to be vigilant, whereas at higher densities, intake rate declines. Interference apparently arises because birds supplant others from feeding sites. Interference only occurs when food is scarce, presumably because it is not cost-effective, or worth the risk of being damaged, for dominants to supplant others when food is abundant (Dolman, 1995).

For understanding density dependence, the important question is how susceptibility to interference varies between individuals. So far, this has only been measured in oystercatchers, *Haematopus ostralegus*, eating mussels, *Mytilus edulis*. Dominants attack subdominants increasingly as bird density rises and often steal their mussels. Intake rate in subdominants decreases with competitor density but, as in Dolman's (1995) study, only above a threshold density. The threshold is unrelated to bird dominance but the slope decreases with dominance and even becomes positive in top dominants (Stillman *et al.*, 1996). Interference does not arise solely because more mussels are stolen as density increases. Rather, it is mainly due to the lower rate at which subdominants capture mussels in the first place (Ens & Goss-Custard, 1984), perhaps because they spend more time avoiding other birds (Vines, 1980). There is a clear need for more empirical studies of how interference really operates (Ens & Cayford, 1996).

Habitat quality

Habitats will, of course, vary in quality with some patches being better than others. The intake rate that individuals experience will depend upon a complex interaction of the quality of the chosen patch, its foraging efficiency, the amount of competition within the patch and its susceptibility to interference. The ideal free distribution (see Fig. 16.3) describes patch choice and incorporates differences in competitive ability. There is a range of models based on the ideal free distribution which make different assumptions about the shape of the interference function (Ruxton *et al.*, 1992; Holmgren, 1995; Moody & Houston, 1995; Goss-Custard *et al.*, 1995a,c; Clarke & Goss-Custard, 1996) and the manner in which individual differences are incorporated. One way to model such differences is to incorporate the decisions of individual animals (Holmgren, 1995; Moody & Houston, 1995; Goss-Custard *et al.*, 1995a). Another is to consider classes of animals (for example, individuals differing in dominance or competitive ability) and determine the strategy by which individuals of each competitive class will obtain the highest intake rate (Sutherland & Parker, 1985; Parker & Sutherland, 1986; Sutherland, 1996a). All individuals of a given class will obtain the same intake rate — otherwise they would move — but individuals of different classes will differ in intake rate. But, however it is modelled, interference and depletion will be greater at high population levels and most or all individuals will obtain a lower intake rate as a result of a combination of both increased interference within each patch and a greater use of poorer patches (Goss-Custard, 1993; Sutherland & Dolman, 1994).

Deriving density dependent functions from ideal free models

Models that predict forager mortality rate assume that individuals die when they fail to achieve a certain minimum intake rate (Sutherland & Dolman,

1994; Goss-Custard *et al.*, 1995a,b) or use up all their energy reserves (Goss-Custard *et al.*, 1995c). Such models generate density dependent mortality functions by calculating mortality at a range of initial population sizes. In an empirical individuals-based model of oystercatchers eating mussels on the Exe estuary in England, each bird has its unique combination of foraging efficiency and dominance (Clarke & Goss-Custard, 1996). Some birds starve late in the non-breeding season because of depleted food stocks and high, temperature-related energy demands. Individuals with quite high rank but low efficiency die, whereas those with above-average efficiency seldom do so (Fig. 16.7). This suggests that efficiency is often more important than dominance in determining a bird's survival chances, as empirical studies have already indicated (Goss-Custard & Durell, 1988). This underlines the need for further studies, like that of Illius *et al.* (1995), on the selective processes determining average efficiency and on the causes of variation between individuals. Surprisingly little is known about the selective constraints on further increases in foraging efficiency and about the causes of individual variation. As population size, when food is limiting, is so sensitive to the foraging efficiency of individuals

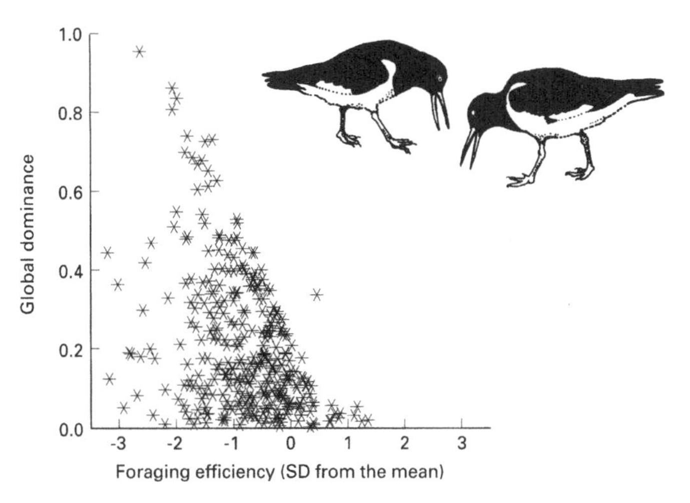

Fig. 16.7 The characteristics of the individuals that starve during the winter, as predicted by the individuals-based model of oystercatchers foraging on the mussel beds of the Exe estuary. Each symbol shows the dominance, and thus susceptibility to interference, and foraging efficiency of a bird that starved to death at some point during the winter when its energy reserves fell to zero. An individual's efficiency is measured in empirically determined standard deviations of the interference-free intake rates of a sample of individuals feeding on the same food supply. Global dominance is a hypothetical, estuary-wide quality that, for example, may reflect an individual's fighting ability. The model assumes that the value for each bird stays the same on all mussel beds but determines its local dominance, and thus its susceptibility to interference, on any one mussel bed. An individual's local dominance increases as, on the mussel bed where it is feeding, the proportion of birds on the same bed that have lower global dominance scores to its own increases. (From Goss-Custard *et al.*, 1995a.)

(Goss-Custard *et al.*, 1996a), more studies on the determinants of foraging efficiency, to parallel the effort invested in studies of dominance, are required if behavioural ecologists are to contribute all they can to predicting population responses to habitat loss and change.

The mortality rate from starvation increases with initial population size in the oystercatcher–mussel model (Fig. 16.8). The density dependence is quite gradual, as would be expected where individuals vary so widely in competitive ability, and patch quality varies threefold. For ecological prediction, this function would only apply if habitat of average quality was lost, because it changes considerably if habitat of above- or below-average quality is lost (Fig. 16.8). As more of the better-quality habitat is removed, density dependence begins at a lower density, and mortality rate rises more steeply. Clearly, it is important to know whether habitat of average or atypical quality is to be lost if population predictions are not to be flawed seriously.

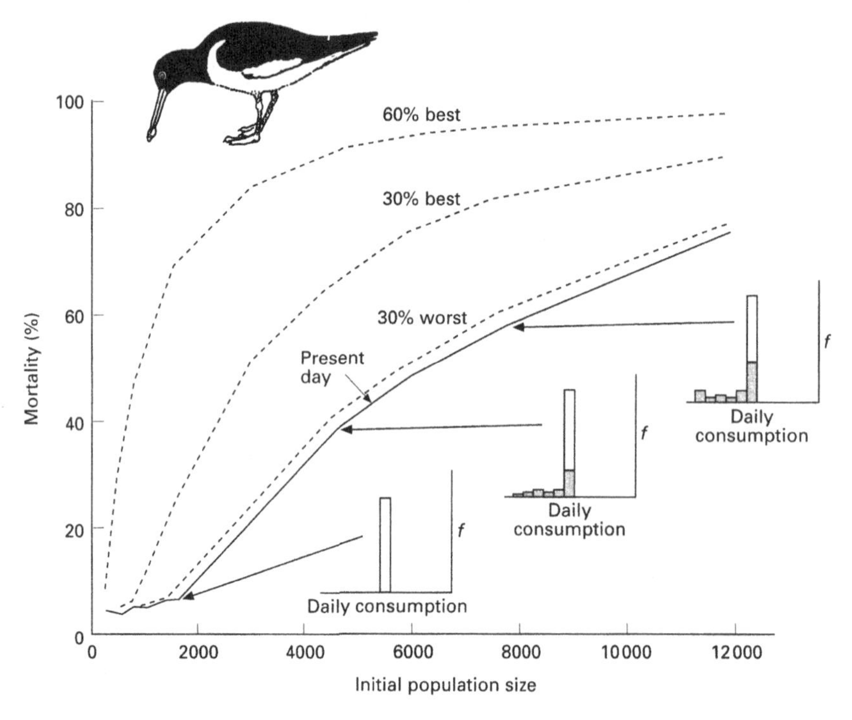

Fig. 16.8 Density dependent starvation functions as predicted by the game-theoretic model of oystercatchers feeding on the mussel beds of the Exe estuary. The solid line shows the percentage of birds that, with different initial population sizes, are predicted to die at some point during the winter with the present-day mussel beds of the estuary. The frequency histograms show the distribution of the daily consumption rates of all individuals in the population, averaged over a full spring–neaps' 14-day cycle in October, before many birds have died. As the initial population size increases, an increasing proportion of birds have low consumption rates in autumn and so fail to accumulate the energy reserves needed to survive the poor feeding conditions that arrive later in the winter; the numbers that die are indicated by the hatching. The dashed lines show the density dependent functions that arise if 60% or 30% of the best feeding areas, or 30% of the worst, are removed. (From Goss-Custard *et al.*, 1995c, 1996a.)

16.3.3 Measuring density dependence in the breeding season

In many species, individuals either defend space and exclude others or exhibit dominance hierarchies within groups. The ideal despotic distribution (see Fig. 16.4) is often an appropriate framework for considering such breeding systems. As densities increase, individuals may: (i) defend smaller territories; (ii) occupy poorer breeding areas; or (iii) refrain from breeding altogether. All of these may result in reduced breeding output *per capita* at high population sizes.

There are field examples of each of these three mechanisms that lead to density dependent production. The red grouse, *Lagopus lagopus*, studied by Watson and Miller (1971) provide an example of territory size decreasing with population size. All the suitable habitat was occupied by grouse at both low and high population densities; thus, at high densities individuals possessed smaller territories than at low densites. Evidence for the second mechanism, the buffer effect (Brown, 1969), comes from a number of species. For example, when nuthatches, *Sitta europea*, are scarce the mean territory quality is greater than when the population is high (Nilsson, 1987). Similarly, female red squirrels, *Sciurus vulgaris*, only occupy poorer quality territories at high population densities, and these territories produce fewer young (Wauters & Dhondt, 1989). A number of studies have shown that juveniles and poorer quality individuals tend to occupy the poorer territories, further reducing *per capita* reproductive rate at high population sizes. For example, blackbirds, *Turdus merula*, breeding in small patches of woodland were more likely to be juveniles and suffered greater nest predation and starvation than the birds breeding in large patches (Møller, 1991). Refraining from breeding at high population sizes has also been recorded (Smith & Arcese, 1989). For example, many species, especially mammals, breed in social groups in which only the highest ranking females breed (Macdonald, 1979), with the result that, as the population increases, a lower proportion of individuals breed.

In many cases, there may be more than one of these mechanisms operating to produce density dependence. In oystercatchers breeding in territories on the island of Schiermonnikoog in the Netherlands (Ens *et al.*, 1992, 1996), both buffering and refraining from breeding occur. There are two main strategies of territorial defence. Some birds, known as residents, defend territories along the edge of the saltmarsh and walk with their chicks straight onto the mudflats to feed. Others, known as leapfrogs, defend two territories: one on the saltmarsh but inland of the resident territories and another on the mudflats but seawards of the resident territories. Leapfrogs thus have to fly back to the nest with each food item to feed the chicks and, as a result, have a considerably lower breeding success (between one-half and one-sixth, depending upon the year) of the residents, many adult oystercatcher birds refrain from breeding even though they are physiologically capable of doing so (Ens *et al.*, 1992). The explanation seems to be that some birds defer breeding in order to obtain a high-quality territory. Ens *et al.* (1996) show that most of these delayed breeders are birds that eventually become residents. The mean expected lifetime reproductive success is very similar for the two strategies. The leapfrogs, on average, breed for more years but with a low average success, whereas the residents usually delay breeding but then, on average, have a higher annual success. These strategies are then frequency dependent. At the evolutionary stable strategy, the mean lifetime reproductive success will be the same for residents and leapfrogs. If a higher proportion of individuals were to wait to become residents, for example, the leapfrog strategy would become more advantageous and more birds would then become leapfrogs.

The data from this study can be used to derive a density dependent breeding function for oystercatchers (Sutherland, 1996a). Assume that there is a fixed number of resident and leapfrog territories. By assuming that individuals adopt the strategy that results in the highest lifetime reproductive success, the number of individuals waiting for an opportunity to gain a territory of either type and to breed, along with the numbers breeding in resident or leapfrog territories, can be calculated for a range of population sizes. At low population sizes, all individuals are residents and thus have a high *per capita* breeding success but, as numbers increase, some individuals become leapfrogs and others defer breeding. Both of these strategies result in a reduction in *per capita* breeding success as numbers increase and thus lead to a density dependent reproductive rate.

16.4 Population size and the effects of habitat loss

As the effect on equilibrium population size of habitat loss in one season depends on the strength of the density dependence in the other, the winter density dependent function does not, of itself, tell us how the population will respond. From the equations in Box 16.1, the theoretical expectation is that the population should decrease in the proportion $d'/(b' + d')$. The values for oystercatchers of d' (for the Exe estuary) and b' (for Schiermonnikoog) in the vicinity of the present-day population size were estimated from the work reviewed above, and are 0.00011 and 0.00005, respectively (Sutherland, 1996b). The effect of the loss of winter habitat of average quality on population size is therefore 0.69 and is thus subproportional to the proportion of the winter area lost. This is confirmed by a simulation study of the world population of the Palearctic subspecies of oystercatcher, *H. ostralegus ostralegus* (Goss-Custard *et al.*, 1995d,e). Under a wide variety of assumptions of winter density dependence, the reduction in equilibrium population size is subproportional to the amount of winter habitat lost across the whole range of habitat removal, at least until all has gone and the population goes extinct (Fig. 16.9). The impact of winter habitat loss is influenced by the strength of the summer density dependence (Fig. 16.9b), but, in these particular simulations, by rather a small amount (Goss-Custard *et al.*, 1995e).

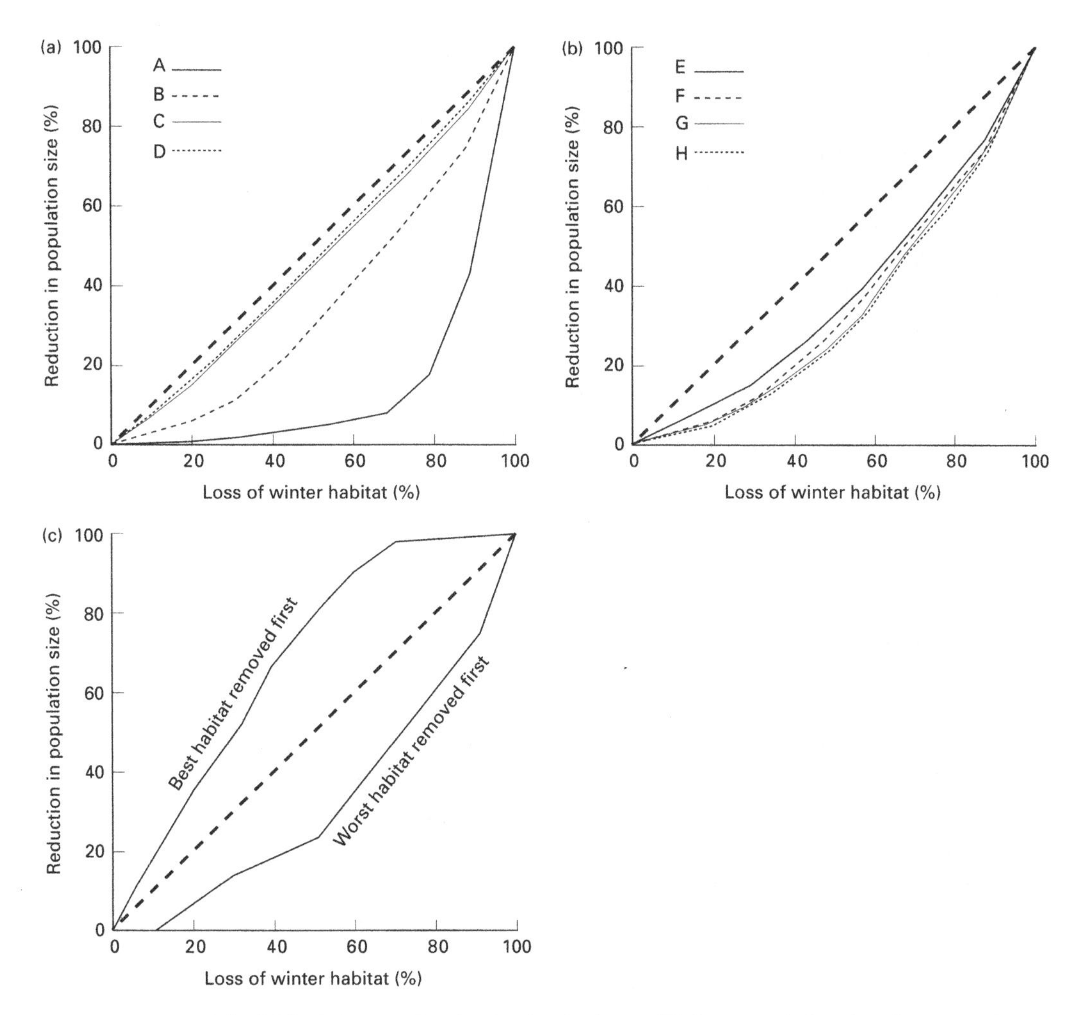

Fig. 16.9 The proportional reduction in the equilibrium population size of oystercatchers resulting from an increasing loss of winter habitat, as predicted by an empirical demographical model of the European population (a,b) and of the Exe estuary (c). The heavy, dashed diagonal line shows the line of proportionality along which a given percentage reduction in habitat area results in the same percentage reduction in equilibrium population size. In (a), summer density dependence, *bT*, is set at 0.5 but the strength of the winter density dependence, *bW*, is increased from 0.001 (A) through 0.01 (B) and 0.05 (C) to 1.0 (D). In (b), the winter density dependence is set at 0.1 and the strength of the summer density dependence is increased from 0.3 (E) through 0.5 (F) and 0.7 (G) to 0.9 (H). In (c) the change in population size resulting from reducing, in successive steps of 10%, the area of the mussel bed feeding grounds, starting with either the worst or the best quality beds. At each step, the appropriate density dependent winter mortality function for that food supply, as derived from the game-theoretic model simulations illustrated in Fig. 16.8, was inserted into a demographical model of the population in which the strength of the summer density dependence, *bT*, was assumed to be 0.5 (From Goss-Custard *et al.*, 1995d,e, 1996b, where model construction, parameter values and the units used in (a–c) are detailed.)

The conclusion that the reduction in equilibrium population size is subproportional to the amount of winter habitat lost depends critically, however, on the assumption that habitat of average quality is removed. The effect of removing habitat of above- or below-average quality was simulated by inserting the density dependent functions derived from the individuals-based model for oystercatchers (see Fig. 16.8) into a demographical population model (Goss-Custard *et al.*, 1996b). With the best habitat being removed first, the reduction in equilibrium population size now becomes supraproportional (Fig. 16.9c). This simulation shows the inadequacy of using present-day density dependent functions to make predictions when habitat other than of average quality is being lost, and underlines the importance of being able to devise a methodology to predict the form and parameter values of density dependent functions in new environments (Goss-Custard *et al.*, 1995c).

16.5 Flexibility and constraint in response to habitat loss

Global environmental change is probably proceeding at an ever increasing rate, resulting in both habitat loss and habitat deterioration. There is also the possibility of latitudinal shifts in ecosystems due to global warming. One key issue is the extent to which populations can respond to these changes and, in the present context in particular, whether populations can readily alter migration routes where necessary. The world environment is in a continual state of gradual flux, for example, due to the movement of continental plates and the succession of ice ages and it is inevitable that species have undergone dramatic changes in distribution and migration routes. The populations of most birds in temperate regions, for example, must have arrived since the last ice age 10 000 years ago, and the current migration routes must have evolved in that time so evolution in migration routes must be possible over such relatively short periods.

There are examples both of very rapid changes in migratory behaviour and of the apparent failure of species to adapt. The clearest example of how a species may respond to environmental change is the study by Berthold *et al.* (1992) of the the blackcap, *Sylvia atricapila*. Berthold *et al.* showed that much of the variation in migratory behaviour between or within populations is determined genetically and thus subject to selection. By an extraordinary coincidence, over the period during which the capacity of blackcaps to change migratory behaviour was being studied, a change did occur. Blackcaps are common breeders in Britain but, until recent decades, were scarce in winter. The wintering blackcaps were initially assumed to be British breeding birds which had become resident, but ringing studies showed that, in fact, they bred in Germany and Austria. By capturing blackcaps wintering in Britain, Berthold *et al.* (1992) showed that their migratory direction differed from those from southwest Germany. They then showed that this orientation behaviour was genetically determined because crossing the two produced offspring that orientated in an intermediate direction (Helbig *et al.*, 1994).

Other species have also changed their migratory behaviour. Over recent years, an increasing proportion of great crested grebes, *Podiceps cristatus*, in the Netherlands have become sedentary (Adriaensen *et al.*, 1993). Serins, *Serinus serinus*, have expanded their range this century from the Mediterranean to central and northern Europe and in doing so have changed from being sedentary to migratory (Berthold, 1993). Such studies may suggest that changes in migratory behaviour can occur both readily and rapidly.

There is, however, evidence that other species may be very slow to alter migration routes. All red-backed shrikes, *Lanius collurio*, migrating from Europe to Africa cross the eastern side of the Mediterranean, including those in Spain that start their migration to Africa by flying northeast! Perhaps there is an adaptive explanation for this migration route but another possibility is that they have simply failed to evolve a new and more direct route. Similarly, all wheatears, *Oenanthe oenanthe*, migrate to Africa in the winter even though the species has spread in both directions from its European breeding ground to eastern Canada in the west and across Asia to western Canada in the east. Were they to migrate directly to Asia or central America, respectively, these newly-established populations could halve their migration distances. Although Africa may be the only continent with suitable wintering habitat for wheatear, perhaps a more likely explanation is that they have not evolved a change in route. We therefore do not have a clear picture yet of how species will respond to changes in the distribution of their habitats at a global scale and it does not necessarily follow that animals will be able to adjust rapidly to a change in the world distribution of their resources.

16.6 Discussion

The science of predicting density dependent functions for the present day and the novel environments brought about by habitat loss and change is still in its infancy and the range of empirical and theoretical studies with which to illustrate the approach is limited. However, the increasing demand from the community at large that ecologists should more willingly address environmental concerns may stimulate more interest in the population consequences of behavioural decision rules, the identification of which has dominated the first 25 years of modern behavioural ecology. This subject began largely because Robert MacArthur, an evolutionary ecologist, wanted to make sense of such ecological concepts as the niche. This required a quantitative understanding of the rules of animal design and their fitness consequences. Since then, much of the ecological talent of a generation has been devoted to pursuing this aim. The increasing interest in using the resulting understanding for the purpose of predicting animal numbers in new environments should bring closer together the two disciplines of population (and community) and behavioural ecology, as MacArthur always intended.

Attempts to do this will undoubtedly run into the major dilemma in ecology. Ecological models that are sufficiently detailed to predict the population consequences of environmental change in a particular case are usually complex, contain estimates of some of their many parameters that are no better than guesses and give predictions that are thought to be situation-specific. Models that are general yet simple and claim to capture the essence of the problem are difficult to test and generate predictions that many do not believe because, they suspect, some vital part of species natural history has been ignored for convenience and analytical tractability. There seems to be no solution to this dilemma other than through the pragmatic and heuristic interchange of results between theoreticians and empiricists. As this chapter has shown, the main conclusions of an empirical model and those of a highly simplified general model can be encouragingly similar.

Although the empirical model summarized here is rather complex compared with many theoretical models, it is nonetheless still very simple in an absolute sense. The need for time-consuming simulations arises from the decision to follow individuals (Goss-Custard *et al.*, 1995a). Individual-based models are becoming popular in population ecology, partly because modern computing power is available and partly because, for many scientists, they do provide the most convincing way in which to represent the real world. Even so, the simulations are cumbersome, slowing down progress. The potential for adding further realism and complicating models still further is considerable. For example, introducing state-dependent decision making in a stochastic environment (Mangel & Clark, 1988) will complicate the models enormously, making their operation and properties difficult to understand. Nonetheless, it will be important in models that predict mortality rates to include the trade-offs animals make between, for example their rate of food intake and risk from predation, and how the trade-off is affected by their body condition (see Chapter 5; Houston, 1993; Goss-Custard *et al.*, 1995c, 1996a). (Clearly, such trade-offs also have important implications for how individual variations in efficiency and susceptibility to interference are measured in the field and included in empirical models.) The solution to this continuing dilemma between simplicity and realism may lie in using behavioural ecological models to derive the decision rules to be used by animals in other models that are designed to predict the population consequences of environmental change. By providing a firm basis for predicting how individual animals would be expected to behave in such new environments, behavioural ecology will be able in this way to contribute significantly to the difficult task of ecological prediction.

However, we finish with a note of warning. Malthus was an early ecological predictor who underestimated the capacity of humans to increase the world food supply. This illustrates one major problem of prediction. It assumes that we understand how individuals will respond to a changed environment. In fact, a species may be able to draw on adaptations that it has not yet displayed. In the 1960s, the population of the dark-bellied brent goose, *Branta bernicla bernicla*, was very low and was restricted to feeding on intertidal plants, especially *Zostera* spp. (Ogilvie & Mathews, 1969). It was argued during the 1970s that the proposed airport on the Foulness intertidal flats in England would have had serious consequences for the population as it would have resulted in the loss of a substantial proportion of the feeding area in Britain. Subsequent work has shown that the population was low at that time because of hunting and it has since increased markedly as a result of conservation measures (Ebbinge, 1991). As the population increased, the geese fed on a wider range of species and have, indeed, started to feed inland. Their ability to do this had not been foreseen. Such an example of unexpected shifts in behaviour may be the exception but it is clear that the approach described in this chapter not only requires appropriate modelling but also a thorough understanding of the adaptive repertoire, or the natural history, of the species.

References

Abrahams, M.V. & Healey, M.C. (1990) Variation in the competitive abilities of fishermen and its influence on the spatial distribution of the British Columbian salmon troll fleet. *Can J Fish Aquat Sci*, **6**, 1116–21.

Abrams, P.A. (1982) Functional responses of optimal foragers. *Am Natur*, **120**, 382–90.

Abrams, P.A. (1993à) Why predation rate should not be proportional to predator density. *Ecology*, **74**, 726–33.

Abrams, P.A. (1993b) Optimal traits when there are several costs: the interaction of mortality and energy costs in determining foraging behavior. *Behav Ecol*, **4**, 246–53.

Abrams, P.A. (1995) Overestimation versus underestimation of predation risk: a reply to Bonskila *et al. Am Natur*, **145**, 1020–4.

Adams, E.S. & Mesterton-Gibbons, M. (1995) The cost of threat displays and the stability of deceptive communication. *J Theor Biol*, **175**, 405–21.

Adriaensen, F., Ulenaers, P. & Dhont, A.A. (1993) Ringing recoveries and the increase in numbers of European great crested grebes (*Podiceps cristatus*). *Ardea*, **81**, 59–70.

Alatalo, R.V. & Lundberg, A. (1984) Density dependence in breeding success of pied flycatcher. *J Anim Ecol*, **53**, 969–77.

Alatalo, R.V., Carlson, A. & Lundberg, A. (1988) The search cost in mate choice of the pied flycatcher. *Anim Behav*, **36**, 289–91.

Alberts, A.C. (1992) Constraints on the design of chemical communication systems in terrestrial vertebrates. *Am Natur*, **139**, S62–S89.

Alcock, J. (1993) *Animal Behavior*, 5th edn. Sinauer, Sunderland, Massachusetts.

Alerstam, T. (1996) The geographical scale factor in orientation of migrating birds. *J Exp Biol*, **199**, 9–19.

Alerstam, T. & Lindström, Å. (1990) Optimal bird migration: the relative importance of time, energy and safety. In: *Bird Migration: Physiology and Ecophysiology* (ed. E. Gwinner), pp. 331–51. Springer-Verlag, Berlin.

Alerstam, T. & Pettersson, S.-G. (1991) Orientation along great circles by migrating birds using a sun compass. *J Theor Biol*, **152**, 191–202.

Alexander, R.D. (1979) *Darwinism and Human Affairs*. University of Washington Press, Seattle, Washington.

Alexander, R.D. (1990) Epigenetic rules and Darwinian algorithms: the adaptive study of learning and development. *Ethol Sociobiol*, **11**, 241–303.

Alexander, R.D. (1991) Social learning and kin recognition: an addendum and reply to Sherman. *Ethol Sociobiol*, **12**, 387–99.

Alexander, R.D. & Borgia, G. (1978) Group selection, altruism, and the levels of organization of life. *Ann Rev Ecol Syst*, **9**, 449–74.

Alexander, R.D. & Sherman, P.W. (1977) Local mate competition and parental investment in social insects. *Science*, **196**, 494–500.

Alexander, R.D., Noonan, K.M. & Crespi, B.J. (1991) The evolution of eusociality. In: *The Biology of the Naked Mole-Rat* (eds P.W. Sherman, J.U.M. Jarvis & R.D. Alexander), pp. 3–44. Princeton University Press, Princeton, New Jersey.

Aloimonos, Y. (1993) *Active Perception*. L. Erlbaum Ass., Hillsdale.

Alonso, J.C., Alonso, J.A., Bautista, L.M. & Muñoz-Pulido, R. (1995) Patch use in cranes: a field test of optimal foraging predictions. *Anim Behav*, **49**, 1367–79.

Altmann, J., Sapolsky, R. & Licht, P. (1995) Baboon fertility and social status. *Nature Lond*, **377**, 688–9.

Ammerman, A.J. & Cavalli-Sforza, L.L. (1984) *The Neolithic Transition and the Genetics of Populations in Europe*. Princeton University Press, Princeton, New Jersey.

Andersson, M. (1984) The evolution of eusociality. *Ann Rev Ecol Syst*, **15**, 165–89.

Andersson, M. (1994) *Sexual Selection*. Princeton University Press, Princeton, NJ.

Andersson, S. (1989) Sexual selection and cues

for female choice in leks of Jackson's Widowbird *Euplectes jacksoni*. *Behav Ecol Sociobiol*, **25**, 403–10.

Andersson, S. (1991) Bowers on the savanna: display courts and mate choice in a lekking widowbird. *Behav Ecol*, **2**, 210–18.

Andersson, S. (1993) Sexual dimorphism and modes of sexual selection in lekking Jackson's Widowbird *Euplectes jacksoni*. *Biol J Linn Soc Lond*, **49**, 1–17.

Andren, C. (1985) Risk of predation in male and female adders, *Vipera berus* (Linne). *Amphib Reptilid*, **6**, 203–6.

Aoki, S. (1977) *Colophina clematis* (Homoptera, Pemphigidae), an aphid species with 'soldiers'. *Kontyû*, **45**, 276–82.

Arak, A. & Enquist, M. (1993) Hidden preferences and the evolution of signals. *Phil Trans Roy Soc Lond B*, **340**, 207–13.

Arak, A. & Enquist, M. (1995) Conflict, receiver bias and the evolution of signal form. *Phil Trans Roy Soc Lond B*, **349**, 337–44.

Arnold, S.J. (1981) The microevolution of feeding behavior. In: *Foraging Behavior: Ecological, Ethological and Psychological Approaches* (eds A.C. Kamil & T.D. Sargent), pp. 409–53. Garland STPM, New York.

Arnold, S.J. (1983) Sexual selection: the interface of theory and empiricism. In: *Mate Choice* (ed. P. Bateson), pp. 67–107. Cambridge University Press, Cambridge.

Aron, S., Passera, L. & Keller, L. (1994) Queen–worker conflict over sex ratio: a comparison of primary and secondary sex ratios in the Argentine ant, *Iridomyrmex humilis*. *J Evol Biol*, **7**, 403–18.

Aron, S., Vargo, E.L. & Passera, L. (1995) Primary and secondary sex ratios in monogyne colonies of the fire ant. *Anim Behav*, **49**, 749–57.

Astheimer, L.B., Buttemer, W.A. & Wingfield, J.C. (1992) Interactions of corticosterone with feeding, activity and metabolism in passerine birds. *Ornis Scand*, **23**, 355–65.

Atkinson, S.N. & Ramsay, M.A. (1995) The effects of prolonged fasting on the body composition and reproductive success of female polar bears (*Ursus maritimus*). *Funct Ecol*, **9**, 559–67.

Au, W.W.L. (1988) Sonar target detection and recognition by odontocetes. In: *Animal Sonar* (eds P.E. Nachtigall & P.W.B. Moore), pp. 451–65. Plenum Press, New York.

Austad, S.N. (1983) A game theoretical interpretation of male combat in the bowl and doily spider, *Frontinella pyramitela*. *Anim Behav*, **31**, 59–73.

Austad, S.N. (1984) Evolution of sperm priority patterns in spiders. In: *Sperm Competition and The Evolution of Animal Mating Systems* (ed. R.L. Smith), pp. 223–49. Academic Press, London.

Avery, M. (1994) Finding good food and avoiding bad food: does it help to associate with experienced flockmates? *Anim Behav*, **48**, 1371–8.

Avise, J.C. (1989) Gene trees and organismal histories: a phylogenetic approach to population biology. *Evolution*, **43**, 1192–1208.

Avise, J.C. (1994) *Molecular Markers, Natural History and Evolution*. Chapman & Hall, New York.

Axelrod, R. & Hamilton, W.D. (1981) The evolution of cooperation. *Science*, **211**, 1390–6.

Backus, V.L. & Herbers, J.M. (1992) Sexual allocation ratios in forest ants: food limitation does not explain observed patterns. *Behav Ecol Sociobiol*, **30**, 425–9.

Baker, R.R. & Bellis, M. (1995) *Human Sperm Competition*. Chapman & Hall, London.

Bakker, T.C.M. (1990) Genetic variation in female mating preference. *Netherl J Zool*, **40**, 617–42.

Bakker, T.C.M. (1993) Positive genetic correlation between female preference and preferred male ornament in sticklebacks. *Nature*, **363**, 255–7.

Bakker, T.C.M. & Pomiankowski, A. (1995) The genetic basis of female mate preferences. *J Evol Biol*, **8**, 129–71.

Balda, R.P. & Kamil, A.C. (1989) A comparative study of cache recovery by three corvid species. *Anim Behav*, **38**, 486–95.

Balda, R.P., Kamil, A.C. & Bednekoff, P.A. (1996) Predicting cognitive capacities from natural histories: examples from four species of corvids. *Curr Ornithol* (in press).

Ball, M.A. & Parker, G.A. (1996) Sperm competition games: external fertilization and 'adaptive' infertility. *J Theor Biol*, **180**, 141–50.

Ballard, D.H. (1991) Animate vision. *Art Intell*, **48**, 57–86.

Barker, D.M. (1994) Copulatory plugs and paternity assurance in the nematode *Caenorhabditis elegans*. *Anim Behav*, **48**, 147–56.

Barlow, H. (1995) The neuron doctrine in perception. In: *The Cognitive Neurosciences* (ed. M.S. Gazzaniga), pp. 415–35. MIT Press, Cambridge, Massachusetts.

Barnard, C.J. (1984) *Producers and Scroungers*. Chapman & Hall, New York.

Barnard, C.J. & Sibly, R.M. (1981) Producers and scroungers: a general model and its application to captive flocks of house sparrows. *Anim Behav*, **29**, 543–50.

Barnett, M., Telford, S.R. & Tibbles, B.J. (1995) Female mediation of sperm competition in the millipede *Alloporus unicatus* (Diplopoda: Spirostreptidae). *Behav Ecol Sociobiol*, **36**, 413–19.

Barraclough, T.G., Harvey, P.H. & Nee, S. (1995) Sexual selection as a cause of taxonomic diversity in passerine birds. *Proc Roy Soc Lond B*, **259**, 211–15.

Barrett, S.C.H. & Harder, L.D. (1996) Ecology and evolution of plant mating. *Trends Ecol Evol*, **11**, 73–9.

Barton, N.H. (1995) A general model for the evolution of recombination. *Genet Res*, **65**, 123–44.

Barton, N.H. & Charlesworth, B. (1984) Genetic revolutions, founder effects and speciation. *Ann Rev Ecol Syst*, **15**, 133–64.

Bartz, S.H. (1979) Evolution of eusociality in termites. *Proc Nat Acad Sci USA*, **76**, 5764–8.

Basolo, A.L. (1990a) Female preference for male sword length in the green swordtail, *Xiphophorus helleri* (Pisces: Poeciliidae). *Anim Behav*, **40**, 332–8.

Basolo, A.L. (1990b) Female preference predates the evolution of the sword in swordtail fish. *Science*, **250**, 808–10.

Basolo, A.L. (1995a) A further examination of a pre-existing bias favouring a sword in the genus *Xiphophorus*. *Anim Behav*, **50**, 365–75.

Basolo, A.L. (1995b) Phylogenetic evidence for the role of a pre-existing bias in sexual selection. *Proc Roy Soc Lond B*, **259**, 307–11.

Bateman, A.J. (1948) Intra-sexual selection in *Drosophilia*. *Heredity*, **2**, 349–68.

Bateson, M. & Kacelnik, A. (1995) Accuracy of memory for amount in the foraging starling, *Sturnus vulgaris*. *Anim Behav*, **50**, 431–43.

Bateson, P.P.G. (1978) Sexual imprinting and optimal outbreeding. *Nature*, **273**, 659–60.

Bateson, P.P.G. (1979) How do sensitive periods arise and what are they for? *Anim Behav*, **27**, 470–86.

Beauchamp, G. & Giraldeau, L.-A. (1996) Group foraging revisited: information-sharing or producer-scrounger game? *Am Natur*, **148**, 738–43.

Beauchamp, G. & Kacelnik, A. (1991) Effects of the knowledge of partners on learning rates in zebra finches *Taeniopygia guttata*. *Anim Behav*, **41**, 247–53.

Beauchamp, G.K., Yamazaki, K., Bard, J. & Boyse, E.A. (1988) Preweaning experience in the control of mating preferences by genes in the major histocompatibility complex of the mouse. *Behav Genet*, **18**, 537–47.

Bednekoff, P.A., Biebach, H. & Krebs, J.R. (1994) Great tit fat reserves under unpredictable temperatures. *J Avian Biol*, **25**, 156–60.

Bednekoff, P.A. & Houston, A.I. (1994) Optimizing fat reserves over the entire winter: a dynamic model. *Oikos*, **71**, 408–15.

Bednekoff, P.A. & Krebs, J.R. (1995) Great tit fat reserves: effects on changing and unpredictable feeding day length. *Funct Ecol*, **9**, 457–62.

Beecher, M.D. (1988) Kin recognition in birds. *Behav Genet*, **18**, 465–82.

Beecher, M.D. (1991) Successes and failures of parent–offspring recognition in animals. In: *Kin Recognition* (ed. P.G. Hepper), pp. 94–124. Cambridge University Press, New York.

Beecher, M.D., Beecher, I.M. & Hahn, S. (1981a) Parent–offspring recognition in bank swallows (*Riparia riparia*): II. Development and acoustic basis. *Anim Behav*, **29**, 95–101.

Beecher, M.D., Beecher, I.M. & Lumpkin, S. (1981b) Parent–offspring recognition in bank swallows *Riparia riparia*: I. Natural history. *Anim Behav*, **29**, 86–94.

Bell, G. (1978) The handicap principle in sexual selection. *Evolution*, **32**, 872–85.

Bell, G. (1982) *The Masterpiece of Nature — the Evolution and Genetics of Sexuality*. University of California Press, Berkeley.

Belovsky, G.E. (1978) Diet optimization in a generalist herbivore: the moose. *Theor Popul Biol*, **14**, 105–34.

Benford, F.A. (1978) Fisher's theory of the sex ratio applied to the social Hymenoptera. *J Theor Biol*, **72**, 701–27.

Bengtsson, B.O. (1978) Avoiding inbreeding: at what cost? *J Theor Biol*, **73**, 439–44.

Benirschke, K., Anderson, J.M. & Brownhill, L.E. (1962) Marrow chimerism in marmosets. *Science*, **138**, 513–15.

Benne, R. (1994) RNA editing in trypanosomes. *Eur J Biochem*, **221**, 9–23.

Bennett, P.M. & Harvey, P.H. (1988) Active and restive metabolism in birds: allometry, phylogeny and ecology. *J Zool (Lond)*, **213**, 327–63.

Bennett-Clark, H.C. (1971) Acoustics of insect song. *Nature*, **234**, 255–9.

Benton, T.G. & Foster, W.A. (1992) Altruistic housekeeping in a social aphid. *Proc Roy Soc Lond B*, **247**, 199–202.

Bercovitch, F.B. (1988) Coalitions, cooperation and reproductive tactics among adult male baboons. *Anim Behav*, **36**, 1198–209.

Berg, E.E. & Hamrick, J.L. (1995) Fine-scale genetic structure of a turkey oak forest. *Evolution*, **49**, 110–20.

Berthold, P. (1993) *Bird Migration*. Oxford University Press, Oxford.

Berthold, P., Helbig, A.J., Mohr, G. & Querner, U. (1992) Rapid microevolution of migratory behaviour in a wild bird species. *Nature*, **360**, 668–70.

Bestor, T.H. (1990) DNA methylation: evolution of a bacterial immune function into a regulator of gene expression and genome structure in higher eukaryotes. *Phil Trans Roy Soc Lond B*, **326**, 179–87.

Bestor, T.H. & Tycko, B. (1996) Creation of genomic methylation patterns. *Nature Genet*, **12**, 363–7.

Bianchi, D.W., Zickwolf, G.K., Weil, G.J., Sylvester, S. & DeMaria, M.A. (1996) Male fetal progenitor cells persist in maternal blood for as long as 27 years postpartum. *Proc Nat Acad Sci USA*, **93**, 705–8.

Bird, A.P. (1993) Functions for DNA methylation in vertebrates. *Cold Spring Harbor Symp Quant Biol*, **58**, 281–5.

Bird, A.P. (1995) Gene number, noise reduction and biological complexity. *Trends Genet*, **11**, 94–100.

Birkhead, T.R. (1995) Sperm competition: evolutionary causes and consequences. In: *'Seventh International Symposium on Spermatology: Plenary Papers'. Reproduction, Fertility and Development*, **7**, 755–75.

Birkhead, T.R. (1996) Sperm competition: evolution and mechanisms. *Curr Topics Develop Biol*, **33**, 103–58.

Birkhead, T.R. & Fletcher, F. (1995a) Male phenotype and ejaculate quality in the zebra finch *Taeniopygia guttata*. *Proc Roy Soc Lond B*, **262**, 329–34.

Birkhead, T.R. & Fletcher, F. (1995b) Depletion determines sperm numbers in male zebra finches. *Anim Behav*, **48**, 451–6.

Birkhead, T.R. & Møller, A.P. (1992) *Sperm Competition in Birds: Evolutionary Causes and Consequences*. Academic Press, London.

Birkhead, T.R. & Møller, A.P. (1993) Why do male birds stop copulating while their partners are still fertile? *Anim Behav*, **45**, 105–18.

Birkhead, T.R. & Møller, A.P. (1996) Monogamy and sperm competition in birds. In: *Partnerships in Birds: the Ecology of Monogamy* (ed. J.M. Black), Oxford University Press, Oxford.

Birkhead, T.R. & Pringle, S. (1986) Multiple mating and paternity in *Gammarus pulex*. *Anim Behav*, **34**, 611–13.

Birkhead, T.R., Pellatt, E.J. & Fletcher, F. (1993) Selection and utilization of spermatozoa in the reproductive tract of the female zebra finch *Taeniopygia guttata*. *J Reprod Fert*, **99**, 593–600.

Birkhead, T.R., Pellatt, J.E. & Hunter, F.M. (1988) Extra-pair copulation and sperm competition in the zebra finch. *Nature Lond*, **334**, 60–2.

Birkhead, T.R., Burke, T., Zann, R., Hunter, F.M. & Krupa, A.P. (1990) Extra-pair paternity and intraspecific brood parasitism in wild zebra finches *Taeniopygia guttata*, revealed by DNA fingerprinting. *Behav Ecol Sociobiol*, **27**, 315–24.

Birkhead, T.R., Wishart, G.J. & Biggins, J.D. (1995a) Sperm precedence in the domestic fowl. *Proc Roy Soc Lond B*, **261**, 285–9.

Birkhead, T.R., Fletcher, F., Pellatt, E.J. & Staples, A. (1995b) Ejaculate quality and the success of extra-pair copulations in the zebra finch. *Nature (Lond)*, **377**, 422–3.

Birks, S.M. (1996) *Reproductive behavior and paternity in the Australian brush turkey,* Alectura lathami. PhD dissertation, Cornell University, Ithaca, New York.

Bjorntörp, P. & Brodoff, B.N. (1992) *Obesity*. J.B. Lippincott Company, Philadelphia.

Black, J.M. & Owen, M. (1989) Agonistic behaviour in barnacle goose flocks: assessment, investment and reproductive success. *Anim Behav*, **37**, 199–209.

Black, J.M., Carbone, C., Wells, R.L. & Owen, M. (1991) Foraging dynamics in goose flocks: the cost of living on the edge. *Anim Behav*, **44**, 41–50.

Blaustein, A.R. & O'Hara, R.K. (1982) Kin recognition in *Rana cascadae* tadpoles: maternal and paternal effects. *Anim Behav*, **30**, 1151–7.

Blaustein, A.R. & Waldman, B. (1992) Kin recognition in anuran amphibians. *Anim Behav*, **44**, 207–21.

Blaustein, A.R., Bekoff, M. & Daniels, T.J. (1987) Kin recognition in vertebrates (excluding primates): empirical evidence. In: *Kin Recognition in Animals* (eds D.J.C. Fletcher & C.D. Michener), pp. 287–331. John Wiley & Sons, Chichester.

Blem, C.R. (1990) Avian energy storage. *Curr Ornithol*, **7**, 59–113.

Boomsma, J.J. (1989) Sex-investment ratios in ants: has female bias been systematically overestimated? *Am Natur*, **133**, 517–32.

Boomsma, J.J. & Grafen, A. (1990) Intraspecific variation in ant sex ratios and the Trivers–Hare hypothesis. *Evolution*, **44**, 1026–34.

Boomsma, J.J. & Grafen, A. (1991) Colony-level sex ratio selection in the eusocial Hymenoptera. *J Evol Biol*, **3**, 383–407.

Boomsma, J.J., Keller, L. & Nielson, M.G. (1995) A comparative analysis of sex ratio investment parameters in ants. *Funct Ecol*, **9**, 743–53.

Boorman, E. & Parker, G.A. (1976) Sperm (ejaculate) competition in *Drosophila melanogaster*, and the reproductive value of females to males in relation to female age and mating status. *Ecol Entomol*, **1**, 145–55.

Borgia, G. (1985) Bower quality, number of decorations and mating success of male satin bowerbirds (*Ptilonorhynchus violaceus*): an experimental analysis. *Anim Behav*, **33**, 266–71.

Borgia, G. & Collis, K. (1989) Female choice for parasite-free male satin bowerbirds and the evolution of bright male plumage. *Behav Ecol Sociobiol*, **25**, 445–54.

Borowsky, R.L., McClelland, M., Cheng, R. & Welsh, J. (1995) Arbitrarily primed DNA fingerprinting for phylogenetic reconstruction in vertebrates: the *Xiphophorus* model. *Mol Biol Evol*, **12**, 1022–32.

Bourke, A.F.G. (1988) Worker reproduction in the higher eusocial Hymenoptera. *Quart Rev Biol*, **63**, 291–311.

Bourke, A.F.G. (1989) Comparative analysis of sex-investment ratios in slave-making ants. *Evolution*, **43**, 913–18.

Bourke, A.F.G. (1994) Worker matricide in social bees and wasps. *J Theor Biol*, **167**, 283–92.

Bourke, A.F.G. & Franks, N.R. (1995) *Social Evolution in Ants*. Princeton University Press, Princeton, New Jersey.

Bourke, A.F.G. & Heinze, J. (1994) The ecology of communal breeding: the case of multiple-queen leptothoracine ants. *Phil Trans Roy Soc Lond B*, **345**, 359–72.

Bourne, G.R. (1993) Proximate costs and benefits of mate acquisition at leks of the frog *Ololygon rubra*. *Anim Behav*, **45**, 1051–9.

Boyce, E.A., Beauchamp, G.K., Yamazaki, K. & Bard, J. (1991) Genetic components of kin recognition in mammals. In: *Kin Recognition* (ed. P.G. Hepper), pp. 148–61. Cambridge University Press, Cambridge.

Boyd, R. & Lorberbaum, J.P. (1987) No strategy is evolutionarily stable in the repeated prisoner's dilemma game. *Nature Lond*, **327**, 58–9.

Boyd, R. & Richerson, P. (1985) *Culture and the Evolutionary Process*. University of Chicago Press, Chicago.

Boyd, R. & Richerson, P. (1988a) An evolutionary model of social learning: the effects of spatial and temporal variation. In: *Social Learning: Psychological and Biological Perspectives* (eds T.R. Zentall & B.G. Galef Jr) pp. 29–48. Lawrence-Erlbaum Associations, Hillsdale, New Jersey.

Boyd, R. & Richerson, P.J. (1988b) The evolution of reciprocity in sizable groups. *J Theor Biol*, **132**, 337–56.

Boyd, R. & Richerson, P.J. (1992) Punishment allows the evolution of cooperation (or anything else) in sizable groups. *Ethol Sociobiol*, **13**, 171–95.

Breden, F. & Stoner, G. (1987) Male predation risk determines female preference in the Trinidad guppy. *Nature*, **329**, 831–3.

Breden, F. & Wade, M.J. (1991) 'Runaway' social evolution: reinforcing selection for inbreeding and altruism. *J Theor Biol*, **153**, 323–37.

Breed, M.D. (1983) Nestmate recognition in honey bees. *Anim Behav*, **31**, 86–91.

Breed, M.D., Garry, M.F., Pearce, A.N., Hibbard, B.E., Bjostad, L.B. & Page, R.E. Jr (1995) The role of wax comb in honey bee nestmate recognition. *Anim Behav*, **50**, 489–96.

Breed, M.D., Snyder, L.E., Lynn, T.L. & Morhart, J.A. (1992) Acquired chemical camouflage in a tropical ant. *Anim Behav*, **44**, 519–23.

Brillard, J.P., Galut, O. & Nys, Y. (1987) Possible causes of subfertility in hens following insemination near the time of ovulation. *Brit Poult Sci*, **28**, 307–18.

Brinkhof, M.W.G. (1995) *Timing of reproduction. An experimental study in coots*. PhD thesis, University of Groningen, Groningen.

Briskie, J.V., Naugler, C.T. & Leech, S.M. (1994) Begging intensity of nestling birds varies with sibling relatedness. *Proc Roy Soc Lond B*, **258**, 73–8.

Brockmann, H.J. (1984) The evolution of social behaviour in insects. In: *Behavioural Ecology: an Evolutionary Approach*, 2nd edn (eds J.R. Krebs & N.B. Davies), pp. 340–61. Blackwell Scientific Publications, Oxford.

Brockmann, H.J., Colson, T. & Potts, W. (1994) Sperm competition in horseshoe crabs (*Limulus polyphemus*). *Behav Ecol Sociobiol*, **35**, 153–60.

Brodbeck, D.R. (1994) Memory for spatial and local cues: a comparison of a storing and nonstoring species. *Anim Learn Behav*, **22**, 119–33.

Brodbeck, D.R. & Shettleworth, S.J. (1995) Matching location and color of a compound stimulus: comparison of a food storing and non-storing bird species. *J Exper Psychol Anim Behav Proces*, **21(1)**, 64–77.

Bronson, F.H. & Kerbeshian, M.C. (1995) Reactions of reproductively photoresponsive versus unresponsive meadow voles to simulated winter conditions. *Can J Zool*, **73**, 1479–88.

Brooks, R. & Caithness, N. (1995) Female choice in a feral guppy population: are there multiple cues? *Anim Behav*, **50**, 301–7.

Brown, J.L. (1969a) The buffer effect and productivity in tit populations. *Am Natur*, **103**, 347–54.

Brown, J.L. (1969b) Territorial behavior and population regulation in birds: a review and re-evaluation. *Wilson Bull*, **81**, 293–329.

Brown, J.L. (1978) Avian communal breeding systems. *Ann Rev Ecol Syst*, **9**, 123–55.

Brown, J.L. (1983) Some paradoxical goals of cells and organisms: the role of MHC. In: *Ethical Questions in Brain and Behavior* (ed. D.W. Pfaff), pp. 111–24. Springer, New York.

Brown, J.L. (1987) *Helping and Communal Breeding in Birds*. Princeton University Press, Princeton, NJ.

Brown, J.L. & Brown, E.R. (1990) Mexican jays: uncooperative breeding. In: *Cooperative Breeding in Birds: Long-term Studies of Ecology and Behavior* (eds P.B. Stacey & W.D. Koenig), pp. 267–88. Cambridge University Press, Cambridge.

Brown, J.L. & Ecklund, A. (1994) Kin recognition and the major histocompatibility complex: an integrative review. *Am Natur*, **143**, 435–61.

Brown, W.L. & Wilson, E.O. (1959) The evolution of

the dacetine ants. *Quart Rev Biol*, **34**, 278–94.

Brunner, D., Kacelnik, A. & Gibbon, J. (1992) Optimal foraging and timing processes in the starling, *Sturnus vulgaris*: effect of inter-capture interval. *Anim Behav*, **44**, 597–613.

Bryant, D.M. & Newton, A.V. (1994) Metabolic costs of dominance in dippers, *Cinclus cinclus*. *Anim Behav*, **48**, 447–55.

Bryant, D.M. & Tatner, P. (1991) Intraspecies variation in energy expenditure: correlates and constraints. *Ibis*, **133**, 236–45.

Buchholz, R. (1995) Female choice, parasite load and male ornamentation in wild turkeys. *Anim Behav*, **50**, 929–43.

Buckle, G.R. & Greenberg, L. (1981) Nestmate recognition in sweat bees (*Lasioglossum zephyrum*): does an individual recognize its own odour or only odours of nestmates? *Anim Behav*, **29**, 802–9.

Bull, C.D., Metcalfe, N.B. & Mangel, M. (1996) Seasonal matching of foraging and anticipated energy requirements in anorexic juvenile salmon. *Proc Roy Soc Lond B*, **263**, 13–18.

Bulmer, M.G. (1983) Sex ratio theory in social insects with swarming. *J Theor Biol*, **100**, 329–39.

Burke, T. & Bruford, M.W. (1987) DNA fingerprinting in birds. *Nature Lond*, **327**, 149–52.

Burrows, M.T. (1994) An optimal foraging and migration model for juvenile plaice. *Evol Ecol*, **8**, 125–49.

Burt, A. (1995) The evolution of fitness. *Evolution*, **49**, 1–8.

Buss, L.W. (1982) Somatic cell parasitism and the evolution of somatic tissue compatibility. *Proc Nat Acad Sci USA*, **79**, 5337–41.

Butcher, G.S. & Rohwer, S. (1989) The evolution of conspicuous and distinctive coloration for communication in birds. *Curr Ornithol*, **6**, 51–108.

Butlin, R.K. (1990) Divergence in emergence time of host races due to differential gene flow. *Heredity*, **65**, 47–50.

Butlin, R.K. (1993) The variability of mating signals and preferences in the brown planthopper, *Nilaparvata lugens* (Homoptera: Delphacidae). *J Insect Behav*, **6**, 125–40.

Butlin, R.K. (1995a) Reinforcement — an idea evolving. *Trends Ecol Evol*, **10**, 432–4.

Butlin, R.K. (1995b) Genetic variation in mating signals and responses. In: *Speciation and the Recognition Concept: Theory and Applications* (eds D.M. Lambert & H.G. Spencer), pp. 327–66. Johns Hopkins University Press, Baltimore and London.

Butlin, R.K. & Hewitt, G.M. (1985a) A hybrid zone between *Chorthippus parallelus parallelus* and *Chorthippus parallelus erythropus* (Orthoptera: Acrididae) I. Morphological and electrophoretic characters. *Biol J Linn Soc*, **26**, 269–85.

Butlin, R.K. & Hewitt, G.M. (1985b) A hybrid zone between *Chorthippus parallelus parallelus* and *Chorthippus parallelus erythropus* (Orthoptera: Acrididae) II. Behavioural characters. *Biol J Linn Soc*, **26**, 287–99.

Butlin, R.K. & Ritchie, M.G. (1991) Variation in female mate preference across a grasshopper hybrid zone. *J Evol Biol*, **4**, 227–40.

Byers, J.A. & Bekoff, M. (1986) What does 'kin recognition' mean? *Ethology*, **72**, 342–5.

Bygott, J.D. (1979) Agonistic behavior, dominance, and social structure in wild chimpanzees of the Gombe National Park. In: *The Great Apes* (eds D.A. Hamburg & E.R. McCown), pp. 405–27. Benjamin/Cummings, Menlo Park, California.

Byrne, R.W. & Tomasello, M. (1995) Do rats ape? *Anim Behav*, **50**, 1417–20.

Cade, W.H. (1979) The evolution of alternative male reproductive strategies in field crickets. In: *Sexual Selection and Reproductive Competition in Insects* (eds M.S. Blum & N.A. Blum), pp. 343–79. Academic Press, London.

Caldwell, R.L. (1992) Recognition, signalling and reduced aggression between former mates in a stomatopod. *Anim Behav*, **44**, 11–19.

Caraco, T. (1981) Risk-sensitivity and foraging groups. *Ecology*, **62**, 527–34.

Caraco, T. & Brown, J.L. (1986) A game between communal breeders: when is food-sharing stable? *J Theor Biol*, **118**, 379–93.

Caraco, T. & Giraldeau, L.-A. (1991) Social foraging: producing and scrounging in a stochastic environment. *J Theor Biol*, **153**, 559–83.

Carlin, N.F. (1989) Discrimination between and within colonies of social insects: two null hypotheses. *Nether J Zool*, **39**, 86–100.

Caro, T.M. (1994) *Cheetahs of the Serengeti Plains: Group Living in an Asocial Species*. University of Chicago Press, Chicago.

Carpenter, J.M. (1989) Testing scenarios: wasp social behaviour. *Cladistics*, **5**, 131–44.

Carson, H.L. (1986) Sexual selection and speciation. In: *Evolutionary Process and Theory* (eds S. Karlin & E. Nevo), pp. 391–409. Academic Press, Orlando.

Carson, H.L. & Templeton, A.R. (1984) Genetic revolutions in relation to speciation phenomena: the founding of new populations. *Ann Rev Ecol Syst*, **15**, 97–131.

Carson, H.L., Lockwood, J.P. & Craddock, E.M. (1990) Extinction and recolonisation of local populations on a growing shield volcano. *Proc Nat Acad Sci USA*, **87**, 7055–7.

Cartwright, B.A. & Collett, T.S. (1983) Landmark learning in bees. Experiments and models. *J Comp Physiol*, **151**, 521–43.

Catchpole, C.K. (1987) Bird song, sexual selection and female choice. *Trends Ecol Evol*, **2**, 94–7.

Catchpole, C.K. & Slater, P.J.B. (1995) *Bird Song. Biological Themes and Variations*. Cambridge University Press, New York.

Cavalli-Sforza, L.L. & Feldman, M.W. (1981) *Cultural Transmission and Evolution: a Quantitative Approach*. Princeton University Press, Princeton.

Chan, G.L. & Bourke, A.F.G. (1994) Split sex ratios in a multiple-queen ant population. *Proc Roy Soc Lond B*, **258**, 261–6.

Chapais, B. (1983) Dominance, relatedness and structure of female rhesus relationships. In: *Primate Social Relationships* (ed. R.A. Hinde), pp. 208–17. Sinauer, Sunderland, Massachusetts.

Chapais, B. (1988a) Rank maintenance in female Japanese macaques: experimental evidence for social dependency. *Behaviour*, **104**, 41–59.

Chapais, B. (1988b) Experimental matrilineal inheritance of rank in female Japanese macaques. *Anim Behav*, **36**, 1025–37.

Chapais, B. (1992) The role of alliances in social inheritance of rank among female primates. In: *Coalitions and Alliances in Humans and other Animals* (eds A.H. Harcourt & F.B.M. de Waal), pp. 29–59. Oxford University Press, Oxford.

Chapais, B., Girard, M. & Primi, G. (1991) Non-kin alliances and the stability of matrilineal dominance relations in Japanese macaques. *Anim Behav*, **41**, 481–91.

Chapman, T., Liddle, L.F., Kalb, J.M., Wolfner, M.F. & Partridge, L. (1995) Cost of mating in *Drosophila melanogaster* females is mediated by male accessory gland products. *Nature Lond*, **373**, 241–4.

Charlesworth, D. (1985) Distribution of dioecy and self-incompatibility in angiosperms. In: *Evolution: Essays in Honor of John Maynard Smith* (eds P.J. Greenwood, P.H. Harvey & M. Slatkin), pp. 237–68. Cambridge University Press, Cambridge.

Charnov, E.L. (1976a) Optimal foraging: attack strategy of a mantid. *Am Natur*, **110**, 141–51.

Charnov, E.L. (1976b) Optimal foraging: the marginal value theorem. *Theor Popul Biol*, **9**, 129–36.

Charnov, E.L. (1982) *The Theory of Sex Allocation*. Princeton University Press, Princeton, New Jersey.

Charnov, E.L. (1993) *Life History Invariants. Some Explorations of Symmetry in Evolutionary Ecology*. Oxford Series in Ecology and Evolution. Oxford University Press, Oxford.

Charnov, E.L. & Parker, G.A. (1995) Dimensionless invariants from foraging theory's marginal value theorem. *Proc Nat Acad Sci USA*, **92**, 1446–50.

Charlesworth, B. (1990) Mutation–selection balance and the evolutionary advantage of sex and recombination. *Genet Res*, **55**, 199–221.

Chase, I.D. (1974) Models of hierarchy formation in animal societies. *Behav Sci*, **19**, 374–82.

Chen, X.F. & Baur, B. (1993) The effect of multiple mating on female reproductive success in the simultaneously hermaphroditic land snail *Arianta arbustorum*. *Can J Zool*, **71**, 2431–6.

Cheney, D.L. (1977) The acquisition of rank and the development of reciprocal alliances among free-ranging immature baboons. *Behav Ecol Sociobiol*, **2**, 303–18.

Cheney, D.L. & Seyfarth, R.M. (1990) *How Monkeys See the World*. University of Chicago Press, Chicago.

Cheng, K.M., Burns, J.T. & McKinney, F. (1983) Forced copulation in captive mallards. III. Sperm competition. *Auk*, **100**, 302–10.

Chivers, D.P. & Smith, J.F. (1994) Fathead minnows, *Pimephales promelas*, acquire predator recognition when alarm substance is associated with the sight of unfamiliar fish. *Anim Behav*, **48**, 597–605.

Chivers, D.P. & Smith, R.J.F. (1995) Fathead minnows (*Pimephales promelas*) learn to recognize chemical stimuli from high-risk habitats by the presence of alarm substance. *Behav Ecol*, **6**, 155–8.

Choe, J.C. (1988) Worker reproduction and social evolution in ants (Hymenoptera: Formicidae). In: *Advances in Myrmecology* (ed. J.C. Trager), pp. 163–87. E.J. Brill, Leiden.

Christenfeld, N.J.S. & Hill, E.A. (1995) Whose baby are you? *Nature*, **378**, 669.

Christy, J.H. (1995) Mimicry, mate choice, and the sensory trap hypothesis. *Am Natur*, **146**, 171–81.

Christy, J.H. & Salmon, M. (1991) Comparative studies of reproductive behavior in mantis shrimp and fiddler crabs. *Am Zool*, **31**, 329–37.

Churchland, P.S., Ramachandran, V.S. & Sejnowski, T.J. (1994) A critique of pure vision. In: *Large-Scale Neuronal Theories of the Brain* (eds C. Koch & J.H.L. Davis), pp. 23–60. MIT Press, Cambridge.

Clark, A.G., Aguade, M., Prout, T., Harshman, L.G. & Langley, C.H. (1995) Variation in sperm displacement and its association with accessory gland protein in *Drosophila melanogaster*. *Genetics*, **139**, 189–201.

Clark, C.W. & Levy, D.A. (1988) Diel vertical migrations by juvenile sockeye salmon and the antipredation window. *Am Natur*, **131**, 271–90.

Clark, C.W. & Mangel, M. (1984) Foraging and flocking strategies: information in an uncertain environment. *Am Natur*, **123**, 626–41.

Clark, C.W. & Mangel, M. (1986) The evolutionary advantages of group foraging. *Theor Popul Biol*, **30**, 45–75.

Clark, D.L. & Uetz, G.W. (1993) Signal efficacy and the evolution of male dimorphism in the jumping spider, *Maevia inclemens*. *Proc Nat Acad Sci USA*, **90**, 11 954–7.

Clarke, M.F. (1984) Co-operative breeding by the Australian Bell miner, *Manorina melanophrys* Latham: a test of kin selection theory. *Behav Ecol Sociobiol*, **14**, 137–46.

Clarke, M.F. (1989) The pattern of helping in the Bell miner *Manorina melanophrys*. *Ethology*, **80**, 292–306.

Clarke, R.T. & Goss-Custard, J.D. (1996) The Exe estuary oystercatcher–mussel model. In: *The Oystercatcher: from Individuals to Populations* (ed. J.D. Goss-Custard), pp. 390–93. Oxford University Press, Oxford.

Clayton, D.L. (1990) Mate choice in experimentally parasitized rock doves: lousy males lose. *Am Zool*, **30**, 251–62.

Clayton, N.S. (1995a) The neuroethological development of food-storing memory: a case of use it or lose it! *Behav Brain Res*, **70**, 95–102.

Clayton, N.S. (1995b) Development of memory and the hippocampus: comparison of food-storing and nonstoring birds on a one-trial associative memory task. *J Neurosci*, **15**, 2796–807.

Clayton, N.S. & Krebs, J.R. (1993) Lateralization in the Paridae: comparison of a storing and a non-storing species on a one-trial associative memory task. *J Comp Physiol A*, **171**, 807–15.

Clayton, N.S. & Krebs, J.R. (1994a) Lateralization and unilateral transfer of spatial memory in marsh tits: are two eyes better than one? *J Comp Physiol A*, **174**, 769–73.

Clayton, N.S. & Krebs, J.R. (1994b) Memory for spatial and object-specific cues in food-storing and non-storing birds. *J Comp Physiol A*, **174**, 371–9.

Clayton, N.S. & Krebs, J.R. (1994c) Hippocampal growth and attrition in birds affected by experience. *Proc Nat Acad Sci USA*, **91**, 7410–14.

Clayton, N.S. & Krebs, J.R. (1994d) One-trial associative memory: comparisons of food-storing and non-storing species of birds. *Anim Learn Behav*, **22**, 366–72.

Clements, K.C. & Stephens, D.W. (1995) Testing models of non-kin cooperation: mutualism and the prisoner's dilemma. *Anim Behav*, **50**, 527–35.

Clutton-Brock, T.H. & Albon, S.D. (1979) The roaring of red deer and the evolution of honest advertisement. *Behaviour*, **69**, 145–70.

Clutton-Brock, T.H. & Harvey, P.H. (1984) Comparative approaches to investigating adaptation. In: *Behavioural Ecology: an Evolutionary Approach* (eds J.R. Krebs & N.B. Davies), 2nd edn, pp. 7–29. Blackwell Scientific Publications, Oxford.

Clutton-Brock, T.H. & Parker, G.A. (1992) Potential reproductive rates and the operation of sexual selection. *Quart Rev Biol*, **67**, 437–55.

Clutton-Brock, T.H. & Parker, G.A. (1995) Sexual coercion in animal societies. *Anim Behav*, **49**, 1345–65.

Clutton-Brock, T.H. & Parker, G.A. (1995) Punishment in animal societies. *Nature Lond*, **373**, 209–16.

Clutton-Brock, T.H. & Vincent, A.C.J. (1991) Sexual selection and the potential reproductive rates of males and females. *Nature Lond*, **352**, 58–60.

Coase, R.H. (1993) The nature of the firm (1937). In: *The Nature of the Firm* (eds O.E. Williamson & S.G. Winter), pp. 18–33. Oxford University Press, New York.

Cohet, Y. & David, J.R. (1980) Geographical divergence and sexual behaviour: comparison of mating systems in French and Afrotropical populations of *Drosophila melanogaster*. *Genetica*, **54**, 161–5.

Cole, B.J. (1986) The social behavior of *Leptothorax allardycei* (Hymenoptera, Formicidae): time budgets and the evolution of worker reproduction. *Behav Ecol Sociobiol*, **18**, 165–73.

Colegrave, N., Birkhead, T.R. & Lessells, C.M. (1995) Sperm precedence in zebra finches does not require special mechanisms of sperm competition. *Proc Roy Soc Lond B*, **259**, 223–8.

Colgan, P. (1983) *Comparative Social Recognition*. John Wiley & Sons, New York.

Collett, T.S. & Land, M.F. (1978) How hoverflies compute interception courses. *J Comp Physiol*, **125**, 191–204.

Collins, S.A. (1994) Male displays: cause or effect of female preference? *Anim Behav*, **48**, 371–5.

Collins, S.A. (1995) The effect of recent experience on female choice in zebra finches. *Anim Behav*, **49**, 479–86.

Collins, S.A., Hubbard, C. & Houtman, A.M. (1994) Female mate choice in the zebra finch: the effect of male beak colour and male song. *Behav Ecol Sociobiol*, **35**, 21–5.

Compton, M.M., Van Krey, H.P. & Siegel, P.B. (1978) The filling and emptying of the uterovaginal sperm-host glands in the domestic hen. *Poult Sci*, **57**, 1696–1700.

Conner, W.E., Roach, B., Benedict, E., Meinwald, J. & Eisner, T. (1990) Courtship pheromone production and body size as correlates of larval diet in males of the arctiid moth, *Utetheisa ornatrix*. *J Chem Ecol*, **16**, 543–52.

Cook, P.A. & Gage, M.J.G. (1995) Effects of risks of sperm competition on the numbers of eupyrene and apyrene sperm ejaculated by the male moth *Plodia interpunctella* (Lepidoptera: Pyralidae). *Behav Ecol Sociobiol*, **36**, 261–8.

Cooper, S.J.B., Ibrahim, K.M. & Hewitt, G.M. (1995) Postglacial expansion and subdivision of the grasshopper *Chorthippus parallelus* in Europe. *Mol Ecol*, **4**, 49–60.

Cooper, W.E., Vitt, L.J., Hedges, R. & Huey, R.B. (1990) Locomotor impairment and defence in gravid lizards (*Eumeces laticeps*): behavioural shift in activity may offset costs of reproduction in an active forager. *Behav Ecol Sociobiol*, **27**, 153–7.

Coopersmith, C.B. & Lenington, S. (1992) Female preferences based on male quality in house mice: interaction between male dominance rank and t-complex genotype. *Ethology*, **90**, 1–16.

Cornell, T.J., Berven, K.A. & Gamboa, G.J. (1989) Kin recognition by tadpoles and froglets of the wood frog *Rana sylvatica*. *Oecologia*, **78**, 312–16.

Cosmides, L.M. & Tooby, J. (1981) Cytoplasmic inheritance and intragenomic conflict. *J Theor Biol*, **89**, 83–129.

Cowie, R.J. (1977) Optimal foraging in great tits (*Parus major*). *Nature*, **268**, 137–9.

Cowie, R.J. & Krebs, J.R. (1979) Optimal foraging in patchy environments. *British Ecological Society Symposium on Population Dynamics* (eds R.M. Anderson, R.D. Turner & L.R. Taylor), pp. 183–205. Blackwell Scientific Publications, Oxford.

Craig, R. (1979) Parental manipulation, kin selection, and the evolution of altruism. *Evolution*, **33**, 319–34.

Creel, S.R. & Waser, P.M. (1991) Failures of reproductive suppression in Dwarf mongooses (*Helogale parvula*): accident or adaptation? *Behav Ecol Sociobiol*, **2**, 7–15.

Creel, S. & Creel, N.M. (1995) Communal hunting and pack size in African wild dogs *Lycaon pictus*. *Anim Behav*, **50**, 1325–39.

Creel, S., Creel, N. & Monfort, S.L. (1996) Social stress and dominance. *Nature Lond*, **379**, 212.

Creel, S., Creel, N., Wildt, D.E. & Monfort, S.L. (1992) Behavioural and endocrine mechanisms of reproductive suppression in Serengeti dwarf mongooses. *Anim Behav*, **43**, 231–45.

Creel, S.R., Monfort, S.L., Wildt, D.E. & Waser, P.M. (1991) Spontaneous lactation is an adaptive result of pseudo-pregnancy. *Nature*, **351**, 660–2.

Crespi, B.J. (1992) Eusociality in Australian gall thrips. *Nature*, **359**, 724–6.

Crespi, B.J. (1994) Three conditions for the evolution of eusociality: are they sufficient? *Insect Soc*, **41**, 395–400.

Crespi, B.J. & Yanega, D. (1995) The definition of eusociality. *Behav Ecol*, **6**, 109–15.

Cresswell, W. (1994) Song as a pursuit–deterrent signal, and its occurrence relative to other anti-predator behaviours of skylark (*Alauda arvensis*) on attack by merlins (*Falco columbarius*) *Behav Ecol Sociobiol*, **34**, 217–3.

Cronin, H. (1991) *The Ant and the Peacock*. Cambridge University Press, Cambridge.

Crook, J.H. (1964) The evolution of social organisation and visual communication in the weaver birds (Ploceinae). *Behaviour*, **10** (Suppl.), 1–178.

Crook, J.H. (1965) The adaptive significance of avian social organisation. *Symp Zool Soc Lond*, **14**, 181–218.

Crook, J.H. & Gartlan, J.S. (1966) Evolution of primate societies. *Nature Lond*, **210**, 1200–3.

Crozier, R.H. (1979) Genetics of sociality. In: *Social Insects*, Vol. I (ed. H.R. Hermann), pp. 223–86. Academic Press, New York.

Crozier, R.H. & Dix, M.W. (1979) Analysis of two genetic models for the innate components of colony odor in social Hymenoptera. *Behav Ecol Sociobiol*, **4**, 217–24.

Crozier, R.H. & Luykx, P. (1985) The evolution of termite eusociality is unlikely to have been based on a male-haploid analogy. *Am Natur*, **126**, 867–9.

Crozier, R.H. & Page, R.E. (1985) On being the right size: male contributions and multiple mating in social Hymenoptera. *Behav Ecol Sociobiol*, **18**, 105–15.

Crozier, R.H. & Pamilo, P. (1996) *Evolution of Social Insect Colonies: Sex Allocation and Kin Selection*. Oxford University Press, Oxford.

Cullen, E. (1957) Adaptations in the kittiwake to cliff nesting. *Ibis*, **99**, 275–302.

Cullen, J.M. (1966) Reduction of ambiguity through ritualization. *Phil Trans Roy Soc Lond B*, **251**, 363–4.

Curio, E. (1988) Cultural transmission of enemy recognition by birds. In: *Social Learning* (eds T.R. Zentall & B.G. Galef, Jr), pp. 75–97. Lawrence Erlbaum Associates, Hillsdale, New Jersey.

Curio, E. (1989) Some aspects of avian mortality patterns. *Mitteil Zool Museum Berl*, **65**(Suppl.), 47–70.

Curry, R.L. (1988) Influence of kinship on helping behavior in Galapagos mockingbirds. *Behav Ecol Sociobiol*, **22**, 141–52.

Cuthill, I.C. & Guilford, T. (1990) Perceived risk and obstacle avoidance in flying birds. *Anim Behav*, **40**, 188–90.

Cuthill, I.C. & Macdonald, W.A. (1990) Experimental manipulation of the dawn and dusk chorus in the blackbird *Turdus merula*. *Behav Ecol Sociobiol*, **26**, 209–16.

Cuthill, I.C., Haccou, P. & Kacelnik, A. (1994) Starlings (*Sturnus vulgaris*) exploiting patches: response to long-term changes in travel time. *Behav Ecol*, **5**, 81–90.

Cuthill, I.C., Kacelnik, A., Krebs, J.R., Haccou, P. & Iwasa, Y. (1990) Starlings exploiting patches: the effect of recent experience on foraging decisions. *Anim Behav*, **40**, 625–40.

Cwynar, L.C. & MacDonald, G.M. (1987) Geographical variation of lodgepole pine in relation to population history. *Am Natur*, **129**, 463–9.

Daan, S. & Aschoff, J. (1982) Circadian contributions to survival. In: *Vertebrate Circadian Systems* (eds J. Aschoff, S. Daan & G. Groos), pp. 305–21. Springer-Verlag, Berlin-Heidelberg.

Daan, S., Deerenberg, C. & Dijkstra, C. (1996) Increased daily work precipitates natural death in the kestrel. *J Anim Ecol*, **65**, 539–44.

Daan, S., Dijkstra, C. & Tinbergen, J.M. (1990a) Family planning in the kestrel (*Falco tinnunculus*): the ultimate control of covariation of laying date and clutch size. *Behaviour*, **114**, 83–116.

Daan, S., Masman, D. & Groenewold, A.A. (1990b) Avian basal metabolic rates: their association with body composition and energy expenditure in nature. *Am J Physiol*, **259**, R333–40.

Daan, S., Dijkstra, C., Drent, R. & Meijer, T. (1989a) Food supply and the annual timing of avian reproduction. In: *Acta XIX Congressus Internationalis Ornithologici*, Vol. I (ed. H. Ouellet), pp. 392–407. University of Ottawa Press, Ottawa.

Daan, S., Masman, D., Strijkstra, A.M. & Verhulst, S. (1989b) Intraspecific allometry of basal metabolic rate: relations with body size, temperature, composition and circadian phase in the kestrel, *Falco tinnunculus*. *J Biol Rhythms*, **4**, 267–83.

Dadour, I.R. & Bailey, W.J. (1990) The acoustic behaviour of males *Mygalopsis marki* (Copiphorinae; Tettigoniidae). In: *The Tettigoniidae: Behaviour, Systematics, Evolution* (eds W.J. Bailey & D.C.F. Rentz), pp. 98–111. Crawford House Press, Bathurst.

Dallas, J.F., Dod, B., Boursot, P., Prager, E.M. & Bonhomme, F. (1995) Population subdivision and gene flow in Danish house mice. *Mol Ecol*, **4**, 311–20.

Darwin, C. (1859) *On the Origin of Species by Means of Natural Selection*. Murray, London.

Darwin, C. (1871) *The Descent of Man, and Selection in Relation to Sex*. John Murray, London.

Darwin, C. (1883) *The Descent of Man and Selection in Relation to Sex*, 3rd edn. Murray, London.

Datta, S.B. (1983) Relative power and the acquisition of rank. In: *Primate Social Relationships* (ed. R.A. Hinde), pp. 93–103. Sinauer, Sunderland, Massachusetts.

Davis, L.I. (1995) The nuclear pore complex. *Ann Rev Biochem*, **64**, 865–96.

Davies, N.B. (1978) Territorial defence in the speckled wood butterfly (*Pararge aegeria*): the resident always wins. *Anim Behav*, **26**, 138–47.

Davies, N.B. (1992) *Dunnock Behaviour and Social Evolution*. Oxford University Press, Oxford.

Davies, N.B. & Brooke, M. de L. (1989) An experimental study of co-evolution between the cuckoo *Cuculus canorus* and its hosts. I. Host egg discrimination. *J Anim Ecol*, **58**, 207–24.

Davies, N.B. & Halliday, T.R. (1978) Deep croaks and fighting assessment in toads *Bufo bufo*. *Nature*, **274**, 683–5.

Davies, N.B., Brooke, M. de L. & Kacelnik, A. (1996) Recognition errors and probability of parasitism determine whether reed warblers should accept or reject mimetic cuckoo eggs. *Proc Roy Soc Lond B*, **263**, 925–31.

Davies, N.B., Hartley, I.R., Hatchwell, B.J. & Langmore, N.E. (1996) Female control of copulations to maximize male help: a comparison of polygynandrous alpine accentors, *Prunella collaris*, and dunnocks *P. modularis*. *Anim Behav*, **51**, 27–47.

Davies, N.B., Hatchwell, B.J., Robson, T. & Burke, T. (1992) Paternity and parental effort in dunnocks *Prunella modularis*: how good are male chick-feeding rules? *Anim Behav*, **43**, 729–45.

Dawkins, M. (1971) Perceptual changes in chicks: another look at the 'search image' concept. *Anim Behav*, **19**, 566–74.

Dawkins, M.S. & Guilford, T.C. (1991) The corruption of honest signalling. *Anim Behav*, **41**, 865–73.

Dawkins, M.S. & Guilford, T. (1995) An exaggerated preference for simple neural network models of signal evolution? *Proc Roy Soc Lond B*, **261**, 357–60.

Dawkins, R. (1976) *The Selfish Gene*. Oxford University Press, Oxford.

Dawkins, R. (1979) Twelve misunderstandings of kin selection. *Z Tierpsychol*, **51**, 184–200.

Dawkins, R. (1982) *The Extended Phenotype*. W.H. Freeman, San Francisco, California.

Dawkins, R. & Krebs, J.R. (1978) Animal signals: information or manipulation? In: *Behavioural Ecology: an Evolutionary Approach*, 1st edn (eds J.R. Krebs & N.B. Davies), pp. 282–309. Blackwell Scientific Publications, Oxford.

Dawson, A., Goldsmith, A.R. & Nicholls, T.J. (1985) Thyroidectomy results in termination of photorefractoriness in starlings (*Sturnus vulgaris*) kept in long day lengths. *J Reprod Fert*, **74**, 527–33.

Dawson, B.V. & Foss, B.M. (1965) Observational learning in budgerigars. *Anim Behav*, **13**, 470–4.

DeChiara, T.M., Robertson, E.J. & Efstratiadis, A. (1991) Parental imprinting of the mouse insulin-like growth factor II gene. *Cell*, **64**, 849–59.

Deerenberg, C. (1996) *Parental energy and fitness costs in birds*. PhD thesis, University of Groningen, Groningen.

Deerenberg, C., Pen, I., Dijkstra, C., Arkies, B.-J., Visser, G.H. & Daan, S. (1995) Parental energy expenditure in relation to manipulated brood size in the European kestrel *Falco tinnunculus*. *Zoology; Analysis Complex Systems* **99**, 39–48.

Dempson, J.B. & Green, J.M. (1985) Life history of anadromous arctic charr, *Salvelinus alpinus*, in the Fraser River, northern Labrador. *Can Zool*, **63**, 315–24.

Dennett, D.C. (1995) *Darwin's Dangerous Idea*. Simon & Schuster, New York.

Denno, R.F. (1994) The evolution of dispersal polymorphisms in insects: the influence of habitats, host plants and mates. *Res Popul Ecol*, **36**, 127–35.

DeSalle, R., Gatesy, J., Wheeler, W. & Grimaldi, D. (1992) DNA sequences from a fossil termite in Oligo-Miocene amber and their phylogenetic implications. *Science*, **257**, 1933–6.

Deslippe, R.J. & Savolainen, R. (1995) Sex investment in a social insect: the proximate role of food. *Ecology*, **76**, 375–82.

Devenport, J.A. & Devenport, L.D. (1993) Time-dependent decisions in dogs (*Canis familiaris*). *J Compar Psychol*, **107**, 169–73.

Devenport, L.D. & Devenport, J.A. (1994) Time-dependent averaging of foraging information in least chipmunks and golden-mantled ground squirrels. *Anim Behav*, **47**, 787–802.

Dewsbury, D.A. (1982) Ejaculate cost and male choice. *Am Natur*, **119**, 601–10.

Dewsbury, D.A. (1984) Sperm competition in murid rodents. In: *Sperm Competition and The Evolution of Animal Mating Systems* (ed. R.L. Smith), pp. 547–71. Academic Press, Orlando.

Diamond, J.M. (1993) Evolutionary physiology. In: *The Logic of Life* (eds C.A.R. Boyd & D. Noble), pp. 89–111. Oxford University Press, Oxford, New York.

Diehl, S.R. & Bush, G.L. (1989) The role of habitat preference in adaptation and speciation. In: *Speciation and its Consequences* (eds D. Otte & J.A. Endler), pp. 345–65. Sinauer, Sunderland, MA.

Diesel, R. (1990) Sperm competition and reproductive success in the decapod *Inachus phalangium* (Majidae): a male ghost spider crab that seals off rivals' sperm. *J Zool Lond*, **220**, 213–23.

Dijkstra, C., Bult, A., Bijlsma, S., Daan, S., Meijer, T. & Zijlstra, M. (1990) Brood size manipulations in the kestrel *Falco tinnunculus*: effects on offspring and adult survival. *J Anim Ecol*, **59**, 269–86.

Dixon, A., Ross, D., O'Malley, S.L.C. & Burke, T. (1994) Paternal investment inversely related to degree of extra-pair paternity in the reed bunting (*Emberiza schoeniclus*). *Nature (Lond)*, **371**, 698–700.

Doherty, J. & Hoy, R. (1985) The auditory behavior of crickets: some views of genetic coupling, song recognition, and predator detection. *Quart Rev Biol*, **60**, 457–72.

Dolman, P.M. (1995) The intensity of interference varies with resource density: evidence from a field study with snow buntings, *Plectrophenax nivalis*. *Oecologia*, **102**, 511–14.

Dominey, W.J. (1984) Effects of sexual selection and life histories on speciation: species flocks in African cichlids and Hawaiian *Drosophila*. In: *Evolution of Fish Species Flocks* (eds A.A. Echelle & I. Kornfield), pp. 231–49. Orono Press, Maine.

Douglas, R.H. & Djamgoz, M.B.A. (1990) *The Visual System of Fish*. Chapman & Hall, London, New York.

Dover, G.A. (1993) Evolution of genetic redundancy for advanced players. *Curr Opin Genet Develop*, **3**, 902–10.

Drent, R.H. & Daan, S. (1980) The prudent parent: energetic adjustments in avian breeding. *Ardea*, **68**, 225–52.

Drickamer, L.C. (1992) Oestrus female house mice discriminate dominant from subordinate males and sons of dominant from sons of subordinate males by odour cues. *Anim Behav*, **43**, 868–70.

Drummond, H. & Osorno, J.L. (1992) Training siblings to be submissive losers: dominance between booby nestlings. *Anim Behav*, **44**, 881–93.

Dugatkin, L.A. (1988) Do guppies play tit-for-tat during predator inspection visits? *Behav Ecol Sociobiol*, **23**, 395–9.

Dugatkin, L.A. (1991) Dynamics of the tit for tat strategy during predator inspection in guppies. *Behav Ecol Sociobiol*, **29**, 127–32.

Dugatkin, L.A. (1992a) Tendency to inspect predators predicts mortality risk in the guppy (*Poecilia reticulata*). *Behav Ecol*, **3**, 124–7.

Dugatkin, L.A. (1992b) Sexual selection and imitation: females copy the mate choice of others. *Am Natur*, **139**, 1384–9.

Dugatkin, L.A. (1996a) Tit for tat, by-product mutualism and predator inspection: a reply to Connor. *Anim Behav*, **51**, 455–7.

Dugatkin, L.A. (1996b) The interface between culturally-based preferences and genetic preferences: female mate choice in *Poecilia reticulata*. *Proc Nat Acad Sci USA*, **93**, 2770–3.

Dugatkin, L.A. & Godin, J.-G.J. (1992) Reversal of female mate choice by copying in the guppy (*Poecilia reticulata*). *Proc Roy Soc Lond B*, **249**, 179–84.

Dugatkin, L.A. & Godin, J.-G.J. (1993) Female mate copying in the guppy (*Poecilia reticulata*): age-dependent effects. *Behav Ecol Sociobiol*, **4**, 289–92.

Dugatkin, L.A., Mesterton-Gibbons, M. & Houston, A.I. (1992) Beyond the prisoner's dilemma: towards models to discriminate among mechanisms of cooperation in nature. *Trends Ecol Evol*, **7**, 202–5.

Duquesnoy, B. & Filipo, R.M. (1994) Polyarthrite rhumatoide et neoplasies. *Rev Rhum Ed Fr*, **61**, 194–7.

Dukas, R. & Clark, C.W. (1995) Searching for cryptic prey: a dynamic model. *Ecology*, **76**, 1320–6.

Dukas, R. & Ellner, S. (1993) Information processing and prey detection. *Ecology*, **74**, 1337–46.

Dukas, R. & Real, L.A. (1993) Learning constraints and floral choice in bumble bees. *Anim Behav*, **46**, 637–44.

Dunbar, R.I.M. (1988) *Primate Social Systems*. Croom Helm, London.

Durell, S.E.A. Le V. dit, Goss-Custard, J.D. & Caldow, R.W.G. (1993) Sex-related differences in diet and feeding method in the oystercatcher *Haematopus ostralegus*. *J Anim Ecol*, **62**, 205–15.

Dusenbery, D.B. (1992) *Sensory Ecology*. Freeman, New York.

Dyer, F.C. (1996) Spatial memory and navigation by honeybees on the scale of the foraging range. *J Exp Biol*, **199**, 147–54.

Ebbinge, B. (1989) A multifactorial explanation for variation in breeding performance of brent geese *Branta bernicla*. *Ibis*, **131**, 196–204.

Ebbinge, B. (1991) The impact of hunting on mortality rates and spatial distribution of geese wintering in the Western Palearctic. *Ardea*, **79**, 197–210.

Eberhard, W.G. (1980) Evolutionary consequences of intracellular organelle competition. *Quart Rev Biol*, **55**, 231–49.

Eberhard, W.G. (1996) *Female Control: Sexual Selection by Cryptic Female Choice*. Princeton University Press, Princeton.

Egid, K. & Brown, J.L. (1989) The major histocompatibility complex and female mating preferences in mice. *Anim Behav*, **38**, 548–50.

Ehlinger, T.J. (1989) Learning and individual variation in bluegill foraging: habitat-specific techniques. *Anim Behav*, **38**, 643–58.

Eisner, T. & Meinwald, J. (1995) The chemistry of sexual selection. *Proc Nat Acad Sci USA*, **92**, 50–5.

Ekman, J.B. (1987) Exposure and time use in willow tit flocks: the cost of subordination. *Anim Behav*, **35**, 445–52.

Ekman, J.B. & Hake, M.K. (1990) Monitoring starvation risk: adjustments of body reserves in greenfinches *Cardulis chloris* during periods of unpredictable foraging success. *Behav Ecol*, **1**, 62–7.

Ekman, J.B. & Lilliendahl, K. (1993) Using priority to food access: fattening strategies in dominance-structured willow tit (*Parus montanus*) flocks. *Behav Ecol*, **4**, 232–8.

Elgar, M.A. & Harvey, P.H. (1987) Basal metabolic rates in mammals: Allometry, phylogeny and ecology. *Funct Ecol*, **1**, 25–38.

Ellis, H.I. (1984) Energetics of free-ranging seabirds. In: *Seabird Energetics*, (eds G.C. Whittow & H. Rahn), pp. 203–34. Plenum Press, New York.

Elton, C. (1927) *Animal Ecology*. Macmillan, New York.

Emlen, S.T. (1978) The evolution of cooperative breeding in birds. In: *Behavioural Ecology: an evolutionary approach*, (eds J.R. Krebs & N.B. Davies), pp. 245–81. Blackwell Scientific Publications, Oxford.

Emlen, S.T. (1982a) The evolution of helping. II. The role of behavioral conflict. *Am Natur*, **119**, 40–53.

Emlen, S.T. (1982b) The evolution of helping. I. An ecological constraints model. *Am Natur*, **119**, 29–39.

Emlen, S.T. (1984) Cooperative breeding in birds and mammals. In: *Behavioral Ecology: an Evolutionary Approach*, 2nd edn (eds J. Krebs & N. Davies), pp. 305–35. Blackwell Scientific Publications, Oxford.

Emlen, S.T. (1990) The white-fronted bee-eater: helping in a colonially nesting species. In: *Cooperative Breeding in Birds: Long-Term Studies of Ecology and Behavior* (eds P. Stacey & W. Koenig), pp. 305–35. Cambridge University Press, Cambridge.

Emlen, S.T. (1991) Evolution of cooperative breeding in birds and mammals. In: *Behavioural Ecology: an Evolutionary Approach* (eds J. Krebs & N.B. Davies), 3rd edn, pp. 301–37. Blackwell Scientific Publications, Oxford.

Emlen, S.T. (1994) Benefits, constraints, and the evolution of the family. *Trends Ecol Evol*, **9**, 282–5.

Emlen, S.T. (1995a) An evolutionary theory of the family. *Proc Nat Acad Sci*, **92**, 8092–9.

Emlen, S.T. (1995b) Can avian biology be useful to the social sciences? *J Avian Biol*, **26**, 273–6.

Emlen, S.T. (1996) Reproductive sharing in different types of kin associations. *Am Natur*, **148**, 756–63.

Emlen, S.T. & Oring, L.W. (1977) Ecology, sexual selection, and the evolution of mating systems. *Science*, **197**, 215–23.

Emlen, S.T. & Vehrencamp, S.L. (1983) Cooperative breeding strategies among birds. In: *Perspectives in*

Ornithology (eds A.H. Brush & G.A. Clark, Jr), pp. 93–133. Cambridge University Press, Cambridge.

Emlen, S.T. & Wrege, P.H. (1988) The role of kinship in helping decisions among white-fronted bee-eaters. *Behav Ecol Sociobiol*, **23**, 305–15.

Emlen, S.T. & Wrege, P.H. (1992a) Parent–offspring conflict and the recruitment of helpers among bee-eaters. *Nature*, **356**, 331–3.

Emlen, S.T. & Wrege, P.H. (1992b) Parent–offspring conflict. *Nature*, **359**, 24.

Emlen, S.T., Demong, N.J. & Emlen, D.J. (1989) Experimental induction of infanticide in female wattled jacanas. *Auk*, **106**, 1–7.

Endler, J.A. (1977) *Geographic Variation, Speciation, and Clines*. Princeton University Press, Princeton, NJ.

Endler, J.A. (1978) A predator's view of animal colour patterns. *Evol Biol*, **11**, 319–64.

Endler, J.A. (1980) Natural selection on colour patterns in *Poecilia reticulata*. *Evolution*, **34**, 76–91.

Endler, J.A. (1983) Natural and sexual selection on color patterns in poeciliid fishes. *Environ Biol Fishes*, **9**, 173–90.

Endler, J.A. (1991a) Interactions between predators and prey. In: *Behavioural Ecology An Evolutionary Approach* (eds J.R. Krebs & N.B. Davies), 3rd edn, pp. 169–96. Blackwell Scientific Publications, Oxford.

Endler, J.A. (1991b) Variation in the appearance of guppy colour patterns to guppies and their predators under different visual conditions. *Vis Res*, **31**, 587–608.

Endler, J.A. (1992) Signals, signal conditions and the direction of evolution. *Am Natur*, **139**, S125–S53.

Endler, J.A. (1993) Some general comments on the evolution and design of animal signalling systems. *Phil Trans Roy Soc Lond B*, **340**, 215–25.

Endler, J.A. & Houde, A.E. (1995) Geographic variation in female preferences for male traits in *Poecilia reticulata*. *Evolution*, **49**, 456–68.

Engels, W. (ed.) (1990) *Social Insects: an Evolutionary Approach to Castes and Reproduction*. Springer-Verlag, Berlin.

Enquist, M. (1985) Communication during aggressive interactions with particular reference to variation in choice of behaviour. *Anim Behav*, **33**, 1152–61.

Enquist, M. & Arak, A. (1993) Selection of exaggerated male traits by female aesthetic senses. *Nature*, **361**, 446–8.

Enquist, M. & Arak, A. (1994) Symmetry, beauty and evolution. *Nature*, **372**, 169–72.

Ens, B.J. & Cayford, J.T. (1996) Feeding with other oystercatchers. In: *The Oystercatcher: from Individuals to Populations* (ed. J.D. Goss-Custard), pp. 77–104. Oxford University Press, Oxford.

Ens, B.J. & Goss-Custard, J.D. (1984) Interference among oystercatchers, *Haematopus ostralegus*, feeding on mussels, *Mytilus edulis*, on the Exe estuary. *J Anim Ecol*, **53**, 217–31.

Ens, B.J., Safriel, U.N. & Harris, M.P. (1993) Divorce in the long-lived and monogamous oystercatcher, *Haemotopus ostralegus*: incompatibility or choosing a better option? *Anim Behav*, **45**, 1199–217.

Ens, B.J., Weissing, F.J. & Drent, R.H. (1995) The despotic distribution and deferred maturity: two sides of the same coin. *Am Natur*, **146**, 536–64.

Ens, B.J., Kersten, M., Brenninkmeijer, A. & Hulscher, J.B. (1992) Territory quality, parental effort and reproductive success of oystercatchers (*Haematopus ostralegus*). *J Anim Ecol*, **61**, 703–15.

Erhardt, C.L. & Bernstein, I.S. (1986) Matrilineal overthrows in rhesus monkey groups. *Int J Primatol*, **7**, 157–81.

Errington, J. (1996) Determination of cell fate in *Bacillus subtilis*. *Trends Genet*, **12**, 31–4.

Eshel, I. (1985) Evolutionary genetic stability of Mendelian segregation and the role of free recombination in the chromosomal system. *Am Natur*, **125**, 412–20.

Espelie, K.E. & Hermann, H.R. (1990) Surface lipids of the social wasp *Polistes annularis* (L.) and its nest and nest pedicel. *J Chem Ecol*, **16**, 1841–52.

Evans, C.S., Macedonia, J.M. & Marler, P. (1993) Effects of apparent size and speed on the response of chickens, *Gallus gallus*, to computer-generated simulations of aerial predators. *Anim Behav*, **46**, 1–11.

Evans, H.E. (1977) Extrinsic versus intrinsic factors in the evolution of insect sociality. *Bioscience*, **27**, 613–17.

Evans, J.D. (1993) Parentage analyses in ant colonies using simple sequence repeat loci. *Molec Ecol*, **2**, 393–7.

Evans, J.D. (1995) Relatedness threshold for the production of female sexuals in colonies of a polygynous ant, *Myrmica tahoensis*, as revealed by microsatellite DNA analysis. *Proc Nat Acad Sci USA*, **92**, 6514–17.

Faaborg, J., Parker, P.G., DeLay, L. *et al.* (1995) Confirmation of cooperative polyandry in the Galapagos hawk (*Buteo galapagoensis*). *Behav Ecol Sociobiol*, **36**, 83–90.

Fabre, J.H. (1897) *Souvenirs Entomologiques* (A.T. de Mattos, Trans). C. Delgrave, Paris.

Farr, J.A. (1980) Social behaviour patterns as determinations of reproductive success in the guppy *Poecilia reticulata* Peters (Pisces, Poeciliidae): An experimental study of the effects of intermale competition, female choice, and sexual selection.

Behaviour, **74**, 38–91.

Feder, J.L. & Bush, G.L. (1989) Gene frequency clines for host races of *Rhagoletis pomonella* (Diptera: Tephritidae) in the midwestern United States. *Heredity*, **63**, 245–66.

Feder, J.L., Hunt, T.A. & Bush, G.L. (1993) The effects of climate, host phenology and host fidelity on the genetics of apple and hawthorn infesting races of *Rhagoletis pomonella*. *Entomol Exp Appl*, **69**, 117–35.

Feder, J.L., Opp, S.B., Wlazlo, B., Reynolds, K., Go, W. & Spisak, S. (1994) Host fidelity is an effective premating barrier between sympatric races of the apple maggot fly. *Proc Nat Acad Sci USA*, **91**, 7990–4.

Felsenstein, J. (1985) Phylogenies and the comparative method. *Am Natur*, **125**, 1–15.

Fisher, J. & Hinde, R.A. (1949) The opening of milk bottle by birds. *Br Birds*, **42**, 347–57.

Fisher, R.A. (1930) *The Genetical Theory of Natural Selection*. Clarendon Press, Oxford.

Fishwild, T.G. & Gamboa, G.J. (1992) Colony defense against conspecifics: caste-specific differences in kin recognition by paper wasps, *Polistes fuscatus*. *Anim Behav*, **43**, 95–102.

Fiske, P. & Kålås, J.A. (1995) Mate sampling and copulation behaviour of great snipe females. *Anim Behav*, **49**, 209–19.

FitzGibbon, C.D. & Fanshawe, J.H. (1988) Stotting in Thomson's gazelles: an honest signal of condition. *Behav Ecol Sociobiol*, **23**, 69–74.

Fleishman, L.J. (1985) Cryptic movement in the vine snake *Oxybelis aeneus*. *Copeia*, **1985**, 242–5.

Fleishman, L.J. (1988a) Sensory influences on physical design of a visual display. *Anim Behav*, **36**, 1420–4.

Fleishman, L.J. (1988b) Sensory and environmental influences on display form in *Anolis auratus*, a grass anole from Panama. *Behav Ecol Sociobiol*, **22**, 309–16.

Fleishman, L.J. (1992) The influence of the sensory system and the environment on motion patterns in the visual displays of anoline lizards and other vertebrates. *Am Natur*, **139**, S36–S61.

Foley, R.A. & Lee, P.C. (1989) Finite social space, evolutionary pathways, and reconstructing hominid behavior. *Science*, **243**, 901–6.

Fonstein, M. & Haselkorn, R. (1995) Physical mapping of bacterial genomes. *J Bacteriol*, **177**, 3361–9.

Forrest, T.G. (1994) From sender to receiver: propagation and environmental effects of acoustic signals. *Am Zool*, **34**, 644–54.

Fournier, F. & Festa-Bianchet, M. (1995) Social dominance in female mountain goats. *Anim Behav*, **49**, 1449–59.

Fowler, K. & Partridge, L. (1989) A cost of mating in female fruitflies. *Nature Lond*, **338**, 760–1.

Fragaszy, D.M. & Visalberghi, E. (1989) Social influences on the acquisition of tool-using behaviors in tufted capuchin monkeys (*Cebus apella*). *J Comp Psychol*, **103(2)**, 159–70.

Fragaszy, D.M. & Visalberghi, E. (1990) Social processes affecting the appearance of innovative behaviors in capuchin monkeys. *Folia Primatol*, **54**, 155–65.

Frank, L.G. (1986) Social organization of the spotted hyena (*Crocuta crocuta*). II. Dominance and reproduction. *Anim Behav*, **34**, 1510–27.

Frank, L.G., Holekamp, K.E. & Smale, L. (1995a) Dominance, demography, and reproductive success of spotted hyenas. In: *Serengeti II: Dynamics, Management, and Conservation of an Ecosystem* (eds A.R.E. Sinclair & P. Arcese), pp. 364–84. University of Chicago Press, Chicago.

Frank, L.G., Weldele, M.L. & Glickman, S.E. (1995b) Masculinization costs in hyenas. *Nature Lond*, **377**, 584–5.

Frank, S.A. (1987) Variable sex ratio among colonies of ants. *Behav Ecol Sociobiol*, **20**, 195–201.

Franks, N.R., Blum, M., Smith, R.-K. & Allies, A.B. (1990) Behavior and chemical disguise of the cuckoo ant *Leptothorax kutteri* in relation to its host *Leptothorax acervorum*. *J Chem Ecol*, **16**, 1431–44.

Frantsevich, L.I. (1982) A compound eye model for the orientation towards polarized light. *Sens Syst (Leningrad)*, **82**, 170–83 [in Russian].

Fretwell, S.D. (1972) *Populations in a Seasonal Environment*. Princeton University Press, Princeton.

Fretwell, S.D. & Lucas, H.L. (1970) On territorial behaviour and other factors influencing habitat distribution in birds. *Acta Biotheoretica*, **19**, 16–36.

Fricke, H.W. (1975) Sozialstruktur und ökologische spezialisierung bei verwandten fischen (Pomacentridae). *Z Tierpsychol*, **39**, 492–520.

Frischknecht, M. (1993) The breeding colouration of male three-spined sticklebacks (*Gasterosteus aculeatus*) as an indicator of energy investment in vigour. *Evol Ecol*, **7**, 439–50.

Futuyma, D.J. & Moreno, G. (1988) The evolution of ecological specialisation. *Ann Rev Ecol Syst*, **19**, 207–33.

Gadagkar, R. (1990) Evolution of eusociality: the advantage of assured fitness returns. *Phil Trans Roy Soc Lond B*, **329**, 17–25.

Gadagkar, R. (1991) Demographic predisposition to the evolution of eusociality: a hierarchy of models. *Proc Nat Acad Sci USA*, **88**, 10 993–7.

Gadagkar, R. (1994) Why the definition of eusociality is not helpful to understand its evolution and what should we do about it. *Oikos*, **70**, 485–8.

Gadgil, M. & Bossert, W. (1970) Life history consequences of natural selection. *Am Natur*, **104**, 1–24.

Gage, M.J.G. (1991) Risk of sperm competition directly affects ejaculate size in the Mediterranean fruit fly. *Anim Behav*, **42**, 1036–7.

Gage, M.J.G., Stockley, P. & Parker, G.A. (1996) Effects of alternative male mating strategies on characteristics of sperm production in the Atlantic salmon (*Salmo salar*): theoretical and empirical investigations. *Phil Trans Roy Soc Lond B*, **350**, 391–9.

Galef, B.G. Jr (1976) Social transmission of acquired behavior: a discussion of tradition and social learning in vertebrates. In: *Advances in the Study of Behavior*, Vol 6, (eds J.S. Rosenblatt, R.A. Hinde, E. Shaw & C. Beer), pp. 77–99. Academic Press, New York.

Galef, B.G. Jr (1988) Imitation in animals: history, definition, and interpretation of data from the psychological laboratory. In: *Social Learning: Psychological and Biological Perspectives* (eds T.R. Zentall & B.G. Galef Jr), pp. 3–28. Lawrence-Erlbaum Associates, New Jersey.

Galef, B.G. Jr (1990) A historical perspective on recent studies of social learning about foods by Norway rats. *Can J Psychol*, **44**, 311–29.

Galef, B.G. Jr (1995) Why behaviour patterns that animals learn socially are locally adaptive. *Anim Behav*, **49**, 1325–34.

Galef, B.G. Jr & Allen, C. (1995) A new model system for studying behavioural traditions in animals. *Anim Behav*, **50**, 705–17.

Galef, B.G. Jr, Manzig, L.A. & Field, R.M. (1986) Imitation learning in budgerigars: Dawson and Foss (1965) revisited. *Behav Process*, **13**, 191–202.

Gallistel, C.R. (1990) *The Organization of Learning*. MIT Press, Cambridge Massachussets.

Gamboa, G.J. (1996) Kin recognition in social wasps. In: *Natural History and Evolution of Paper Wasps* (eds S. Turillazzi & M.J. West-Eberhard), pp. 161–77. Oxford University Press, Oxford.

Gamboa, G.J., Reeve, H.K., Ferguson, I.D. & Wacker, T.L. (1986a) Nestmate recognition in social wasps: the origin and acquisition of recognition odours. *Anim Behav*, **34**, 685–95.

Gamboa, G.J., Reeve, H.K. & Pfennig, D.W. (1986b) The evolution and ontogeny of nestmate recognition in social wasps. *Ann Rev Entomol*, **31**, 431–54.

Gamboa, G.J., Berven, K.A., Schemidt, R.A., Fishwild, T.G. & Jankens, K.M. (1991a) Kin recognition by larval wood frogs (*Rana sylvatica*): effects of diet and prior exposure to conspecifics. *Oecologia*, **86**, 319–24.

Gamboa, G.J., Foster, R.L., Scope, J.A. & Bitterman, A.M. (1991b) Effects of stage of colony cycle, context, and intercolony distance on conspecific tolerance by paper wasps (*Polistes fuscatus*). *Behav Ecol Sociobiol*, **29**, 87–94.

Gamboa, G.J., Reeve, H.K. & Holmes, W.G. (1991c) Conceptual issues and methodology in kin-recognition research: a critical discussion. *Ethology*, **88**, 109–27.

Gamboa, G.J., Grudzien, T.A., Espelie, K.E. & Bura, E.A. (1996) Kin recognition pheromones in social wasps: combining chemical and behavioral evidence. *Anim Behav*, **51**, 625–9.

Garland, T.J. (1992) Rate tests for phenotypic evolution using phylogenetically independent contrasts. *Am Natur*, **140**, 509–19.

Gaulin, S.J.C. & FitzGerald, W. (1986) Sex differences in spatial ability: an evolutionary hypothesis and test. *Am Natur*, **127**, 74–88.

Gaulin, S.J.C. & FitzGerald, W. (1989) Sexual selection for spatial-learning ability. *Anim Behav*, **37**, 322–31.

Gebhardt, M.D. & Stearns, S.C. (1988) Reaction norms for developmental time and weight at eclosion in *Drosophila meractorum*. *J Evol Biol*, **1**, 335–54.

Gendron, R.P. (1986) Searching for cryptic prey: evidence for optimal search rates and the formation of search image in quail. *Anim Behav*, **34**, 898–912.

Gendron, R.P. & Staddon, J.E.R. (1983) Searching for cryptic prey: the effects of search rate. *Am Natur*, **121**, 172–86.

Gerhardt, H.C. (1983) Communication and the environment. In: *Communication* (eds T.R. Halliday & P.J.B. Slater), pp. 82–113. Blackwell Scientific Publications, Oxford.

Gerhardt, H.C. (1992) Conducting playback experiments and interpreting their results. In: *Playback and Studies of Animal Communication* (ed. P.K. McGregor), pp. 59–77. Plenum Press, New York.

Gerhardt, H.C. (1994) Reproductive character displacement of female mate choice in the grey treefrog, *Hyla chrysoscelis*. *Anim Behav*, **47**, 959–69.

Getty, T. (1985) Discriminability and the sigmoid functional response: how optimal foragers can stabilize model-mimic systems. *Am Natur*, **125**, 239–56.

Getty, T. (1993) Search tactics and frequency-dependent prey detection. *Am Natur*, **141**, 804–11.

Getty, T. & Krebs, J.R. (1985) Lagging partial preferences for cryptic prey: a signal detection analysis of great tit foraging. *Am Natur*, **125**, 39–60.

Getty, T. & Pulliam, H.R. (1993) Search and prey detection by foraging sparrows. *Ecology*, **74**, 734–42.

Getty, T., Kamil, A.C. & Real, P.G. (1987) Signal detection theory and foraging for cryptic or mimetic prey. In: *Foraging Behavior* (eds A.C. Kamil, J.R. Krebs & H.R. Pulliam), pp. 525–49. Plenum Press, New York.

Getz, W.M. (1981) Genetically based kin recognition systems. *J Theor Biol*, **92**, 209–26.

Getz, W.M. (1991) The honey bee as a model kin recognition system. In: *Kin Recognition* (ed. P.G. Hepper), pp. 358–412. Cambridge University Press, Cambridge.

Getz, W.M. & Chapman, R.F. (1987) An odor discrimination model with application to kin recognition in social insects. *Int J Neurosci*, **32**, 963–78.

Getz, W.M. & Smith, K.B. (1983) Genetic kin recognition: honeybees discriminate between full and half sisters. *Nature*, **302**, 147–8.

Getz, W.M. & Smith, K.B. (1986) Honeybee kin recognition: learning self and nestmate phenotypes. *Anim Behav*, **34**, 1617–26.

Gibbon, J. (1977) Scalar expectancy theory and Weber's law in animal timing. *Psychol Rev*, **84**, 279–325.

Gibbon, J. & Church, R.M. (1990) Representation of time. *Cognition*, **37**, 23–54.

Gibbon, J., Church, R.M., Fairhurst, S. & Kacelnik, A. (1988) Scalar expectancy theory and choice between delayed rewards. *Psychol Rev*, **95**, 102–14.

Gibson, R.M. & Höglund, J. (1992) Copying and sexual selection. *Trends Ecol Evol*, **7**, 229–32.

Gibson, R.M., Bradbury, J.W. & Vehrencamp, S.L. (1991) Mate choice in lekking sage grouse revisited: the roles of vocal display, female site fidelity, and copying. *Behav Ecol*, **2**, 165–80.

Gibson, W. & Garside, L. (1990) Kinetoplast DNA minicircles are inherited from both parents in genetic hybrids of *Trypanosoma brucei*. *Molec Biochem Parasitol*, **42**, 45–54.

Gilliam, J.F. & Fraser, D.F. (1987) Habitat selection under predation hazard: test of a model with foraging minnows. *Ecology*, **68**, 1856–62.

Gingerich, P.D. (1993) Quantification and comparison of evolutionary rates. *Am J Sci*, **293-A**, 453–78.

Ginsberg, J.R. & Huck, U.W. (1989) Sperm compoetition in mammals. *Trends Ecol Evol*, **4,** 74–9.

Giraldeau, L.-A. (1984) Group foraging: the skill pool effect and frequency-dependent learning. *Am Natur*, **124**, 72–9.

Giraldeau, L.-A. & Livoreil, B. (1997) Game theory and social foraging. In: *Advances in Game Theory and the Study of Animal Behaviour* (eds L.A. Dugatkin & H.K. Reeve), in press. Oxford University Press, Oxford.

Giraldeau, L.-A. & Lefebvre, L. (1986) Exchangeable producer scrounger roles in a captive flock of feral pigeons. *Anim Behav*, **34**, 797–803.

Giraldeau, L.-A. & Lefebvre, L. (1987) Scrounging prevents cultural transmission of food-finding behaviour in pigeons. *Anim Behav*, **35**, 387–94.

Giraldeau, L.-A. & Templeton, J.J. (1991) Food scrounging and diffusion of foraging skills in pigeons, *Columba livia*: the importance of tutor and observer rewards. *Ethology*, **89**, 63–72.

Giraldeau, L.-A., Soos, C. & Beauchamp, G. (1994a) A test of the producer–scrounger foraging game in captive flocks of spice finches, *Lonchura punctulata*. *Behav Ecol Sociobiol*, **34**, 251–6.

Giraldeau, L.-A., Caraco, T. & Valone, T. (1994b) Social foraging: individual learning and cultural transmission of innovations. *Behav Ecol*, **5**, 35–43.

Gittleman, J.L. (1981) The phylogeny of parental care in fishes. *Anim Behav*, **29**, 936–41.

Gittleman, J.L., Anderson, C.G., Kot, M. & Luh, H.-K. (1996) Comparative tests of evolutionary lability and rates using molecular phylogenies. In: *New Uses for New Phylogenies* (eds P.H. Harvey, A.J. Leigh Brown, J. Maynard Smith & S. Nee), pp. 289–307. Oxford University Press, Oxford.

Gladstein, D.S., Carlin, N.F. & Austad, S.N. (1991) The need for sensitivity analyses of dynamic optimization models. *Oikos*, **60**, 121–6.

Godfray, H.C.J. (1991) Signalling of need between parents and offspring. *Nature Lond*, **352**, 328–30.

Godfray, H.C.J. (1995a) Evolutionary theory of parent–offspring conflict. *Nature Lond*, **376**, 133–8.

Godfray, H.C.J. (1995b) Signalling of need between parents and young: parent–offspring conflict and sibling rivalry. *Am Natur*, **146**, 1–24.

Godin, J.-G.J. & Briggs, S.E. (1996) Female mate choice under predation risk in the guppy. *Anim Behav*, **51**, 117–30.

Godin, J.-G.J. & Davis, S.A. (1995a) Who dares, benefits: predator approach behavior in the guppy (*Poecilia reticulata*) deters predator pursuit. *Proc Roy Soc Lond B*, **259**, 193–200.

Godin, J.-G.J. & Davis, S.A. (1995b) Boldness and predator deterrence: a reply to Milinski and Boltshauser. *Proc Roy Soc Lond B*, **262**, 107–12.

Goldizen, A.W. (1988) Tamarin and marmoset mating systems; unusual flexibility. *Trends Ecol Evol*, **3**, 36–40.

Goldsmith, T.H. (1991) The evolution of visual pigments and colour vision. In: *Vision and Visual Dysfunction*, Vol. 6 (ed. P. Gouras), pp. 62–89. Macmillan Press, Houndmills, London.

Goodnight, C.J. (1988) Epistasis and the effect of founder events on the additive genetic variance. *Evolution*, **42**, 441–54.

Gosler, A.G. (1987) Pattern and process in the bill morphology of the great tit *Parus major*. *Ibis*, **129**, 451–76.

Gosler, A.G. (1996) Environmental and social determinants of winter fat storage in the great tit *Parus major*. *J Anim Ecol*, **65**, 1–17.

Gosler, A.G., Greenwood, J.J.D. & Perrins, C.M. (1995) Predation risk and the cost of being fat. *Nature*, **377**, 621–3.

Goss-Custard, J.D. (1977) The ecology of the Wash. III. Density-related behaviour and the possible effects of a loss of feeding grounds on wading birds (Charadrii). *J Appl Ecol*, **14**, 721–39.

Goss-Custard, J.D. (1980) Competition for food and interference among waders. *Ardea*, **68**, 31–52.

Goss-Custard, J.D. (1985) Foraging behaviour of wading birds and the carrying capacity of estuaries. In: *Behavioural Ecology: Ecological Consequences of Adaptive Behaviour* (eds R.M. Sibly & R.H. Smith), pp. 169–88. Blackwell Scientific Publications, Oxford.

Goss-Custard, J.D. (1993) The effect of migration and scale on the study of bird populations: 1991 Witherby Lecture. *Bird Study*, **40**, 81–96.

Goss-Custard, J.D. & Durell, S.E.A. Le V. dit. (1988) The effect of dominance and feeding method on the intake rates of oystercatchers, *Haematopus ostralegus*, feeding on mussels, *Mytilus edulis*. *J Anim Ecol*, **57**, 827–44.

Goss-Custard, J.D., Caldow, R.W.G., Clarke, R.T., Durrell, S.E.A. Le V. dit & Sutherland, W.J. (1995a) Deriving population parameters from individual variations in foraging behaviour. 1. Empirical game theory distribution model of oystercatchers *Haematopus ostralegus* feeding on mussels *Mytilus edulis*. *J Anim Ecol*, **64**, 265–76.

Goss-Custard, J.D., Caldow, R.W.G., Clarke, R.T. & West, A.D. (1995b) Deriving population parameters from individual variations in foraging behaviour. II Model tests and population parameters. *J Anim Ecol*, **64**, 277–89.

Goss-Custard, J.D., Caldow, R.W.G., Clarke, R.T., Durell, S.E.A. Le V. dit, Urfi, A.J. & West, A.D. (1995c) Consequences of habitat loss and change to populations of wintering migratory birds: predicting the local and global effects from studies of individuals. *Ibis*, **137**, S56–66.

Goss-Custard, J.D., Clarke, R.T., Briggs, K.B. *et al.* (1995d) Population consequences of winter habitat loss in a migratory shorebird: I. Estimating model parameters. *J Appl Ecol*, **32**, 317–33.

Goss-Custard, J.D., Clarke, R.T., Durell, S.E.A. Le V. dit, Caldow, R.W.G. & Ens, B.J. (1995e) Population consequences of winter habitat loss in a migratory shorebird: II. Model predictions. *J Appl Ecol*, **32**, 334–48.

Goss-Custard, J.D., West, A.D., Caldow, R.W.G., Clarke, R.T. & Durell, S.E.A. le V. dit (1996a) The carrying capacity of coastal habitats for oystercatchers. In: *The Oystercatcher: from Individuals to Populations* (ed. J.D. Goss-Custard), pp. 327–51. Oxford University Press, Oxford.

Goss-Custard, J.D., Durell, S.E.A. le V. Dit, Clarke, R.T. *et al.* (1996b) Population dynamics of the Oystercatcher. In: *The Oystercatcher: from Individuals to Populations* (ed. J.D. Goss-Custard), pp. 352–83. Oxford University Press, Oxford.

Gould, J.L. (1986) The locale map of honey bees: do insects have cognitive maps? *Science*, **232**, 861–3.

Gould, S.J. (1989) *Wonderful Life*. Hutchinson, London.

Gould, S.J. & Lewontin, R.C. (1979) The spandrels of San Marco and the Panglossian paradigm: a critique of the adaptationist programme. *Proc Roy Soc Lond B*, **205**, 581–98.

Grafen, A. (1984) Natural selection, kin selection and group selection. In: *Behavioural Ecology: an Evolutionary Approach*, 2nd edn, (eds J.R. Krebs & N.B. Davies), pp. 62–84. Blackwell Scientific Publications, Oxford.

Grafen, A. (1985) A geometric view of relatedness. In: *Oxford Surveys in Evolutionary Biology*, Vol. 2 (eds R. Dawkins & M. Ridley), pp. 28–89. Oxford University Press, Oxford.

Grafen, A. (1986) Split sex ratios and the evolutionary origins of eusociality. *J Theor Biol*, **122**, 95–121.

Grafen, A. (1987) The logic of divisively asymmetric contests: respect for ownership and the desperado effect. *Anim Behav*, **35**, 462–7.

Grafen, A. (1988) On the use of data on lifetime reproductive success. In: *Reproductive Success: Studies of Individual Variation in Contrasting Breeding Systems* (ed. T.H. Clutton-Brock), pp. 454–71. The University of Chicago Press, Chicago.

Grafen, A. (1990a) Do animals really recognize kin? *Anim Behav*, **39**, 42–54.

Grafen, A. (1990b) Sexual selection unhandicapped by the Fisher process. *J Theor Biol*, **144**, 473–516.

Grafen, A. (1990c) Biological signals as handicaps. *J Theor Biol*, **144**, 517–46.

Grafen, A. (1991) Modelling in behavioural ecology. In: *Behavioural Ecology: an Evolutionary Approach*, 3rd edn, (eds J.R. Krebs & N.B. Davies), pp. 5–31. Blackwell Scientific Publications, Oxford.

Grafen, A. (1992) Of mice and the MHC. *Nature*, **360**, 530.

Graham, D.S. & Middleton, A.L.A. (1989) Conspecific recognition by juvenile brown-headed cowbirds. *Bird Behav*, **8**, 14–22.

Graveland, J. & Vangijzen T. (1994) Arthropods and seeds are not sufficient as calcium sources for shell formation and skeletal growth in passerines. *Ardea*, **82**, 299–314.

Green, R.F. (1980) Bayesian birds: a simple example of Oaten's stochastic model of optimal foraging. *Theor Popul Biol*, **18**, 244–56.

Green, R.F. (1984) Stopping rules for optimal foragers. *Am Natur*, **123**, 80–90.

Greenfield, M.D. (1994a) Cooperation and conflict in the evolution of signal interactions. *Ann Rev Ecol Syst*, **25**, 97–126.

Greenfield, M.D. (1994b) Synchronous and alternating choruses in insects and anurans: common mechanisms and diverse functions. *Am Zool*, **34**, 605–15.

Greenfield, M.D. & Roizen, I. (1993) Katydid synchronous chorusing is an evolutionarily stable outcome of female choice. *Nature Lond*, **364**, 618–20.

Greenwood, P.J. (1980) Mating systems, philopatry and dispersal in birds and mammals. *Anim Behav*, **28**, 1140–62.

Gridley, G., McLaughlin, J.K., Ekbom, A. *et al.* (1993) Incidence of cancer among patients with rheumatoid arthritis. *J Nat Cancer Inst*, **85**, 307–11.

Grinnell, J., Packer, C. & Pusey, A.E. (1995) Cooperation in male lions: kinship, reciprocity or mutualism? *Anim Behav*, **49**, 95–105.

Gromko, M.H., Gilbert, D.G. & Richmond, R.C. (1984) Sperm transfer and use in the multiple mating system of *Drosophila*. In: *Sperm Competition and the Evolution of Animal Mating Systems* (ed. R.L. Smith), pp. 371–426. Academic Press, Orlando.

Grosberg, R.K. (1988) The evolution of alleorecognition specificity in clonal invertebrates. *Quart Rev Biol*, **63**, 377–412.

Grosberg, R.K. & Quinn, J.F. (1986) The genetic control and consequences of kin recognition by the larvae of a colonial marine invertebrate. *Nature*, **322**, 457–9.

Gross, M.R. (1994) The evolution of behavioural ecology. *Trends Ecol Evol*, **9**, 358–60.

Gross, M.R. (1996) Alternative reproductive strategies and tactics: diversity within sexes. *Trends Ecol Evol*, **11**, 92–8.

Guhl, A.M., Collias, N.E. & Allee, W.C. (1945) Mating behavior and the social hierarchy in small flocks of white leghorns. *Physiol Zool*, **18**, 365–90.

Guilford, T. & Dawkins, M.S. (1987) Search images not proven: a reappraisal of recent evidence. *Anim Behav*, **35**, 1838–45.

Guilford, T. & Dawkins, M.S. (1991) Receiver psychology and the evolution of animal signals. *Anim Behav*, **42**, 1–14.

Guilford, T. & Dawkins, M.S. (1993) Receiver psychology and the design of animal signals. *Trends Neurosci*, **16**, 430–6.

Guilford, T. & Dawkins, M.S. (1995) What are conventional signals? *Anim Behav*, **49**, 1689–95.

Gustafsson, L. & Sutherland, W. J. (1988) The costs of reproduction in the collared flycatcher *Ficedula albicollis*. *Nature*, **335**, 813–15.

Gwinner, E. (1990) Circannual rhythms in bird migration: control of temporal patterns and interaction with photoperiod. In: *Bird Migration. Physiology and Ecology* (ed. E. Gwinner), pp. 257–68. Springer-Verlag, Berlin.

Gwynne, D.T. (1984) Courtship feeding increases female reproductive success in bush-crickets. *Nature*, **307**, 361–3.

Gwynne, D.T. (1984) Male mating effort, confidence of paternity, and insect sperm competition. In: *Sperm Competition and the Evolution of Animal Mating Systems* (ed. R.L. Smith), pp. 117–49. Academic Press, Orlando.

Gwynne, D.T. (1988) Courtship feeding and the fitness of female katydids (Orthoptera: Tettigoniidae). *Evolution*, **42**, 545–55.

Hahn, P.D. & Stuart, A.M. (1987) Sibling interactions in two species of termites: a test of the haplodiploid analogy (Isoptera: Kalotermitidae; Rhinotermitidae). *Sociobiology*, **13**, 83–92.

Haig, D. (1992) Genomic imprinting and the theory of parent–offspring conflict. *Semin Develop Biol*, **3**, 153–60.

Haig, D. (1993) The evolution of unusual chromosomal systems in sciarid flies: intragenomic conflict and the sex ratio. *J Evol Biol*, **6**, 249–61.

Haig, D. (1996) Gestational drive and the green-bearded placenta. *Proc Nat Acad Sci USA* **93**, 6547–51.

Haig, D. & Grafen, A. (1991) Genetic scrambling as a defence against meiotic drive. *J Theor Biol*, **153**, 531–58.

Haig, D. & Graham, C. (1991) Genomic imprinting and the strange case of the insulin-like growth factor-II receptor. *Cell*, **64**, 1045–6.

Hake, M. & Ekman, J. (1988) Finding and sharing depletable patches: when group foraging decreases intake rates. *Ornis Scand*, **19**, 275–9.

Haldane, J.B.S. (1949) Suggestions as to quantitative measurement of rates of evolution. *Evolution*, **3**, 51–6.

Halliday, T.R. & Verrell, P.A. (1984) Sperm competition in amphibians. In: *Sperm Competition and The Evolution of Animal Mating Systems* (ed. R.L. Smith), pp. 487–508. Academic Press, Orlando.

Halpin, Z.T. (1991) Kin recognition cues of vertebrates. In: *Kin Recognition* (ed. P.G. Hepper), pp. 220–58. Cambridge University Press, Cambridge.

Hamaguchi, K., Itô, Y. & Takenaka, O. (1993) GT dinucleotide repeat polymorphisms in a polygynous ant, *Leptothorax spinosior* and their use for measurement of relatedness. *Naturwissenschaften*, **80**, 179–81.

Hamilton, W.D. (1964a,b) The genetical evolution of social behaviour. I, II. *J Theor Biol*, **7**, 1–52.

Hamilton, W.D. (1967) Extraordinary sex ratios. *Science*, **156**, 477–88.

Hamilton, W.D. (1972) Altruism and related phenomena, mainly in social insects. *Ann Rev Ecol Syst*, **3**, 193–232.

Hamilton, W.D. (1979) Wingless and fighting males in fig wasps and other insects. In: *Sexual Selection and Reproductive Competition in Insects* (eds M.S. Blum & N.A. Blum), pp. 167–220. Academic Press, New York.

Hamilton, W.D. (1987a) Discriminating nepotism: expectable, common, overlooked. In: *Kin Recognition in Animals* (eds D.J.C. Fletcher & C.D. Michener), pp. 417–37. John Wiley & Sons, New York.

Hamilton, W.D. (1987b) Kinship, recognition, disease, and intelligence: constraints of social evolution. In: *Animal Societies: Theories and Facts* (eds Y. Itô, J.L. Brown & J. Kikkawa), pp. 81–102. Japan Scientific Societies Press, Tokyo.

Hamilton, W.D. & Zuk, M. (1982) Heritable true fitness and bright birds: a role for parasites? *Science*, **218**, 384–6.

Hamilton, W.D., Axelrod, R. & Tanese, R. (1990) Sexual reproduction as an adaptation to resist parasites (a review). *Proc Nat Acad Sci USA*, **87**, 3566–73.

Hammerstein, P. (1981) The role of asymmetries in animal contests. *Anim Behav*, **29**, 193–205.

Hammerstein, P. & Parker, G.A. (1982) The asymmetric war of attrition. *J Theor Biol*, **96**, 647–82.

Hanken, J. & Sherman, P.W. (1981) Multiple paternity in Belding's ground squirrel litters. *Science Wash*, **212**, 351–3.

Hannon, S.J., Mumme, R.L., Koenig, W.D. & Pitelka, F.A. (1985) Replacement of breeders and within-group conflict in the cooperatively breeding Acorn woodpecker. *Behav Ecol Sociobiol*, **17**, 303–12.

Harley, C.B. (1981) Learning the evolutionarily stable strategy. *J Theor Biol*, **89**, 611–33.

Harrison, R.G. (ed.) (1993) *Hybrid Zones and the Evolutionary Process*. Oxford University Press, New York.

Harrison, S. & Hastings, A. (1996) Genetic and evolutionary consequences of metapopulation structure. *Trends Ecol Evol*, **11**, 180–3.

Harshman, L.G. & Prout, T. (1994) Sperm displacement without sperm transfer in *Drosophila melanogaster*. *Evolution*, **48**, 758–66.

Hartl, D.L. & Clark, A.G. (1989) *Principles of Population Genetics*. Sinauer Associates, Sunderland, MA.

Hartley, I.R., Davies, N.B., Hatchwell, B.J., Desrochers, A., Nebel, D. & Burke, T. (1995) The polygynandrous mating system of the alpine accentor, *Prunella collaris*. II. Multiple paternity and parental effort. *Anim Behav*, **49**, 789–803.

Hartshorne, C. (1956) The monotony threshold in singing birds. *Auk*, **95**, 758–60.

Harvey, P.H. (1996) Phylogenies for ecologists. *J Anim Ecol*, **65**, 255–63.

Harvey, P. & Bradbury, J.W. (1991) Sexual selection. In: *Behavioural Ecology, An Evolutionary Approach*, 3rd edn. (eds J.R. Krebs & N.B. Davies), pp. 203–33. Blackwell Scientific Publications, Oxford.

Harvey, P.H. & Krebs, J.R. (1990) Comparing brains. *Science*, **249**, 140–6.

Harvey, P.H. & Pagel, M.D. (1991) *The Comparative Method in Evolutionary Biology*. Oxford University Press, Oxford.

Harvey, P.H. & Paxton, R.J. (1981) The evolution of aposematic coloration. *Oikos*, **37**, 391–3.

Harvey, P.H. & Purvis, A. (1991) Comparative methods for explaining adaptations. *Nature*, **351**, 619–24.

Hasegawa, E. (1994) Sex allocation in the ant *Colobopsis nipponicus* (Wheeler). I. Population sex ratio. *Evolution*, **48**, 1121–9.

Hasegawa, E. & Yamaguchi, T. (1995) Population structure, local mate competition, and sex-allocation pattern in the ant *Messor aciculatus*. *Evolution*, **49**, 260–5.

Hasselquist, D., Bensch, S. & Von Schantz, T. (1996) Correlation between male song repertoire, extra-pair paternity and offspring survival in the great reed warbler. *Nature*, **381**, 229–32.

Hastings, I.M. (1992) Population genetic aspects of deleterious cytoplasmic genomes and their effect on the evolution of sexual reproduction. *Genet Res*, **59**, 215–25.

Hausfater, G. & Hrdy, S.B. (1984) *Infanticide: Comparative and Evolutionary Perspectives*. Aldine Press, NY.

Hausfater, G., Altmann, J. & Altmann, S. (1982) Long-term consistency of dominance relations among female baboons (*Papio cynocephalus*). *Science*, **217**, 752–5.

Healy, D.D., Clayton, N.S. & Krebs, J.R. (1995) Development of hippocampal specialization in two species of tit (*Parus* spp.). *Behav Brain Res*, **61**, 23–8.

Healy, S. & Guilford, T. (1990) Olfactory bulb size and nocturnality in birds. *Evolution*, **44**, 339–46.

Heard, S.B. & Hauser, D.L. (1995) Key innovations and their ecological mechanisms. *Histor Biol*, **10**, 151–73.

Hedenström, A. (1992) Flight performance in relation to fuel load in birds. *J Theor Biol*, **158**, 535–7.

Hedenström, A. & Alerstam, T. (1995) Optimal flight speed of birds. *Phil Trans Roy Soc Lond B*, **348**, 471–87.

Hedrick, A.V. & Dill, L.M. (1993) Mate choice by female crickets is influenced by predation risk. *Anim Behav*, **46**, 193–6.

Heinsohn, R. & Packer, C. (1995) Complex cooperative strategies in group-territorial African lions. *Science*, **269**, 1260–2.

Heinze, J. (1995) Reproductive skew and genetic relatedness in *Leptothorax* ants. *Proc Roy Soc Lond B*, **261**, 375–9.

Heinze, J. & Buschinger, A. (1988) Polygyny and functional monogyny in *Leptothorax* ants (Hymenoptera: Formicidae). *Psyche*, **95**, 309–25.

Heinze, J., Lipski, N., Schlehmeyer, K. & Hölldobler, B. (1995) Colony structure and reproduction in the ant, *Leptothorax acervorum*. *Behav Ecol*, **6**, 359–67.

Heisler, I.L. (1985) Quantitative genetic models of female choice based on 'arbitrary' male characters. *Heredity*, **55**, 187–98.

Helbig, A.J., Bertold, P., Mohr, G. & Querner, U. (1994) Inheritance of a novel migratory direction in central European blackcaps. *Naturwissenschaften*, **81**, 184–6.

Heldmaier, G. (1989) Seasonal acclimatization of energy requirements in mammals: functional significance of body weight control, hypothermia, torpor and hibernation. In: *Energy Transformation in Cells and Organisms* (eds W. Wieser & E. Gnaiger), pp. 130–9. Thieme, Stuttgart.

Heldmaier, G. & Steinlechner, S. (1981) Seasonal control of energy requirements for thermoregulation in the Djungarian hamster (*Phodopus sungorus*) living in natural photoperiod. *J Comp Physiol*, **142**, 429–37.

Heldmaier, G., Klaus, S., Wiesinger, H., Friedrichs, U. & Wenzel, M. (1989) Cold acclimation and thermogenesis. In: *Living in the Cold II* (eds A. Malan & B. Canguilhem), pp. 347–58. John Libbey, Eurotext Ltd London–Paris.

Hemelrijk, C.K. (1994) Support for being groomed in long-tailed macaques, *Macaca fascicularis*. *Anim Behav*, **48**, 479–81.

Hemelrijk, C.K. (1996) Reciprocation in apes: from complex cognition to self-structuring. In: *Great Ape Societies* (eds W.C. McGrew, L.F. Marchant & T. Nishida), pp. 185–95. University Press, Cambridge.

Henderson, M.T.G., Merriam, G. & Wegner, J. (1985) Patchy environments and species survival: chipmunks in an agricultural mosaic. *Biol Conserv*, **31**, 95–105.

Hengeveld, R. (1989) *Dynamics of Biological Invasions*. Chapman & Hall, London.

Hepper, P.G. & Waldman, B. (1992) Embryonic olfactory learning in frogs. *Quart J Exp Psychol B*, **44**, 179–97.

Herbers, J.M. (1990) Reproductive investment and allocation ratios for the ant *Leptothorax longispinosus*: sorting out the variation. *Am Natur*, **136**, 178–208.

Herbers, J.M. (1993) Ecological determinants of queen number in ants. In: *Queen Number and Sociality in Insects* (ed. L. Keller), pp. 262–93. Oxford University Press, Oxford.

Heusner, A.A. (1987) What does the power function reveal about structure and function in animals of different size? *Am Rev Physiol*, **49**, 121–33.

Hewitt, G.M. (1988) Hybrid zones — natural laboratories for evolutionary studies. *Trends Ecol Evol*, **3**, 158–67.

Hewitt, G.M. (1993a) After the ice — *parallelus* meets *erythropus* in the Pyrenees. In: *Hybrid Zones and the Evolutionary Process* (ed. R.G. Harrison), pp. 140–64. Oxford University Press, New York.

Hewitt, G.M. (1993b) Postglacial distribution and species substructure: lessons from pollen, insects and hybrid zones. In: *Evolutionary Patterns and Processes* (eds D.R. Lees & D. Edwards), pp. 97–123. Linnaean Society of London, Academic Press, London.

Hewitt, G.M. (1996) Some genetic consequences of ice ages, and their role in divergence and speciation. *Biol J Linn Soc*, **58**, 247–76.

Heyes, C.M. (1993) Imitation, culture and cognition. *Anim Behav*, **46**, 999–1010.

Heyes, C.M. & Dawson, G.R. (1990) A demonstration of observational learning using a bidirectional control. *Quart J Exp Psychol*, **42B**, 59–71.

Hickey, D.A. (1982) Selfish DNA: a sexually transmitted nuclear parasite. *Genetics*, **101**, 519–31.

Hicks, G.R. & Raikhel, N.V. (1995) Protein import into the nucleus: an integrated view. *Ann Rev Cell Develop Biol*, **11**, 155–88.

Higashi, M., Abe, T. & Burns, T.P. (1992) Carbon–nitrogen balance and termite ecology. *Proc Roy Soc Lond B*, **249**, 303–8.

Higashi, M., Yamamura, N., Abe, T. & Burns, T.P. (1991) Why don't all termite species have a sterile worker caste? *Proc Roy Soc Lond B*, **246**, 25–9.

Hill, G.E. (1990) Female house finches prefer colourful males: sexual selection for a condition-dependent trait. *Anim Behav*, **40**, 563–72.

Hill, G.E. (1991) Plumage colouration is a sexually selected indicator of male quality. *Nature Lond*, **350**, 337–9.

Hill, G.E. (1992) The proximate basis of variation in

carotenoid pigmentation in male house finches. *Auk*, **109**, 1–12.

Hill, G.E. (1994) Geographic variation in male ornamentation and female mate preference in the house finch: a comparative test of models of sexual selection. *Behav Ecol*, **5**, 64–73.

Hill, G.E. & Montgomerie, R. (1994) Plumage colour signals nutritional condition in the house finch. *Proc Roy Soc Lond B*, **258**, 47–52.

Hillis, D.M., Moritz, C. & Mable, B.K. (1996) *Molecular Systematics*, 2nd edn. Sinauer, Sunderland.

Hinde, R.A. (1983) A conceptual framework. In: *Primate Social Relationships* (ed. R.A. Hinde), pp. 1–7. Sinauer, Sunderland, Massachusetts.

Högstedt, G. (1980) Evolution of clutch size in birds: adaptive variation in relation to territory quality. *Science*, **210**, 1148–50.

Hogendoorn, K. & Velthuis, H.H.W. (1993) The sociality of *Xylocopa pubescens*: does a helper really help? *Behav Ecol Sociobiol*, **32**, 247–57.

Hogendoorn, K. & Velthuis, H.H.W. (1995) The role of young guards in *Xylocopa pubescens*. *Insect Soc*, **42**, 427–48.

Hoggren, M. (1995) *Mating strategies and sperm competition in the adder (Vipera berus)*. PhD thesis, Uppsala University.

Höglund, J. (1989) Size and plumage dimorphism in lek-breeding birds: a comparative analysis. *Am Natur*, **134**, 72–87.

Höglund, J. & Alatalo, R.V. (1995) *Leks*. Princeton University Press, Princeton, New Jersey.

Höglund, J. & Sillén-Tullberg, B. (1996) Does lekking promote the evolution of male biased size dimorphism in birds? On the use of comparative approaches. *Am Natur* (in press).

Höglund, J., Alatalo, R.V. & Lundberg, A. (1990) Copying the mate choice of others, observation on female Black grouse. *Behaviour*, **114**, 221–31.

Hogstad, O. (1987) It is expensive to be dominant. *Auk*, **104**, 333–6.

Hoikkala, A. & Kaneshiro, K.Y. (1993) Change in the signal–response sequence responsible for asymmetric isolation between *Drosophila planitibia* and *Drosophila silvestris*. *Proc Nat Acad Sci USA*, **90**, 5813–17.

Hoikkala, A. & Welbergen, P. (1995) Signals and responses of females and males in successful and unsuccessful courtships of three Hawaiian lek-mating *Drosophila* species. *Anim Behav*, **50**, 177–90.

Hokit, D.G., Walls, S.C. & Blaustein, A.R. (1996) Context-dependent kin discrimination in larvae of the marbled salamander, *Ambystoma opacum*. *Anim Behav*, **52**, 17–31.

Hölldobler, B. & Wilson, E.O. (1990) *The Ants*. Springer–Verlag, Berlin.

Holmes, W.G. (1984) Ontogeny of dam-young recognition in captive Belding's ground squirrels. *J Comp Psychol*, **98**, 246–56.

Holmes, W.G. (1986a) Kin recognition by phenotype matching in female Belding's ground squirrels. *Anim Behav*, **34**, 38–47.

Holmes, W.G. (1986b) Identification of paternal half siblings by captive Belding's ground squirrels. *Anim Behav*, **34**, 321–7.

Holmes, W.G. (1994) The development of littermate preferences in juvenile Belding's ground squirrels. *Anim Behav*, **48**, 1071–84.

Holmes, W.G. & Sherman, P.W. (1982) The ontogeny of kin recognition in two species of ground squirrels. *Am Zool*, **22**, 491–517.

Holmgren, N. (1995) The ideal free distribution of unequal competitors: predictions from a behaviour-based functional response. *J Anim Ecol*, **64**, 197–212.

Holmgren, N. & Hedenström, A. (1995) The scheduling of molt in migratory birds. *Evol Ecol*, **9**, 354–68.

Homann, H. (1928) Beiträge zur Physiologie der Springspinnen. *Z Vergl Physiol*, **7**, 201–68.

Honda, K. (1995) Chemical basis of differential oviposition by lepidopterous insects. *Arch Insect Biochem Physiol*, **30**, 1–23.

Hoogland, J.L. & Sherman, P.W. (1976) Advantages and disadvantages of bank swallow (*Riparia riparia*) coloniality. *Ecol Monogr*, **46**, 33–58.

Houde, A.E. (1987) Mate choice based upon naturally occurring colour pattern variation in a guppy population. *Evolution*, **41**, 1–10.

Houde, A.E. (1993) Evolution by sexual selection: what can population comparisons tell us? *Am Natur*, **141**, 796–803.

Houde, A.E. & Endler, J.A. (1990) Correlated evolution of female mating preferences and male color patterns in the Guppy *Poecilia reticulata*. *Science*, **248**, 1405–8.

Houde, A.E. & Torio, A.J. (1992) Effect of parasitic infection on male color pattern and female choice in guppies. *Behav Ecol*, **3**, 346–51.

Houston, A.I. (1986) The optimal flight velocity for a bird exploiting patches of food. *J Theor Biol*, **119**, 345–62.

Houston, A.I. (1993) The efficiency of mass-loss in breeding birds. *Proc Roy Soc Lond B*, **254**, 221–5.

Houston, A.I. (1993) The importance of state. In: *Diet Selection: an Interdisciplinary Approach to Foraging Behaviour* (ed. R.N. Hughes), pp. 10–31. Blackwell Scientific Publications, Oxford.

Houston, A.I. (1995) Energetic constraints and foraging efficiency. *Behav Ecol*, **6**, 393–6.

Houston, A.I. & McNamara, J.M. (1988) Fighting for food: a dynamic version of the Hawk–Dove game. *Evol Ecol*, **2**, 51–64.

Houston, A.I. & McNamara, J.M. (1988) There's no such thing as a free lunch. *Behav Brain Sci*, **11**, 154–9.

Houston, A.I. & McNamara, J.M. (1992) Phenotypic plasticity as a state-dependent life-history decision. *Evol Ecol*, **6**, 243–53.

Houston, A.I. & McNamara, J.M. (1993) A theoretical investigation of the fat reserves and mortality levels of small birds in winter. *Ornis Scand*, **24**, 205–19.

Houston, A.I. & Sumida, B.H. (1987) Learning rules, matching and frequency-dependence. *J Theor Biol*, **126**, 289–308.

Houston, A.I., McNamara, J.M. & Hutchinson, J.M.C. (1993) General results concerning the trade-off between gaining energy and avoiding predation. *Phil Trans Roy Soc Lond B*, **341**, 375–97.

Houston, A.I., McNamara, J.M. & Thomson, W.A. (1992) On the need for a sensitive analysis of optimization models, or 'the simulation is not as the former'. *Oikos*, **63**, 513–17.

Houston, A.I., Schmid-Hempel, P. & Kacelnik, A. (1988) Foraging strategy, worker mortality, and the growth of the colony in social insects. *Am Natur*, **131**, 107–14.

Houston, A.I., Clark, C.W., McNamara, J.M. & Mangel, M. (1988) Dynamic models in behavioural and evolutionary ecology. *Nature*, **332**, 29–34.

Houston, D.C., Donnan, D. & Jones, P.J. (1995a) Use of labelled methionine to investigate the contribution of muscle proteins to egg production in zebra finches. *J Comp Physiol B*, **165**, 161–4.

Houston, D.C., Donnan, D., Jones, P., Hamilton, I. & Osborn, D. (1995b) Changes in the muscle condition of female Zebra Finches *Poephila guttata* during egg laying and the role of protein storage in bird skeletal muscle. *Ibis*, **137**, 322–8.

Howard, R.D., Whiteman, H.H. & Schueller, T.I. (1994) Sexual selection in American toads: a test of a good-genes hypothesis. *Evolution*, **48**, 1286–1300.

Hoy, R.R., Hahn, J. & Paul, R.C. (1977) Hybrid cricket auditory behavior: evidence for genetic coupling in animal communication. *Science*, **195**, 82–4.

Hoy, R.R., Hoikkala, A. & Kaneshiro, K.Y. (1988) Hawaiian courtship songs: evolutionary innovation in communication signals in *Drosophila*. *Science*, **240**, 217–19.

Hubbell, S.P. & Johnson, L.K. (1987) Environmental variance in lifetime mating success, mate choice and sexual selection. *Am Natur*, **130**, 91–112.

Huck, U., Labov, J. & Lisk, R. (1987) Food-restricting first generation juvenile female hamsters (*Mesocricetus auratus*) affects sex ratio and growth of third generation offspring. *Biol Reprod*, **37**, 612–17.

Huelsenbeck, J.P., Bull, J.J. & Cunningham, C.W. (1996) Combining data in phylogenetic analysis. *Trends Ecol Evol*, **11**, 152–8.

Hughes-Schrader, S. (1948) Cytology of coccids (Coccoidea — Homoptera). *Adv Genet*, **2**, 127–203.

Hunte, W. & Horrocks, J. (1986) Kin and non-kin interventions in the aggressive disputes of vervet monkeys. *Behav Ecol Sociobiol*, **20**, 257–63.

Hunte, W., Myers, R.A. & Doyle, R.W. (1985) Bayesian mating decisions in an amphipod, *Gammarus lawrencianus* Bousfield. *Anim Behav*, **33**, 366–72.

Hunter, F.M., Burke, T.A. & Watts, S.E. (1992) Frequent copulation as a method of paternity assurance in the Northern Fulmar. *Anim Behav*, **44**, 149–56.

Hunter, F.M., Petrie, M., Otronen, M., Birkhead, T.R. & Møller, A.P. (1993) Why do females copulate repeatedly with one male? *Trends Ecol Evol*, **8**, 21–6.

Hunter-Jones, P. (1960) Fertilization of eggs of the desert locust by spermatozoa from successive copulations. *Nature (Lond)*, **185**, 336.

Huntingford, F.A. & Turner, A.K. (1987) *Animal Conflict*. Chapman & Hall, London.

Huntingford, F.A., Lazarus, J., Barrie, B.D. & Webb, S. (1994) A dynamic analysis of cooperative predator inspection in sticklebacks. *Anim Behav*, **47**, 413–23.

Hurly, T.A. (1992) Energetic reserves of marsh tits (*Parus palustris*): food and fat storage in response to variable food supply. *Behav Ecol*, **3**, 181–8.

Hurst, L.D. (1990) Parasite diversity and the evolution of diploidy, multicellularity and anisogamy. *J Theor Biol*, **144**, 429–3.

Hurst, L.D. & Hamilton, W.D. (1992) Cytoplasmic fusion and the nature of sexes. *Proc Roy Soc Lond B*, **247**, 189–94.

Hurst, L.D., Atlan, A. & Bengtsson, B.O. (1996) Genetic conflicts *Quart Rev Biol*, **71**, 317–64.

Hutchinson, J., McNamara, J.M. & Cuthill, I.C. (1993) Song, sexual selection, starvation and strategic handicaps. *Anim Behav*, **45**, 1153–77.

Huxley, J.S. (1912) A 'disharmony' in the reproductive habits of the wild duck (*Anas boschas* L.). *Biol Zentralbl*, **32**, 621–3.

Huxley, J.S. (1938) Darwin's theory of sexual selection and the data subserved by it, in the light of recent research. *Am Natur*, **72**, 416–33.

Huxley, J.S. (1940) *The New Systematics*. Clarendon Press, Oxford.

Ibrahim, K.M., Nichols, R.A. & Hewitt, G.M. (1996) Spatial patterns of genetic variation generated by different forms of dispersal during range expansion. *Heredity*, **77**, 282–91.

Illius, A.W., Albon, S.D., Pemberton, J.M., Gordon,

I.J. & Clutton-Brock, T.H. (1995) Selection for foraging efficiency during a population crash in Soay sheep. *J Anim Ecol*, **64**, 481–92.

Inman, A., Lefebvre, L. & Giraldeau, L.-A. (1988) Individual diet differences in feral pigeons: evidence for resource partitioning. *Anim Behav*, **35**, 1902–3.

Irons, W (1990) Let's make our perspective broader rather than narrower. *Ethol Sociobiol*, **11**, 361–74.

Itô, Y. (1993) *Behaviour and Social Evolution of Wasps: the Communal Aggregation Hypothesis*. Oxford University Press, Oxford.

Iwasa, Y. & Pomiankowski, A. (1994) The evolution of mate preferences for multiple sexual ornaments. *Evolution*, **48**, 853–67.

Iwasa, Y., Higashi, W. & Yamamura, N. (1981) Prey distribution as a factor determining the choice of optimal foraging strategy. *Am Natur*, **117**, 710–23.

Jackson, R.R. & Blest, A.D. (1982) The distances at which a primitive jumping spider, *Portia fimbriata*, makes visual discriminations. *J Exp Biol*, **97**, 441–5.

Jacobs, L. (1995) The ecology of spatial cognition. In: *Behavioral Brain Research in Naturalistic and Semi-naturalistic Settings* (eds E. Alleva, A. Fasolo, H.P. Lipp, L. Nadel & L. Ricceri), pp. 301–22. NATO ASI Series, Kluwer Academic Press, Dordrecht.

Jacobs, L.F., Gaulin, S.J.C., Sherry, D.F. & Hoffman, G. (1990) Evolution of spatial cognition: sex-specific patterns of spatial behavior predict hippocampal size. *Proc Nat Acad Sci USA*, **87**, 6349–52.

Jacobs, M.E. (1955) Studies on territorialism and sexual selection in dragonflies. *Ecology*, **36**, 566–86.

Jaenike, J. & Holt, R.D. (1991) Genetic variation for habitat preference: evidence and explanations. *Am Natur*, **137**(Suppl.), S67–S90.

Jamieson, I.G., Quinn, J.S., Rose, P.A. & White, B.N. (1994) Shared paternity among non-relatives is a result of an egalitarian mating system in a communally breeding bird, the pukeko. *Proc R Soc Lond B*, **257**, 271–7.

Janetos, A.C. (1980) Strategies of female mate choice: a theoretical analysis. *Behav Ecol Sociobiol*, **7**, 107–12.

Janis, C.M. (1996) Do legs support the arms race hypothesis in mammalian predator/prey relationships. In: *Vertebrate Behavior as Derived from the Fossil Record* (eds J.R. Horner & L. Ellis), (in press). Columbia University Press, New York.

Jarman, P.J. (1974) The social organisation of antelope in relation to their ecology. *Behaviour*, **48**, 215–67.

Jeffreys, A.J., Wilson, V. & Thein, S.L. (1985) Hypervariable 'minisatellite' regions in human DNA. *Nature Lond*, **314**, 67–73.

Jenni, D.A. (1974) Evolution of polyandry in birds. *Am Zool*, **14**, 129–44.

Jennions, M.D. & Oakes, E.J. (1994) Symmetry and sexual selection. *Trends Ecol Evol*, **9**, 440.

Jennions, M.D. & Passmore, N.I. (1993) Sperm competition in frogs: testis size and a 'sterile male' experiment on *Chiromantis xerampelina* (Rhacophoridae). *Biol J Linn Soc*, **50**, 211–20.

Jimenez, J.A., Hughes, K.A., Alaks, G., Graham, L & Lacy, R.C. (1994) An experimental study of inbreeding depression in a natural habitat. *Science*, **266**, 271–3.

Johnson, J. (1987) Dominance rank in juvenile olive baboons, *Papio anubis*: the influence of gender, size, maternal rank and orphaning. *Anim Behav*, **35**, 1694–1708.

Johnson, M.L. & Gaines, M.S. (1990) Evolution of dispersal: theoretical models and empirical tests using birds and mammals. *Ann Rev Ecol Syst*, **21**, 449–80.

Johnson, M.S. & Black, R. (1995) Neighborhood size and the importance of barriers to gene flow in an intertidal snail. *Heredity*, **75**, 142–54.

Johnson, M.S., Murray, J. & Clarke, B. (1993) The ecological genetics and adaptive radiation of *Partula* on Moorea. *Oxf Surv Evol Biol*, **9**, 167–238.

Johnstone, R.A. (1994) Female preference for symmetrical males as a by-product of selection for mate recognition. *Nature*, **372**, 172–5.

Johnstone, R.A. (1995a) Sexual selection, honest advertisement and the handicap principle: reviewing the evidence. *Biol Rev*, **70**, 1–65.

Johnstone, R.A. (1995b) Honest advertisement of multiple qualities using multiple signals. *J Theor Biol*, **177**, 87–94.

Johnstone, R.A. (1996a) Game theory and communication. In: *Game Theory and the Study of Animal Behaviour* (eds L.A. Dugatkin & H.K. Reeve). Oxford University Press, Oxford (in press).

Johnstone, R.A. (1996b) Multiple displays in animal communication: 'backup signals' and 'multiple messages'. *Phil Trans Roy Soc Lond B*, **351**, 329–38.

Johnstone, R.A. & Grafen, A. (1992) The continuous Sir Philip Sidney game: a simple model of biological signalling. *J Theor Biol*, **156**, 215–34.

Johnstone, R.A. & Grafen, A. (1993) Dishonesty and the handicap principle. *Anim Behav*, **46**, 759–64.

Johnstone, R.A., Reynolds, J.D. & Deutsch, J.C. (1996) Mutual mate choice and sex differences in choosiness. *Evolution*, **50**, 1382–91.

Jones, G. (1986) Sexual chases in sand martins (*Riparia riparia*): cues for males to increase their reproductive success. *Behav Ecol Sociobiol*, **19**, 179–85.

Jones, M.M. (1991) Muscle protein loss in laying House sparrows *Passer domesticus*. *Ibis*, **133**, 193–8.

Kacelnik, A. (1989) Short-term adjustment of parental effort in starlings. In: *Acta XIX Congressus Internationalis Ornithologici*, Vol. II, (ed. H. Ouellet), pp. 1843–56. University of Ottawa Press, Ottawa.

Kacelnik, A. & Krebs, J.R. (1985) Learning to exploit patchily distributed prey. In: *Behavioural Ecology*. British Ecological Society Symposium (eds R.M. Sibly & R.H. Smith), pp. 189–205. Blackwell Scientific Publications, Oxford.

Kacelnik, A. & Todd, I.A. (1992) Psychological mechanisms and the marginal value theorem: effect of variability in travel time on patch exploitation. *Anim Behav*, **43**, 313–22.

Kacelnik, A., Brunner, D. & Gibbon, J. (1990) Timing mechanisms in optimal foraging: some applications of scalar expectancy theory. In: *Behavioural Mechanisms of Food Selection*, NATO ASI series Vol. G20 (ed. R.N. Hughes), pp. 61–82. Springer-Verlag, Berlin.

Kacelnik, A., Krebs, J.R. & Ens, B. (1987) Foraging in a changing environment: an experiment with starlings (*Sturnus vulgaris*). In: *Harvard Symposium on the Quantitative Analysis of Behavior Vol. 6 Foraging* (eds M.L. Commons, A. Kacelnik & S.J. Shettleworth), pp. 63–87. Lawrence-Erlbaum Associates, Hillsdale, NJ.

Kacelnik, A., Cotton, P.A., Stirling, L. & Wright, J. (1995) Food allocation among nestling starlings: sibling competition and the scope of parental choice. *Proc Roy Soc Lond B*, **259**, 259–63.

Kaiser, D. (1986) Control of multicellular development: *Dictyostelium* and *Myxococcus*. *Ann Rev Genet*, **20**, 539–66.

Kalmus, H. (1932) Ueber die Erhaltungswert der Phanotypischen (Morphologischen) Anisogamie und die Entstehung der Erstern Geslechtsunterschiede. *Biol Zentralblatt*, **552**, 716–26.

Kambhampati, S. (1995) A phylogeny of cockroaches and related insects based on DNA sequence of mitochondrial ribosomal RNA genes. *Proc Nat Acad Sci USA*, **92**, 2017–20.

Kamil, A.C., Balda, R.P. & Olson, D. (1994) Performance of four seed-caching corvid species in the radial-arm maze analog. *J Comp Psychol*, **108**, 385–93.

Kaneshiro, K.Y. (1976) Ethological isolation and phylogeny in the planitibia subgroup of Hawaiian *Drosophila*. *Evolution*, **30**, 740–5.

Kaneshiro, K.Y. (1989) Dynamics of sexual selection and founder effects in species formation. In: *Genetics, Speciation and the Founder Principle* (eds L.V. Giddings, K.Y. Kaneshiro & W.W. Anderson), pp. 279–96. Oxford University Press, New York.

Kareem, A.M. & Barnard, C.J. (1986) Kin recognition in mice: age, sex, and parental effects. *Anim Behav*, **34**, 1814–24.

Kasuya, E., Hibino, Y. & Itô, Y. (1980) On 'intercolonial' cannibalism in Japanese paper wasps, *Polistes chinensis antennalis* Perez and *P. jadwigae* Dalla Torre (Hymenoptera: Vespidae). *Res Popul Ecol (Kyoto)*, **22**, 255–62.

Kavaliers, M. & Colwell, D.D. (1995) Odours of parasitized males induce aversive responses in female mice. *Anim Behav*, **50**, 1161–9.

Kawai, M. (1958) In the system of social ranks in a natural troop of Japanese monkeys (1): basic rank and dependent rank. *Primates*, **1**, 111–48.

Kawai, M. (1965) Newly acquired pre-cultural behavior of the natural troop of Japanese monkeys on Koshima Islet. *Primates*, **6**, 1–30.

Keane, B., Waser, P.M., Creel, S.M., Creel, N.M., Elliott, L.F. & Minchella, D.J. (1994) Subordinate reproduction in dwarf mongooses. *Anim Behav*, **47**, 65–75.

Keller, L. (1995) Social life: the paradox of multiple-queen colonies. *Trends Ecol Evol*, **10**, 355–60.

Keller, L. & Nonacs, P. (1993) The role of queen pheromones in social insects: queen control or queen signal? *Anim Behav*, **45**, 787–94.

Keller, L. & Reeve, H.K. (1994) Partitioning of reproduction in animal societies. *Trends Ecol Evol*, **9**, 98–102.

Keller, L. & Vargo, E.L. (1993) Reproductive structure and reproductive roles in colonies of eusocial insects. In: *Queen Number and Sociality in Insects* (ed. L. Keller), pp. 16–44. Oxford University Press, Oxford.

Keller, L.F., Arcese, P., Smith, J.N.M., Hochacka, W.M. & Stearns, S.C. (1994) Selection against inbred song sparrows during a natural population bottleneck. *Nature*, **372**, 356–7.

Kempenaers, B. & Sheldon, B.C. (1996) Why do male birds not discriminate between their own and extra-pair offspring? *Anim Behav*, **51**, 1165–73.

Kempenaers, B., Verheyen, G.R., Broeck, M.V.D., Burke, T., Broeckhoven, C.V. & Dhondt, A.A. (1992) Extra-pair paternity results from female preference for high-quality males in the blue tit. *Nature*, **357**, 494–6.

Kendeigh, S.C., Kontogiannis, J.E., Maza, A. & Roth, R. (1969) Environmental regulation of food intake by birds. *Comp Biochem Physiol*, **31**, 941–57.

Kennedy, C.E.J., Endler, J.A., Poyton, S.L. & McMinn, H. (1987) Parasite load predicts mate choice in guppies. *Behav Ecol Sociobiol*, **17**, 199–206.

Kennedy, J.S. (1992) *The New Anthropomorphism*.

Cambridge University Press, Cambridge.

Kent, D.S. & Simpson, J.A. (1992) Eusociality in the beetle *Austroplatypus incompertus* (Coleoptera: Curculionidae). *Naturwissenschaften*, **79**, 86–7.

Kersten, M. & Piersma, T. (1987) High levels of energy expenditure in shorebirds: metabolic adaptations to an energetically expensive way of life. *Ardea*, **75**, 175–88.

Kilner, R. (1995) When do canary parents respond to nestling signals of need? *Proc Roy Soc Lond B*, **260**, 343–8.

King, J.R. & Murphy, M.E. (1985) Periods of nutritional stress in the annual cycles of endotherms: fact or fiction? *Am Zool*, **25**, 955–64.

Kinlen, L.J. (1992) Malignancy and autoimmune diseases. *J Autoimmun*, **5(A)**, 363–71.

Kirkpatrick, M. (1982) Sexual selection and the evolution of female choice. *Evolution*, **36**, 1–12.

Kirkpatrick, M. & Dugatkin, L.A. (1994) Sexual selection and the evolutionary effects of copying mate choice. *Behav Ecol Sociobiol*, **34**, 443–9.

Kirkpatrick, M. & Barton, N.H. (1996) The strength of indirect selection on female mating preferences, *Proc Nat Acad Sci (USA)*, in press.

Kirkpatrick, M. & Ryan, M.J. (1991) The paradox of the lek and the evolution of mating preferences. *Nature*, **350**, 33–8.

Kirkwood, J.K. (1983) A limit to metabolisable energy intake in mammals and birds. *Comp Biochem Physiol*, **75A**, 1–3.

Klahn, J.E. (1988) Intraspecific comb usurpation in the social wasp *Polistes fuscatus*. *Behav Ecol Sociobiol*, **23**, 1–8.

Kleiber, M. (1947) Body size and metabolic rate. *Physiol Rev*, **27**, 511–41.

Klomp, H. (1970) The determination of clutch-size in birds. A review. *Ardea*, **58**, 2–123.

Klopfer, P. (1961) Observational learning in birds: the establishment of behavioral modes. *Behaviour*, **17**, 71–9.

Knowlton, N. (1974) A note on the evolution of gamete dimorphism. *J Theor Biol*, **46**, 283–5.

Knowlton, N., Weigt, L.A., Solórzano, L.A., Mills, D.K. & Bermingham, E. (1993) Divergence in proteins, mitochondrial DNA, and reproductive compatibility across the Isthmus of Panama. *Science*, **260**, 1629–32.

Kodric-Brown, A. (1989) Dietary carotenoids and male mating success in the guppy: an environmental component to female choice. *Behav Ecol Sociobiol*, **25**, 393–401.

Kodric-Brown, A. (1985) Female preference and sexual selection for male colouration in the guppy *Poecilia reticulata*. *Behav Ecol Sociobiol*, **25**, 393–401.

Kodric-Brown, A. (1993) Female choice of multiple male criteria in guppies: interacting effects of dominance, coloration and courtship. *Behav Ecol Sociobiol*, **32**, 415–20.

Koenig, W.D. (1990) Opportunity of parentage and nest destruction in polygynandrous acorn woodpeckers. *Behav Ecol*, **1**, 55–61.

Koenig, W.D. & Mumme, R.L. (1987) *Population ecology of the cooperative breeding acorn woodpecker*. Monograph in *Population Biology*, No. 24. Princeton University Press, Princeton, New Jersey.

Koenig, W.D., Mumme, R.L., Stanback, M.T. & Pitelka, F.A. (1995) Patterns and consequences of egg destruction among joint-nesting acorn woodpeckers. *Anim Behav*, **50**, 607–21.

Koenig, W.D., Pitelka, F.A., Carmen, W.J., Mumme, R.L. & Stanback, M.T. (1992) The evolution of delayed dispersal in cooperative breeders. *Quart Rev Biol*, **67**, 111–50.

Koford, R.R., Bowen, B.S. & Vehrencamp, S.L. (1990) Groove-billed Anis: joint-nesting in a tropical cuckoo. In: *Cooperative Breeding in Birds: Long-term Studies in Ecology and Behavior* (eds P.B. Stacey & W.D. Koenig), pp. 335–55. Cambridge University Press, Cambridge.

Komdeur, J. (1992) Importance of habitat saturation and territory quality for evolution of cooperative breeding in the Seychelles warbler. *Nature*, **358**, 493–5.

Komdeur, J. (1994) The effect of kinship on helping in the cooperative breeding Seychelles warbler (*Acrocephalus seychellensis*). *Proc Roy Soc Lond B*, **256**, 47–52.

Konarzewski, M. & Diamond, J. (1994) Peak sustained metabolic rate and its individual variation in cold-stressed mice. *Physiol Zool*, **67**, 1186–212.

Kontogiannis, J.E. (1967) Day and night changes in body weight in the white-crowned sparrow (*Zonotrichia albicollis*). *Auk*, **84**, 390–5.

Koops, M. & Giraldeau, L.-A. (1996) Producer–scrounger foraging games in starlings: a test of mean-maximizing and risk-minimizing foraging models. *Anim Behav*, **51**, 773–83.

Kortner, G. & Heldmaier, G. (1995) Body weight cycles and energy balance in the alpine marmot (*Marmota marmota*). *Physiol Zool*, **68**, 149–63.

Kozłowski, J. (1992) Optimal allocation of resources to growth and reproduction: implications for age and size at maturity. *Trends Ecol Evol*, **7**, 15–19.

Kozłowski, J. (1996) Optimal allocation of resources explains interspecific life-history patterns in animals with indeterminate growth. *Proc Roy Soc Lond B*, **263**, 559–66.

Kozłowski, J. & Uchmanski, J. (1987) Optimal

individual growth and reproduction in perennial species with indeterminate growth. *Evol Ecol*, **1**, 214–30.

Kozłowski, J. & Weiner, J. (1996) Interspecific allometries are byproducts of body size optimization. *Am Natur* (in press).

Kozłowski, J. & Wiegert, R. (1986) Optimal allocation of energy to growth and reproduction. *Theor Popul Biol*, **29**, 16–37.

Kozłowski, J. & Wiegert, R.G. (1987) Optimal age and size at maturity in annuals and perennials with determinate growth. *Evol Ecol*, **1**, 231–44.

Krakauer, D.C. & Johnstone, R.A. (1995) The evolution of exploitation and honesty in animal communication: a model using artificial neural networks. *Phil Trans Roy Soc Lond B*, **348**, 355–61.

Krebs, J.R. (1970) The efficiency of courtship feeding in blue tits. *Ibis*, **112**, 108–10.

Krebs, J.R. (1973) Social learning and the significance of mixed-species flocks of chickadees. *Can J Zool*, **51**, 1275–88.

Krebs, J.R. (1982) Territorial defence in the great tit *Parus major*: do residents always win? *Behav Ecol Sociobiol*, **11**, 185–94.

Krebs, J.R. (1990) Food storing birds: adaptive specialization in brain and behaviour. *Phil Trans Roy Soc Lond B*, **329**, 153–60.

Krebs, J.R. & Davies, N.B. (1993) *An Introduction to Behavioural Ecology*, 3rd edn. Blackwell Scientific Publications, Oxford.

Krebs, J.R. & Dawkins, R. (1984) Animal signals: mind-reading and manipulation. In: *Behavioural Ecology: an Evolutionary Approach*, 2nd edn (eds J.R. Krebs & N.B. Davies), pp. 380–402. Blackwell Scientific Publications, Oxford.

Krebs, J.R. & Inman, A.J. (1992) Learning and foraging: individuals, groups and populations. *Am Natur*, **140**, S63–S84.

Krebs, J.R. & Kacelnik, A. (1991) Decision-making. In: *Behavioural Ecology. An Evolutionary Approach* (eds J.R. Krebs & N.B. Davies), 3rd edn, pp. 105–36. Blackwell Scientific Publications, Oxford.

Krebs, J.R., Healy, S.D. & Shettleworth, S.J. (1990) Spatial memory of Paridae: comparison of storing and non-storing species, the coal tit, *Parus ater*, and the great tit, *P. major*. *Anim Behav*, **39**, 1127–37.

Krebs, J.R., MacRoberts, M.H. & Cullen, J.M. (1972) Flocking and feeding in the great tit *Parus major* — an experimental study. *Ibis*, **114**, 507–30.

Krebs, J.R., Clayton, N.S., Hampton, R.R. & Shettleworth, S.J. (1995) Effects of photoperiod on food-storing and the hippocampus in birds. *Neuroreport*, **6**, 1701–4.

Krebs, J.R., Sherry, D.F., Healy, S.D., Perry, V.H. & Vaccarino, A.L. (1989) Hippocampal specialization of food storing birds. *Proc Nat Acad Sci USA*, **86**, 1388–92.

Kreitman, M. & Akashi, H. (1995) Molecular evidence for natural selection. *Ann Rev Ecol Syst*, **26**, 403–22.

Krementz, D.G. & Ankney, C.D. (1995) Changes in total-body calcium and diet of breeding house sparrows. *J Avian Biol*, **26**, 162–7.

Kruuk, H. (1975) Functional aspects of social hunting by carnivores. In: *Function and Evolution in Behaviour: Essays in honour of Professor Niko Tinbergen F.R.S.* (eds G. Baerends, C. Beer & A. Manning), pp. 119–41. Oxford University Press, Oxford.

Kusano, K., Naito, T., Handa, N. & Kobayashi, I. (1995) Restriction–modification systems as genomic parasites in competition for specific sequences. *Proc Nat Acad Sci USA*, **92**, 11 095–9.

Kuusaari, M., Nieminen, M. & Hanski, I. (1996) An experimental study of migration in the Glanville fritillary butterfly, *Melitaea cinxia*. *J Anim Ecol*, **65**, 791–801.

Kyriacou, C.P. & Hall, J.C. (1986) Interspecific genetic control of courtship song production and reception in *Drosophila*. *Science*, **232**, 494–7.

Lack, D. (1950) The breeding seasons of European birds. *Ibis*, **92**, 288–316.

Lack, D. (1966) *Population Studies of Birds*. Clarendon Press, Oxford.

Lack, D. (1968) *Ecological Adaptations for Breeding in Birds*. Methuen, London.

Lacy, R.C. (1980) The evolution of eusociality in termites: a haplodiploid analogy? *Am Natur*, **116**, 449–51.

Lacy, R.C. & Sherman, P.W. (1983) Kin recognition by phenotype matching. *Am Natur*, **121**, 489–512.

Lagercrantz, U. & Ryman, N. (1990) Genetic structure of Norway Spruce (*Picea abies*): concordance of morphological and allozymic variation. *Evolution*, **44**, 38–53.

Lambrechts, M.M. & Dhondt, A.A. (1995) Individual voice discrimination in birds. *Curr Ornithol*, **12**, 115–39.

Land, M.F. (1969) Structure of the retinae of the principal eyes of jumping spiders (Salticidae: Dendryphantinae) in relation to visual optics. *J Exp Biol*, **51**, 443–70.

Land, M.F. (1981) Optics and vision in invertebrates. In: *Handbook of Sensory Physiology*, Vol. VII/6B (ed. H. Autrum), pp. 471–592. Springer-Verlag, Berlin, New York.

Land, M.F. (1989) Variations in the structure and design of compound eyes. In: *Facets of Vision* (eds D.G. Stavenga & R.C. Hardie), pp. 90–111. Springer-Verlag, Berlin.

Land, M.F. & Fernald, R.D. (1992) The evolution of eyes. *Ann Rev Neurosci*, **15**, 1–29.

Landau, H.G. (1951) On dominance relations and the structure of animal societies. I. Effect of inherent characteristics. *Bull Math Biophysics*, **13**, 1–19.

Lande, R. (1981) Models of speciation by sexual selection of polygenic traits. *Proc Nat Acad Sci USA*, **78**, 3721–5.

Landwer, A.J. (1994) Manipulation of egg production reveals cost of reproduction in the tree lizard (*Urosaurus ornatus*). *Oecologia*, **100**, 243–9.

Langen, T.A. & Rabenold, K.E. (1994) Dominance and diet selection in juncos. *Behav Ecol*, **5**, 334–8.

Lank, D.B., Smith, C.M., Hanotte, O., Burke, T. & Cooke, F. (1995) Genetic polymorphism for alternative mating behaviour in lekking male ruff *Philomachus pugnax*. *Nature Lond*, **378**, 59–62.

Lau, M.M.H., Stewart, C.E.H., Liu, Z., Bhatt, H., Rotwein, P. & Stewart, C.L. (1994) Loss of the imprinted IGF2/cation-independent mannose 6-phosphate receptor results in fetal overgrowth and perinatal lethality. *Genes Develop*, **8**, 2953–63.

Lauder, G.V. (1981) Form and function: structural analysis in evolutionary morphology. *Paleobiology*, **7**, 430–42.

Laughlin, S.B. (1995) Towards the cost of seeing. In: *Nervous Systems and Behaviour* (eds M. Burrows, T. Matthews, P.L. Newland & H.J. Schuppe), p. 290. Thieme, Stuttgart, New York.

Laughlin, S.B. (in press) Observing design with compound eyes. In: *Optimal Animal Design: Fact or Fancy?* (eds E.R. Weibel & L. Bolis). University Press, Cambridge.

Law, R. & Hutson, V. (1992) Intracellular symbionts and the evolution of uniparental cytoplasmic inheritance. *Proc Roy Soc Lond B*, **248**, 69–77.

Lawrence, E.S. & Allen, J.A. (1983) On the term 'search image'. *Oikos*, **40**, 313–14.

Lazarus, J. & Metcalfe, N. (1990) Tit for tat cooperation in sticklebacks: a critique of Milinski. *Anim Behav*, **39**, 987–8.

Lazcano, A., Guerrero, R., Margulis, L. & Oró, J. (1988) The evolutionary transition from RNA to DNA in early cells. *J Molec Evol*, **27**, 283–90.

Lee, R.B. & DeVore, I. (eds) (1968) *Man the Hunter*. Aldine Publishing Company, Chicago.

Lee, S.J., Witter, M.S., Cuthill, I.C. & Goldsmith, A.R. (1996) Reduction in escape performance as a cost of reproduction in gravid starlings (*Sturnus vulgaris*). *Proc Roy Soc Lond B*, **263**, 619–23.

Lefebvre, L. (1986) Cultural diffusion of a novel food-finding behaviour in urban pigeons: an experimental field test. *Ethology*, **71**, 295–304.

Lefebvre, L. (1995a) The opening of milk bottles by birds: evidence for accelerating learning rates, but against the wave-of-advance model of cultural transmission. *Behav Process*, **34**, 43–54.

Lefebvre, L. (1995b) Culturally-transmitted feeding behaviour in primates: evidence for accelerating learning rates. *Primates*, **36(2)**, 227–39.

Lefebvre, L. & Giraldeau, L.-A. (1994) Cultural transmission in pigeons is affected by the number of tutors and bystanders present during demonstrations. *Anim Behav*, **47**, 331–7.

Lefebvre, L., Whittle, P., Lascaris, E. & Finkelstein, A. (1996a) Feeding innovations and forebrain size in birds. *Anim Behav* (in press).

Lefebvre, L., Palameta, B. & Hatch, K.K. (1996b) Is group-living associated with social learning? A comparative test of a gregarious and a territorial columbid. *Behaviour*, **133**, 241–61.

Lehnherr, H., Maguin, E., Jafri, S. & Yarmolinsky, M.B. (1993) Plasmid addiction genes of bacteriophage P1: *doc*, which causes cell death on curing of prophage, and *phd*, which prevents host death when prophage is retained. *J Molec Biol*, **233**, 414–28.

Leigh, E.G. (1971) *Adaptation and Diversity*. Freeman, Cooper & Company, San Francisco.

Leigh, E.G. (1990) Community diversity and environmental stability: a re-examination. *Trends Ecol Evol*, **5**, 340–4.

Lemon, W.C. (1993a) The energetics of lifetime reproductive success in the zebra finch *Taeniopygia guttata*. *Physiol Zool*, **66**, 946–63.

Lemon, W.C. (1993b) Heritability of selectively advantageous foraging behaviour in a small passerine. *Evol Ecol*, **7**, 421–8.

Lenington, S. (1991) The t-complex: a story of genes, behavior, and populations. *Adv Stud Behav*, **20**, 51–86.

Lenington, S., Egid, K. & Williams, J. (1988) Analysis of a genetic recognition system in wild house mice. *Behav Genet*, **18**, 549–64.

Lenington, S., Coopersmith, C.B. & Williams, J. (1992) Genetic basis of mating preferences in wild house mice. *Am Zool*, **32**, 40–7.

Lenington, S., Coopersmith, C.B. & Erhardt, M. (1994) Female preference and variability among *t*-haplotypes in wild house mice. *Am Natur*, **143**, 766–84.

Lenski, R.E., Simpson, S.C. & Nguyen, T.T. (1994) Genetic analysis of a plasmid-encoded, host genotype-specific enhancement of bacterial fitness. *J Bacteriol*, **176**, 3140–7.

Lessells, C.M. (1991) The evolution of life histories. In: *Behavioural Ecology. An Evolutionary Approach* (eds J. Krebs & N. Davies), pp. 32–68. Blackwell Scientific

Publications, Oxford.

Lessells, C.M. & Birkhead, T.R. (1990) Mechanisms of sperm competition in birds: mathematical models. *Behav Ecol Sociobiol*, **27**, 325–37.

Levitan, D.R. & Petersen, C. (1995) Sperm limitation in the sea. *Trends Ecol Evol*, **10**, 228–31.

Lewis, D., Verma, S.C. & Zuberi, M.I. (1988) Gametophytic–sporophytic incompatibility in the Cruciferae *Raphanus sativus*. *Heredity*, **61**, 355–66.

Lewontin, R.C. (1991) *Biology as Ideology*. House of Anansi Press, Concord, Ontario.

Lewontin, R.C., Rose, S. & Kanin, L.J. (1984) *Not in our Genes*. Pantheon, New York.

Liem, K.F. (1980) Adaptive significance of intra- and interspecific differences in the feeding repertoires of cichlid fishes. *Am Zool*, **20**, 295–314.

Liem, K.F. (1973) Evolutionary strategies and morphological innovations: cichlid pharyngeal jaws. *System Zool*, **22**, 425–41.

Lima, S.L. (1986) Predation risk and unpredictable feeding conditions: determinants of body mass in wintering birds. *Ecology*, **67**, 377–85.

Lima, S.L. (1989) Iterated prisoner's dilemma: an approach to evolutionarily stable cooperation. *Am Natur*, **134**, 828–44.

Lima, S.L. & Zollner, P.A. (1996) Towards a behavioral ecology of ecological landscapes. *Trends Ecol Evol*, **11**, 131–5.

Lin, N. & Michener, C.D. (1972) Evolution of sociality in insects. *Quart Rev Biol*, **47**, 131–59.

Linsenmair, K.E. (1987) Kin recognition in subsocial arthropods, in particular in the desert isopod *Hemilepistus reaumuri*. In: *Kin Recognition in Animals* (eds D.J.C. Fletcher & C.D. Michener), pp. 121–208. John Wiley & Sons, New York.

Lindstedt, S.L. & Calder, W.A. (1981) Body size, physiological time, and longevity of homeothermic animals. *Quart Rev Biol*, **56**, 1–16.

Lindström, Å. & Alerstam, T. (1992) Optimal fat load in migrating birds: a test of the time minimisation hypothesis. *Am Natur*, **140**, 477–91.

Lindström, Å. & Piersma, T. (1993) Mass changes in migrating birds: the evidence for fat and protein storage re-examined. *Ibis*, **135**, 70–8.

Locket, N.A. (1977) Adaptations to the deep-sea environment. In: *Handbook of Sensory Physiology*, Vol. VII/5 (ed. F. Crescitelli), pp. 67–192. Springer-Verlag, Berlin, New York.

Loehle, C. (1995) Social barriers to pathogen transmission in wild animal populations. *Ecology*, **76**, 326–35.

Loftus-Hills, J.J. & Littlejohn, M.J. (1992) Reinforcement and reproductive character displacement in *Gastrophryne carolinensis* and *G. olivacea* (Anura: Microhylidae): a reexamination. *Evolution*, **46**, 896–906.

Loher, W. & Dambach, M. (1989) Reproductive behavior. In: *Cricket Behavior and Neurobiology* (eds F. Huber, T.E. Moore & W. Loher), pp. 43–82. Cornell University Press, Ithaca, New York.

Long, K.D. & Houde, A.E. (1989) Orange spots as visual cues for female mate choice in the guppy (*Poecilia reticulata*). *Ethology*, **82**, 316–24.

Lord, E.M. & Eckard, K.J. (1984) Incompatibility between the dimorphic flowers of *Collomia grandiflora*, a cleistogamous species. *Science*, **223**, 695–6.

Lord, R.D. (1960) Litter size and latitude in North American mammals. *Am Midland Natur*, **63**, 488–99.

Lorenz, M.G. & Wackerknagel, W. (1994) Bacterial gene transfer by natural genetic transformation in the environment. *Microbiol Rev*, **58**, 563–602.

Losos, J. (1992) The evolution of convergent structure in Caribbean *Anolis* communities. *System Biol*, **401**, 403–20.

Losos, J.B. (1996) Community evolution in Greater Antillean *Anolis* lizards: phylogenetic patterns and experimental tests. In: *New Uses for New Phylogenies* (eds P.H. Harvey, A.J. Leigh Brown, J. Maynard Smith & S. Nee), pp. 308–21. Oxford University Press, Oxford.

Lotem, A. (1993) Learning to recognize nestlings is maladaptive for *Cuculus canorus* hosts. *Nature*, **362**, 743–5.

Lotem, A., Nakamura, H. & Zahavi, A. (1995) Constraints on egg discrimination in cuckoo-host evolution. *Anim Behav*, **49**, 1185–1209.

Lovejoy, C.O. (1981) The origin of man. *Science*, **211**, 341–50.

Lucas, J.R. & Howard, R.D. (1995) On alternative reproductive tactics in anurans: dynamic games with density and frequency dependence. *Am Natur*, **146**, 365–97.

Lucas, J.R. & Walter, L.R. (1991) When should chickadees hoard food? Theory and experimental results. *Anim Behav*, **41**, 579–601.

Lucas, J.R., Peterson, L.J. & Boudinier, R.L. (1993) The effects of time constraints and changes in body mass and satiation on the simultaneous expression of caching and diet-choice decisions. *Anim Behav*, **45**, 639–58.

Luykx, P. & Syren, R.M. (1979) The cytogenetics of *Incisitermes schwarzi* and other Florida termites. *Sociobiology*, **4**, 191–209.

Luykx, P., Michel, J. & Luykx, J. (1986) The spatial distribution of the sexes in colonies of the termite *Incisitermes schwarzi* Banks (Isoptera: Kalotermitidae).

Insect Soc, **33**, 406–21.

Lythgoe, J.N. (1988) Light and vision in the aquatic environment. In: *Sensory Biology of Aquatic Animals* (eds J. Atema, R.R. Fay, A.N. Popper & N. Tavoga), pp. 57–82. Springer-Verlag, New York.

Lythgoe, J.N., Muntz, W.R.A., Partridge, J.C., Shand, J. & Williams, D.McB. (1994) The ecology of the visual pigments of snappers (Lutjanidae) on the Great Barrier Reef. *J Comp Physiol A*, **174**, 461–7.

MacArthur, R.H. & Wilson, E.O. (1967) *The Theory of Island Biogeography*. Princeton University Press, Princeton, NJ.

MacArthur, R.H. (1972) *Geographical Ecology*. Harper & Row, New York.

MacArthur, R.H. & Pianka, E.R. (1966) On the optimal use of a patchy environment. *Am Natur*, **100**, 603–9.

McBeath, M.K., Shaffer, D.M. & Kaiser, M.K. (1995) How baseball outfielders determine where to run to catch fly balls. *Science*, **268**, 569–73.

McCauley, D.E. (1991) Genetic consequences of local population extinction and recolonization. *Trends Ecol Evol*, **6**, 5–8.

McComb, K.E., Packer, C. & Pusey, A.E. (1994) Roaring and numerical assessment in contests between groups of female lions, *Panthera leo*. *Anim Behav*, **47**, 379–87.

McDonald, D.B. & Potts, W.K. (1994) Cooperative display and relatedness among males in a lek-mating bird. *Science*, **266**, 1030–2.

Macdonald, D.W. (1979) Helpers in fox society. *Nature*, **282**, 69–71.

Macdonald, D.W. & Moehlman, P.D. (1982) Cooperation, altruism, and restraint in the reproduction of carnivores. *Perspec Ethol*, **5**, 433–69.

McGregor, P.K. (1993) Signalling in territorial systems: a context for individual identification, ranging and eavesdropping. *Phil Trans Roy Soc Lond B*, **340**, 237–44.

McGregor, P.K. & Krebs, J.R. (1982) Song types in a population of great tits (*Parus major*): their distribution, abundance, and acquisition by individuals. *Behaviour*, **79**, 126–52.

Mackeney, P.A. & Hughes, R.N. (1995) Foraging behaviour and memory window in sticklebacks. *Behaviour*, **132**, 1241–53.

McKinney, F., Cheng, K.M. & Bruggers, D.J. (1984) Sperm competition in apparently monogamous birds. In: *Sperm Competition and the Evolution of Animal Mating Systems* (ed R.L. Smith), pp. 523–45. Academic Press, Orlando.

McKinney, F., Derrickson, S.R. & Mineau, P. (1983) Forced copulation in waterfowl. *Behaviour*, **86**, 250–94.

McLaughlin, R.L. & Montgomerie, R.D. (1990) Flight speeds of parent birds feeding nestlings: maximization of foraging efficiency or food delivery rate? *Can J Zool*, **68**, 2269–74.

McLennan, D.A. & McPhail, J.D. (1990) Experimental investigations of the evolutionary significance of sexually dimorphic nuptial colouration in *Gasterosteus aculeatus* (L.): the relationship between male colour and female behaviour. *Can J Zool*, **68**, 482–92.

McLeod, P. & Dienes, Z. (1993) Running to catch the ball. *Nature*, **362**, 23.

McMahon, T. (1973) Size and shape in biology. *Science*, **179**, 1201–4.

McMinn, H. (1990) Effects of the nematode parasite *Camallanus cotti* in sexual and non-sexual behaviours in the guppy (*Poecilia reticulata*). *Am Zool*, **30**, 245–9.

McNamara, J.M. (1990) The starvation–predation trade-off and some behavioural and ecological consequences. In: *Behavioural Mechanisms of Food Selection, NATO ASI Series A, Life Sciences* (ed. R.N. Hughes), pp. 39–59. Springer-Verlag, New York.

McNamara, J.M. (1994) Timing of entry into diapause: optimal allocation to 'growth' and 'reproduction' in a stochastic environment. *J Theor Biol*, **168**, 201–9.

McNamara, J.M. & Houston, A.I. (1980) The application of statistical decision theory to animal behaviour. *J Theor Biol*, **85**, 673–90.

McNamara, J.M. & Houston, A.I. (1986) The common currency for behavioral decisions. *Am Natur*, **127**, 358–78.

McNamara, J.M. & Houston, A.I. (1987) Starvation and predation as factors limiting population size. *Ecology*, **68**, 1515–19.

McNamara, J.M. & Houston, A.I. (1990) The value of fat reserves and the trade-off between starvation and predation. *Acta Biotheor*, **38**, 37–61.

McNamara, J.M. & Houston, A.I. (1992) State-dependent life-history theory and its implications for optimal clutch size. *Evol Ecol*, **6**, 170–85.

McNamara, J.M. & Houston, A.I. (1994) The effect of a change in foraging options on intake rate and predation rate. *Am Natur*, **144**, 978–1000.

McNamara, J.M., Houston, A.I. & Krebs, J.R. (1990) Why hoard? The economics of food storing in tits, *Parus* spp. *Behav Ecol*, **1**, 12–23.

McNamara, J.M., Houston, A.I. & Lima, S.L. (1994) Foraging routines of small birds in winter: a theoretical investigation. *J Avian Biol*, **25**, 287–302.

McNamara, J.M., Mace, R.H. & Houston, A.I. (1987) Optimal daily routines of singing and foraging. *Behav Ecol Sociobiol*, **20**, 399–405.

McQuoid, L.M. & Galef, B.G. Jr (1992) Social influences of feeding site selection by Burmese fowl *Gallus gallus*. *J Compar Psychol*, **106**, 137–41.

Maddison, W.P. (1990) A method for testing the correlated evolution of two binary characters: are gains or losses concentrated on certain branches of a phylogenetic tree? *Evolution*, **44**, 539–57.

Maddison, W.P. & Maddison, D.R. (1992) *MacClade: Analysis of Phylogeny and Character Evolution. Version 3*. Sinauer, Sunderland.

Madsen, T. (1987) Cost of reproduction and female life-history tactics in a population of grass snakes, *Natrix natrix*, in southern Sweden. *Oikos*, **49**, 129–32.

Madsen, T., Shine, R., Loman, J. & Hakansson, T. (1992) Why do female adders copulate so frequently? *Nature Lond*, **355**, 440–1.

Malcolm, J.R. & Marten, K. (1982) Natural selection and the communal rearing of pups in African wild dogs (*Lycaon pictus*). *Behav Ecol Sociobiol*, **10**, 1–13.

Malthus, T.R. (1798) *An Essay on the Principle of Population*. Reprinted by Macmillan, New York.

Mangel, M. & Clark, C.W. (1986) Towards a unified foraging theory. *Ecology*, **67**, 1127–38.

Mangel, M. & Clark, C.W. (1988) *Dynamic Modelling in Behavioural Ecology*. Princeton University Press, Princeton.

Manning, C.J., Wakeland, E.K. & Potts, W.K. (1992) Communal nesting patterns in mice implicate MHC genes in kin recognition. *Nature*, **360**, 581–3.

Manning, C.J., Dewsbury, D.A., Wakeland, E.K. & Potts, W.K. (1995) Communal nesting and communal nursing in house mice, *Mus musculus domesticus*. *Anim Behav*, **50**, 741–51.

Marchetti, K. (1993) Dark habitats and bright birds illustrate the role of the environment in species divergence. *Nature Lond*, **362**, 149–52.

Marchetti, K. & Price, T. (1989) Differences in the foraging of juvenile and adult birds: the importance of developmental constraints. *Biol Rev*, **64**, 51–70.

Markl, H. (1983) Vibrational communication. In: *Neuroethology and Behavioral Physiology* (eds F. Huber & H. Markl), pp. 332–53. Springer-Verlag, Berlin, Heidelberg.

Marler, P. (1991) Song-learning behavior: the interface with neuroethology. *Trends Neurosci*, **14**, 199–206.

Marr, D. (1982) *Vision*. W.H. Freeman, San Francisco.

Marsh, R.L. (1984) Adaptations of the grey catbird *Dumetella carolinensis* to long-distance migration: flight muscle hypertrophy associated with elevated body mass. *Physiol Zool*, **57**, 105–117.

Martin, P.A., Reimers, T.J., Lodge, J.R. & Dziuk, P.J. (1974) The effect of ratios and numbers of spermatozoa mixed from two males on proportions of offspring. *J Reprod Fert*, **39**, 251–8.

Martins, E.P. & Garland, T.H. (1991) Phylogenies and the evolution of continuous characters. *Evolution*, **45**, 534–57.

Marzluff, J.M. & Balda, R.P. (1990) Pinyon jays: making the best of a bad situation by helping. In: *Cooperative Breeding in Birds: Long-term Studies of Ecology and Behavior* (eds P.B. Stacey & W.D. Koenig), pp. 197–237. Cambridge University Press, Cambridge.

Masman, D., Daan, S. & Dijkstra, C. (1988) Time allocation in the kestrel (*Falco tinnunculus*), and the principle of energy minimization. *J Anim Ecol*, **57**, 411–32.

Mason, R.J. (1988) Direct and observational learning by redwinged blackbirds (*Agelaius phoeniceus*): the importance of complex visual stimuli. In: *Social Learning: Psychological and Biological Perspectives* (eds T.R. Zentall & B.G. Galef Jr), pp. 99–115. Lawrence-Erlbaum Associates, New Jersey.

Masters, W. & Waite, T. (1990) Tit-for-tat during predator inspection, or shoaling? *Anim Behav*, **39**, 603–4.

Mateo, J. (1996) The development of alarm-call response behaviour in free-living juvenile Belding's ground squirrels. *Anim Behav*, **52**, 489–505.

Mayfield, H. (1965) The brown-headed cowbird, with old and new hosts. *Living Bird*, **4**, 13–28.

Maynard Smith, J. (1972) *On Evolution*. Edinburgh University Press, Edinburgh.

Maynard Smith, J. (1974) The theory of games and the evolution of animal conflicts. *J Theor Biol*, **47**, 209–21.

Maynard Smith, J. (1977) Parental investment — a prospective analysis. *Anim Behav*, **25**, 1–9.

Maynard Smith, J. (1978) *The Evolution of Sex*. Cambridge University Press, Cambridge.

Maynard Smith, J. (1978) Optimization theory in evolution. *Ann Rev Ecol Syst*, **9**, 31–56.

Maynard Smith, J. (1979) Game theory and the evolution of behaviour. *Proc Roy Soc Lond B*, **205**, 475–88.

Maynard Smith, J. (1982) *Evolution and the Theory of Games*. Cambridge University Press, Cambridge.

Maynard Smith, J. (1991) Honest signalling: the Philip Sidney game. *Anim Behav*, **42**, 1034–5.

Maynard Smith, J. (1991) Theories of sexual selection. *Trends Ecol Evol*, **6**, 146–51.

Maynard Smith, J. & Parker, G.A. (1976) The logic of asymmetric contests. *Anim Behav*, **24**, 159–75.

Maynard Smith, J. & Price, G.R. (1973) The logic of animal conflict. *Nature Lond*, **246**, 15–18.

Maynard Smith, J. & Riechert, S.E. (1984) A conflicting-tendency model of spider agonistic behavior: hybrid-pure population line comparisons. *Anim Behav*, **32**, 564–78.

Maynard Smith, J. & Szathmáry, E. (1993) The origin of chromosomes. I. Selection for linkage. *J Theor Biol*, **164**, 437–46.

Maynard Smith, J. & Szathmáry, E. (1995) *The Major Transitions in Evolution*. W.H. Freeman, Oxford.

Mayr, E. (1970) *Populations, Species, and Evolution*. Belknap Press of Harvard University Press, Cambridge, MA.

Meddis, R. (1983) The evolution of sleep. In: *Sleep Mechanisms and Functions* (ed. A. Mayes), pp. 57–106. Van Nostrand Reinhold, London.

Meffert, L.M. (1995) Bottleneck effects on genetic variance for courtship repertoire. *Genetics*, **139**, 365–74.

Mesterton-Gibbons, M. & Dugatkin, L.A. (1995) Toward a theory of dominance hierarchies: effects of assessment, group size, and variation in fighting ability. *Behav Ecol*, **6**, 416–23.

Metcalf, R.A. & Whitt, G.S. (1977a) Relative inclusive fitness in the social wasp *Polistes metricus*. *Behav Ecol Sociobiol*, **2**, 353–60.

Metcalf, R.A. & Whitt, G.S. (1977b) Intra-nest relatedness in the social wasp *Polistes metricus*. A genetic analysis. *Behav Ecol Sociobiol*, **2**, 339–51.

Metcalfe, N.B. & Ure, S.E. (1995) Diurnal variation in flight performance and hence predation risk in small birds. *Proc Roy Soc Lond B*, **261**, 395–400.

Meyer, A., Morrissey, J.M. & Schartl, M. (1994) Recurrent origin of a sexually selected trait in *Xiphophorus* fishes inferred from a molecular phylogeny. *Nature*, **368**, 539–42.

Michelsen, A. (1983) Hearing and sound communication in small animals: evolutionary adaptations to the laws of physics. In: *The Evolutionary Biology of Hearing* (eds D.B. Webster, R.R. Fay & A.N. Popper), pp. 61–77. Springer-Verlag, New York, Berlin.

Michelsen, A., Towne, W.F., Kirchner, W.H., Kryger, P. (1987) The acoustic near field of a dancing honeybee. *J Comp Physiol A*, **161**, 633–43.

Michener, C.D. & Smith, B.H. (1987) Kin recognition in primitively eusocial insects. In: *Kin Recognition in Animals* (eds D.J.C. Fletcher & C.D. Michener), pp. 209–42. John Wiley & Sons, New York.

Michod, R.E. (1982) The theory of kin selection. *Ann Rev Ecol Syst*, **13**, 23–55.

Milinski, M. (1984) Competitive resource sharing: an experimental test of a learning rule for ESSs. *Anim Behav*, **32**, 233–42.

Milinski, M. (1984) A predator's cost of overcoming the confusion effect of swarming prey. *Anim Behav*, **32**, 233–42.

Milinski, M. (1987) Tit for tat and the evolution of cooperation in sticklebacks. *Nature Lond*, **325**, 433–7.

Milinski, M. (1992) Predator inspection: cooperation or 'safety in numbers'? *Anim Behav*, **43**, 679–80.

Milinski, M. & Bakker, T.C.M. (1990) Female sticklebacks use male coloration in mate choice and hence avoid parasitized males. *Nature*, **344**, 330–3.

Milinski, M. & Bakker, T.C.M. (1993) Costs influence sequential mate choice in sticklebacks. *Proc Roy Soc Lond B*, **250**, 229–33.

Milinski, M. & Boltshauser, P. (1995) Boldness and predator deterrence: a critique of Godin and Davis. *Proc Roy Soc Lond B*, **262**, 103–5.

Milinski, M., Pfluger, D., Kulling, D. & Kettler, R. (1990) Do sticklebacks cooperate repeatedly in pairs? *Behav Ecol Sociobiol*, **27**, 17–23.

Mills, J.A. (1994) Extra-pair copulations in the red-billed gull: females with high quality, attentive males resist. *Behaviour*, **128**, 41–64.

Mineka, S. & Cook, M. (1988) Social learning and the acquisition of snake fear in monkeys. In: *Social Learning: Psychological and Biological Perspectives* (eds T.R. Zentall & B.G. Galef, Jr), pp. 51–75. Lawrence Erlbaum Associates, New Jersey.

Mittelstaedt, H. (1985) Analytical cybernetics of spider navigation. In: *Neurobiology of Arachnids* (ed. F.G. Barth), pp. 298–316. Springer-Verlag, Berlin.

Mitter, C., Farrell, B. & Wiegmann, B. (1988) The phylogenetic study of adaptive zones: has phytophagy promoted diversification? *Am Natur*, **132**, 107–28.

Møller, A.P. (1988) Female choice selects for male sexual tail ornaments in the monogamous swallow. *Nature Lond*, **332**, 640–2.

Møller, A.P. (1989) Viability costs of male tail ornaments in a swallow. *Nature Lond*, **339**, 132–5.

Møller, A.P. (1990a) Effects of a haematophagous mite on the barn swallow (*Hirundo rustica*): a test of the Hamilton and Zuk hypothesis. *Evolution*, **44**, 771–84.

Møller, A.P. (1990b) Parasites and sexual selection: current status of the Hamilton and Zuk hypothesis. *J Evol Biol*, **3**, 319–28.

Møller, A.P. (1991) Clutch size, nest predation and distribution of avian unequal competitors in a patchy environment. *Ecology*, **72**, 228–34.

Møller, A.P. (1992) Parasites differentially increase the degree of fluctuating asymmetry in secondary sexual characters. *J Evol Biol*, **5**, 691–9.

Møller, A.P. (1993) Developmental stability, sexual selection, and the evolution of secondary sexual characters. *Ecologia*, **3**, 199–208.

Møller, A.P. (1994) Male ornament size as a reliable cue to enhanced offspring viability in the barn swallow. *Proc Nat Acad Sci USA*, **91**, 6926–32.

Møller, A.P. (1994) *Sexual Selection and the Barn Swallow*. Oxford University Press, Oxford.

Møller, A.P. & Birkhead, T.R. (1989) Copulation behaviour of mammals: evidence that sperm competition is widespread. *Biol J Linn Soc*, **38**, 119–31.

Møller, A.P. & Birkhead, T.R. (1992) A pairwise comparative method as illustrated by copulation frequency in birds. *Am Natur*, **139**, 644–56.

Møller, A.P. & Birkhead, T.R. (1994) The evolution of plumage brightness in birds is related to extra-pair paternity. *Evolution*, **48**, 1089–100.

Møller, A.P. & de Lope, F. (1995) Differential costs of a secondary sexual character: an experimental test of the handicap principle. *Evolution*, **48**, 1676–83.

Møller, A.P. & Höglund, J. (1991) Patterns of fluctuating asymmetry in avian feather ornaments: implications for models of sexual selection. *Proc Roy Soc Lond B*, **245**, 1–5.

Møller, A.P. & Pomiankowski, A. (1993a) Fluctuating asymmetry and sexual selection. *Genetica*, **89**, 267–79.

Møller, A.P. & Pomiankowski, A. (1993b) Why have birds got multiple sexual ornaments? *Behav Ecol Sociobiol*, **32**, 167–76.

Møller, A.P., Dufva, R. & Allander, K. (1993) Parasites and the evolution of host social behavior. *Adv Study Behav*, **22**, 65–102.

Møller, A.P., Soler, M. & Thornhill, R. (1995) Breast asymmetry, sexual selection, and human reproductive success. *Ethol Sociobiol*, **16**, 207–19.

Moody, A.L. & Houston, A.I. (1995) Interference and the ideal free distribution. *Anim Behav*, **49**, 1065–72.

Moore, T. & Haig, D. (1991) Genomic imprinting in mammalian development: a parental tug-of-war. *Trends Genet*, **7**, 45–9.

Morton, N.E. (1991) Parameters of the human genome. *Proc Nat Acad Sci USA*, **88**, 7474–6.

Mrosovsky, N. & Sherry, D.F. (1980) Animal anorexias. *Science*, **207**, 837–42.

Mueller, U.G. (1991) Haplodiploidy and the evolution of facultative sex ratios in a primitively eusocial bee. *Science*, **254**, 442–4.

Mueller, U.G., Eickwort, G.C. & Aquadro, C.F. (1994) DNA fingerprinting analysis of parent–offspring conflict in a bee. *Proc Nat Acad Sci USA*, **91**, 5143–7.

Mulder, R.A., Dunn, P.O., Cockburn, A., Lazenby-Cohen, K.A. & Howell, M.J. (1994) Helpers liberate female fairy-wrens from constraints on extra-pair mate choice. *Proc Roy Soc Lond B*, **255**, 223–9.

Mumme, R.L. (1992a) Delayed dispersal and cooperative breeding in the Seychelles warbler. *Trends Ecol Evol*, **7**, 330–1.

Mumme, R.L. (1992b) Do helpers increase reproductive success? An experimental analysis in the Florida scrub jay. *Behav Ecol Sociobiol*, **31**, 319–8.

Mumme, R.L., Koenig, W.D. & Pitelka, F.A. (1983) Reproductive competition in the communal acorn woodpecker: sisters destroy each other's eggs. *Nature*, **306**, 583–4.

Munz, F.W. & McFarland, W.N. (1977) Evolutionary adaptations of fishes to the photic environment. In: *Handbook of Sensory Physiology*, Vol. VII/5 (ed. F. Crescitelli), pp. 193–274. Springer-Verlag, Berlin, New York.

Murphy, M.E. (1994) Dietary complementation by wild birds — considerations for field studies. *J Biosci*, **19**, 355–68.

Murton, R.K. (1971) The significance of a specific search image in the feeding behaviour of the wood-pigeon. *Behaviour*, **40**, 10–42.

Myles, T.G. & Nutting, W.L. (1988) Termite eusocial evolution: a re-examination of Bartz's hypothesis and assumptions. *Quart Rev Biol*, **63**, 1–23.

Nachtigall, P.E. & Moore, P.W.B. (1988) *Animal Sonar*. Plenum Press, New York.

Nagy, K. (1980) CO_2 production in animals: analysis of potential errors in the doubly labeled water method. *Am J Physiol*, **238**, R466–R73.

Nalepa, C.A. (1984) Colony composition, protozoan transfer and some life history characteristics of the woodroach *Cryptocercus punctulatus* Scudder (Dictyoptera: Cryptocercidae). *Behav Ecol Sociobiol*, **14**, 273–9.

Nalepa, C.A. (1994) Nourishment and the origin of termite eusociality. In: *Nourishment and Evolution in Insect Societies* (eds J.H. Hunt & C.A. Nalepa), pp. 57–104. Westview Press, Boulder, Colorado.

Nalepa, C.A. & Jones, S.C. (1991) Evolution of monogamy in termites. *Biol Rev*, **66**, 83–97.

Nee, S., Read, A.F., Greenwood, J.J.D. & Harvey, P.H. (1991) The relationship between abundance and body size in British birds. *Nature*, **351**, 312–13.

Neems, R.M. & Butlin, R.K. (1994) Variation in cuticular hydrocarbons across a hybrid zone in the grasshopper *Chorthippus parallelus*. *Proc Roy Soc Lond B*, **257**, 135–40.

Nelson, W.A. (1984) Effects of nutrition of animals on their ectoparasites. *J Med Entomol*, **21**, 621–35.

Netto, W.J. & van Hooff, J.A.R.A.M. (1986) Conflict interference and the development of dominance relationships in immature *Macaca fascicularis*. In: *Primate Ontogeny, Cognition and Social Behaviour* (eds J.G. Else & P.C. Lee), pp. 291–300. Cambridge University Press, Cambridge.

Neukirch, A. (1982) Dependence of the life span of the honeybee (*Apis mellifera*) upon flight performance and energy consumption. *J Comp Physiol*, **146**, 35–40.

Neuweiler, G., Singh, S. & Sripathi, K. (1984) Audiograms of a South Indian bat community. *J Comp Physiol A*, **154**, 133–42.

Newton, I. (ed.) (1989) *Lifetime Reproduction in Birds*. Academic Press, London.

Nichols, R.A. & Beaumont, M.A. (1996) Is it ancient or modern history that we can read in our genes? In: *Aspects of the Genesis and Maintenance of Biological Diversity* (eds M.E. Hochberg, J. Clobert & R. Barbault), pp. 69–87. Oxford University Press, Oxford.

Nicol, C.J. & Pope, S.J. (1994) Social learning in small flocks of laying hens. *Anim Behav*, **47**, 1289–96.

Nicoletto, P.F. (1993) Female sexual response to condition-dependent ornaments in the guppy, *Poecilia reticulata*. *Anim Behav*, **46**, 441–50.

Nilsson, S.G. (1987) Limitation and regulation of population density in the nuthatch *Sitta europaea* (Aves) breeding in natural cavities. *J Anim Ecol*, **56**, 921–37.

Nishida, T. (1983) Alpha status and agonistic alliance in wild chimpanzees (*Pan troglodytes scheinfurthii*). *Primates*, **24**, 318–36.

Noble, G.K. & Bradley, H.T. (1933) The mating behavior of lizards; its bearing on the theory of sexual selection. *Ann N Y Acad Sci*, **35**, 35–100.

Noë, R. (1990) A veto game played by baboons: a challenge to the use of the prisoner's dilemma as a paradigm for reciprocity and cooperation. *Anim Behav*, **39**, 78–90.

Noirot, C. (1989) Social structure in termite societies. *Ethol Ecol Evol*, **1**, 1–17.

Noirot, C. & Pasteels, J.M. (1987) Ontogenetic development and evolution of the worker caste in termites. *Experientia*, **43**, 851–60.

Noirot, C. & Pasteels, J.M. (1988) The worker caste is polyphyletic in termites. *Sociobiology*, **14**, 15–20.

Nonacs, P. (1986) Ant reproductive strategies and sex allocation theory. *Quart Rev Biol*, **61**, 1–21.

Nonacs, P. (1991) Alloparental care and eusocial evolution: the limits of Queller's head-start advantage. *Oikos*, **61**, 122–5.

Nonacs, P. (1993) Male parentage and sexual deception in the social Hymenoptera. In: *Evolution and Diversity of Sex Ratio in Insects and Mites* (eds D.L. Wrensch & M.A. Ebbert), pp. 384–401. Chapman & Hall, New York.

Nonacs, P. & Reeve, H.K. (1995) The ecology of cooperation in wasps: causes and consequences of alternative reproductive decisions. *Ecology*, **76**, 953–67.

Noonan, K.M. (1981) Individual strategies of inclusive-fitness-maximizing in *Polistes fuscatus* foundresses. In: *Natural Selection and Social Behavior* (eds R.D. Alexander & D.W. Tinkle), pp. 18–44. Chiron Press, New York.

Norberg, R.Å. (1981) Optimal flight speeds in parent birds when feeding young. *J Anim Ecol*, **50**, 473–7.

Norberg, U.M. (1990) *Vertebrate Flight*. Springer-Verlag, Berlin.

Nordeen, E.J. & Nordeen, K.W. (1990) Neurogenesis and sensitive periods in avian song learning. *Trends Neurosci*, **13**, 31–6.

Nordström, K. & Austin, S.J. (1989) Mechanisms that contribute to the stable segregation of plasmids. *Ann Rev Genet*, **23**, 37–69.

Norris, K. (1993) Heritable variation in a plumage indicator of viability in male great tits *Parus major*. *Nature*, **362**, 537–9.

Norris, K., Anwar, M. & Read, A.F.J. (1994) Reproductive effort influences the prevalence of haematozoan parasites in great tits. *J Anim Ecol*, **63**, 601–10.

Nose, A., Nagafuchi, A. & Takeichi, M. (1988) Expressed recombinant cadherins mediate cell sorting in model systems. *Cell*, **54**, 993–1001.

Nottebohm, F. (1991) Reassessing the mechanisms and origins of vocal learning in birds. *Trends Neurosci*, **14**, 206–10.

Nowak, M. & Sigmund, K. (1993) A strategy of win–stay, lose–shift that outperforms tit-for-tat in the prisoner's dilemma game. *Nature Lond*, **364**, 56–8.

Nowak, M. & Sigmund, K. (1994) The alternating prisoner's dilemma. *J Theor Biol*, **168**, 219–26.

Oakes, E.J. (1992) Lekking and the evolution of sexual dimorphism in birds: comparative approaches. *Am Natur*, **140**, 665–94.

O'Day, D.H. (1979) Aggregation during sexual development in *Dictyostelium discoideum*. *Can J Microbiol*, **25**, 1416–26.

O'Donald, P. (1962) The theory of sexual selection. *Heredity*, **17**, 541–52.

Ogilvie, M.A. & Matthews, G.V.T. (1969) Brent geese, mudflats and Man. *Wildfowl*, **20**, 119–25.

Oldroyd, B.P., Rinderer, T.E., Schwenke, J.R. & Buco, S.M. (1994) Subfamily recognition and task specialisation in honey bees (*Apis mellifera* L.) (Hymenoptera: Apidae). *Behav Ecol Sociobiol*, **34**, 169–73.

Oldroyd, B.P., Smolenski, A.J., Cornuet, J.-M. & Crozier, R.H. (1994) Anarchy in the beehive. *Nature*, **371**, 749.

Olivieri, I., Michalakis, Y. & Gouyon, P.-H. (1995) Metapopulation genetics and the evolution of dispersal. *Am Natur*, **146**, 202–28.

Olsson, M. (1995) Forced copulation and costly female

resistance behavior in the Lake Eyre Dragon *Ctenophorus maculosus*. *Herpetologica*, **51**, 19–24.

Olsson, M., Gullberg, A. & Tegelstrom, H. (1995) Sperm competition in the sand lizard, *Lacerta agilis*. *Anim Behav*, **48**, 193–200.

Olsson, M., Gullberg, A., Tegelstrom, H., Madsen, T. & Shine, R. (1994) Can female adders multiply? *Nature*, **369**, 528.

Orians, G.H. & Wittenberger, J.F. (1991) Spatial and temporal scales in habitat selection. *Am Natur*, **137**, S29–49.

Oster, G.F. & Wilson, E.O. (1978) *Caste and Ecology in the Social Insects*. Princeton University Press, Princeton, New Jersey.

Owen, D.D. & Owen, M.J. (1984) Helping behaviour in brown hyaenas. *Nature*, **308**, 843–6.

Owen, R.E. & Plowright, R.C. (1982) Worker–queen conflict and male parentage in bumble bees. *Behav Ecol Sociobiol*, **11**, 91–9.

Owens, I.P.F., Dixon, A., Burke, T. & Thompson, D.B.A. (1995) Strategic paternity assurance in the sex-role reversed Eurasian dotterel (*Charadrius morinellus*): behavioral and genetic evidence. *Behav Ecol*, **6**, 14–21.

Packer, C. (1979) Inter-troop transfer and inbreeding avoidance in *Papio anubis*. *Anim Behav*, **27**, 1–36.

Packer, C. & Pusey, A.E. (1982) Cooperation and competition within coalitions of male lions: kin selection or game theory? *Nature Lond*, **296**, 740–2.

Packer, C. & Pusey, A.E. (1983) Adaptations of female lions to infanticide by incoming males. *Am Natur*, **121**, 716–28.

Packer, C. & Pusey, A.E. (1985) Asymmetric contests in social mammals: respect, manipulation and age-specific aspects. In: *Evolution: Essays in Honour of John Maynard Smith* (eds P.J. Greenwood & M. Slatkin), pp. 173–86. Cambridge University Press, Cambridge.

Packer, C. & Ruttan, L.M. (1988) The evolution of cooperative hunting. *Am Natur*, **132**, 159–98.

Packer, C., Scheel, D. & Pusey, A.E. (1990) Why lions form groups: food is not enough. *Am Natur*, **136**, 1–19.

Packer, C., Collins, D.A., Sindimwo, A. & Goodall, J. (1995) Reproductive constraints on aggressive competition in female baboons. *Nature Lond*, **373**, 60–3.

Packer, C., Gilbert, D., Pusey, A.E. & O'Brien, S.J. (1991) A molecular genetic analysis of kinship and cooperation in African lions. *Nature Lond*, **351**, 562–5.

Packer, C., Herbst, L., Pusey, A.E. *et al.* (1988) Reproductive success of lions. In: *Reproductive Success* (ed. T.H. Clutton-Brock), pp. 363–83. University of Chicago Press, Chicago.

Page, R.E. (1986) Sperm utilization in social insects. *Ann Rev Entomol*, **31**, 297–320.

Page, R.E., Robinson, G.E. & Fondrk, M.K. (1989) Genetic specialists, kin recognition and nepotism in honey bee colonies. *Nature*, **338**, 576–9.

Pagel, M.D.. & Harvey, P.H. (1989) Taxonomic differences in the scaling of brain on body weight among mammals. *Science*, **244**, 1589–93.

Palameta, B. & Lefebvre, L. (1985) The social transmission of a food-finding technique in pigeons: what is learned? *Anim Behav*, **33**, 892–6.

Pamilo, P. (1990) Sex allocation and queen–worker conflict in polygynous ants. *Behav Ecol Sociobiol*, **27**, 31–6.

Pamilo, P. (1991a) Evolution of the sterile caste. *J Theor Biol*, **149**, 75–95.

Pamilo, P. (1991b) Evolution of colony characteristics in social insects. I. Sex allocation. *Am Natur*, **137**, 83–107.

Pancer, Z., Gershon, H. & Rinkevich, B. (1995) Coexistence and possible parasitism of somatic and germ cell lines in chimeras of the colonial urochordate *Botryllus schlosseri*. *Biol Bull*, **189**, 106–12.

Parker, G.A. (1970a) Sperm competition and its evolutionary consequences in the insects. *Biol Rev*, **45**, 525–67.

Parker, G.A. (1970b) The reproductive behaviour and the nature of sexual selection in *Scatophaga stercoraria* L. (Diptera: Scatophagidae). VII. The origin and evolution of the passive phase. *Evolution*, **24**, 774–88.

Parker, G.A. (1970c) The reproductive behaviour and the nature of sexual selection in *Scatophaga stercoraria* L. (Diptera: Scatophagidae). II. The fertilization rate and spatial and temporal relationships of each sex around the site of mating and oviposition. *J Anim Ecol*, **39**, 205–28.

Parker, G.A. (1974a) Assessment strategy and the evolution of fighting behavior. *J Theor Biol*, **47**, 223–43.

Parker, G.A. (1974b) The reproductive behaviour and the nature of sexual selection in *Scatophaga stercoraria*. IX. Spatial distribution of fertilisation rates and evolution of male search strategy within the reproductive area. *Evolution*, **28**, 93–108.

Parker, G.A. (1978a) Searching for mates. In: *Behavioural Ecology: an Evolutionary Approach* (eds J.R. Krebs & N.B. Daries). Blackwell Scientific Publications, Oxford.

Parker, G.A. (1978b) Selection on non-random fusion of gametes during the evolution of anisogamy. *J Theor Biol*, **73**, 1–28.

Parker, G.A. (1979) Sexual selection and sexual conflict. In: *Sexual Selection and Reproductive Competition in Insects* (eds M.S. Blum & N.A. Blum), pp. 123–66. Academic Press, New York.

Parker, G.A. (1982) Why are there so many tiny sperm? Sperm competition and the maintenance of two sexes. *J Theor Biol*, **96**, 281–94.

Parker, G.A. (1983) Mate quality and mating decisions. In: *Mate Choice* (ed. P. Bateson), pp. 141–66. Cambridge University Press, Cambridge.

Parker, G.A. (1984a) Evolutionarily stable strategies. In: *Behavioural Ecology: an Evolutionary Approach* (eds J.R. Krebs & N.B. Davies), 2nd edn, pp. 30–61. Blackwell Scientific Publications, Oxford.

Parker, G.A. (1984b) Sperm competition and the evolution of animal mating strategies. In: *Sperm Competition and the Evolution of Animal Mating Systems* (ed. R.L. Smith), pp. 1–60. Academic Press, Orlando.

Parker, G.A. (1990a) Sperm competition games: raffles and roles. *Proc Roy Soc Lond B*, **242**, 120–6.

Parker, G.A. (1990b) Sperm competition games: sneaks and extra-pair copulations. *Proc Roy Soc Lond B*, **242**, 127–33.

Parker, G.A. (1992) Marginal value theorem with exploitation time costs: diet, sperm reserves, and optimal copula duration in dung flies. *Am Natur*, **139**, 1237–56.

Parker, G.A. (1993) Sperm competition games: sperm size and sperm number under adult control. *Proc Roy Soc Lond B*, **253**, 245–54.

Parker, G.A. & MacNair, M.R. (1978) Models of parent–offspring conflict. I. Monogamy. *Anim Behav*, **26**, 97–110.

Parker, G.A. & Simmons, L.W. (1991) A model of constant random sperm displacement during mating: evidence from *Scatophaga*. *Proc Roy Soc Lond B*, **246**, 107–15.

Parker, G.A. & Simmons, L.W. (1994) Evolution of phenotypic optima and copula duration in dungflies. *Nature Lond*, **370**, 53–6.

Parker, G.A. & Smith, J.L. (1975) Sperm competition and the evolution of the precopulatory passive phase behaviour in *Locusta migratoria migratorioides*. *J Entomol (A)*, **49**, 155–71.

Parker, G.A. & Stuart, R.A. (1976) Animal behaviour as a strategy optimizer: evolution of resource assessment strategies and optimal emigration thresholds. *Am Natur*, **110**, 1055–76.

Parker, G.A. & Sutherland, W.J. (1986) Ideal free distribution when individuals differ in competitive ability: phenotype-limited ideal free models. *Anim Behav*, **34**, 1222–42.

Parker, G.A., Baker, R.R. & Smith, V.G.F. (1972) The origin and evolution of gamete dimorphism and the male–female phenomenon. *J Theor Biol*, **36**, 529–53.

Parker, G.A., Simmons, L.W. & Kirk, H. (1990) Analysing sperm competition data: simple models for predicting mechanisms. *Behav Ecol Sociobiol*, **27**, 55–65.

Partridge, L. (1976) Individual differences in feeding efficiencies and feeding preferences of captive great tits. *Anim Behav*, **24**, 230–40.

Partridge, L. & Green, P. (1987) An advantage for specialist feeding in jackdaws *Corvus monedula*. *Anim Behav*, **35**, 982–90.

Paterson, H.E.H. (1985) The recognition concept of species. In: *Species and Speciation* (ed. E.S. Vrba), pp. 21–9. Transvaal Museum, Pretoria.

Payne, R.B. (1985) Behavioral continuity and change in local song populations of village indigobirds *Vidua chalybeata*. *Z Tierpsychol*, **70**, 1–44.

Payne, R.S. & Webb, D. (1971) Orientation by means of long range acoustic signalling in Baleen whales. *Ann N Y Acad Sci*, **188**, 110–41.

Pearl, R. (1928) *The Rate of Living*. Alfred A. Knopf, New York.

Pearson, B., Raybould, A.F. & Clarke, R.T. (1995) Breeding behaviour, relatedness and sex-investment ratios in *Leptothorax tuberum* Fabricius. *Entomol Exper Appl*, **75**, 165–74.

Pemberton, J.M., Albon, S.D., Guinness, F.E., Clutton-Brock, T.H. & Dover, G.A. (1992) Behavioural estimates of male mating success tested by DNA fingerprinting in a polygynous mammal. *Behav Ecol*, **3**, 66–75.

Pennycuick, C.J. (1989) *Bird Flight Performance: a Practical Calculation Manual*. Oxford Scientific Publications, Oxford.

Pennycuick, C.J. (1992) *Newton Rules Biology*. Oxford University Press, Oxford.

Peters, A., Streng, A. & Michiels, N. (1996) Mating behaviour in a hermaphroditic flatworm with reciprocal insemination: do they assess their mates during copulation? *Ethology*, **102**, 236–51.

Peters, J.M., Queller, D.C., Strassmann, J.E. & Solís, C.R. (1995) Maternity assignment and queen replacement in a social wasp. *Proc Roy Soc Lond B*, **260**, 7–12.

Peters, R.H. (1983) *The Ecological Implications of Body Size*. Cambridge University Press, Cambridge.

Petrie, M. (1983) Female moorhens compete for small fat males. *Science*, **220**, 413–15.

Petrie, M. (1994) Improved growth and survival of offspring of peacocks with more elaborate trains. *Nature*, **371**, 598–9.

Petrie, M., Halliday, T. & Sanders, C. (1991) Peahens prefer peacocks with elaborate trains. *Anim Behav*,

41, 323–31.

Pfennig, D.W. (1990) 'Kin recognition' among spadefoot toad tadpoles: a side-effect of habitat selection? *Evolution*, **44**, 785–98.

Pfennig, D.W. (1992) Polyphenism in spadefoot toad tadpoles as a locally-adjusted evolutionarily stable strategy. *Evolution*, **46**, 1408–20.

Pfennig, D.W. & Collins, J.P. (1993) Kinship affects morphogenesis in cannibalistic salamanders. *Nature*, **362**, 836–8.

Pfennig, D.W. & Sherman, P.W. (1992) Identifying relatives. *Science*, **255**, 217–18.

Pfennig, D.W. & Sherman, P.W. (1995) Kin recognition. *Sci Am*, **272**, 98–103.

Pfennig, D.W., Loeb, M.L.G. & Collins, J.P. (1991) Pathogens as a factor limiting the spread of cannibalism in tiger salamanders. *Oecologia*, **88**, 161–6.

Pfennig, D.W., Reeve, H.K. & Sherman, P.W. (1993) Kin recognition and cannibalism in spadefoot toad tadpoles. *Anim Behav*, **46**, 87–94.

Pfennig, D.W., Sherman, P.W. & Collins, J.P. (1994) Kin recognition and cannibalism in polyphenic salamanders. *Behav Ecol*, **5**, 225–32.

Pfennig, D.W., Gamboa, G.J., Reeve, H.K., Reeve, J.S. & Ferguson, I.D. (1983) The mechanism of nestmate discrimination in social wasps (*Polistes*, Hymenoptera: Vespidae). *Behav Ecol Sociobiol*, **13**, 299–305.

Piersma, T. (1988) Breast muscle atrophy and constraints on foraging during the flightless period of wing moulting Great Crested Grebes. *Ardea*, **76**, 96–106.

Piersma, T. (1990) Pre-migratory 'fattening' usually involves more than the deposition of fat alone. *Ring Migr*, **11**, 113–15.

Pietrewicz, A.T. & Kamil, A.C. (1979) Search image formation in the blue jay (*Cyanocitta cristata*). *Science*, **204**, 1332–3.

Pinxten, R., Hanotte, O., Eens, M., Verheyen, R.F., Dhondt, A.A. & Burke, T. (1993) Extra-pair paternity and intraspecific brood parasitism in the European starling, *Sturnus vulgaris*: evidence from DNA fingerprinting. *Anim Behav*, **45**, 795–809.

Piper, W.H. & Slater, G. (1993) Polyandry and incest avoidance in the cooperative stripe-backed wren of Venezuela. *Behaviour*, **124**, 227–47.

Pitcher, T.J., Green, D. & Magurran, A.E. (1986) Dicing with death: predator inspection behavior. *J Fish Biol*, **28**, 1439–48.

Pitcher, T.J., Magurran, A.E. & Winfield, I.J. (1982) Fish in larger shoals find food faster. *Behav Ecol Sociobiol*, **10**, 149–51.

Pitelka, F.A., Holmes, R.T. & Maclean, S.F.J. (1974) Ecology and evolution of social organisation in Arctic sandpipers. *Am Zool*, **14**, 185–204.

Pitnick, S., Spicer, G.S. & Markow, T.A. (1995) How long is a giant sperm? *Nature*, **375**, 109.

Pøldmaa, T. (1995) *Parentage, kinship and the significance of helping behaviour in the cooperatively breeding noisy minor, Manorina melanocephala*. PhD thesis, Queens University, Kingston, Ontario, Canada.

Pomerantz, J.R. & Kubovy, M. (1986) Theoretical approaches to perceptual organization, simplicity and likelihood principles. In: *Handbook of Perception and Human Performance*, Vol. II, *Cognitive Processes and Performance* (eds K.R. Boff, L. Kaufman & J.P. Thomas), pp. (36)1–(36)46. John Wiley and Sons, New York.

Pomiankowski, A.N. (1988) The evolution of female mate preferences for male genetic quality. *Oxford Surv Evol Biol*, **5**, 136–84.

Pomiankowski, A. & Iwasa, Y. (1993) Evolution of multiple sexual ornaments by Fisher's process of sexual selection. *Proc Roy Soc Lond B*, **253**, 173–81.

Pomiankowski, A. & Møller, A.P. (1995) A resolution of the lek paradox. *Proc Roy Soc Lond B*, **260**, 21–9.

Pond, C.M. (1978) Morphological aspects and the ecological and mechanical consequences of fat deposition in wild vertebrates. *Ann Rev Ecol Syst*, **9**, 519–70.

Porter, R.H. (1991) Mutual mother–infant recognition in humans. In: *Kin Recognition* (ed. P.G. Hepper), pp. 413–32. Cambridge University Press, Cambridge.

Potts, W.K., Manning, C.J. & Wakeland, E.K. (1991) Mating patterns in seminatural populations of mice influenced by MHC genotype. *Nature*, **352**, 619–21.

Potts, W.K., Manning, C.J. & Wakeland, E.K. (1994) The role of infectious disease, inbreeding, and mating preferences in maintaining MHC genetic diversity: an experimental test. *Phil Trans Roy Soc Lond B*, **346**, 369–78.

Prestwich, K.N. (1994) The energetics of acoustic signaling in anurans and insects. *Am Natur*, **34**, 625–3.

Price, M.V. & Waser, N.M. (1979) Pollen dispersal and optimal outcrossing in *Delphinium nelsoni*. *Nature*, **277**, 294–7.

Price, K., Harvey, H. & Ydenberg, R. (1996) Begging tactics of nestling yellow-headed blackbirds, *Xanthocephalus xanthocephalus*, in relation to need. *Anim Behav*, **51**, 421–35.

Primack, R.B. (1993) *Essentials of Conservation Biology*. Sinauer, Sunderland, MA.

Proctor, H.C. (1991) Courtship in the water mite *Neumania papillator*: males capitalize on female adaptations for predation. *Anim Behav*, **42**, 589–98.

Proctor, H.C. (1992) Sensory exploitation and the

evolution of male mating behaviour: a cladistic test using water mites (Acari: Parasitengona). *Anim Behav*, **44**, 745–52.

Prop, J. & Deerenberg, C. (1991) Feeding constraints in spring staging brent geese and the impact of diet on the accumulation of body reserves. *Oecologia*, **87**, 19–28.

Pruett-Jones, S. (1992) Independent versus non-independent mate choice: do females copy each other? *Am Natur*, **140**, 1000–9.

Pruett-Jones, S.G. & Lewis, M.J. (1990) Habitat limitation and sex ratio promotes delayed dispersal in Superb fairy-wrens. *Nature*, **348**, 541–2.

Pruett-Jones, S.G. & Wade, M.J. (1990) Female copying increases the variance in male mating success. *Proc Nat Acad Sci USA*, **87**, 5749–53.

Pusey, A.E. & Packer, C. (1994) Non-offspring nursing in social carnivores: minimizing the costs. *Behav Ecol*, **5**, 362–74.

Queller, D.C. (1989) The evolution of eusociality: reproductive head starts of workers. *Proc Nat Acad Sci USA*, **86**, 3224–6.

Queller, D.C. (1994) Extended parental care and the origin of eusociality. *Proc Roy Soc Lond B*, **256**, 105–11.

Queller, D.C. & Strassmann, J.E. (1989) Measuring inclusive fitness in social wasps. In: *The Genetics of Social Evolution* (eds M.D. Breed & R.E. Page), pp. 103–22. Westview Press, Boulder, Colorado.

Queller, D.C., Strassmann, J.E. & Hughes, C.R. (1993) Microsatellites and kinship. *Trends Ecol Evol*, **8**, 285–8.

Quinn, T.W., Quinn, J.S., Cooke, F. & White, B.N. (1987) DNA marker analysis detects multiple maternity and paternity in single broods of the lesser snow goose. *Nature Lond*, **326**, 392–4.

Rabenold, K.N. (1985) Cooperation in breeding by nonreproductive wrens: kinship, reciprocity and demography. *Behav Ecol Sociobiol*, **17**, 1–17.

Rabenold, P.P., Rabenold, K.N., Piper, W.H., Haydock, J. & Zack, S.W. (1990) Shared paternity revealed by genetic analysis in cooperatively breeding tropical wrens. *Nature*, **348**, 538–40.

Radwan, J. & Siva-Jothy, M.T. (1996) The function of postinsemination mate association in the bulb mite, *Rhizoglyphus robini*. *Anim Behav*, **52**, 651–7.

Ralls, K., Harvey, P.H. & Lyles, A.M. (1986) Inbreeding in natural populations of birds and mammals. In: *Conservation Biology: the Science of Scarcity and Diversity* (ed. M. Soule), pp. 35–56. Sinauer Associates, Sunderland, Massachusetts.

Ralls, K., Ballou, J.D. & Templeton, A. (1988) Estimates of lethal equivalents and the cost of inbreeding in mammals. *Conserv Biol*, **2**, 185–93.

Rand, A.S., Ryan, M.J. & Wilczynski, W. (1992) Signal redundancy and receiver permissiveness in acoustic mate recognition by the tungara frog, *Physalaemus pustulosus*. *Am Zool*, **32**, 81–90.

Ranta, E., Rita, H. & Lindström, K. (1993) Competition versus cooperation: success of individuals foraging alone and in groups. *Am Natur*, **142**, 42–58.

Ratnieks, F.L.W. (1988) Reproductive harmony via mutual policing by workers in eusocial Hymenoptera. *Am Natur*, **132**, 217–36.

Ratnieks, F.L.W. (1993) Egg-laying, egg-removal, and ovary development by workers in queenright honey bee colonies. *Behav Ecol Sociobiol*, **32**, 191–8.

Ratnieks, F.L.W. (1995) Evidence for a queen-produced egg-marking pheromone and its use in worker policing in the honey bee. *J Apicult Res*, **34**, 31–7.

Ratnieks, F.L.W. & Reeve, H.K. (1991) The evolution of queen-rearing nepotism in social Hymenoptera: effects of discrimination costs in swarming species. *J Evol Biol*, **4**, 93–115.

Ratnieks, F.L.W. & Reeve, H.K. (1992) Conflict in single-queen Hymenopteran societies: the structure of conflict and processes that reduce conflict in advanced eusocial species. *J Theoret Biol*, **158**, 33–65.

Ratnieks, F.L.W. & Visscher, P.K. (1989) Worker policing in the honeybee. *Nature*, **342**, 796–7.

Raubenheimer, D. & Simpson, S.J. (1995) A multilevel analysis of feeding behaviour: the geometry of nutritional decisions. *Phil Trans Roy Soc Lond B*, **342**, 381–402.

Rayner, J.M.V. (1990) The mechanics of flight and bird migration performance. In: *Bird Migration. Physiology and Ecology* (ed. E. Gwinner), pp. 283–99. Springer-Verlag, Berlin.

Read, A.F. & Harvey, P.H. (1989) Reassessment of comparative evidence for Hamilton and Zuk theory on the evolution of secondary sexual characters. *Nature*, **339**, 618–20.

Read, A.F. & Nee, S. (1995) Inference from binary comparative data. *J Theor Biol*, **173**, 99–108.

Real, L.A. (1990) Search theory and mate choice. I. Models of single-sex discrimination. *Am Natur*, **136**, 376–404.

Real, L.A. (1991) Animal choice behavior and the evolution of cognitive architecture. *Science*, **253**, 980–6.

Reboreda, J. & Kacelnik, A. (1990) On cooperation, tit-for-tat and mirrors. *Anim Behav*, **40**, 1188–9.

Reboreda, J.C. & Kacelnik, A. (1991) Risk sensitivity in starlings: variability in food amount and food delay. *Behav Ecol*, **2**, 301–8.

Redfield, R.J. (1993) Genes of breakfast: the have-your-cake-and-eat-it-too of bacterial transformation. *J Hered*, **84**, 400–4.

Redondo, T. (1993) Exploitation of host mechanisms for parental care by avian brood parasites. *Etología*, **3**, 235–97.

Redondo, T. & Castro, F. (1992) Signalling of nutritional need by magpie nestlings. *Ethology*, **92**, 193–204.

Reeve, H.K. (1989) The evolution of conspecific acceptance thresholds. *Am Natur*, **133**, 407–35.

Reeve, H.K. (1993) Haplodiploidy, eusociality and absence of male parental and alloparental care in Hymenoptera: a unifying genetic hypothesis distinct from kin selection theory. *Phil Trans Roy Soc Lond B*, **342**, 335–52.

Reeve, H.K. (1997a) Game theory, reproductive skews and nepotism. In: *Game Theory and Behavior* (eds L.A. Dugatkin & H.K. Reeve). Oxford University Press, Oxford.

Reeve, H.K. (1997b) Acting for the good of others: kinship and reciprocity, with some new twists. In: *Evolution and Human Behavior* (eds C. Crawford & D. Krebs), in press. Lawrence Erlbaum Associates, Hillsdale. New Jersey.

Reeve, H.K. & Keller, L. (1995) Partitioning of reproduction in mother–daughter versus sibling associations: a test of optimal skew theory. *Am Natur*, **145**, 119–32.

Reeve, H.K. & Nonacs, P. (1992) Social contracts in wasp societies. *Nature*, **359**, 823–5.

Reeve, H.K. & Ratnieks, F.L.W. (1993) Queen–queen conflicts in polygynous societies: mutual tolerance and reproductive skew. In: *Queen Number and Sociality in Insects* (ed. L. Keller), pp. 45–85. Oxford University Press, Oxford.

Reeve, H.K. & Sherman, P.W. (1993) Adaptation and the goals of evolutionary research. *Quart Rev Biol*, **68**, 1–32.

Reid, M.L. (1987) Costliness and reliability in the singing vigour of Ipswich sparrows. *Anim Behav*, **35**, 1735–43.

Reid, P.J. & Shettleworth, S.J. (1992) Detection of cryptic prey: search image or search rate? *J Exper Psychol Animal Behav Process*, **18**, 273–86.

Reilly, L.M. (1987) Measurements of inbreeding and average relatedness in a termite population. *Am Natur*, **130**, 339–49.

Reimann, J.G., Moen, D.O. & Thorson, B.J. (1967) Female monogamy and its control in the housefly, *Musca domestica* L. *J Insect Physiol*, **13**, 407–18.

Rensch, I. & Rensch, B. (1956) Relative organmasze bei tropischen warmblütern. *Zool Anz*, **156**, 106–24.

Reynolds, J.D. (1996) Animal breeding systems. *Trends Ecol Evol*, **11**, 68–72.

Reynolds, J.D. & Gross, M.R. (1992) Female mate preference enhances offspring growth and reproduction in a fish, *Poecilia reticulata*. *Proc Roy Soc Lond B*, **250**, 57–62.

Rheinländer, J. & Römer, H. (1986) Insect hearing in the field. *J Comp Physiol A*, **158**, 647–51.

Rice, W.R. (1987) The accumulation of sexually antagonistic genes as a selective agent promoting the evolution of reduced recombination between primitive sex chromosomes. *Evolution*, **41**, 911–14.

Rice, W.R. & Hostert, E.E. (1994) Perspective: laboratory experiments on speciation: what have we learned in forty years? *Evolution*, **47**, 1637–53.

Richman, A.D. & Price, T. (1992) Evolution of ecological differences in the Old World leaf warblers. *Nature*, **355**, 817–21.

Richner, H., Christe, P. & Oppliger, A. (1995) Paternal investment affects prevalence of malaria. *Proc Nat Acad Sci USA*, **92**, 1192–4.

Ridley, M. (1983) *The Explanation of Organic Diversity: the Comparative Method and Adaptations for Mating*. Oxford University Press, Oxford.

Ridley, M. & Grafen, A. (1981) Are green beard genes outlaws? *Anim Behav*, **29**, 954–5.

Ridley, M. & Grafen, A. (1996) How to study discrete comparative methods. In: *Phylogenies and the Comparative Method in Animal Behavior* (ed. E.P. Martins), pp. 76–103. Oxford University Press, New York.

Rinkevich, B., Porat, R. & Goren, M. (1995) Allorecognition elements on a urochordate histocompatibility locus indicate unprecedented extensive polymorphism. *Proc Roy Soc Lond B*, **260**, 319–24.

Ritchie, M. (1996) What is 'the paradox of the lek'? *Trends Ecol Evol*, **11**, 175.

Ritchie, M.E. (1988) Individual variation in the ability of Columbian ground squirrels to select an optimal diet. *Evol Ecol*, **2**, 232–52.

Rivier, D.H. & Pillus, L. (1994) Silencing speaks up. *Cell*, **76**, 963–6.

Robertson, J.G.M. (1990) Female choice increases fertilization success in the Australian frog *Uperolia laevigata*. *Anim Behav*, **39**, 639–45.

Robinson, S.K. & Terborgh, J. (1995) Interspecific aggression and habitat selection by Amazonian birds. *J Anim Ecol*, **64**, 1–11.

Rodriguez-Gironés, M.A., Drummond, H. & Kacelnik, A. (1996) Effect of food deprivation on dominance status in blue-footed booby (*Sula nebouxii*) broods. *Behav Ecol*, **7**, 82–8.

Roff, D.A. (1984) The evolution of life history parameters in teleosts. *Can J Fish Aquat Sci*, **41**, 989–1000.

Roff, D.A. (1994) Why is there so much genetic variation for wing dimorphism? *Res Popul Ecol*, **36**, 145–50.

Rogers, C.M. (1987) Predation risk and fasting capacity: do winter birds maintain optimal body mass? *Ecology*, **68**, 1051–61.

Rogers, C.M. & Smith, J.N.M. (1993) Life-history theory in the non-breeding period: trade-offs in avian fat reserves. *Ecology*, **74**, 419–26.

Rogers, C.M., Nolan, V. & Ketterson, E.D. (1994) Winter fattening in the dark-eyed junco: plasticity and possible interaction with migration trade-offs. *Oecologia*, **97**, 526–32.

Rogers, C.M., Ramenofsky, M., Ketterson, E.D., Nolan, V. & Wingfield, J.C. (1993) Plasma-corticosterone, adrenal mass, winter weather, and season in nonbreeding populations of dark-eyed juncos (*Junco hyemalis*). *Auk*, **110**, 279–85.

Rohwer, S. (1986) Selection for adoption versus infanticide by replacement mates in birds. *Curr Ornithol*, **3**, 253–336.

Roisin, Y. (1994) Intragroup conflicts and the evolution of sterile castes in termites. *Am Natur*, **143**, 751–65.

Römer, H. (1993) Environmental and biological constraints for the evolution of long-range signalling and hearing in acoustic insects. *Phil Trans Roy Soc Lond B*, **340**, 179–85.

Römer, H. & Bailey, W.J. (1990) Insect hearing in the field. *Comp Biochem Physiol A*, **97**, 443–7.

Rose, M.R. (1991) *Evolutionary Biology of Aging*. Oxford University Press, Oxford.

Ross, K.G. & Matthews, R.W. (eds) (1991) *The Social Biology of Wasps*. Comstock, Ithaca.

Rossel, S. (1979) Regional differences in photoreceptor performance in the eye of the praying mantis. *J Comp Physiol*, **131**, 95–112.

Roubik, D.W. (1989) *Ecology and Natural History of Tropical Bees*. Cambridge University Press, Cambridge.

Rowe, L. Ludwig, D. & Schluter, D. (1994) Time, condition and the seasonal decline of avian clutch size. *Am Natur*, **143**, 698–722.

Rowley, A., Dowell, S.J. & Diffley, J.F.X. (1994) Recent developments in the initiation of chromosomal DNA replication: a complex picture emerges. *Biochim Biophys Acta*, **1217**, 239–56.

Rowley, I. (1965) The life history of the Superb blue wren, *Malurus cyaneus*. *Emu*, **64**, 251–97.

Rowley, I. & Russell, E. (1990) Splendid fairy-wrens: demonstrating the importance of longevity. In: *Cooperative Breeding in Birds: Long-term Studies of Ecology and Behavior*, (eds P.B. Stacey & W.D. Koenig), pp. 1–30. Cambridge University Press, Cambridge.

Royer, M. (1975) Hermaphroditism in insects. Studies on *Icerya purchasi*. In: *Intersexuality in the Animal Kingdom* (ed. R. Reinboth), pp. 135–45. Springer-Verlag, Berlin.

Rubner, M. (1893) Über den einfluss der körpergrösse auf stoff- und kraftwechsel. *Zeitsch Biol*, **19**, 535–62.

Russon, A.E. & Galdikas, B.M.F. (1993) Imitation in free-ranging rehabilitant orangutans (*Pongo pygmaeus*) *J Comp Psychol*, **107**, 147–61.

Russon, A.E. & Galdikas, B.M.F. (1995) Constraints on great apes' imitation: model and action selectivity in rehabilitant orangutan (*Pongo pygmaeus*) imitation. *J Comp Psychol*, **109**, 5–17.

Ruxton, G.D., Gurney, W.S.C. & de Roos, A.M. (1992) Interference and generation cycles. *Theor Popul Biol*, **42**, 235–53.

Ruxton, G.D., Hall, S.J. & Gurney, W.S.C. (1995) Attraction toward feeding conspecifics when food patches are exhaustible. *Am Natur*, **145**, 653–60.

Ryan, M.J. (1983) Sexual selection and communication in a neotropical frog, *Physalaemus pustulosus*. *Evolution*, **37**, 261–72.

Ryan, M.J. (1985) *The Túngara Frog, a Study in Sexual Selection and Communication*. Chicago University Press, Chicago.

Ryan, M.J. (1988) Energy, calling and selection. *Am Natur*, **28**, 885–98.

Ryan, M.J. (1990) Sensory systems, sexual selection, and sensory exploitation. *Oxford Surv Evol Biol*, **7**, 157–95.

Ryan, M.J. (1994) Mechanistic studies in sexual selection. In: *Behavioral Mechanisms in Evolutionary Ecology* (ed. L. Real), pp. 190–215. University of Chicago Press, Chicago.

Ryan, M.J. (1996) Phylogenetics and behavior: some cautions and expectations. In: *Phylogenies and the Comparative Method in Animal Behavior* (ed. E. Martins), pp. 1–21. Oxford University Press, Oxford.

Ryan, M.J. & Keddy-Hector, A. (1992) Directional patterns of female mate choice and the role of sensory biases. *Am Natur*, **139**, S4–S35.

Ryan, M.J. & Rand, A.S. (1990) The sensory basis of sexual selection for complex calls in the túngara frog, *Physalaemus pustulosus* (sexual selection for sensory exploitation). *Evolution*, **44**, 305–14.

Ryan, M.J. & Rand, A.S. (1993a) Sexual selection and signal evolution: the ghost of biases past. *Phil Trans Roy Soc Lond B*, **340**, 187–95.

Ryan, M.J. & Rand, A.S. (1993b) Species recognition and sexual selection as a unitary problem in animal communication. *Evolution*, **47**, 647–57.

Ryan, M.J. & Rand, A.S. (1995) Female responses to ancestral advertisement calls in the túngara frog. *Science*, **269**, 390–2.

Ryan, M.J. & Wagner, W. (1987) Asymmetries in mating preferences between species: female swordtails prefer heterospecific mates. *Science*, **236**, 595–7.

Ryan, M.J., Hews, D.K. & Wagner, W.E. Jr (1990b)

Sexual selection on alleles that determine body size in the swordtail *Xiphophorus nigrensis*. *Behav Ecol Sociobiol*, **26**, 231–7.

Ryan, M.J., Fox, J.H., Wilczynski, W. & Rand, A.S. (1990a) Sexual selection for sensory exploitation in the frog *Physalaemus pustulosus*. *Nature*, **343**, 66–7.

Ryan, M.J., Warkentin, K.M., McClelland, B.E. & Wilczynski, W. (1995) Fluctuating asymmetries and advertisement call variation in the cricket frog, *Acris crepitans*. *Behav Ecol*, **6**, 124–31.

Saether, B.-E. (1994) Reproductive strategies in relation to prey size in altricial birds: homage to Charles Elton. *Am Natur*, **144**, 285–99.

Saetre, G.-P. & Slagsvold, T. (1992) Evidence for sex recognition from plumage colour by the pied flycatcher, *Ficedula hypoleuca*. *Anim Behav*, **44**, 293–9.

Saino, N., Primmer, C., Ellegren, H. & Møller, A.P. (1996) Sexual ornamentation affects paternity in the socially monogamous barn swallow *Hirundo rustica*. *Evolution* (in press).

Sakaluk, S.K. & Belwood, J.J. (1984) Gecko phonotaxis to cricket calling song — a case of satellite predation. *Anim Behav*, **32**, 659–62.

Sakaluk, S.K. & Eggert, A.-K. (1996) Female control of sperm transfer and intraspecific variation in sperm precedence: antecedents to the evolution of a courtship food gift. *Evolution*, **50**, 694–703.

Salmon, M.A., Van Melderen, L., Bernard, P. & Couturier, M. (1994) The antidote and autoregulatory functions of the F plasmid CcdA protein: a genetic and biochemical survey. *Molec Gen Genet*, **244**, 530–8.

Salvini-Plawen, L.V. & Mayr, E. (1977) On the evolution of photoreceptors and eyes. *Evol Biol*, **10**, 207–63.

Samuels, A., Silk, J. & Altmann, J. (1987) Continuity and change in dominance relationships among female baboons. *Anim Behav*, **35**, 785–93.

Sanderson, N., Szymura, J.M. & Barton, N.H. (1992) Variation in mating call across the hybrid zone between the fire-bellied toads *Bombina bombina* and *B. variegata*. *Evolution*, **46**, 595–607.

Saumitou-Laprade, P., Cuguen, J. & Vernet, P. (1994) Cytoplasmic male sterility in plants: molecular evidence and the nucleocytoplasmic conflict. *Trends Ecol Evol*, **9**, 431–5.

Scheel, D. & Packer, C. (1991) Group hunting behaviour of lions: a search for cooperation. *Anim Behav*, **41**, 697–710.

Scheffer, S.J., Uetz, G.W. & Stratton, G.E. (1996) Sexual selection, male morphology, and the efficacy of courtship signalling in two wolf spiders (Araneae: Lycosidae). *Behav Ecol Sociobiol*, **38**, 17–23.

Schierwater, B., Streit, B., Wagner, G.P. & DeSalle, R. (Eds) (1994) *Molecular Ecology and Evolution: Approaches and Applications*. Birkhauser Verlag, Basel.

Schildberger, K., Huber, F. & Wohlers, D.W. (1989) Central auditory pathway: neuronal correlates of phonotactic behavior. In: *Cricket Behavior and Neurobiology* (eds F. Huber, T.E. Moore & W. Loher), pp. 423–58. Cornell University Press, Ithaca, New York.

Schjelderup-Ebbe, T. (1935) Social behavior of birds. In: *Handbook of Social Psychology* (ed. C. Murchison), pp. 947–72. Clark University Press, Worcester, Massachusetts.

Schlupp, I., Marler, C.A. & Ryan, M.J. (1994) Benefit to male sail fin mollies of mating with heterospecific females. *Science*, **263**, 373–4.

Schluter, D. & Nagel, L.M. (1995) Parallel speciation by natural selection. *Am Natur*, **146**, 292–301.

Schluter, D. & Price, T. (1993) Honesty, perception and population divergence in sexually selected traits. *Proc Roy Soc Lond B*, **253**, 117–22.

Schmid-Hempel, P. (1994) Infection and colony variability in social insects. *Phil Trans Roy Soc Lond B*, **346**, 313–21.

Schmid-Hempel, P. & Schmid-Hempel, R. (1993) Transmission of a pathogen in *Bombus terrestris*, with a note on division of labour in social insects. *Behav Ecol Sociobiol*, **33**, 319–27.

Schmid-Hempel, P., Kacelnik, A. & Houston, A.I. (1985) Honeybees maximize efficiency by not filling their crop. *Behav Ecol Sociobiol*, **17**, 61–6.

Schmidt, J.M. & Smith, J.J.B. (1986) Correlations between body angles and substrate curvature in the parasitoid wasp *Trichogramma minutum*: a possible mechanism of host radius measurements. *J Exp Biol*, **125**, 271–85.

Schmidt-Nielsen, K. (1984) *Scaling*. Cambridge University Press, Cambridge.

Schoener, T.W. (1971) Theory of feeding strategies. *Ann Rev Ecol Syst*, **2**, 369–404.

Schwagmeyer, P.L. & Parker, G.A. (1987) Queuing for mates in thirteen-lined ground squirrels. *Anim Behav*, **35**, 1015–25.

Schwagmeyer, P.L. & Parker, G.A. (1990) Male mate choice as predicted by sperm competition in thirteen-lined ground squirrels. *Nature (Lond)*, **348**, 62–4.

Schwagmeyer, P.L. & Parker, G.A. (1994) Mate-quitting rules for male thirteen-lined ground squirrels. *Behav Ecol*, **5**, 142–50.

Schwarzkopf, L. & Shine, R. (1992) Costs of reproduction in lizards: escape tactics and susceptibility to predation. *Behav Ecol Sociobiol*, **31**, 17–25.

Scudo, F.M. (1967) The adaptive value of sexual dimorphism. I. Anisogamy. *Evolution*, **21**, 285–91.

Searcy, W.A. (1992) Song repertoire and mate choice in birds. *Am Zool*, **32**, 71–80.

Searcy, W.A. & Andersson, M.B. (1986) Sexual selection and the evolution of song. *Ann Rev Ecol Syst*, **17**, 507–33.

Seeley, T.D. (1985) *Honeybee Ecology. A Study of Adaptation in Social Life*. Princeton University Press, Princeton, New Jersey.

Seelinger, G. & Seelinger, U. (1983) On the social organisation, alarm and fighting in the primitive cockroach *Cryptocercus punctulatus* Scudder. *Z Tierpsychol*, **61**, 315–33.

Seger, J. (1983) Partial bivoltinism may cause alternating sex-ratio biases that favour eusociality. *Nature*, **301**, 59–62.

Seger, J. (1991) Cooperation and conflict in social insects. In: *Behavioural Ecology: an Evolutionary Approach*, 3rd edn, (eds J.R. Krebs & N.B. Davies), pp. 338–73. Blackwell Scientific Publications, Oxford.

Selker, E.U. (1990) Premeiotic instability of repeated sequences in *Neurospora crassa*. *Ann Rev Genet*, **24**, 579–613.

Shapiro, D.Y., Marconato, A. & Yoshikawa, T. (1994) Sperm economy in a coral reef fish *Thalassoma bifasciatum*. *Ecology*, **75**, 1334–44.

Shaw, K. (1995) Phylogenetic tests of the sensory exploitation model of sexual selection. *Trends Ecol Evol*, **10**, 117–20.

Sheldon, B.C. (1994) Male phenotype, fertility, and the pursuit of extra-pair copulations by female birds. *Proc Roy Soc Lond B*, **257**, 25–30.

Shellman, J.S. & Gamboa, G.J. (1982) Nestmate discrimination in social wasps: the role of exposure to nest and nestmates (*Polistes fuscatus*, Hymenoptera: Vespidae). *Behav Ecol Sociobiol*, **11**, 51–3.

Sherman, P.W. (1977) Nepotism and the evolution of alarm calls. *Science*, **197**, 1246–53.

Sherman, P.W. (1980) The limits of ground squirrel nepotism. In: *Sociobiology: Beyond Nature/Nurture?* (eds G.W. Barlow & J. Silverberg), pp. 505–44. Westview Press, Boulder, Colorado.

Sherman, P.W. (1981a) Reproductive competition and infanticide in Belding's ground squirrels and other organisms. In: *Natural Selection and Social Behavior* (eds R.D. Alexander & D.W. Tinkle), pp. 311–31. Chiron Press, New York.

Sherman, P.W. (1981b) Kinship, demography, and Belding's ground squirrel nepotism. *Behav Ecol Sociobiol*, **8**, 251–9.

Sherman, P.W. (1985) Alarm calls of Belding's ground squirrels to aerial predators: nepotism or self-preservation? *Behav Ecol Sociobiol*, **17**, 313–23.

Sherman, P.W. (1989) Mate guarding as paternity insurance in Idaho ground squirrels. *Nature Lond*, **338**, 418–20.

Sherman, P.W. (1991) Multiple mating and kin recognition by self-inspection. *Ethol Sociobiol*, **12**, 377–86.

Sherman, P.W. & Holmes, W.G. (1985) Kin recognition: issues and evidence. In: *Experimental Behavioral Ecology and Sociobiology* (eds B. Hölldobler & M. Lindauer), pp. 437–60. Sinauer, Sunderland, Massachusetts.

Sherman, P.W., Seeley, T.D. & Reeve, H.K. (1988) Parasites, pathogens, and polyandry in social Hymenoptera. *Am Natur*, **131**, 602–10.

Sherman, P.W., Lacey, E.A., Reeve, H.K. & Keller, L. (1995) The eusociality continuum. *Behav Ecol*, **6**, 102–8.

Sherry, D.F. (1985) Food storage by birds and mammals. *Adv Stud Behav*, **15**, 153–88.

Sherry, D.F. & Galef, B.G. Jr (1984) Cultural transmission without imitation: milk bottle opening by birds. *Anim Behav*, **32**, 937–8.

Sherry, D.F. & Vaccarino, A.L. (1989) Hippocampus and memory for caches in the black-capped chickadee. *Behav Neurosci*, **103**, 308–18.

Sherry, D.F., Jacobs, L.F. & Gaulin, S.J.C. (1992) Spatial memory and adaptive specialization of the hippocampus. *Trends Neurosci*, **15**, 298–303.

Sherry, D.F., Krebs, J.R. & Cowie, R.J. (1981) Memory for the location of stored food in marsh tits. *Anim Behav*, **29**, 1260–6.

Sherry, D.F., Mrosovsky, N. & Hogan, J.A. (1980) Weight loss and anorexia during incubation in birds. *J Comp Physiol Psychol*, **94**, 89–98.

Sherry, D.F., Vaccarino, A.L., Buckenham, K. & Herz, R. (1989) The hippocampal complex of food storing birds. *Brain Behav Evol*, **34**, 308–17.

Shettleworth, S.J. (1990) Spatial memory in food storing birds. *Phil Trans Roy Soc Lond B*, **329**, 143–51.

Shettleworth, S.J. (1993) Where is the comparison in comparative cognition? Alternative research programs. *Psychol Sci*, **4**, 179–84.

Shettleworth, S.J. (1995) Comparative studies of memory in food storing birds: from the field to the Skinner box. In: *Behavioral Brain Research in Naturalistic and Semi-naturalistic Settings* (eds E. Alleva, A. Fasolo, H.P. Lipp, L. Nadel & L. Ricceri), pp. 159–92. NATO ASI Series, Kluwer Academic Press, Dordrecht.

Shettleworth, S.J. & Plowright, C.M.S. (1992) How pigeons estimate rates of prey encounter. *J Exper Psychol*, **18**, 219–35.

Shettleworth, S.J., Krebs, J.R., Healy, S.D. & Thomas, C.M. (1990) Spatial memory of food-storing tits

(*Parus ater* and *P. atricapillus*): comparison of storing and non-storing tasks. *J Compar Psychol*, **104**, 71–81.

Shettleworth, S.J., Krebs, J.R. Stephens, D.W. & Gibbon, J. (1988) Tracking a fluctuating environment: a study of sampling. *Anim Behav*, **36**, 87–105.

Shields, W.M. (1982) *Philopatry, Inbreeding, and the Evolution of Sex*. State University of New York Press, Albany, New York.

Shimkets, L.J. (1990) Social and developmental biology of the myxobacteria. *Microbiol Rev*, **54**, 473–501.

Shine, R. (1980) 'Costs' of reproduction in reptiles. *Oecologia*, **46**, 92–100.

Short, R.V. (1977) Sexual selection and the descent of man. In: *Proceedings of the Canberra Symposium on Reproduction and Evolution. Australian Academy of Science*, pp. 3–19. Canberra.

Shuster, S.M. & Wade, M.J. (1991) Equal mating success among male reproductive strategies in a marine isopod. *Nature Lond*, **350**, 608–10.

Sibley, C.G. & Ahlquist, J.E. (1990) *Phylogeny and Classification of Birds*. Yale University Press, New Haven.

Siegel, R.A., Huggins, M.M. & Ford, N.B. (1987) Reduction in locomotor ability as a cost of reproduction in gravid snakes. *Oecologia*, **73**, 481–5.

Sigurjonsdottir, H. & Parker, G.A. (1981) Dung fly struggles: evidence for assessment strategy. *Behav Ecol Sociobiol*, **8**, 219–30.

Sillén-Tullberg, B. (1988) Evolution of gregariousness in aposematic butterfly larvae: a phylogenetic analysis. *Evolution*, **42**, 293–305.

Sillén-Tullberg, B. (1993) The effect of biased inclusion of taxa on the correlation between discrete characters in phylogenetic trees. *Evolution*, **47**, 1182–91.

Simmons, L.W. (1986) Female choice in the field cricket, *Gryllus bimaculatus* (De Geer). *Anim Behav*, **34**, 1463–70.

Simmons, L.W. (1989) Kin recognition and its influence on mating preferences of the field cricket, *Gryllus bimaculatus* (de Geer). *Anim Behav*, **38**, 68–77.

Simmons, L.W. (1991) Female choice and the relatedness of mates in the field cricket, *Gryllus bimaculatus*. *Anim Behav*, **41**, 493–501.

Simmons, L.W., Stockley, P., Jackson, R.L. & Parker, G.A. (1996) Sperm competition or sperm selection: no evidence for female influence over paternity in yellow dung flies *Scatophaga stercoraria*. *Behav Ecol Sociobiol*, **38**, 199–206.

Simmons, L.W., Llorens, T., Schinzig, M., Hosken, D. & Craig, M. (1994) Courtship role reversal in bush crickets: another role for parasites? *Anim Behav*, **47**, 117–22.

Sinclair, A.R.E. (1989) Population regulation in animals. In: *Ecological Concepts* (ed. J.M. Cherrett), pp. 197–241. Blackwell Scientific Publications, Oxford.

Sinervo, B. & Lively, C.M. (1996) The rock–paper–scissors game and the evolution of alternative male strategies. *Nature Lond*, **380**, 240–3.

Sinervo, B., Hedges, R. & Adolph, S.C. (1991) Decreased sprint speed as a cost of reproduction in the lizard *Sceloporus occidentalis*: variation among populations. *J Exp Biol*, **155**, 323–36.

Sinervo, B., Doughty, P., Huey, R.B. & Zamudio, K. (1992) Allometric engineering: a causal analysis of natural selection on offspring size. *Science*, **258**, 1927–30.

Singer, T.L. & Espelie, K.E. (1992) Social wasps use nest paper hydrocarbons for nestmate recognition. *Anim Behav*, **44**, 63–8.

Slatkin, M. (1985) Gene flow in natural populations. *Ann Rev Ecol Syst*, **16**, 393–430.

Slatkin, M. (1993) Isolation by distance in equilibrium and non-equilibrium populations. *Evolution*, **47**, 264–79.

Slatkin, M. (1995) A measure of population subdivision based on microsatellite allele frequencies. *Genetics*, **139**, 457–62.

Smale, L., Nunes, S. & Holekamp, K.E. (1997) Sexually dimorphic dispersal in mammals: patterns, primal causes and consequences. In: *Advances in the Study of Behavior*, Vol. 26 (eds P.J.B. Slater, J.S. Rosenblatt, C.T. Snowdon & M. Milinski), pp. 181–250. Academic Press, London (in press).

Smith, E.A. & Winterhalder, B. (eds) (1992) *Evolutionary Ecology and Human Behavior*. Aldine De Gruyter, New York.

Smith, G.T., Wingfield, J.C. & Veit, R.R. (1994) Adrenocortical response to stress in the common diving petrel, *Pelecanoides urinatrix*. *Physiol Zool*, **67**, 526–37.

Smith, J.N.M. & Arcese, P. (1989) How fit are floaters? Consequences of alternative territorial behaviour in a nonmigratory sparrow. *Am Natur*, **133**, 830–45.

Smith, R.L. (1979) Repeated copulation and sperm precedence: paternity assurance for a male brooding water bug. *Science*, **205**, 1029–31.

Smith, R.L. (ed.) (1984) *Sperm Competition and the Evolution of Animal Mating Systems*. Academic Press, Orlando.

Smith, S.M. (1975) Innate recognition of coral snake patterns by a possible avian predator. *Science*, **187**, 759–60.

Smuts, B.B., Cheney, D.L., Seyfarth, R.M., Wrangham, R.W. & Struhsaker, T.T. (1987) *Primate Societies*. University of Chicago Press, Chicago.

Snell, H., Jennings, R.D., Snell, H.M. & Harcourt, S. (1988) Intra-population variance in predator-avoidance performance of Galapagos lava lizards: the interaction of sexual and natural selection. *Evol Ecol*, **2**, 353–69.

Sokal, R.R., Harding, R.M. & Oden, N.L. (1989) Spatial patterns of human gene frequencies in Europe. *Am J Phys Anthrop*, **80**, 267–94.

Solomon, N. & French, J.A. (1996) *Cooperative Breeding in Mammals*. Cambridge University Press, Cambridge.

Soulé, M. (1982) Allometric variation. I. The theory and some consequences. *Am Natur*, **120**, 751–64.

Soulé, M. & Cuzin-Roudy, J. (1982) Allometric variation: 2. Development instability of extreme phenotypes. *Am Natur*, **120**, 765–80.

Southwood, T.R.E. (1973) The insect/plant relationship — an evolutionary perspective. *Symp Roy Entomol Soc Lond*, **6**, 3–30.

Spiess, E.B. (1987) Discrimination among prospective mates in *Drosophila*. In: *Kin Recognition in Animals* (eds D.J.C. Fletcher & C.D. Michener), pp. 75–119. John Wiley & Sons, New York.

Squire, L.R. (1992) Memory and the hippocampus: a synthesis from findings with rats, monkeys, and humans. *Psychol Rev*, **99**, 195–231.

Stacey, P.B. (1979) Kinship, promiscuity and communal breeding in the acorn woodpecker. *Behav Ecol Sociobiol*, **6**, 53–66.

Stacey, P.B. & Edwards, T.C., Jr (1983) Possible cases of infanticide by immigrant females in a group-breeding bird. *Auk*, **100**, 731–3.

Stacey, P.B. & Koenig, W.D. (1990) *Cooperative Breeding in Birds: Long-term Studies of Ecology and Behavior*. Cambridge University Press, Cambridge.

Stacey, P. & Ligon, J.D. (1987) Territory quality and dispersal options in the Acorn woodpecker, and a challenge to the habitat-saturation model of cooperative breeding. *Am Natur*, **130**, 654–76.

Stacey, P. & Ligon, J.D. (1991) The benefits of philopatry hypothesis for the evolution of cooperative breeding: variation in territory quality and group size effects. *Am Natur*, **137**, 831–46.

Staddon, J.E.R. (1975) A note on the evolutionary significance of supernormal stimuli. *Am Natur*, **109**, 541–5.

Stahl, W. (1962) Similarity and dimensional methods in biology. *Science*, **137**, 205–12.

Stamps, J.A. (1991) The effect of conspecifics on habitat selection in territorial species. *Behav Ecol Sociobiol*, **28**, 29–36.

Stander, P.E. (1992) Cooperative hunting in lions: the role of the individual. *Behav Ecol Sociobiol*, **29**, 445–54.

Stark, R.E. (1989) Beobachtungen zur Nestgründung und Brutbiologie der Holzbiene *Xylocopa sulcatipes* Maa (Apoidea: Anthophoridae). *Mitteilungen der Deutschen Gesellschaft für allgemeine und angewandte Entomologie*, **7**, 252–6.

Stark, R.E. (1990) Intraspecific nest usurpation in the large carpenter bee *Xylocopa sulcatipes* Maa (Apoidea: Anthophoridae). In: *Social Insects and the Environment* (eds G.K. Veeresh, B. Mallik & C.A. Viraktamath), pp. 165–6. Oxford & IBH Publishing Co. Pvt. Ltd, New Delhi.

Stark, R.E. (1992a) Cooperative nesting in the multivoltine large carpenter bee *Xylocopa sulcatipes* Maa (Apoidea: Anthophoridae): do helpers gain or lose to solitary females? *Ethology*, **91**, 301–10.

Stark, R.E. (1992b) Sex ratio and maternal investment in the multivoltine large carpenter bee *Xylocopa sulcatipes* (Apoidea: Anthophoridae). *Ecol Entomol*, **17**, 160–6.

Stark, R.E., Hefetz, A., Gerling, D. & Velthuis, H.H.W. (1990) Reproductive competition involving oophagy in the socially nesting bee *Xylocopa sulcatipes*. *Naturwissenschaften*, **77**, 38–40.

Starr, C.K. (1979) Origin and evolution of insect sociality: a review of modern theory. In: *Social Insects*, Vol. I (ed. H.R. Hermann), pp. 35–79. Academic Press, New York.

Starr, C.K. (1984) Sperm competition, kinship, and sociality in the aculeate Hymenoptera. In: *Sperm Competition and the Evolution of Animal Mating Systems* (ed. R.L. Smith), pp. 427–64. Academic Press, Orlando.

Stearns, S.C. (1992) *The Evolution of Life Histories*. Oxford University Press, Oxford.

Stearns, S.C. & Koella, J. (1986) The evolution of phenotypic plasticity in life-history traits: predictions of reaction norms for age and size in maturity. *Evolution*, **40**, 893–913.

Stearns, S.C. & Crandall, R.E. (1981) Quantitative predictions of delayed maturity. *Evolution*, **35**, 455–63.

Stearns, S.C. & Crandall, R.E. (1984) Plasticity for age and size at sexual maturity: a life-history adaptation to unavoidable stress. In: *Fish Reproduction* (eds G. Potts & R. Wootton), pp. 13–33. Academic Press, New York.

Stephens, D.W. (1981) The logic of risk-sensitive foraging preferences. *Anim Behav*, **29**, 628–9.

Stephens, D.W. & Krebs, J.R. (1986) *Foraging Theory*. Princeton University Press, Princeton.

Stephens, D.W., Anderson, J.P. & Toyer, K.B. (1996) On the spurious occurrence of tit-for-tat in pairs of predator-approaching fish. *Anim Behav* (in press).

Stephens, D.W., Nishimura, K. & Toyer, K.B. (1995) Error and discounting in the iterated prisoner's dilemma. *J Theor Biol*, **176**, 457–69.

Stern, D.L. (1994) A phylogenetic analysis of soldier evolution in the aphid family Hormaphididae. *Proc Roy Soc Lond B*, **256**, 203–9.

Stern, D.L. & Foster, W.A. (1996) The evolution of soldiers in aphids. *Biol Rev*, **71**, 27–79.

Stevens, T.A. & Krebs, J.R. (1986) Retrieval of stored seeds by marsh tits, *Parus palustris*, in the field. *Ibis*, **128**, 513–25.

Stewart, G.J. & Carlson, C.A. (1986) The biology of natural transformation. *Ann Rev Microbiol*, **40**, 211–35.

Stillman, R.A., Goss-Custard, J.D., Clarke, R.T. & Durell, S.E.A. le V. dit (1996) Shape of the interference function in a foraging vertebrate. *J Anim Ecol*, **65**, in press.

Stockley, P. & Purvis, A. (1993) Sperm competition in mammals: a comparative study of male roles and relative investment in sperm production. *Funct Ecol*, **7**, 560–70.

Stone, G.N. & Sunnocks, P. (1993) Genetic consequences of an invasion through a patchy environment — the cynipid gallwasp *Andricus quercuscalis* (Hymenoptera: Cynipidae). *Mol Ecol*, **2**, 251–68.

Stoner, G. & Breden, F. (1988) Phenotypic differentiation in female preference related to geographic variation in male predation risk in the Trinidad guppy (*Poecilia reticulata*). *Behav Ecol Sociobiol*, **22**, 285–91.

Strong, D.R., Lawton, J.H. & Southwood, T.R.E. (1984) *Insects on Plants: Community Patterns and Mechanisms*. Harvard Univesity Press, Cambridge, Massachusetts.

Stuebe, M.M. & Ketterson, E.D. (1982) A study of fasting in tree sparrows (*Spizella arborea*) and dark-eyed juncos (*Junco hyemalis*): ecological implications. *Auk*, **99**, 299–308.

Su, T.T., Follette, P.J. & O'Farrell, P.H. (1995) Qualifying for the license to replicate. *Cell*, **81**, 825–8.

Subowski, M.D. (1990) Releaser-induced recognition learning. *Psychol Rev*, **97(2)**, 271–84.

Suga, N. (1995) Processing of auditory information carried by species-specific complex sounds. In: *The Cognitive Neurosciences* (ed. M.S. Gazzaniga), pp. 295–313. MIT Press, Cambridge, Massachusetts.

Sundström, L. (1994) Sex ratio bias, relatedness asymmetry and queen mating frequency in ants. *Nature*, **367**, 266–8.

Sundström, L., Chapuisat, M. & Keller, L. (1996) Conditional manipulation of sex ratios by ant workers: a test of kin selection theory. *Science*, **274**, 993–5.

Sutherland, W.J. (1992) Game theory models of functional and aggregative responses. *Oecologia*, **90**, 150–2.

Sutherland, W.J. (1996a) *From Individual Behaviour to Population Ecology*. Oxford University Press, Oxford.

Sutherland, W.J. (1996b) Predicting the consequences of habitat loss for migrating populations. *Proc Roy Soc Lond B*, **263**, 1325–7.

Sutherland, W.J. & Dolman, P.M. (1994) Combining behaviour and population dynamics with applications for predicting the consequences of habitat loss. *Proc Roy Soc B*, **255**, 133–8.

Sutherland, W.J. & Moss, D. (1985) The inactivity of animals: influence of stochasiticity and prey size. *Behaviour*, **92**, 1–8.

Sutherland, W.J. & Parker, G.A. (1985) Distribution of unequal competitors. In: *Behavioural Ecology: Ecological Consequences of Adaptive Behaviour* (eds R.M. Sibly & R.H. Smith), pp. 255–74. Blackwell Scientific Publications, Oxford.

Sutherland, W.J., Jones, D.W.F. & Hadfield, R.W. (1986) Age differences in the feeding abilities of moorhens *Gallinula chloropus*. *Ibis*, **128**, 414–18.

Székely, T. (1992) Reproduction of Kentish Plover, *Charadrius alexandrinus*, in grasslands and fishponds: habitat mal-assessment hypothesis. *Aquila*, **99**, 59–68.

Székely, T. & Reynolds, J.D. (1995) Evolutionary transitions of parental care in shorebirds. *Proc Roy Soc Lond B*, **262**, 57–64.

Szymura, J.M. (1993) Analysis of hybrid zones with *Bombina*. In: *Hybrid Zones and the Evolutionary Process* (ed. R.G. Harrison), pp. 261–89. Oxford University Press, New York.

Taylor, C.R., Heglund, N.C. & Maloiy, G.M.O. (1982) Energetics and mechanics of terrestrial locomotion. I. Metabolic energy consumption as a function of speed and body size in birds and mammals. *J Exp Biol*, **97**, 1–21.

Taylor, C.R., Maloiy, G.M.O., Weibel, E.R. *et al.* (1980) Design of the mammalian respiratory system. 3. Scaling maximum aerobic capacity to body mass: wild and domestic animals. *Resp Physiol*, **44**, 25–37.

Temeles, E.J. (1994) The role of neighbours in territorial systems: when are they 'dear enemies?' *Anim Behav*, **47**, 339–50.

Templeton, A.R., Routman, E. & Phillips, C.A. (1995) Separating population structure from population history: a cladistic analysis of the geographical distribution of mitochondrial DNA haplotypes in the tiger salamander, *Ambystoma tigrinum*. *Genetics*, **140**, 767–82.

Templeton, J.J. (1993) *The use of personal and public information in foraging flocks of European starlings*. PhD thesis, Concordia University.

Templeton, J.J. & Giraldeau, L.-A. (1995) Patch

assessment in foraging flocks of European starlings: evidence for public information use. *Behav Ecol*, **6**, 65–72.

Templeton, J.J. & Giraldeau, L.-A. (1996) Vicarious sampling: the use of personal and public information by starlings foraging in a simple patchy environment. *Behav Ecol Sociobiol*, **38**, 105–13.

ten Cate, C. (1985) On sex differences in sexual imprinting. *Anim Behav*, **33**, 1310–17.

Thisted, T., Sørensen, N.S., Wagner, E.G.H. & Gerdes, K. (1994) Mechanism of post-segregational killing: Sok antisense RNA interacts with Hok mRNA via its 5′-end single-stranded leader and competes with the 3′-end of Hok mRNA for binding to the *mok* translation initiation region. *EMBO J*, **13**, 1960–8.

Thorne, B.L. & Carpenter, J.M. (1992) Phylogeny of the Dictyoptera. *System Entomol*, **17**, 253–67.

Thornhill, N.W. (ed.) (1993) *The Natural History of Inbreeding and Outbreeding: Theoretical and Empirical Perspectives*. University of Chicago Press, Chicago.

Thornhill, R. (1976) Sexual selection and nuptial feeding behavior in *Bittacus apicalis* (Insecta: Mecoptera). *Am Natur*, **110**, 529–48.

Thornhill, R. (1980) Sexual selection in the black-tipped hangingfly. *Sci Am*, **242**, 162–72.

Thornhill, R. (1981) *Panorpa* (Mecoptera: Panorpidae) scorpionflies: systems for understanding resource-defense polygyny and alternative male reproductive efforts. *Ann Rev Ecol Syst*, **12**, 355–86.

Thornhill, R. (1983) Cryptic female choice and its implications in the scorpionfly *Harpovittacus nigriceps*. *Am Natur*, **122**, 765–88.

Thornhill, R. (1992) Female preference for the pheromone of males with low fluctuating asymmetry in the Japanese scorpionfly (*Panorpa japonica*: Mecoptera). *Behav Ecol*, **3**, 277–83.

Thornhill, R. & Alcock, J. (1983) *The Evolution of Insect Mating Systems*. Harvard University Press, Cambridge, Massachusetts and London.

Thorpe, W.H. (1956) *Learning and Instinct in Animals*. Methuen & Co., London.

Tinbergen, J.M. (1981) Foraging decisions in starlings (*Sturnus vulgaris* L.). *Ardea*, **69**, 1–67.

Tinbergen, J.M. & Boerlijst, M. (1990) Nestling weight and survival in individual great tits (*Parus major*). *J Anim Ecol*, **59**, 1113–27.

Tinbergen, J.M. & Daan, S. (1990) Family planning in the Great tit (*Parus major*): optimal clutch size as integration of parent and offspring fitness. *Behaviour*, **114**, 161–90.

Tinbergen, L. (1960) The natural control of insects in pine woods. I. Factors influencing the intensity of predation by songbirds. *Arch Neerland Zool*, **13**, 265–343.

Tinbergen, N. (1953) *Social Behaviour in Animals*. Butler and Tanner, London.

Tinbergen, N. (1953) *The Herring Gull's World*. New Naturalist Series, Collins, London.

Tinbergen, N. (1963) On aims and methods of ethology. *Z Tierpsychol*, **20**, 410–33.

Tinbergen, N. (1964) *Social Behaviour in Animals*, 2nd edn. Chapman & Hall, London.

Tinbergen, N. (1972) *The Animal in its World*. Vol. I. *Field Studies*. George Allen & Unwin, London.

Tinbergen, N., Brockhuysen, G.J., Feekes, F., Houghton, J.C.W., Kruuk, H. & Szulc, E. (1963) Egg shell removal by the black headed gull, *Larus ridibundus*: a behaviour component of camouflage. *Behaviour*, **19**, 74–117.

Tomlinson, I. & O'Donald, P. (1989) The co-evolution of multiple female mating preferences and preferred male characters: the 'gene-for-gene' hypothesis of sexual selection. *J Theor Biol*, **139**, 219–38.

Tracy, N.D. & Seaman, J.W. Jr (1995) Properties of evolutionarily stable learning rules. *J Theor Biol*, **177**, 193–8.

Trail, P.W. & Adams, E.S. (1989) Active mate choice at cock-of-the-rock leks: tactics of sampling and comparison. *Behav Ecol Sociobiol*, **25**, 283–92.

Trivers, R.L. (1971) The evolution of reciprocal altruism. *Quart Rev Biol*, **46**, 35–57.

Trivers, R.L. (1972) Parental investment and sexual selection. In: *Sexual Selection and the Descent of Man 1871–1971* (ed. B. Campbell), pp. 136–79. Aldine, Chicago.

Trivers, R.L. (1974) Parent–offspring conflict. *Am Zool*, **14**, 249–64.

Trivers, R. (1985) *Social Evolution*. Benjamin/Cummings, Menlo Park, California.

Trivers, R.L. & Hare, H. (1976) Haplodiploidy and the evolution of the social insects. *Science*, **191**, 249–63.

Turner, G.F. & Robinson, R.L. (1992) Milinski's tit-for-tat hypothesis: do fish preferentially inspect in pairs? *Anim Behav*, **43**, 677–9.

Tuttle, M.D. & Ryan, M.J. (1981) Bat predation and the evolution of frog vocalisations in the Neotropics. *Science*, **214**, 677–8.

Ueno, T. (1994) Self-recognition by the parasitic wasp *Itoplectis naranyae* (Hymenoptera: Ichneumonidae). *Oikos*, **70**, 333–9.

Valone, T.J. (1989) Group foraging, public information and patch estimation. *Oikos*, **56**, 357–63.

Valone, T.J. (1991) Bayesian and prescient assessment: foraging with pre-harvest information. *Anim Behav*, **41**, 569–77.

Valone, T.J. (1992a) Information for patch assessment: a field investigation with black-chinned hummingbirds. *Behav Ecol*, **3**, 211–22.

Valone, T.J. (1992b) Patch estimation via memory windows and the effect of travel time. *J Theor Biol*, **157**, 243–51.

Valone, T.J. & Brown, J.S. (1989) Measuring patch assessment abilities of desert granivores. *Ecology*, **70**, 1800–10.

Valone, T.J. & Giraldeau, L.-A. (1993) Patch estimation in group foragers: what information is used? *Anim Behav*, **45**, 721–8.

Van der Have, T.M., Boomsma, J.J. & Menken, S.B.J. (1988) Sex-investment ratios and relatedness in the monogynous ant *Lasius niger* (L.). *Evolution*, **42**, 160–72.

Vander Meer, R.P. & Wojcik, D.K. (1982) Chemical mimicry in the myrmechophilus beetle *Myrmecaphodius excavaticollis*. *Science*, **218**, 806–8.

Vander Wall, S.B. (1990) *Food Hoarding In Animals*. Chicago University Press, Chicago.

van Dijk, B.A., Boomsma, D.I. & de Man, A.J.M. (1996) Blood group chimerism in multiple human births is not rare. *Am J Med Genet*, **61**, 264–8.

Van Honk, C.G.J., Röseler, P.-F., Velthuis, H.H.W. & Hoogeveen, J.C. (1981) Factors influencing the egg laying of workers in a captive *Bombus terrestris* colony. *Behav Ecol Sociobiol*, **9**, 9–14.

Van Noordwijk, A.J., VanBalen, J.H. & Scharloo, W. (1981) Genetic variation in the timing of reproduction in the Great tit. *Oecologia*, **49**, 158–66.

van Rhijn, J.G. (1985) A scenario for the evolution of social organisation in ruffs *Philomachus pugnax* and other Charadriiform species. *Ardea*, **73**, 25–37.

van Rhijn, J.G. (1990) Unidirectionality in the phylogeny of social organisation, with special reference to birds. *Behaviour*, **115**, 153–74.

Vega-Redondo, F. & Hasson, O. (1993) A game–theoretic model of predator–prey signalling. *J Theor Biol*, **162**, 309–19.

Vehrencamp, S.L. (1979) The roles of individual, kin, and group selection in the evolution of sociality. In: *Handbook of Behavioral Neurobiology* Vol. 3 (eds P. Marler & J.G. Vandenbergh), pp. 351–79. Plenum Press, New York.

Vehrencamp, S.L. (1983a) A model for the evolution of despotic versus egalitarian societies. *Anim Behav*, **31**, 667–82.

Vehrencamp, S.L. (1983b) Optimal degree of skew in cooperative societies. *Am Zool*, **23**, 327–35.

Verhulst, S. (1995) *Reproductive decisions in Great tits. An optimality approach*. PhD thesis, University of Groningen, Groningen.

Vickery, W.L., Giraldeau, L.-A., Templeton, J.J., Kramer, D.L. & Chapman, C.A. (1991) Producers, scroungers, and group foraging. *Am Natur*, **137**, 847–63.

Vines, G. (1980) Spatial consequences of aggressive behaviour in flocks of oystercatchers *Haematopus ostralegus*. *Anim Behav*, **28**, 1175–83.

Visscher, P.K. (1986) Kinship discrimination in queen rearing by honey bees (*Apis mellifera*). *Behav Ecol Sociobiol*, **18**, 453–60.

Visscher, P.K. (1989) A quantitative study of worker reproduction in honey bee colonies. *Behav Ecol Sociobiol*, **25**, 247–54.

Visscher, P.K. & Dukas, R. (1995) Honey bees recognize development of nestmates' ovaries. *Anim Behav*, **49**, 542–4.

Vogel, S. (1988) *Life's Devices*. Princeton University Press, Princeton.

von Boehmer, H. & Kisielow, P. (1991) How the immune system learns about self. *Sci Am*, **265**(4), 74–81.

von Frisch, K. (1967) *The Dance Language and Orientation of Bees*. Harvard University Press, Cambridge, MA.

Vos, D.R. (1994) Sex recognition in zebra finch males results from early experiences. *Behaviour*, **128**, 1–14.

Vos, D.R. (1995a) The role of sexual imprinting for sex recognition in zebra finches: a difference between males and females. *Anim Behav*, **50**, 645–53.

Vos, D.R. (1995b) Sexual imprinting in zebra finch females: do females develop a preference for males that look like their father? *Ethology*, **99**, 252–62.

Waage, J.K. (1979) Dual function of the damselfly penis: sperm removal and transfer. *Science Wash*, **203**, 916–18.

Waage, J.K. (1988) Confusion over residence and the escalation of damselfly territorial disputes. *Anim Behav*, **36**, 586–95.

de Waal, F. (1982) *Chimpanzee Politics: Power and Sex Among Apes*. Harper and Row, New York.

de Waal, F.B.M. & Luttrell, L.M. (1988) Mechanisms of social reciprocity in three primate species: symmetrical relationship characteristics or cognition? *Behav Ecol Sociobiol*, **9**, 101–18.

Waddell, D.R. (1982) A predatory slime mould. *Nature*, **298**, 464–6.

Waldman, B. (1981) Sibling recognition in toad tadpoles: the role of experience. *Z Tierpsychol*, **56**, 341–58.

Waldman, B. (1986) Chemical ecology of kin recognition in anuran amphibians. In: *Chemical Signals in Vertebrates*, Vol.4 (eds D. Duval, D. Müller-Schwarze & R.M. Silverstein), pp. 225–42. Plenum, New York.

Waldman, B. (1987) Mechanisms of kin recognition. *J Theor Biol*, **128**, 159–85.

Waldman, B. (1991) Kin recognition in amphibians. In: *Kin Recognition* (ed. P. Hepper), pp. 162–219.

Cambridge University Press, Cambridge.

Waldman, B., Frumhoff, P.C. & Sherman, P.W. (1988) Problems of kin recognition. *Trends Ecol Evol*, **3**, 8–13.

Waldman, B., Rice, J.E. & Honeycutt, R.L. (1992) Kin recognition and incest avoidance in toads. *Am Zool*, **32**, 18–30.

Walkowiak, W. (1988) Neuroethology of anuran call recognition. In: *The Evolution of the Amphibian Auditory System* (eds B. Fritzsch, M.J. Ryan, W. Wilczynski, T.E. Hetherington & W. Walkowiak), pp. 485–509. John Wiley & Sons, New York.

Wallace, A.R. (1889) *Darwinism*, 2nd edn. MacMillan, London.

Walters, J.R. (1990) Red-cockaded woodpeckers: a 'primitive' cooperative breeder. In: *Cooperative Breeding in Birds: Long-term Studies of Ecology and Behavior* (eds P.B. Stacey & W.D. Koenig), pp. 67–102. Cambridge University Press, Cambridge.

Walters, J.R. & Seyfarth, R.M. (1987) Conflict and cooperation. In: *Primate Societies* (eds B.B. Smuts, D.L. Cheney, R.M. Seyfarth, R.W. Wrangham & T.T. Struhsaker), pp. 306–17. University of Chicago Press, Chicago.

Walters, J.R., Copeyon, C.K. & Carter, I.J.H. (1992) Test of the ecological basis of cooperative breeding in Red-cockaded woodpecker. *Auk*, **109**, 90–7.

Ward, P.I. (1986) Prey availability increases less quickly than nest size in the social spider *Stegodyphus mimosarum*. *Behaviour*, **97**, 213–25.

Warner, R.R. (1987) Female choice of sites versus mates in a coral reef fish, *Thalassoma bifasciatum*. *Anim Behav*, **35**, 1470–8.

Warner, R.R., Shapiro, D.Y., Marcanato, A. & Petersen, C.W. (1995) Sexual conflict: males with highest mating success convey the lowest fertilization benefits to females. *Proc Roy Soc Lond B*, **262**, 135–9.

Warner, R.R., Wernerus, F., Lejeune, P. & van den Berghe, E. (1995) Dynamics of female choice for parental care in a fish species where care is facultative. *Behav Ecol*, **6**, 73–81.

Waser, N.M. (1993) Population structure, optimal outbreeding, and assortative mating in angiosperms. In: *The Natural History of Inbreeding and Outbreeding* (ed. N.W. Thornhill), pp. 173–99. University of Chicago Press, Chicago.

Waser, N.M. & Price, M.V. (1985) Reciprocal transplant experiments with *Delphinium nelsonii* (Ranunculaceae): evidence for local adaptation. *Am J Bot*, **72**, 1726–32.

Waser, N.M. & Price, M.V. (1993) Crossing distance effects on prezygotic performance in plants: an argument for female choice. *Oikos*, **68**, 303–8.

Waser, N.M. & Price, M.V. (1994) Crossing-distance effects in *Delphinium nelsonii*: outbreeding and inbreeding depression in progeny fitness. *Evolution*, **48**, 842–52.

Waser, P.M., Austad, S.N. & Kena, B. (1986) When should animals tolerate inbreeding? *Am Natur*, **128**, 529–37.

Wasser, S.K. (1995) Costs of conception in baboons. *Nature Lond*, **376**, 219–20.

Watson, A. & Miller, G.R. (1971) Territory size and aggression in a fluctuating red grouse population. *J Anim Ecol*, **40**, 367–83.

Watson, P.J. & Thornhill, R. (1994) Fluctuating asymmetry and sexual selection. *Trends Ecol Evol*, **9**, 21–5.

Watts, B.D. (1990) Cover use and predator-related mortality in song and savannah sparrows. *Auk*, **107**, 775–8.

Wauters, L. & Dhondt, A.A. (1989) Body weight, longevity and reproductive success in red squirrels (*Sciurus vulgaris*). *J Anim Ecol*, **58**, 637–51.

Weary, D.M., Guilford, T.C. & Weisman, R.G. (1993) A product of discriminative learning may lead to female preferences for elaborate males. *Evolution*, **47**, 333–6.

Weathers, W.W. (1980) Seasonal and geographic variation in avian standard metabolic rate. *Proc Int Ornithol Congress Berl*, 283–6.

Wedekind, C., Seebeck, T., Bettens, F. & Paepke, A.J. (1995) MHC-dependent mate preferences in humans. *Proc Roy Soc Lond B*, **261**, 245–9.

Wedell, N. (1994) Variation in nuptial gift quality in bush crickets (Orthoptera: Tettigoniidea). *Behav Ecol*, **5**, 418–25.

Wehner, R. (1981) Spatial vision in arthropods. In: *Handbook of Sensory Physiology*, Vol. VII/6C (ed. H. Autrum), pp. 287–616. Springer-Verlag, Berlin, New York.

Wehner, R. (1987) 'Matched filters' — neural models of the external world. *J Comp Physiol A*, **161**, 511–31.

Wehner, R. (1994) The polarization-vision project: championing organismic biology. In: *Neural Basis of Behavioural Adaptation* (eds K. Schildberger & N. Elsner), pp. 103–43. G. Fischer, Stuttgart, New York.

Wehner, R. & Menzel, R. (1990) Do insects have cognitive maps? *Ann Rev Neurosci*, **13**, 403–14.

Wehner, R. & Wehner, S. (1990) Insect navigation: use of maps or Ariadne's thread? *Ethol Ecol Evol*, **2**, 27–48.

Wehner, R. & Srinivasan, M.V. (1984) The world as the insect sees it. In: *Insect Communication* (ed. T. Lewis), pp. 29–47. Academic Press, London.

Wehner, R., Michel, B. & Antonsen, P. (1996) Visual

navigation in insects: coupling of egocentric and geocentric information. *J Exp Biol*, **199**, 129–40.

Weiner, J. (1992) Physiological limits to sustainable energy budgets in birds and mammals: ecological implications. *Trends Ecol Evol*, **7**, 384–8.

Welham, C.V.J. & Ydenberg, R.C. (1993) Efficiency-maximizing flight speeds in parent black terns. *Ecology*, **74**, 1893–1901.

Went, F.W. (1968) The size of man. *Am Sci*, **56**, 400–13.

West, M.J. & King, A.P. (1988) Female visual displays affect the development of male song in the cowbird. *Nature*, **334**, 244–6.

West-Eberhard, M.J. (1975) The evolution of social behavior by kin selection. *Quart Rev Biol*, **50**, 1–33.

West-Eberhard, M.J. (1979) Sexual selection, social competition and evolution. *Proc Am Philosoph Soc*, **123**, 222–34.

West-Eberhard, M.J. (1989) Phenotypic plasticity and the origins of diversity. *Ann Rev Ecol Syst*, **20**, 249–78.

Westneat, D.F. (1987) Extra-pair fertilizations in a predominantly monogamous bird: genetic evidence. *Anim Behav*, **35**, 877–86.

Westneat, D.F. & Sargent, R.C. (1996) Sex and parenting: the effects of sexual conflict and parentage on parental strategies. *Trends Ecol Evol*, **11**, 87–91.

Westneat, D.F. & Sherman, P.W. (1993) Parentage and the evolution of parental behavior. *Behav Ecol*, **4**, 66–77.

Westneat, D.F. & Webster, M.S. (1994) Molecular analysis of kinship in birds: interesting questions and useful techniques. In: *Molecular Ecology and Evolution: Approaches and Applications* (eds B. Schierwater, B. Streit, G.P. Wagner & R. DeSalle), pp. 91–126. Birhauser Verlag, Basel.

Westneat, D.F., Clark, A.B. & Rambo, K.C. (1995) Within-brood patterns of paternity and paternal behavior in red-winged blackbirds. *Behav Ecol Sociobiol*, **37**, 349–56.

Wetton, J.H. & Parkin, D.T. (1991) An association between fertility and cuckoldry in the house sparrow *Passer domesticus*. *Proc Roy Soc Lond B*, **245**, 227–33.

Wetton, J.H., Carter, R.E., Parkin, D.T. & Walters, D. (1987) Demographic study of a wild house sparrow population by DNA fingerprinting. *Nature Lond*, **327**, 147–9.

Whitehead, H. & Hope, P.L. (1991) Sperm whalers off the Galapagos Islands and in the Western North Pacific, 1830–1850: ideal free whalers. *Ethol Sociobiol*, **12**, 147–61.

Whiten, A. & Ham, R. (1992) On the nature and evolution of imitation in the animal kingdom: reappraisal of a century of research. *Adv Study Behav*, **21**, 239–83.

Whitlock, M.C. (1995) Variance-induced peak shifts. *Evolution*, **49**, 252–9.

Whittingham, L.A. & Lifjeld, J.T. (1995) High paternal investment in unrelated young: extra-pair paternity and male parental care in house martins. *Behav Ecol Sociobiol*, **37**, 103–8.

Wiegmann, D.D., Real, L.A., Capone, T.A. & Ellner, S. (1996) Some distinguishing features of models of search behavior and mate choice. *Am Natur*, **147**, 188–204.

Wiggins, D.A. & Morris, R.D. (1986) Criteria for female choice of mates: courtship feeding and paternal care in the common tern. *Am Natur*, **128**, 126–9.

Wilcove, D.S. (1985) Nest predation in forest tracts and the decline of migratory songbirds. *Ecology*, **66**, 1211–14.

Wildhaber, M.L., Green, R.F. & Crowder, L.B. (1994) Bluegill continuously update patch giving up times based on foraging experience. *Anim Behav*, **47**, 501–13.

Wiley, R.H. (1983) The evolution of communication: information and manipulation. In: *Communication* (eds T.R. Halliday & P.J.B. Slater), pp. 82–113. Blackwell Scientific Publications, Oxford.

Wiley, R.H. (1991) Associations of song properties with habitats for territorial oscine birds of eastern North America. *Am Natur*, **138**, 973–93.

Wiley, R.H. (1994) Errors, exaggeration and deception in animal communication. In: *Behavioural Mechanisms in Evolutionary Ecology* (ed. L.A. Real), pp. 157–89. University of Chicago Press, London.

Wiley, R.H. & Richards, D.G. (1978) Physical constraints on acoustic communication in the atmosphere: implications for the evolution of animal vocalization. *Behav Ecol Sociobiol*, **3**, 69–94.

Wilkinson, G.S. & Reillo, P.R. (1994) Female choice response to artificial selection on an exaggerated male trait in a stalk-eyed fly. *Proc Roy Soc Lond B*, **255**, 1–6.

Williams, A.F. & Barclay, A.N. (1988) The immunoglobulin superfamily — domains for cell surface recognition. *Ann Rev Immunol*, **6**, 381–405.

Williams, D. & McIntyre, P. (1980) The principal eyes of a jumping spider have a telephoto component. *Nature*, **288**, 578–80.

Williams, G.C. (1966a) *Adaptation and Natural Selection, a Critique of Some Current Evolutionary Thought*. Princeton University Press, Princeton.

Williams, G.C. (1966b) Natural selection, the costs of reproduction, and a refinement of Lack's principle. *Am Natur*, **100**, 687–90.

Williams, G.C. (1975) *Sex and Evolution*. Princeton University Press, Princeton.

Williams, J.B. (1987) Field metabolism and food consumption of savannah sparrows during the breeding season. *Auk*, **104**, 277–89.

Williams, T.C. & Williams, J.M. (1978) Orientation of transatlantic migrants. In: *Animal Migration, Navigation, and Homing* (eds K. Schmidt-Koenig & W.T. Keeton), pp. 239–51. Springer-Verlag, Berlin.

Wilson, D.S. & Sober, E. (1994) Reintroducing group selection to the human behavioral sciences. *Behav Brain Sci*, **17**, 585–654.

Wilson, E.O. (1971) *The Insect Societies*. Harvard University Press, Cambridge, Massachusetts.

Wilson, E.O. (1975) *Sociobiology*. Harvard University Press, Cambridge, Mass.

Wilson, K. & Gatehouse, A.G. (1993) Seasonal and geographical variation in the migratory potential of outbreak populations of the African armyworm moth, *Spodoptera exempta*. *J Anim Ecol*, **62**, 169–81.

Wingfield, J.C., Schwabb, H. & Mattocks, P.W. Jr (1990) Endocrine mechanisms of migration. In: *Bird Migration. Physiology and Ecology* (ed. E. Gwinner), pp. 232–56. Springer-Verlag, Berlin.

Wingfield, J.C., Suydam, R. & Hunt, K. (1994b) The adrenocortical responses to stress in snow buntings (*Plectrophenax nivalis*) and lapland longspurs (*Calcarius lapponicus*) at Barrow, Alaska. *Comp Biochem Physiol C*, **108**, 299–306.

Wingfield, J.C., Vleck, C.M. & Moore, M.C. (1992) Seasonal changes of the adrenocortical response to stress in birds of the Sonoran desert. *J Exp Zool*, **264**, 419–28.

Wingfield, J.C., Deviche, P., Sharbaugh, S. *et al.* (1994a) Seasonal changes of the adrenocortical responses to stress in redpolls, *Acanthis flammea*, in Alaska. *J Exp Zool*, **270**, 372–80.

Winn, H.E. (1958) The comparative ecology and reproductive behaviour of fourteen species of darter (Percidae). *Ecol Monogr*, **28**, 155–91.

Winston, M.L. (1987) *The Biology of the Honey Bee*. Harvard University Press, Cambridge, Massachusetts.

Winter, U. & Buschinger, A. (1983) The reproductive biology of a slavemaker ant, *Epimyrma ravouxi*, and a degenerate slavemaker, *E. kraussei* (Hymenoptera: Formicidae). *Entomol General*, **9**, 1–15.

Winterhalder, B. (1981) Foraging strategies in the boreal forest: an analysis of Cree hunting and gathering. In: *Hunter–Gatherer Foraging Strategies* (eds B. Winterhalder & E.A. Smith), pp. 66–98. University Chicago Press, Chicago.

Wishart, G.J. (1987) Regulation of the length of the fertile period in the domestic fowl by numbers of oviductal spermatozoa as reflected by those trapped in laid eggs. *J Reprod Fert*, **80**, 493–8.

Witter, M.S. & Cuthill, I.E. (1993) The ecological costs of avian fat storage. *Phil Trans Roy Soc Lond B*, **340**, 73–90.

Witter, M.S. & Swaddle, J.P. (1995) Dominance, competition and energy reserves in the European starling, *Sturnus vulgaris*. *Behav Ecol*, **6**, 343–8.

Witter, M.S., Cuthill, I.C. & Bonser, R.H.C. (1994) Experimental investigations of mass-dependent predation risk in the European starling, *Sturnus vulgaris*. *Anim Behav*, **48**, 201–22.

Witter, M.S., Swaddle, J.P. & Cuthill, I.C. (1995) Periodic food availability and strategic regulation of body mass in the European starling, *Sturnus vulgaris*. *Funct Ecol*, **9**, 568–74.

Woodhead, A.P. (1985) Sperm mixing in the cockroach *Diploptera punctata*. *Evolution*, **39**, 159–64.

Woolfenden, G. & Fitzpatrick, J. (1984) *The Florida Scrub Jay: Demography of a Cooperative Breeding Bird*. Princeton University Press, Princeton, NJ.

Woyciechowski, M. & Lomnicki, A. (1987) Multiple mating of queens and the sterility of workers among eusocial Hymenoptera. *J Theor Biol*, **128**, 317–27.

Wright, R. (1994) *The Moral Animal: the New Science of Evolutionary Psychology*. Pantheon Books, New York.

Wright, S. (1978) *Evolution and the Genetics of Populations*, Vol. 4. *Variability within and among Populations*. University of Chicago Press, Chicago.

Würsig, B. (1989) Cetaceans. *Science*, **244**, 1550–7.

Wynne-Edwards, V.C. (1962) *Animal Dispersion in Relation to Social Behaviour*. Oliver and Boyd, Edinburgh & London.

Yamazaki, K., Boyse, E.A., Mike, V., Thaler, H.T., Mathieson, B.J., Abbott, J., Boyse, J. & Zayas, Z.A. (1976) Control of mating preferences in mice by genes in the major histocompatibility complex. *J Exp Med*, **144**, 1324–35.

Ydenberg, R.C., Welham, C.V.J., Schmid Hempel, R., Schmid-Hempel, P. & Beauchamp, G. (1994) Time and energy constraints and the relationship between currencies in foraging theory. *Behav Ecol*, **5**, 28–34.

Yoerg, S.I. (1991) Social feeding reverses learned flavor aversions in spotted hyenas (*Crocuta crocuta*). *J Compar Psychol*, **105**, 185–9.

Young, R.A. (1976) Fat, energy and mammalian survival. *Am Zool*, **16**, 699–710.

Zack, S. (1990) Coupling delayed breeding with short-distance dispersal in cooperatively breeding birds. *Ethology*, **86**, 265–86.

Zack, S. & Rabenold, K. (1989) Assessment, age, and proximity in dispersal contests among cooperative wrens: field experiments. *Anim Behav*, **38**, 235–47.

Zack, S. & Stutchbury, B. (1993) Delayed breeding in avian social systems: the role of territory quality and 'floater' tactics. *Behaviour*, **123**, 194–219.

Zahavi, A. (1975) Mate selection — a selection for a handicap. *J Theor Biol*, **53**, 205–14.

Zahavi, A. (1977a) The cost of honesty (further remarks on the handicap principle). *J Theor Biol*, **67**, 603–5.

Zahavi, A. (1977b) Reliability in communication systems and the evolution of altruism. In: *Evolutionary Ecology* (eds B. Stonehouse & C. Perrins), pp. 253–9. Macmillan, London.

Zahavi, A. (1987) The theory of signal selection and some of its implications. In: *International Symposium of Biological Evolution* (ed. V.P. Delfino), pp. 305–27. Adriatica Editrice, Bari.

Zahavi, A. (1990) Arabian babblers: the quest for social status in a cooperative breeder. In: *Cooperative Breeding in Birds: Long-term Studies of Ecology and Behavior* (eds P.B. Stacey & W.D. Koenig), pp. 103–30. Cambridge University Press, Cambridge.

Zhivotovsky, L.A., Feldman, M.W. & Christiansen, F.B. (1994) Evolution of recombination among multiple selected loci: a generalized reduction principle. *Proc Nat Acad Sci USA*, **91**, 1079–83.

Zimmerer, E.J. & Kallman, K.D. (1989) Genetic basis for alternative reproductive tactics in the pygmy swordtail, *Xiphophorus nigrensis*. *Evolution*, **43**, 1298–1307.

Ziołko, M. & Kozłowski, J. (1983) Evolution of body size: an optimization model. *Mathemat Biosci*, **64**, 127–43.

Zuk, M., Ligon, J.D. & Thornhill, R. (1992) Effects of experimental manipulation of male secondary sexual characters on female mate preference in red jungle fowl. *Anim Behav*, **44**, 999–1006.

Zuk, M., Simmons, L.W. & Cupp, L. (1993) Calling characteristics of parasitized and unparasitized populations of the field cricket *Teleogryllus oceanicus*. *Behav Ecol Sociobiol*, **33**, 339–43.

Zuk, M., Thornhill, R., Ligon, J.D. *et al.* (1990) The role of male ornaments and courtship behaviour in female mate choice of red jungle fowl. *Am Natur*, **136**, 459–73.

Zyskind, J.W. & Smith, D.W. (1992) DNA replication, the bacterial cell cycle, and cell growth. *Cell*, **69**, 5–8.

Index

Note: page numbers in *italics* refer to figures, those in **bold** to tables.

acceptance errors 72, 92–3
acceptance thresholds
  context-dependent (shifting) 87–8, 89–90, 95–6
  optimal 86, 88, 90, 91–2
Accipiter nisus (sparrowhawk) *110*, 111
acorn woodpeckers *see Melanerpes formicivorus*
acoustic communication 30
acoustic signalling 159–60, 162–4
Acrocephalus arundinaceus (warbler, great reed) 78
Acrocephalus sechellensis (warbler, Seychelles) 233–4, 239
action, recognition component 71–2, 86–92
adaptive behaviour 16
additive genetic variance, in courtship 368–9
adrenal stress hormones 119
Agelaius phoeniceus (blackbird, red-winged) 59, 93, *173*
aggression, sexually-related 237, 251, 252
air oscillation, near-field 30
Alauda arvensis (skylark) 170
Alectura lathami (brush turkey) 90
alerting components, assertion display 158
alleles, recognition promoting 94
allelic diversity, loss of 356
alliances 258–62
  bridging and revolutionary 262
  eukaryotic 299–300
  non-kin 261
Alpheus (shrimp, snapping) 347
altruism 9, 152, 204, 207–8, 228–9
  evolution of 249
  reciprocal and true 208
  reproductive 150, 203
Ambystoma tigrinum (salamander, tiger) 360
Ambystoma tigrinum nebulosum 70, 71
animals, as decision-makers 19
anisogamy 122, 123, 124
Anolis auratus 156–8
Anolis spp. (lizard) 343, 348
Anser anser (goose, domestic) 127
Anser cynoides (goose, Chinese) 127
anthropomorphism 11
Aphelocoma coerulescens (jay, scrub) 57, 58, 239
Aphelocoma ultramarina (jay, Mexican) 57, 58, 235
Apis mellifera (bee, honey) 30, 83, 99, 164–5, 227
apostatic prey selection 51–2
Archilocus alexandri (humming-bird, black-chinned) 47
area copying 59–60
'armpit effect' 83
assertion display 156–8
assortative mating 190, 368
auditory communications 24, *25*
Augochlorella striata (bee, halictine) 224

Bacillus subtilis 293
'back-up signal' hypothesis 175
Bayesian models 45–8
behavioural models 379–80, 386, 389
  ideal free-distribution models 379–80, 386
behavioural routines, investigations 117–18
behaviour
  anti-predatory 109–10, 111
  complex 15–16
  constraints imposed by physical environment 20–4, *26*
  and ecology 5–6
  economic models of 6–7, 17
  effects of last male precedence mechanisms 139–42
  evolution of in small isolated populations 368–72
  female, and sperm competition 141
  interactions with population structure 363–8
  males seeking extrapair copulations 141–2
  selfish 266, 267
  short-term
    consequences of 17
    and lifetime reproductive success 111–18
  vigilance behaviour 110
  see also social behaviour
behaviour copying 61–2, 66
behavioural decision rules, population consequences of 393
behavioural ecological models 394
Belding's ground squirrel *see Spermophilus beldingi*
Bembicium vittatum (snail) **357**, 361
bird navigation 31–3
blue jays 281–3
body mass
  changing prior to migration 104–5
  as function of age *316*, 316
  increased, and flight 107–8
  and photoperiod 102–3
  power functions of 318
  production and age of maturity 318–19, *319*, 320, *321*
body size
  constraints due to 26–30
  and sound–source detection 30
Bombina bombina (toad) **357**, 368
Bombina bombina/variegata (toad, fire-bellied) 355
Bombus (bee, bumble) 226
Botryllus schlosseri (urochordate) 82, 295
Branta bernicla bernicla (goose, dark-bellied brent) 395
breeders
  dominant, harassment by 248

early breeders 324, 332
replacement breeders 241–2, 252
breeding
communal 216
delayed 231, 389–90
multiple 216
optimal timing 325
shared 239–40
see also cooperative breeding
breeding season, measurement of density dependence in 389–90
breeding season habitats, not easily evaluated 380–1
breeding success 381
breeding units, closed, and gene flow 362
brood size 330–1
brood size manipulations 322–4
buffer effect 389
Buteo galapagoensis (hawk, Galapagos) 245
Bufo americanus (toad, American) 74, 194

Calcarius lapponicus (longspur, Lapland) 99, *173*
Calcarius pictus *173*
Camallanus cotti (nematode) 176
Campylorhynchus nuchalis (wren, stripe-backed) 237
Carduelis chloris (greenfinch) 60, 103
Carduus spp. (thistle) 366
Carpodacus mexicanus (house finch) 166
Cataglyphis ants, navigation by 33–6, 40
Cataglyphis fortis (desert ant) *34*, *38*
categorical variables, methods for 345–6
causation 4, 11
Cebus apella (monkey, capuchin) 66
cell-surface recognition 294
central place foragers, navigation of 33–6
Ceratitis capitata (fruit fly, mediterranean) 141
Cercopithecus aethiops (monkey, vervet) 79
chain procedures, maintaining cultural traditions 66
challenge display 156
character change 336–40
combined with tree structure 347–9
correlated 340–7
types of 341
character evolution
rates of 339–40
tracing of 337–8
Charadrius alexandrinus (plover, Kentish) 381
cheating 149
chick begging 167–9, 171, 172, 177
chick position, and food allocation 177–8
chimerism 295
Chiroxiphia linearis (manakin, long-tailed) 163
Chlidonias nigra (tern, black) 99
choice, self-enforcing 190
Chorthippus parallelus (grasshopper, meadow) 354–5, 368
chorusing, synchronous 163–4
chromosomes
bacterial 291, 292, 303, 304
sex 300–1
Clamator glandarius (cuckoo, great spotted) 171
clutch laying dates, life-history consequences of changes 327–8
clutch size, variation in 330, *331*
clutch-date combinations, optimal 326
coercion 273–4, 275
Collomia grandiflora 75
Colobopsis nipponicus *223*
colonies, eusocial and parasocial 224–5
colonization
changes in selection pressures 370–1
of new territory 355
progressive 353
repeated, and role of selection 371–2
colony fission 223
Columba livia (pigeon) 52, 60
Columba palambus (woodpigeon) 60
Columbina inca (dove, inca) 47
communication signals 149
competition 254
among signallers during communication 177–8
competitive ability, in foraging 382–6
competitive relationships 254–65
alternatives to dominance 263–4
benefits and costs of dominance 262–3
dominance hierarchies 258–62
pairwise contests 255–8
computational capabilities, constraints set by 30–40
conflict
and cooperation 7, 229
in family groupings 9, 237–41
intragenomic 285, 290
organellar/nuclear genes 299–300
over who reproduces 241–6, 252
parent–offspring 302
sexual 142–4
conflicts of interest
evolutionary 240
genetic 229
conjugation 291, 292
Conocephalus nigropleurum (katydid) 78
conservation 308, 373
consortship 142
conspicuousness 156–7, 158, 159, 193
constraints 313–14
cooperation 254, 265–83, 284
coerced 273–4, 275
and conflict 7, 229
mutualism in 152
short- and long-term 271–2
producers vs scroungers 272–3, 275
reciprocity 265–70
cooperative breeding 234, 249, 251
changing scope of research 228–30
preferential assistance to closest relatives 235
coordination 284
Cope's law 341
copula duration, male dungflies 132–4
copulation, females with multiple males, benefits **125**, 125–6
coreplicons 290, 303–4
corticosterone 119
Corvus monedula (jackdaw) 66
courtship, additive genetic variance in 368–9
courtship trembling 161
Crenicichla alta 160
Crocuta crocuta (hyena, spotted) 61
Crotophaga sulcirostris (anis, groove-billed) 245
cryptic female choice 141
cryptic prey, detection of 51–2
Cryptocercus (woodroach) 214
cub-rearing, communal 277–8
Cuculus canorus (cuckoo) 4, 85
cues 71
chemical 74
kin recognition 74–7
optimal cue system 77
phenotypic and non-phenotypic 88
predictive, short- and long-term 101

recognition cues 72–4
temporal 88
cultural diffusion, regulation of 65–6
cultural transmission 63–6, *67*
population approach 63, *64*
Cyrtodiopsis dalmanni (stalk-eyed fly) 80
cytoplasmic commons, management of 290
cytoplasmic conflict, nuclear-enforced suppression of 300

D-present cues 72, 74, 75, 81
daily energy expenditure (DEE) 322–3
Daphnia (water flea) 384
dawn chorus 116
deceit, in signalling 166–71
decision rules 313
defence mechanisms 5
Delichon urbica (house martin) 93
Delphinium nelsonii (larkspur) 71, 367
demes 351
Dendroica coronata (warbler) **357**
density dependence
behavioural approach to 379–81
determining strength of 378–90
importance of 374–7, *378*
and macroptery 365
measurement of
in the breeding season 389–90
in the non-breeding season 381–8
determinism 10
development 4
Dictyostelium caveatum 294
Dictyostelium discoideum 294–5
dimensionality theories 321–2
dioecious species 121
disassortative mating 185, 196
discrimination, absent, effects of 93
dispersal, delayed 230–1, 243
parental influence 243
dispersal variation 364–6
display 156–8, 169–70, 180
and the physical environment 158–60
visual 159, 162
dizygotic twins, chimerism between 295
DNA, compartmentalized in vertebrates 296–7
DNA markers 362–3
Dollo's law 341
dominance
benefits and costs of 262–3
carpenter bee pairs 212–13
and mate replacement conflict 242
and models of reproductive skew 264–5
ownership an alternative 263–4
and susceptibility to interference 383–4, 385–6
see also social dominance
dominance hierarchies 258–62, 273
dominance relationships 255–6
dominance variance 369
Drosophila (fruit fly) 30, 144, 351, 353, 359, 369
mating in 144
Drosophila bifurca 140
Drosophila differens 370
Drosophila heteroneura 370
Drosophila melanogaster 371
Drosophila planitibia 370
Drosophila pseudoobscura **357**
Drosophila silvestris 370
Drosophila willistoni **357**
dungfly, yellow *see Scatophaga stercoraria*
dynamic programming 17, 114, 115–16, 118

E-vector patterns 34, 35
early experience (dominance), importance of 256
ecological constraints 248, 249
and formation of family groups 230–4
and reproductive sharing 245
severe, effect of 247
and skew theory 215–18
ecological constraints theory 228–9
ecological diversification 349
ecological models, dilemma of 394
ecology, and behaviour 5–6
ecomorphs, a classificatory tool 348
ecosystems, latitudinal shift in 392
Ectatomma ruidum (ant, neotropical) 73
effective population size 362–3
egg size, and post-hatching survival 329–3
ejaculation, multiple 140
energetic efficiency 19–20
energy, constraints on acquisition and spending 99–100
energy allocation 322–4
plasticity of 324
energy efficiency 99
energy reserves 119
and avian song 115–16
benefits of 100–4
costs of 104–11
environmental change 308, 392
environmental constraints, adaptations to 20–1
Eopsaltria georgiana (robin, white-breasted) 242
Epimyrma kraussei 223
epistasis 369
escape performance, and increased body mass 109–10
Escherichia coli 295
eukaryotic cell cycle, replication in 296
Eulamprus tympanum (water skink) 110
Eumomota superciliosa 79
Euplectes jacksoni (Jackson's widowbird) 158, 175
eusocial insects 204, **205–6**
eusociality 151, 203
Hymenopteran, and the haplodiploidy hypothesis 209–14
origin and evolution of 208–15
evolutionarily stable strategies (ESSs) 7–8
cooperation as 274
stable coexistence of male and female 123
evolutionary change, models of 343–4
evolutionary history 4
evolutionary inertia, rod pigments 21–2
evolutionary rates, measurement of 339–40
evolutionary social theory, unified 249–51
extrapair copulation 137–8, 140, 143–4
sought by males 141–2
extrapair partners, in family groups 236
eyes, compound and single-lens 15, 27–9

Falco columbarius (merlin) 170
Falco tinnunculus (kestrel) 98, *322*, 326
families/family group(ing)s 251, 350–1
changing composition, causing conflict 237–41
conflicts in 9
definitions 229
extrapair partners 236
formation of
and dissolution 231–4, 245
economic model of 231
multigenerational 230–1, 235

stable, myth of 246–9
unstable 231
see also replacement families; social groups
family resources, inheritance of 234
fasting, enforced 100–1
fat storage 119
avian, and feeding interruptions 101–3
costs of 106–11
and social dominance 104
fecundity effects, consequence of mate choice 184–6, 201
feedback 367
negative 380
feeding
acquisition/storage costs 105–6
and energy storage 119–20
patterns 97
feeding areas, evaluation of easier 380–1
feeding interruptions 101–3
female mating preferences 179–82, 282
direct selection on 184–90
null model for 182–4
females
masculinization 263
and offspring paternity 141
Ficedula hypoleuca (flycatcher, pied) 77
fighting ability
and skew 218
and social dominance 246
Fisherian runaway evolution 175
fitness 184, 312–13, *314*
consequences of crosses to male parents 367
costs and benefits 333
direct and indirect 9, 150
and foraging efficiency 384–5
flight
and increased body mass 107–8
speed of, costs and benefits 98–9
fluctuating asymmetry (FA) 199
indicating genetic quality 195
food acquisition *see* feeding; foraging
food allocation, and chick begging 167–9, 171, 172, 177, 178
food caching
costs of storing energy 105–6
use of spatial memory for 53–9
food choice, and object copying 60–1
food stealing (kleptoparasitism) 383–4
foraging
foraging options 112–14
models 98–9
predation risk 105
social, use of public information 50–1
foraging behaviour, and food availability 101
foraging decisions, energetic gain, rate of (τ) 98
foraging efficiency 134, 382, 383, 387–8
individual efficiency in 384–5
foraging theory 42–3
ability to measure and remember time 43–4
Formica podzolica 222
Formica truncorum (ant, wood) *223*, 225
founder events 355, 370
role of 368, 369
frequency filtering 24
frog, tungara *see Physalaemus pustulosus*
Fulica atra (coot, European) 326, *328*
function 4, 11

Gallinago media (snipe, great) 90
Gallus gallus 79
Gallus gallus domesticus (chicken, domestic) 66
Gallus gallus spadiceus (junglefowl, Burmese) 59
gamete competition 122
Gammarus lawrencianus 47
Gammarus pulex 142
Gasterosteus aculeatus 372
Gastrophryne carolinensis (toad, narrow-mouthed) 162–3
Gastrophryne olivacea 162–3
gene conflicts 152
gene flow 332, 333, 358–9
disrupted 347
limit on 361
non-random 363–4
and population structure 361–2
and population subdivision 362–3
genes 126
autosomal 301
chromosomal 291
dominant, for sibling care 211
high-viability 78
material and informational 286, *287*
nuclear and organellar 299–300
selfish 82, 152
social gene 284–304
strategic 286, 287–9
as strategists 285
successful, phenotypic effects of 286
variant 286–7
see also good genes
genetic correlation
and good genes hypothesis 193–5
trait and preference 191–2, 201–2
genetic diversity, loss of 355
genetic drift 189, 356, 358–9
genetic endogenous labels 74–5, 77
genetic interaction 288–9
genetic markers 351
genetic variation 351, 371, 374
Chorthippus parallelus 354–5
genomes, refugial 354–5
genomic imprinting 152, 302–3
genotypes 350
genotypic variation 350
glacials and interglacials 353–4
good genes 126, 180, 198, 199
hypotheses 192–7
selection for 197, 201
good genes-parasite hypothesis 185
green beard effects 288, 292, 294, 298
group hunting 276–7
group selection 8–9
group territoriality 278–9
group-living 244, 254, 272
promoting social learning 61
growth, and maturation 314–17
Grus grus (crane) 47
Gryllus bimaculatus (cricket, field) 82, 131
guarding 131, 142, 213
guppies *see Poecilia reticulata*
guppies
courtship display 176–7
display design and sensory capacities 160–1
Gymnorhinus cyanocephalus (jay, pinyon) 57, 58, 235
Gyrodactylus turnbulli 177

habitat fidelity, and genetic divergence 363–4
habitat loss
consequences of 374–7, *378*
effects on population size 390–2
flexibility and constraint in response to 392–3
habitats 20, 372
variable quality of 386, 388
habituation 186, 197
Haematopus ostralegus (oystercatcher) 386, *391*
Haematopus ostralegus ostralegus 390
Hamilton's rule 9, 204–8, 212–14
handicap principle 8, 167–9

condition-dependent handicap 192
haplodiploidy 150–1
and Hymenopteran eusociality 209–14
Helogale parvula (mongoose, dwarf) 235, 248–9
'helping' behaviour 213, 228–9, *236*, *239*, 243
Hemilepistus reaumeri (woodlouse, desert) 73
hermaphrodites 121–2
hidden preferences 161–2, 180, 189
hierarchies
linear 258, 261–2
matrilineal 259–62
in population structure 350–1
hippocampus 54–5, 56, 68
Hipposideros 24
Hirundo daurica 347
Hirundo rustica (swallow, barn) 79, 169, *173*, 193–4, 346–7
Homo sapiens 296
honesty 149
advertising quality/need 167–9
in display *168*
maintenance of 166, 169
in signalling 165
honey bees, dance language 164
hunting, flight or perch 98
Hyaena brunnea (hyena, brown) 235
hybrid zones 351, 355, 368
Hyla chrysoscelis (frog, grey tree) 83
Hyla versicolor 83
Hylobittacus apicalis (hanging fly) 78
Hymenoptera 209–14
eusocial, split sex ratios 224–6
facultatively social 211
relatedness levels 209–10

Icerya purchasi (scale insect) 295
ideal despotic distribution model 380, 389
ideal free-distribution models 379–80, 386
deriving density-dependent functions from 386–8
imitation *see* behaviour copying
imprinting *see* template learning
inbreeding 70
avoidance 83–4, 236
reduction of 82
inbreeding depression 364, 366, 367, 369
incest avoidance 236–7, 247
inclusive fitness (theory) 204, 234, 285
independent contrasts, method of 308, 341–5
indeterminate growers 316
information, outdated, discarded or devalued 48–9
information processing ability, importance of 67
information sharing *see* area copying
inhibitory resetting 163
insect navigation 33–8
insects, eusocial **205–6**
interception courses, computation of 38–40
interference, susceptibility to 383
individual variation in 385–6
inversions, chromosomal 289
'island models' 357
isogamy 122, 123
isolation
asymmetric 370, 371
by distance 358, 362
reproductive 371

Junco hyemalis (junco, dark-eyed) 119

'Kaneshiro' hypothesis 370–1
key innovation, and phylogenetic tree structure 335–6
kin, forming revolutionary alliances 260
kin conflict 203, 208
over worker reproduction 226–7
kin recognition
action component 89–90
failures of 92–3
fitness consequences of 69–70
misunderstandings concerning 94–5
perception component 81–3
production component 74–7
vertebrate 75–7
kin selection
and Hamilton's rule 204–8
necessity of 208–10
and sociality 203–27
and strategic genes 288
theory 150, 204, 234, 249
and skew theory 151, 216
kinship 245–6
and tendency to cooperate 234–7

labels *see* cues
'Lack date' 325, 326
Lagopus lagopus (grouse, red) 389
landmarks/landmark panoramas, in insect navigation 37–8
Lanius collurio (shrike, red-backed) 393
Larus ridibundus (gull, black-headed) 6
Lasius niger 222, *223*
last male precedence, models and mechanisms of 130–8
bird models 134–8
insect models 130–4
learning
non-social 63
via self-inspection 82, 85
lek, paradox of 126, 179, 185
Lepomis macrochirus (sunfish, bluegill) 47, 384
leptothoracine ants, skew evolution in 219
Leptothorax 219
Leptothorax longispinosus 222
Leptothorax tuberum 223
life history 311, 351
studies 311, 332
theory 311
concepts in 312–14
lifetime reproductive success (LRS) 313, 315, 316, 331
and short-term behaviour 111–18
linear operator models 48–9
linear regression models 344
linkage disequilibrium 184, 190, 289, 297–8, 300
lions 276–9
cooperation in 276–9
local mate competition, and female bias in sex ratios 222–3
local resource competition, effects of 223–4
locomotion
and body size 26
costs of fat storage 107
effects of mass increase 108
loxodrome (constant-angle) routes 32–3

Macaca fuscata (macaque, Japanese) 63, *64*
Macrobrachium crenulatum (prawn, freshwater) 160
macroptery, and habitat persistence 365
major histocompatibility complex (MHC) 196, 294
male display 150
and resistance to parasites *183*
male pursuits 38–9
male sterility, cytoplasmic 300
male traits, variation in and female mate choice 180–2
Malurus cyaneus (fairy-wren) 233
Manorina melanocephala (miner, noisy) 235
Manorina melanophrys (miner, bell) 235
Marmota marmota (marmot, alpine) 101

matching-to-memory 37, 38
mate choice 138
  copying 92
  for correct species 187
  decision rules 90–1
  and sexual selection 179–202
  and signalling 149–50
  and speciation, in passerine birds 336
mate copying 200–1
mate quality, decision rules 90
mate quality recognition 78
mate quality recognition templates 83–4, 86
  learned 83–4
mate recognition 70
  action component 90–2
  perception component 83–6
  production component 77–9
mate recognition cues, and benefits of discrimination 78–9
mate replacement 237–8
mate sharing 247
mate switching 128, 137
mate-acceptance thresholds, optimal 91–2
mate-guarding 142, 143
mate-resource recognition 78
mating
  conflicts 142–3
  incestuous, rare in family groups 236–7
  see also multiple mating
mating preferences
  conspecific 197
  influence of social cues 200
  pleiotropic effects on 186–7
mating systems 9
  lek (lek-like) 179
  polygynandrous 144
  resource-based 184–5
  and sexual dimorphism 56
  and sperm competition 121–45
maturation, and growth 314–17
Megaderma lyra (vampire, false) 23, *24*
meiotic drive 299, 300, 301
Melanerpes formicivorus (woodpecker, acorn) 88, 231–2
Melitaea cinxia (fritillary, Glanville) 364
Melopsittacus undulatus (budgerigar) 47, 62
Melospiza melodia 173
memes 300
memory parameters 67–8
memory window models 48, 68
Merops bullockoides (bee-eater, white-fronted) 235, *236*, 237–8, 239, 248
Messor aciculatus (ant, harvester) 223
Metaohidippus aeneolus (spider, jumping) *29*
metapopulation dynamics 358
microgametes 122
Microtus ochrogaster (vole, prairie) 56
Microtus pennsylvanicus (vole, meadow) 56, 103
Microtus pinetorum (vole, pine) 56
migrants, mating success of 366–8
migration, selective 363
migration routes
  evolution in 392–3
  see also navigation
migratory species
  changes in migratory behaviour 392
  consequences of habitat loss 375–7
molecular interactions 294
Molothrus ater (cowbird, brown-headed) 85, *173*
monocular exclusion experiments 54, 58–9
mortality
  and brood size manipulations 323–4
  and mean age at maturity 320
multicellular organisms, somatic security systems 293–4
multiple mating 9, 18, 144
  by queens 226–7
  see also sperm competition
Mus musculus (mouse, house) 75, **357**
Musca domestica (housefly) 369
mutations 285, 358
  repeat-induced point (RIP) 296
mutualism 152, 208, 276–7, 279, 281
  short- and long-term 271–2B
mutualistic advantages 274, 278
Myrmica tahoensis 225
Mytilus edulis (mussel, edible) **357**, 386
Myxococcus xanthum 293, 294

natural selection, against traits 189
navigation
  by insects 33–8
  dissection into subroutines 33, 40–1
navigational errors 35–6
neighbourhoods, size and structure 360–1
Neoconocephalus spiza (katydid) 164
nepotism 70
nervous systems 19, 20, 36–8
Nesomimus parvulus (mockingbird, Galapagos) 235
Neumannia papillator (water mite) 161
neural mechanisms, and female mate choice 181
neurogenesis, seasonal 56
Neurospora crassa (bread mould) 296
niche partitioning, in adaptive radiations 348–9
non-breeding season, measurement of density dependence in 381–8
Nucifraga columbiana (nutcracker, Clark's) 54, 57, 58

object copying (stimulus enhancement) 60–1
Oenanthe oenanthe (wheatear) 393
offspring, size and numbers 329–32
offspring viability 193–4
Ololygon rubra 185
optimal behaviour rule 116
optimality models 6–7, 97
optimization 313, 333
  of age and size at maturity *315*, 315–16
  individual 324, 326, 330–1
origin recognition complex (ORC) 296
orthodrome (great-circle) routes 32
oscines 159–60
outbreeding 70
  optimal 367–8
outbreeding depression 366, 367
outlaw genes 82
Ovis aries (sheep, Soay) 385
Oxybelis aeneus (snake, vine) 185

P2 values 129–30, 131, 134
pairbonds 230
pairwise comparisons 308, 346–7
pairwise contests, competition in 255–8
Pan troglodytes (chimpanzee) 62, *64*
panglossianism 10–11
panoramic vision 27–9
Panthera leo (lion) 235
paper wasps 17, *89*, 89, 94
parasite load, influencing female mating preferences 194–5
parasitism 4–5
  brood parasitism 93, 171
parent–offspring associations 246
parent–offspring communication 170–1
parental care
  conflict over 143
  evolution of 337–8
parental effort and investment

322–4
parental manipulation 208
parental repairing 252
change in sexual dynamics 237–8
Partula suturalis (snail, land) 368
Parus ater (tit, coal) 57
Parus atricapillus (chickadee, black-capped) 54
Parus major (tit, great) 82, 101, *102*, 108, *110*, *322*, 323–4, 330, *331*
Parus montanus (tit, willow) 104
Parus palustris (tit, marsh) 54, 58–9
Passerculus sandwichensis (sparrow, savanna) *322*
Passerculus sandwichensis princeps (sparrow, Ipswich) 116
Passerina cyanea *173*
passive sperm loss model 135, *136*, 137, 138
patch quality 16
and Bayesian estimation 45–8
patches 351
patchiness
and genetic drift 356–7
and population structure 364
paternity
extrapair 127–8, 128–9
multiple 17–18, 128
and paternal care 127
patterns of 129–38
paternity assignment 128
paternity guards 142
paternity insurance 346–7
path integration *see* insect navigation
Pavlov strategy 269, 270, 277
pay-offs, short-term, role of 281–3
PBS 122–3
depends on sperm competition 123–4
peak shift learning 189
perception, recognition component 71, 79–86
Peromyscus californicus (mouse) **357**
phenotype matching, self-referent 96
phenotypic plasticity 307, 314
pheromones 185
queen, and honest signal 226
photoperiod, and body mass 102–3
photoreceptors 21
photorefractoriness 103
Phylloscopus 174
phylogenetic constraint, adaptive or genetic 340
phylogenetic trees 6, 334
mapping character change onto 337
and phylogeographies 359–60
structure of 335–6
as working hypotheses 335
phylogenies 307–8
bifurcating, independent contrasts 342
can mislead 344–5
phylogeographies 355, 359–60
Physalaemus coloradorum 187, 188
Physalaemus petersi 187
Physalaemus pustulosus (frog, tungara) 174, 187, 188
phytophagy, and insect diversification 336
Pica pica 171
Picoides borealis (woodpecker, red-cockaded) 233
Pipistrellus *24*
plasmids 291
protection rackets 291–2
Plectrophenax nivalis (snow bunting) 385
pleiotropy, affecting female mate preferences 186–7
Plethodon cinereus (salamander) **357**
Plethodon ouachitae (salamander) **357**
plumage brightness/patterns 194–5
and signalling 159, 173–4
Podiceps cristatus (grebe, great crested) 393
Podisma pedestris (grasshopper, brown mountain) 351, *352*, **357**, 359
Poecilia formosa (molly, Amazon) 200–1
Poecilia latipinna (molly, sailfin) 200–1
Poecilia reticulata (guppy) 160, 161, 176
poison–antidote mechanisms 291–2, 299
Polistes dominulus (paper wasp) 219
Polistes fuscatus (polistine wasp) 218
polistine wasps, skew evolution in 218–19
polymerases, DNA and RNA 290
polymorphism 8
and dispersal costs/benefits 364–5
genetic, balanced 333
Pongo pygmaeus (orangutan) 62
population genetic models 357–8
population size, and habitat loss 376–7, 390–2
population structure 350–3
current vs historical effects 359–60
and dispersal behaviour 364–6
interactions with behaviour 363–8
shaped by the Pleistocene period 353–6
shaped by present interactions 356–9
testing predictions 360–3
gene flow 361–2
neighbourhood size and structure 360–1
population subdivision and gene flow 362–3
populations
genetic history, and pairwise comparison 347
genetic structure 308
parapatric 355
small, isolated, evolution of behaviour in 368–72
Porphyrio porphyrio (pukeko) 245
predation risk, mass-dependent 107–8, *110*, 111, 114
predator inspection 279
Priapella 189
prisoner's dilemma 152, 266–70, 280, 281–3
production, recognition component 71, 72–9
protected invasion hypothesis 211
Prunella collaris (accentor, alpine) 93, 144
Prunella modularis (dunnock) 93, *173*
pseudopregnancy 248–9
public information 50–1, 59
pursuit deterrence 279

quality estimation 42–51
in social environments 50–1
in unchanging environments 45–8
Quercus laevis (oak, Turkey) 360

raffles, fair and loaded 130, 131, 135, 139–40
Rana cascadae 75
Rana sylvatica (frog, wood) 75, *76*
Rana temporaria (frog, common) 75
range changes
due to climatic/environmental fluctuation 353–6
range expansion 360
rank
determination experiments 259–61
high, and reduced reproductive success 262–3
maintenance of 260
rank inheritance, maternal 259, 259–60

'rate of living', allometry of *317*, 318
Rattus norvegicus (rat, Norway) 61, *67*
reaction norms 313, 332
rearing, communal 249
receivers
  influencing display design 160–3
  selective pressure on 155, 164–72
reciprocity 265–70, 275–83
  basic problem 266–7
  blue jays, cooperative key pecking 281–3
  cooperation in lions 276–9
  predator approach behaviour 279–81, *281*
recognition
  forms and functions of 69–70
  parent–offspring 75
  research 92–5
recognition systems 16–17, 69–96
  components of 70–92
  evolution of *73*
recombination 289, 290–1, 298
  bacterial 297, 304
  chromosomal 292
  creating relatives 304
  eukaryotic 304
  meiotic 297
reduced search rate hypothesis 52
reduction principle 298
redundancy, and signal detection 158
refrainers 276
refugia (kipuka) 353
  ice-age *354*, 355
rejection errors 72, 92–3
relatedness 204, 215
  affecting skew 217–18
  genetic, and skew models 245
  levels of, Hymenoptera 209–10
  in a social insect colony *207*
relatedness asymmetry 220–1, 222, 224, 225, 303
relationships 301–3
    ecology of 254–83
relative payoff sum (RPS) 48–9
repetition 157–8
replacement breeders 241–2, 252
replacement families 237, *240*
  less stable 240
  reduction in indirect benefits 238–9
replication 295–6
  genetic 289–90
reproduction 314
  alternating with growth 316
  conflict over 241–6
  costs of 322, 323–4
  and ecological constraints 252–3
  independent, probable success 244–5
  seasonal timing of 324–9
reproductive sharing 242–4, 245, 246, 249, 250, 252–3
reproductive skew
  and dominance 264–5
  stable 203, 215–19
  theory 151, 229, 242–4, 247–9
reproductive success 381
  future 98
  reduced, with high rank 262–3
  see also lifetime reproductive success
reproductive value 323
Requena verticalis (cricket, bush) 185
resource allocation 307, 316–17
resource holding power (RHP), and competition 255, 256, 258
resource-management strategies 308, 373
resources, high-quality 251
Rhagoletis pomonella (fly, apple maggot) 363
Rhinopoma 24
Riparia riparia (swallow, bank) 88, 94, 108
risk-spreading theorem 114
Rissa tridactyla (gull, kittiwake) 5
ritualization *see* signal evolution
Rivulus hartii 160
RNA editing 292
runaway sexual selection (Fisher) 180, *183*, 190–2, 197, 199, 201
Rupicola rupicola (cock-of-the-rock) 90

Salmo salar (salmon) 113
Salvelinus alpinus (charr, Arctic) *316*
scalar expectancy theory (SET) 16, 43–5
scaling, of time and energy 317–22
scaling relationships 307
Scaphiopus bombifrons (toad, spadefoot) 90
Scatophaga 133, 142
Scatophaga stercoraria (dungfly, yellow) 7, 18, 131
scatter-hoarding 105
Sceloporus occidentalis 108
Schizocosa ocreata (spider, bush-legged wolf) 176
Schizocosa rovneri 176
Sciurus vulgaris (squirrel, red) 389
scrounging, preventing skills spread 65–6
search costs 185–6
search image hypothesis 51–2
selection
  direct 181, *183*, 184, 197, 201
    on female mating preferences 184–90, 195
    subtle 189–90
  and fitness 4
  indirect 181, 184
    on female mating preferences 192–7
    on mating preferences 190–2
  and 'lying' mutants 166
  see also sexual selection
self-aggrandizement 290
self-displacement 131
self-inspection 82–3, 96
  mediating disassortative mating 85
self-peptide 196
self-recognition 75
  cost of using U-absent cues 74
  genetic *see* green beard effects
self-rejection 74
Semotilus atromaculatus 113
sensory bias 186, 196
sensory capacity, and display design 160–2
sensory exploitation (sensory trap) 161–2, 174, 186, 187–9
sensory systems 15, 20, 186
  bias in 181–2
Serinus canarius (canary) 177–8
Serinus serinus (serin) 393
sex allocation
  manipulation of 225–6
  queen-controlled 222
  worker control 221, 225
sex ratio theory, social Hymenoptera 220–1
  tests of 221–6
sex ratios
  biased 301
  and PBS 123–4
  sex investment ratios, effect of resource levels 222
  sex ratio evolution 203
  see also split sex ratios
sexual competition 17
  absent in intact families 235–6
sexual conflicts, and mating systems 142–4
sexual dimorphism *181*
  and sexual selection 198–9
sexual display 168
sexual selection 17, 78, 121, 124
  and mate choice 179–202
shoaling 280
Sialia sialis 173
sibling competition, and food

allocation 178
sibling–sibling associations 245
signal detection theory 51
signal evolution 155–78
cooperation and conflict in 164–7
selective pressure
on receivers 155, 164–72
on signallers 155, 156–64, 173
signaller interaction effects 162–4
signallers
multiple 177–8
selective pressure on 155, 156–64, 173
signalling
assessment of immigrant genetic contributions 368
costs of 156
deceit in
physical constraints 166–8
possibility rare 170–1
strategic constraints 167–70
and mate choice 149–50
and plumage brightness/ patterns 159, 173–4
sexual 165
see also acoustic signalling
signalling systems 8, 164–5
signals
diversity of, strategic explanations 171–2, *173*
effective 156–8
efficient and reliable 173–4
multiple 175–7
as source of information 164–5
'silver spoon' effects 332
single nesting vs helping 212–14
Sitta europaea (nuthatch) 389
skylight compass 34
skylight pattern, in insect navigation 34–5
slave-makers 221–2
sneak-guarder situations 140
social actions 204
social behaviour 8
evolution of 150–2
social dominance 104, 247, 249, 253
and fighting ability 246
social groups, stability of 152
social insects
monogyny or polygyny 215–16
sex ratio evolution and kin conflict in 220–7
social learning 16, 59–62
and trait acquisition 66
social predispositions, heritable, understanding of 250
social relationships 254
social tensions 151
social vertebrates, predicting family dynamics in 228–53
sociality
affording protection 212–14
and kin selection 203–27
societies, subsocial and semisocial 217–18
soma-germ line, physical cohesion 293–4
somatic exploitation, of slime moulds 294
somatic specialization 293
sonar 23
song
and escape ability 170
temporal structure of 159–60
songbirds 197
singing vs foraging 115–16
sound–source detection, and body size 30
spatial memory 16, 68
as an adaptive cognitive specialization 53–9
speciation 347, 364
by reinforcement 367
parallel 371–2
in passerine birds, and mate choice 336
species diversity, concern over 373
sperm competition 9, 17–18, 121, 123–4
detection and incidence of 126–9
female behaviour 141
sperm displacement
with instant mixing 130, 131
mixing after displacement 130, 135
sperm expenditure theory 139–40
sperm polymorphisms 140
Spermophilus beldingi (squirrel, Belding's ground) 75, 80, 82, 88, 94
Spermophilus columbianus (squirrel, Columbian ground) 384
Spermophilus tridecemlineatus (squirrel, 13-lined ground) 139
spherical geometry, coping with 31–3
split sex ratios
in eusocial Hymenoptera 224–6
and origin of eusociality 210–11
squirrels, red and grey 289
starvation, mortality rate from 387, *388*, 388
state variables 111, 119–20
'stepping stone' models 357–8, 361
strategic genes 286, 287–9
Sturnus vulgaris (starling) 3, 44, 50–1, 103, 108, 109
subordinates, as breeder replacements 241–2
survival, efficiency vs dominance 387–8
suspicious TFT (STFT) 268
Sylvia atricapilla (blackcap) 392
symmetry, sensory bias towards 196
Symphodus tinca (wrasse, peacock) 92
synchronization 351

t-complex, and mate choice 197
Tachycineta bicolor 173
Tadarida 24
Taeniopygia guttata (finch, zebra) 17, 18, 66, 92, 109, *135*, 137–8, 141–2, 324, 384
temperate zones, increased productivity and metabolism of animals 320–1
template learning 82
timing of 80
templates 71, 79–81
genetically encoded 80, 83, 85
kin-recognition 82, 83
learned 80, 81–2, 83–5
mate-quality recognition 83–4
mate-recognition 77, 85
neural 35
sex-recognition 84–5
temporal memory 44
temporal variation 351–3
Tenodera australasiae (praying mantis) *29*
terminal reward 111, 112–14
termites, origin of eusociality in 214–15
territorial defence strategies 389–90
territorial quality, and family group formation 231–2, 233, 234
territory size, density dependent 389
Tettigonia viridissima (bushcricket) *25*
Thalassoma bifasciatum (wrasse, blue-headed) 78, 141
Thamnophis sirtalis (snake, garter) 80
Thomonys bottae (gopher) **357**
threat diplays 165
time and energy, scaling of 317–22

time and energy management 97–120
time estimation, scalar expectancy theory (SET) 43–5
Tinbergen's four questions 4–5, 11
tit-for-tat strategy (TFT) 266–7, 267–8, 277
  and shoaling 280
  weakness of 268–9, 268–70
tit-for-two-tats strategy (TF2T) 268
tits *see Parus* spp.
trade-offs 17, 23, 97, 307, 313, 394–5
  current and future reproduction 322, 329
  dispersal and fecundity 366
  fecundity and survival 314–15
traditions, longevity of 66, *67*
traits 312, *314*
  cultural, spread of 63, *64*
  fitness-related, courtship characters 369
  manipulation of 332–3
  multiple, and their preferences 198–200
  and preferences, genetic correlation of 191–2
  sexually-selected 126
transduction 292
transformation 292–3
transitions, evolutionary 337–8
Trichogramma minutum (wasp, ichneumonid) *31*
Turdoides squamiceps (babbler, Arabian) 242, 247–8
Turdus merula (blackbird) 116, 389

U-absent cues 72, 74, 77, 81
Ultethasia 185
ultrasound 30
underwater hearing 22–3
underwater vision 21–2
updating
  Bayesian 45–7
  and relative payoff sum (RPS) 48–9
  of templates 80–1, 92
Uperolia laevigata 185
Ursus maritimus (polar bear) 106
Uta stansburiana (lizard, side-blotched) 329

variability
  in behaviour 7–8
  environmental 311–12, *314*, 316, 318 19, 332
  of the experienced *S* 44–5
  genetic 311, 333
vervet monkeys *see Cercopithecus aethiops*
viability 193–4, 300
vibration, substrate-borne 30
Vidua chalybeata (indigobird, village) 80–1
vigilance, cooperative 272
vigilance behaviour 110
Vipera berus (adder) 129
vision
  and spatial information 21
  visual acuity 27–9
  visual imitation 62
vocal learning, birds 61
voles 54, 103
Volucella pellucens (hoverfly) 38–9, *39*

white-fronted bee-eaters *see Merops bullockoides*
window-shopping tests 58
wing dimorphism 365
winner–loser effects 256–8
winter, as a foraging environment 101
winter habitat, consequence of loss of 376–7, 390–2
worker policing 226–7
worker reproduction, kin conflict over 226–7
worker–queen conflict 226
  sex allocation 221
workers, assessing their relatedness asymmetry 225

Xanthocephalus xanthocephalus (blackbird, yellow-headed) 178, 381
Xiphophorus maculatus (swordfish) 188–9
Xiphophorus multilineatus 189
Xiphophorus nigrensis 189
Xiphophorus pigmaeus 189
Xiphophorus variatus (swordfish) 188–9
Xylocopa sulcatipes (bee, carpenter) 212, *213*, 216

zebra finches *see Taeniopygia guttata*
Zenaida aurita (dove, Barbados zenaida) 61
Zonotrichia leucophrys 173
Zostera spp. 395

Lightning Source UK Ltd.
Milton Keynes UK
UKOW07f0317310315

248814UK00002B/8/P